The major histocompatibility complex (MHC) genes are involved in the immune system's response to tumour and infected cells and in generating an immune response. This book brings together basic aspects of the regulation of MHC antigens with important clinical applications (in viral infection, viral oncology, cancer biology and autoimmunity). There is a strong emphasis on situations where MHC expression is modulated (either stimulated or repressed). The book's major themes are: the mechanisms of MHC expression, which is explored at several levels including the transcription and translation of MHC genes and the insertion of MHC protein molecules into plasma membranes; the effect of cytokines on MHC expression, both in the aetiology of certain diseases and in possible immunotherapeutic approaches to disease; and the use of gene therapy to modify MHC expression in cancer cells and thereby cause tumour rejection.

# MODULATION OF MHC ANTIGEN EXPRESSION AND DISEASE

# MODULATION OF MHC ANTIGEN EXPRESSION AND DISEASE

*Edited by*

G. ERIC BLAIR
*Senior Lecturer, University of Leeds*

CRAIG R. PRINGLE
*Professor of Biological Sciences, University of Warwick*

&

D. JOHN MAUDSLEY
*Senior Research Fellow, University of Warwick*

Published by the Press Syndicate of the University of Cambridge
The Pitt Building, Trumpington Street, Cambridge CB2 1RP
40 West 20th Street, New York, NY 10011–4211, USA
10 Stamford Road, Oakleigh, Melbourne 3166, Australia



First published 1995

Printed in Great Britain at the University Press, Cambridge

*A catalogue record for this book is available from the British Library*

*Library of Congress cataloguing in publication data*

Modulation of MHC antigen expression and disease / edited by G. E. Blair, C. R. Pringle and D. J. Maudsley.
p. cm.
Includes index.
ISBN 0–521–49578–4 (hardback). – ISBN 0–521–49921–6 (pbk.)
1. Major histocompatibility complex. 2. Immunopathology. 3. Immunotherapy. 4. Cancer – immunological aspects. I. Blair, G. E. (G. Eric) II. Pringle, Craig R. III. Maudsley, D. John.
[DNLM: 1. Histocompatibility Antigens – immunology. 2. Histocompatibility Antigens – genetics. 3. Major Histocompatibility Complex – immunology. 4. Major Histocompatibility Complex – genetics. 5. Gene Expression Regulation. QW 573.5.H6 M692 1995]
QR184.315.M63 1995
616.07′92 – dc20 95–1214 CIP
DNLM/DLC
for Library of Congress

ISBN 0 521 49578 4 hardback
ISBN 0 521 49921 6 paperback

WV

# Contents

*List of contributors* — *page* ix
*Preface* — xiii
*List of abbreviations* — xv
1 General introduction to the MHC *Lynda G. Tussey and Andrew J. McMichael* — 1
2 Organization of the MHC *Jiannis Ragoussis* — 27
3 Interactions of cytokines in the regulation of MHC class I and class II antigen expression *Ian Todd* — 43
4 Control of MHC class I gene expression *Keiko Ozato* — 80
5 Control of MHC class II gene expression *Jenny Pan-Yun Ting and Barbara J. Vilen* — 103
6 Modulation of MHC antigen expression by viruses *D. John Maudsley* — 133
7 Modulation of MHC antigen expression by retroviruses *Douglas V. Faller* — 150
8 Modulation of MHC class I antigen expression in adenovirus infection and transformation *G. Eric Blair, Joanne L. Proffitt and Maria E. Blair Zajdel* — 192
9 MHC expression in HPV-associated cervical cancer *Jennifer Bartholomew, John M. Tinsley and Peter L. Stern* — 233
10 Inhibition of the cellular response to interferon by hepatitis B virus polymerase *Graham R. Foster* — 251
11 Cellular adhesion molecules and MHC antigens in cells infected with Epstein–Barr virus: implications for immune recognition *Martin Rowe and Maria G. Masucci* — 261
12 Effect of human cytomegalovirus infection on the expression of MHC class I antigens and adhesion molecules: potential role in immune evasion and immunopathology *Jane E. Grundy* — 278

13 Oncogenes and MHC class I expression *Lucy T. C. Peltenburg and Peter I. Schrier* 295
14 Mechanisms of tumour cell killing and the role of MHC antigens in experimental model systems *Roger F. L. James* 315
15 Manipulation of MHC antigens by gene transfection and cytokine stimulation: a possible approach for pre-selection of suitable patients for cytokine therapy *Ahmad M. E. Nouri* 335
16 Overexpression of MHC proteins in pancreatic islets: a link between cytokines, viruses, the breach of tolerance and insulin-dependent diabetes mellitus? *Marta Vives-Pi, Nuria Somoza, Francesca Vargas and Ricardo Pujol-Borrell* 361
17 The role of cytokines in contributing to MHC antigen expression in rheumatoid arthritis *Fionula M. Brennan* 390
18 Expression of an MHC antigen in the central nervous system: an animal model for demyelinating diseases *Lionel Feigenbaum, Taduru Sreenath and Gilbert Jay* 409
*Index* 424

# Contributors

Jennifer Bartholomew
*Cancer Research Campaign Department of Immunology, Paterson Institute for Cancer Research, Christie Hospital NHS Trust, Manchester M20 9BX, UK*

G. Eric Blair
*Department of Biochemistry and Molecular Biology, University of Leeds, Leeds LS2 9JT, UK*

Maria E. Blair Zajdel
*Division of Biomedical Sciences, School of Science, Sheffield Hallam University, Sheffield S1 1WB, UK*

Fionula M. Brennan
*The Kennedy Institute of Rheumatology, Sunley Building, 1 Lurgen Avenue, Hammersmith, London W6 8LW, UK*

Douglas V. Faller
*Cancer Research Center, Divisions of Medicine, Biochemistry, Pediatrics, Pathology and Laboratory Medicine, Boston University School of Medicine, Boston, MA, USA*

Lionel Feigenbaum
*Department of Virology, Jerome H. Holland Laboratory, Rockville, MD, USA*

Graham R. Foster
*Department of Medicine, St. Mary's Hospital, Imperial College of Science, Technology & Medicine, London, UK*

Jane E. Grundy
*Department of Clinical Immunology, Royal Free Hospital School of Medicine, London NW3 2QG, UK*

Roger F. L. James
*Department of Surgery, University of Leicester, Clinical Sciences Building, Royal Infirmary, Leicester LE1 7LX, UK*

Gilbert Jay
*Department of Virology, Jerome H. Holland Laboratory, Rockville, MD, USA*

Maria G. Masucci
*Department of Tumour Biology, Karolinska Institute, Box 6040, S-104 01 Stockholm, Sweden*

D. John Maudsley
*Department of Biological Sciences, University of Warwick, Coventry CV4 7AL, UK*

Andrew J. McMichael
*Institute of Molecular Medicine, John Radcliffe Hospital, University of Oxford, Oxford OX3 9DU, UK*

Ahmad M. E. Nouri
*Department of Medical Oncology, The Royal London Hospital, Whitechapel, London E1 1BB, UK*

Keiko Ozato
*Laboratory of Molecular Growth Regulation, National Institute of Child Health and Human Development, National Institutes of Health, Bethesda, MD 20892, USA*

Jenny Pan-Yun Ting
*Lineberger Comprehensive Cancer Center and Department of Microbiology-Immunology, CB 7295, University of North Carolina at Chapel Hill, Chapel Hill, NC 27599–7295, USA*

Lucy T. C. Peltenburg
*Department of Clinical Oncology, University Hospital, PO Box 9600, 2300 RC Leiden, The Netherlands*

Joanne L. Proffitt
*Department of Biochemistry and Molecular Biology, University of Leeds, Leeds LS2 9JT, UK*

Ricardo Pujol-Borrell
*Immunology Unit, Hospital Germans Trias I Pujol, Universitat Autònoma De Barcelona, Badalona, Spain*

Jiannis Ragoussis
*Division of Medical and Molecular Genetics, UMDS – Guy's Hospital, London Bridge, London, SE1 9RD, UK*

Martin Rowe
*Department of Cancer Studies, University of Birmingham Medical School, Birmingham B15 2TJ, UK*

Peter I. Schrier
*Department of Clinical Oncology, University Hospital, PO Box 9600, 2300 RC Leiden, The Netherlands*

Nuria Somoza
*Immunology Unit, Hospital Germans Trias I Pujol, Universitat Autònoma De Barcelona, Badalona, Spain*

Taduru Sreenath
*Department of Virology, Jerome H. Holland Laboratory, Rockville, MD, USA*

Peter L. Stern
*Cancer Research Campaign Department of Immunology, Paterson Institute for Cancer Research, Christie Hospital NHS Trust, Manchester M20 9BX, UK*

John M. Tinsley
*Molecular Genetics Group, Institute of Molecular Medicine, John Radcliffe Hospital, Oxford OX3 9DU, UK*

Ian Todd
*Department of Immunology, University Hospital, Queen's Medical Centre, Nottingham NG7 2UH, UK*

Lynda G. Tussey
*Institute of Molecular Medicine, John Radcliffe Hospital, University of Oxford, Oxford OX3 9DU, UK*

Francesca Vargas
*Immunology Unit, Hospital Germans Trias I Pujol, Universitat Autònoma De Barcelona, Badalona, Spain*

Barbara J. Vilen
*Lineberger Comprehensive Cancer Center and Department of Microbiology-Immunology, CB 7295, University of North Carolina at Chapel Hill, Chapel Hill, NC 27599–7295, USA*

Marta Vives-Pi
*Immunology Unit, Hospital Germans Trias I Pujol, Universitat Autònoma De Barcelona, Badalona, Spain*

# Preface

While the association of particular major histocompatibility complex (MHC) haplotypes with human diseases has been extensively reviewed, we felt that the area of modulation of MHC antigen expression and its association with human and animal diseases was less well explored. These ideas were initially exchanged between the editors and some of the authors of this book at a most interesting meeting organized by the British Society for Immunology in Warwick in 1991 entitled 'Viruses, Cytokines and the MHC'. Following this meeting, we realised that there was a need for a broad, interdisciplinary treatment of the area of MHC modulation that would be useful for both basic medical researchers and clinicians.

Expression of MHC class I and class II antigens follows a complex pathway from gene transcription to plasma membrane insertion and many steps can be stimulated or repressed leading to altered levels of cell surface MHC molecules. Therefore, to provide basic background information on the MHC, the genomic organization, antigen structure, biosynthesis and function and control of transcription of class I and II are considered first along with the important effects which cytokines and other extracellular agents can have on MHC antigen expression. Infection and oncogenic transformation of mammalian cells by viruses have provided powerful systems for analysing precise mechanisms of MHC antigen modulation and the relationship of this process to disease. Major virus/host cell systems described here are the retroviruses, the adenoviruses and the herpesviruses as well as other viruses which are also important human pathogens, such as hepatitis B virus and the papillomaviruses. Loss of MHC antigen expression in human and rodent cancers is then considered, including the interesting effects of a cellular oncogene, *c-myc*, on MHC class I expression in human melanoma cells. Aberrant MHC antigen expression is described in three autoimmune diseases: diabetes, rheumatoid arthritis and multiple sclerosis. The overall aim has been to highlight possible common mechanisms of MHC modulation in systems which are highly diverse.

It is a great pleasure to acknowledge the help of colleagues who participated in the production of this book and gave us support in many different ways. In Leeds, Diane Baldwin and Mani Tummala performed what seemed an impossible secretarial task with great patience and skill. Matthew Hope, Neil Maughan, Andrew Booth and Aruna Asipu were excellent sources of advice about the intricacies of word-processing. David Waller kindly provided the front cover illustration. We also thank Tim Benton, Jane Ward and Robert Harington of Cambridge University Press for their help and sound advice.

*G. Eric Blair*, Leeds
*Craig R. Pringle*, Warwick
*D. John Maudsley*, Warwick

# Abbreviations

| | |
|---|---|
| 5-HT | 5-hydroxytryptamine (serotonin) |
| Ab | antibody |
| Ad | adenovirus |
| ADCC | antibody-dependent cell cytoxicity |
| AIDS | acquired immune deficiency syndrome |
| AMLR | autologous mixed lymphocyte reaction |
| APC | antigen-presenting cell |
| ATP | adenosine triphosphate |
| $\beta_2$-m | $\beta_2$-microglobulin |
| BB rat | Biobreeding rat |
| BCG | bacille Calmette–Guérin |
| BCF | B cell factor |
| BL | Burkitt's lymphoma |
| BLS | bare lymphocyte syndrome |
| cAMP | cyclic AMP |
| CAT | chloramphenicol acetyl transferase |
| CD(antigen) | cluster of differentiation(antigen) |
| cDNA | complementary DNA |
| CIN | cervical intraepithelial neoplasia |
| cM | centimorgan |
| CM | conditioned medium |
| CMV | cytomegalovirus |
| CNS | central nervous system |
| CRE | cyclic AMP response element *also* class I regulatory element |
| CSF | colony-stimulating factor |
| CTL | cytotoxic T lymphocyte |
| DB | dot blot |
| dbcAMP | dibutyryl cAMP |

| | |
|---|---|
| DMSO | dimethylsulphoxide |
| DZ | dizygotic |
| EAE | experimental allergic encephalomyelitis |
| EBV | Epstein–Barr virus |
| EC | embryonal carcinoma (cells) |
| EGF | epidermal growth factor |
| EMCV | encephalomyocarditis virus |
| endo H | endoglycosidase H |
| ER | endoplasmic reticulum |
| FACS | fluorescent-activated cell sorter |
| FCS | foetal calf serum |
| GAD | glutamic acid decarboxylase |
| GM-CSF | granulocyte-macrophage colony-stimulating factor |
| GP | glycoprotein |
| HBV | hepatitis B virus |
| HCG | haemochromatosis candidate gene |
| HCMV | human cytomegalovirus |
| HEV | high endothelial venules |
| HIV | human immunodeficiency virus |
| HLA | human leukocyte antigen |
| HMG | high mobility group |
| HPV | human papillomavirus |
| HSP | heat shock proteins |
| HTH | helix-turn-helix |
| HTLV-I | human T cell leukaemia virus, type I |
| ICAM-1 | intercellular adhesion molecule-1 |
| ICS | interferon consensus sequence |
| ICSBP | ICS-binding protein |
| IDDM | insulin-dependent diabetes mellitus |
| IEF | isoelectric focussing |
| IFN | interferon |
| IFNEX | interferon-α-enhanced X-binding protein |
| Ig | immunoglobulin |
| Ii | invariant chain |
| IL | interleukin |
| IL-2R | IL-2 receptor |
| INR | initiation sequence (for mRNA transcription) |
| IP | immunoprecipitation |
| Ir(gene) | immune response (gene) |
| IRE | interferon response element |
| IRF | interferon (gene) regulatory factor |
| ISG | interferon-stimulated gene |
| ISRE | interferon-stimulated response element |

| | |
|---|---|
| kb | kilobase (pairs) |
| kDa | kilodalton |
| LAK | lymphokine-activated killer (cell) |
| LAM-1 | leukocyte adhesion molecule 1 |
| LCL | lymphoblastoid cell lines |
| LCMV | lymphocytic choriomeningitis virus |
| LFA-1//3 | lymphocyte function associated antigen 1//3 |
| LFB | Luxol fast blue |
| LGL | large granular lymphocyte |
| LMP | low molecular weight protein |
| LPS | lipopolysaccharide |
| LT | lymphotoxin |
| LTR | long terminal repeat |
| Mb | megabase (pairs) |
| MBP | myelin basic protein |
| mAb | monoclonal antibody |
| MCF | mink cell focus-inducing (virus) |
| MCP-1 | monocyte chemoattractant protein-1 |
| MFI | mean fluorescence index |
| MHC | major histocompatibility complex |
| mIg | membrane immunoglobulin |
| MNC | mononuclear cells |
| MoMuLV | Moloney murine leukaemia virus |
| MuLV | murine leukaemia virus |
| MuSV | murine sarcoma virus |
| MS | multiple sclerosis |
| MZ | monozygotic |
| NK | natural killer (cell) |
| NOD (mice) | non-obese diabetic (mice) |
| NON (mice) | non-obese non-diabetic (mice) |
| NP | nucleoprotein (of influenza virus) |
| NRE | negative regulatory element |
| NRD I | negative regulatory domain (of the IFN-β gene) |
| OA | osteoarthritis |
| ORF | open reading frame |
| p24 | protein of 24 kDa molecular weight |
| PAP | peroxidase anti-peroxidase |
| PBL | peripheral blood lymphocyte |
| PCR | polymerase chain reaction |
| PE | phycoerythrin |
| PFGE | pulsed-field gel electrophoresis |
| PGE | prostaglandins of the E series |
| PHA | phytohaemagglutinin |

| | |
|---|---|
| pInspro | (human) insulin promoter |
| PLP | proteolipid protein |
| PNS | peripheral nervous system |
| PPAR | peroxisome proliferation activation receptor |
| PRDI | positive regulatory domain I (of the IFN-β gene) |
| RA | rheumatoid arthritis |
| RadLV | radiation leukaemia virus |
| RAR | retinoic acid receptor |
| RD | radioactive surface binding |
| RER | rough endoplasmic reticulum |
| RIP | rat insulin promoter |
| RSV | Rous sarcoma virus |
| RXR | retinoid X receptor |
| SEA, SEB | staphylococcal enterotoxin A, B |
| SFV | Semliki forest virus |
| SH | *src* homology |
| SIV | simian immunodeficiency virus |
| SLE | systemic lupus erythematosus |
| STZ | streptozotocin |
| SV40 | simian virus 40 |
| T(cell) | thymus-derived (cell) |
| TAP | transporter associated with antigen presentation |
| TATA | tumour-associated transplantation antigen |
| TBP | TATA-binding protein |
| TCR | T cell receptor |
| tg | transgenic mice |
| TGF | transforming growth factor |
| $T_H$(cell) | T helper (cell) |
| THR | thyroid hormone receptor |
| TIL | tumour-infiltrating lymphocyte |
| TNF | tumour necrosis factor |
| TRA | tumour rejection antigen |
| TRE | TPA response element |
| $T_S$(cell) | T suppressor (cell) |
| TSH | thyroid stimulatory hormone |
| TSST | toxic shock syndrome toxin |
| TT | tetanus toxin |
| V-CAM | vascular cell adhesion molecule |
| VDR | vitamin D receptor |
| VSV | vesicular stomatitis virus |
| $X^+$ | cell bearing antigen X, e.g. $CD8^+$ cell bearing CD8 antigen |
| YAC | yeast artificial chromosome |

# 1

# General introduction to the MHC

LYNDA G. TUSSEY and ANDREW J. McMICHAEL
*University of Oxford*

## 1.1 Introduction

To protect the individual from foreign agents, such as viruses and bacteria, mammals have evolved a sophisticated system that allows them to distinguish self from non-self. Self/non-self discrimination was first demonstrated in mammals by the rejection of foreign tissue grafts in mice (Gorer, 1936; Snell, 1958; Klein, 1975). The genetic loci involved in graft rejection were subsequently mapped to a region on chromosome 17 (Klein, 1975), which became known as the major histocompatibility complex (MHC). The human MHC, also known as the human leukocyte antigen (HLA) system, is located on chromosome 6 (van Someren *et al.*, 1974).

The MHC occupies some 3.5 megabase pairs (Mb) of the genome and, in humans, approximately 75 genes (many still of unknown function) have been identified in this region (Trowsdale, Ragoussis & Campbell, 1991). The organization of the MHC is reviewed in detail in Chapter 2. The complex is often divided into three different classes of gene: I, II and III. There are multiple class I loci but the classical 'transplantation antigens' fall into three positions termed HLA-A, HLA-B and HLA-C in humans. The class II genes, encoded in the HLA-D region, encode proteins that help regulate the immune response to different antigens (Fig. 1.1). In early experiments that studied the genetic control of immunity to certain protein antigens, the immune response, whether strong or weak, was found to depend on particular alleles at loci within the MHC. These loci, originally called the immune response (Ir) genes, are now known to correspond directly to the class II genes (Benacerraf, Paul & Green, 1967; McDevitt & Chinitz, 1972). The class III region encodes components of the complement system as well as a diverse collection of at least 20 other genes (Bird, 1987; Spies, Bresnahan & Strominger, 1989; Sargent, Dunham & Campbell, 1989). Little structural and functional similarities have been established between the class III gene products and the class I and

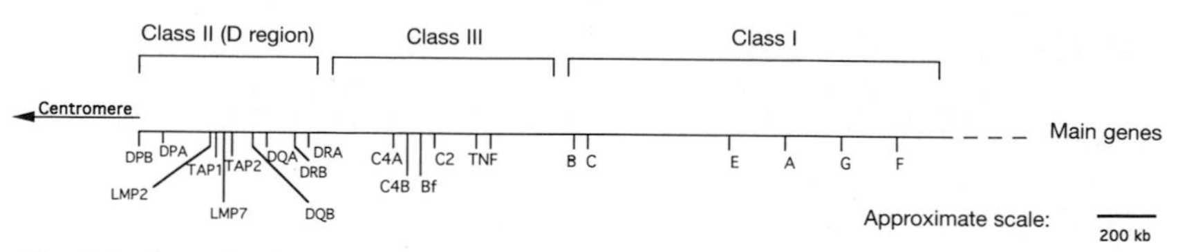


Fig. 1.1. Organization of the human MHC, which is located on chromosome 6. Locations of only the major loci of the class I, class II and class III regions are shown. (Adapted from Campbell & Trowsdale, 1993.)

class II gene products and, therefore, it is still not clear whether the MHC has encompassed this region by accident or if there is, in fact, some selective advantage. This chapter will focus on those MHC molecules involved in triggering T lymphocytes, namely the class I and class II gene products.

Both class I and class II proteins bind antigens that have been processed within the cell and present them at the cell surface to T cells. They are responsible, however, for stimulating different populations of T cells, with class I proteins presenting antigens primarily to CD8 antigen positive ($CD8^+$ CD, cluster of differentiation), cytotoxic T lymphocytes (CTLs) and class II proteins usually stimulating CD4 antigen positive ($CD4^+$) T helper cells ($T_H$). CTLs have long been recognized as an important component of the defence against viral infections (Zinkernagel & Althage, 1977; Yap & Ada, 1978) and they also play an important role in the rejection of foreign tissue grafts (Fischer Lindahl & Wilson, 1977a,b). CTLs therefore function to 'check' cells for the intracellular synthesis of foreign proteins and then to destroy such cells to prevent further infection. Class I antigens play a pivotal role in this process by displaying a 'sampling' of the intracellular protein content at the cell surface. $T_H$ cells, by comparison, are an important component of the defence against extracellular antigens. $T_H$ cells secrete a vast array of cytokines that both drive the differentiation of B cells into antibody-secreting plasma cells and accelerate the proliferation of other T cells. As part of this device for the detection of foreign, extracellular antigens, class II antigens display a sampling of extracellular proteins which have been endocytosed.

Both $T_H$ cells and CTLs express an antigen-specific T cell receptor (TCR). TCRs are multimeric proteins that consist of a variable $\alpha/\beta$ (or $\gamma/\delta$) heterodimer and also of the invariable components of CD3; together they are known as the TCR/CD3 complex. There is a division of labour within the TCR/CD3 complex, with the heterodimer interacting with antigen and the CD3 proteins regulating assembly of the entire complex as well as transduction of any signal resulting from occupancy of the TCR. The heterodimer is encoded by rearranging gene elements similar to the V, D, J elements in immunoglobulins. Many conserved amino acid sequences are also shared with the immunoglobulins and this provides a domain structure for the TCR that is similar to that of the immunoglobulins (Kronenberg *et al.*, 1986; Chothia, Boswe & Lesk, 1988; Davis & Bjorkman, 1988). The rearranging gene elements create TCRs with different binding surfaces and, therefore, different binding specificities. Thus, different T cells express TCRs with unique antigen specificities. Different T cell clones 'probe' for their specific antigen within the organism and only those encountering an antigen that fits its TCR will be activated. The challenge for this system, of course, is to know when not to react; as in the case of antigens normally expressed within and on the cells of various tissues. To avoid this, T cells must be 'educated' before they enter the fray so that T cells bearing receptors that can bind self antigens are either eliminated in the thymus during ontogeny or inactivated in the periphery. In this way both the CTL and $T_H$ cell populations are 'tolerized' to self antigens.

The intact antigen is not presented to T cells. Instead, class I and class II proteins bind fragments of antigens, in the form of short, intracellularly generated peptides, and display these at the cell surface to T cells (Babbitt *et al.*, 1985; Buus *et al.*, 1987; Townsend, Gotch & Davey, 1985; Townsend *et al.*, 1986b). Although class I and class II antigens have similar structures and peptide specificities, they actually sample different intracellular compartments for peptides. It is differences between class I and class II antigens in both their biosynthetic assembly and in the intracellular pathways they travel that account for their ability to bind antigenic peptides generated in different cellular compartments.

The building blocks of the class I antigens, the heavy chain and $\beta_2$-microglobulin ($\beta_2$-m), are translocated co-translationally into the endoplasmic reticulum (ER). It is in this compartment that the subunits are thought to assemble with a suitable peptide and form a functional complex (Nuchtern *et al.*, 1989; Yewdell & Bennink, 1989; Ljunggren *et al.*, 1990; Baas *et al.*, 1992; Lapham *et al.*, 1993); complexes devoid of peptide are unstable. However, while class I antigens are thought to bind their peptides in the ER, most of the proteins from which these peptides are generated reside in the cytoplasm throughout their lifespan. How do peptides generated in the cytosol cross the membrane to associate with class I antigens? The current view is that class I antigens recruit their peptides from the cytosol via an active transport process that uses ATP hydrolysis as the driving force to translocate peptides across the ER membrane (reviewed in DeMars & Spies, 1992). Although direct evidence for this transport process is still limited, it is believed to be mediated by the products of genes that encode peptide transporters associated with antigen presentation (TAP1 and TAP2) (DeMars & Spies, 1992). Once functional complexes are formed, they are transported to the cell surface via the Golgi complex and post-Golgi vesicles, where they present their peptides to circulating CTLs.

In contrast to class I, MHC class II antigens predominantly sample endocytic compartments for peptides (Neefjes & Ploegh, 1992b). Peptides generated in these compartments are derived from extracellular proteins that have been endocytosed or from internalized self proteins and are subsequently degraded by proteases that thrive in these acidified endosomes or lysosomes. Class II antigens avoid binding peptides during their initial assembly in the ER because their $\alpha$ and $\beta$ subunits co-assemble with a third polypeptide called the invariant chain, which serves not only to prevent peptide binding in this compartment but also to direct class II molecules to the cellular compartments associated with the endocytic pathway (Bakke & Dobberstein, 1990; Lotteau *et al.*, 1990; Roche & Cresswell, 1990; 1991; Teyton *et al.*, 1990; Lamb *et al.*, 1991). Within these compartments, the invariant chain is degraded and the class II antigens become competent to bind peptides. Whether the proteases involved in the breakdown of invariant chain are also involved in the generation of presentable peptides remains to be established, but it seems likely. Class II binding of antigenic peptides generated in the endocytic pathway is followed by release of class II from the compartment and expression on the cell surface.

Thus, intracellular and extracellular foreign antigens present different challenges for the immune system, both in terms of its ability to recognize their presence and in its ability to respond appropriately. Not surprisingly, parallel systems have evolved to meet these challenges: with intracellular antigens generally being presented to $CD8^+$ CTLs by class I proteins, while extracellular antigens are presented to $CD4^+$ $T_H$ cells. Differences in the biosynthetic assembly and intracellular trafficking of class I and class II molecules results in their acquisition of peptides from different intracellular sources and, therefore, ensures the segregation of their function.

## 1.2 Class I antigens

### *1.2.1 Structure and function*

Class I antigens are composed of a 45 kDa heavy chain and a 12 kDa light chain called $\beta_2$-m. For classical class I antigens, the heavy chain consists of three extracellular domains, $\alpha_1$, $\alpha_2$ and $\alpha_3$, a transmembrane region and a

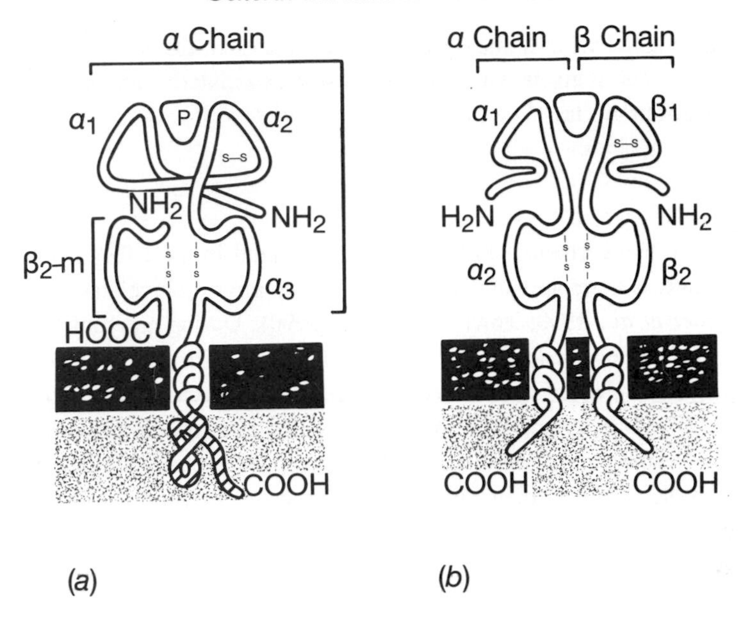


Fig. 1.2. Primary structure of class I and class II histocompatibility antigens.
(*a*) Class I antigens are composed of a 45 kDa heavy chain and the 12 kDa $\beta_2$-m. The peptide-binding site is formed by the $\alpha_1$ and $\alpha_2$ helices of the heavy chain.
(*b*) Class II antigens are composed of a 33–35 kDa α chain and a 25–29 kDa β chain. Each chain contributes an α helix to form the peptide-binding site.

cytoplasmic tail at the carboxy-terminus (Fig. 1.2). The light chain, $\beta_2$-m, is non-covalently attached to the heavy chain and is completely extracellular. The $\alpha_1$ and $\alpha_2$ domains fold to form the peptide-binding site. This site, as revealed by the crystal structure now resolved at 2.5 Å (0.25 nm), is a groove formed by two α helices lying across an eight-stranded β pleated sheet (Saper, Bjorkman & Wiley, 1991). It accommodates peptides, generally 8–10 amino acids residues long, in an extended conformation (Madden *et al.*, 1991; Saper *et al.*, 1991; Silver *et al.*, 1992). Six depressions or 'pockets', designated A to F, are found along the length of the groove and within these pockets specific interactions can occur between class I and peptide side chains (Garrett *et al.*, 1989; Madden *et al.*, 1991; Saper *et al.*, 1991; Silver *et al.*, 1992). The A and F pockets, located at the ends of the groove, are conserved among class I heavy chains and bind the amino- and carboxy-termini of the peptide, respectively (Madden *et al.*, 1991). The B, C, D and E pockets located in the centre of the groove, however, are polymorphic (Parham *et al.*, 1988) and this polymorphism, by modulating the size, shape and polarity of the pockets, determines allele-specific differences in the fine structure of the groove. Thus, there is a unique chemistry associated with the binding groove of different class I alleles.

### *1.2.2 Assembly*

MHC class I heavy chains and $\beta_2$-m are synthesized in the ER, where they assemble into a heterodimer that is unstable at physiological temperature (Schumacher *et al.*, 1990; Townsend *et al.*, 1990). During assembly, MHC class I proteins are transiently associated with an 88 kDa ER protein called calnexin that is released upon binding of peptide (Degen, Cohen Doyle & Williams, 1992). It appears that other molecules also transiently associate with class I during its assembly, but these are poorly characterized. These accessory molecules may play a role in the correct folding of class I antigens.

### *1.2.3 Generation of peptides*

The mechanisms responsible for the generation of antigenic peptides are poorly understood. Most peptides presented by class I antigens are generated from proteins located in either the cytoplasm or nucleus, although the association of peptides derived from mitochondrial or ER resident proteins with class I antigens has been demonstrated (Loveland *et al.*, 1990; Henderson *et al.*, 1992; Wei & Cresswell, 1992). Nevertheless, the existence of a cytosolic or nuclear degradation system in which proteins are degraded to peptide fragments rather than amino acids is suggested. The degradation of most cellular proteins starts with their covalent conjugation with ubiquitin (Rechsteiner, 1987; Hershko & Ciechanover, 1992). This labels proteins for rapid hydrolysis to oligopeptides by an ATP-dependent, large proteolytic complex called the proteasome (Goldberg, 1992; Goldberg & Rock, 1992) (Fig. 1.3*a*). Early experiments in this field did suggest a role for a ubiquitin-dependent pathway in the generation of antigenic peptides (Townsend *et al.*, 1988). In support of these early findings, recent experiments using cells that exhibit a temperature-sensitive defect in ubiquitin conjugation showed that class I presentation of a specific antigen was in fact inhibited at the non-permissive temperature (Michalek *et al.*, 1993). Two genes in the MHC, LMP2 (for low molecular weight protein) and LMP7, specify proteins with homology to subunits of multicatalytic protease complexes, or proteasomes (Monaco & McDevitt, 1986; Brown, Driscoll & Monaco, 1991; Glynne *et al.*, 1991; Kelly *et al.*, 1991; Martinez & Monaco, 1991; Ortiz Navarrete *et al.*, 1991). It has been tempting, primarily because of their location but also because of the sugges-

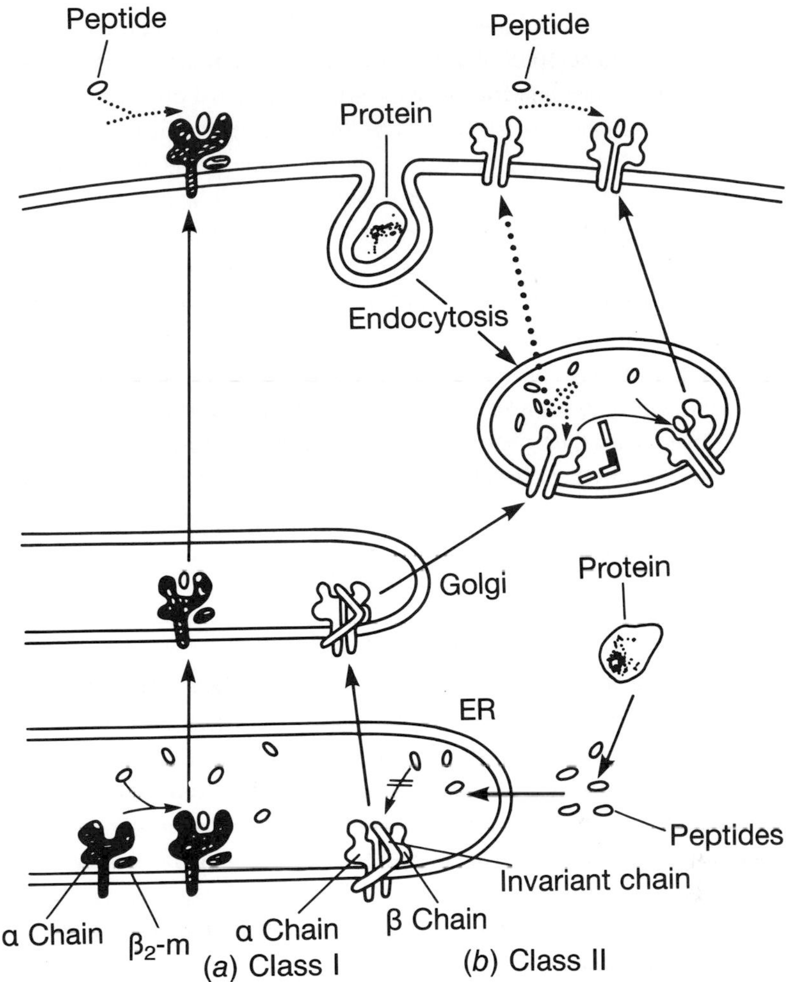


Fig. 1.3. Schematic models of class I and class II antigen presentation. For class I, cytosolic and nuclear proteins are thought to be degraded by large, ATP-dependent proteasomes. The resulting peptides are then translocated to the lumen of the ER by the TAP1 and TAP2 heterodimer. At the same time, class I heavy and $\beta_2$-m chains are inserted into the ER where they associate non-covalently and bind peptide. The resulting heterotrimer is transported through the Golgi to the cell surface. For class II, the $\alpha$ and $\beta$ chains are inserted into the ER where they associate with each other and with the invariant chain. Although it is not shown here, a nonamer consisting of three invariant chains and three $\alpha/\beta$ dimers is actually formed. Complexes are transported through the Golgi complex where they are sorted to a compartment in the endocytic pathway. Here the invariant chain is proteolytically removed, the class II molecules bind degraded antigens and the complexes are released from the endocytic pathway. Class II complexes are then displayed on the cell surface.

tion of this pathway's involvement, to think that these are proteases that are involved in the generation of antigenic peptides, although there is little evidence for this. In fact, genetic evidence argues against a critical role of LMP2 and LMP7 in antigen presentation. The human mutant cell line 721.174 and its derivative T2 have a large deletion in the class II region which not only spans these genes but also two genes that are important for the transport of peptides into the ER. However, in transfection experiments, the antigen-processing defect in these cells appears to be completely reversed by introduction of the transporter genes alone (Arnold *et al.*, 1992; Momburg *et al.*, 1992). It has also been observed that mice in which the LMP2 gene has been deleted express normal levels of class I antigens, whereas transporter gene-deficient mice express at least a 20-fold reduction in surface levels of class I. It is clear, then, that expression of LMP2 and LMP7 proteins is not an absolute requirement for antigen presentation; however, a more subtle role in processing, such as the fine specificity of cleavage, is by no means precluded by these experiments.

### *1.2.4 Peptide translocation*

For virally infected cells, CTL recognition was originally thought to involve recognition of viral glycoproteins expressed on the surface of virally infected cells. However, a critical experiment, demonstrating that cells transfected with only fragments of influenza nucleoprotein (NP) were recognized by NP-specific CTLs, showed that viral proteins are recognized as peptide fragments (Townsend *et al.*, 1986a,b). When these experiments were extended to show that the actual epitopes involved shared no signal sequence that could be involved in translocation across a membrane, the existence of a transport system was proposed.

Experimental support for a transport system came when the mutant murine cell line RMA-S was shown to be capable of presenting externally delivered peptides to CTLs but not endogenously delivered antigen derived from infecting virus (Townsend *et al.*, 1989). These cells characteristically fail to express class I on the cell surface in a stable form but this can be reversed either by the addition of peptide epitopes or by lowered temperatures (Townsend *et al.*, 1989; Ljunggren *et al.*, 1990). A similar defect has been demonstrated in the human cell lines 721.134, 721.174 and the derivative T2 (Cerundolo *et al.*, 1990; Hosken & Bevan, 1990). As class I antigens depend on bound peptides for stability, it was argued that these cell lines were defective in the generation and/or transport of peptides into the ER.

Each of the human cell lines thought to be defective in peptide translocation has a deletion in the MHC class II region, and this suggested that the gene(s) required for transport were located in this region. Mapping of this region, in both humans and rodents, soon revealed two genes encoding proteins of the ATP-binding cassette family, which was already known to be involved in cellular transport processes (Deverson *et al.*, 1990; Monaco, Cho & Attaya, 1990; Spies *et al.*, 1990; Trowsdale *et al.*, 1990). While, to date, there is no biochemical assay for transport function, this process is believed to be mediated by the products of these two genes, which are now named TAP1 (for transporter associated with antigen presentation) and TAP2 (Fig. 1.3*a*).

Two lines of evidence indicate that the products of the TAP1 and TAP2 genes form heterodimers. Firstly, TAP2 protein co-precipitates in immunoprecipitations with anti-TAP1 antibody (Spies *et al.*, 1992). Secondly, two of the cell lines with antigen processing-defective phenotypes (RMA-S and 721.134) have defects in only one or the other gene, suggesting that no functional homodimers exist. Transfections of the TAP1 and/or TAP2 genes do in fact correct the defect in the mutant cell lines (Powis *et al.*, 1991; Spies & DeMars, 1991). The TAP1 protein has been localized to the membranes of the ER and cis-Golgi (Kleijmeer *et al.*, 1992) and is consistent with a role for TAP1 and TAP2 in peptide translocation. Both TAP1 and TAP2 are thought to span the membrane six to eight times, with a putative cytosolic domain providing the ATP-binding site. Using ATP as the driving force, the TAP proteins are then presumed to translocate peptides across the ER membrane. Whether they transport peptides precisely trimmed to fit the binding cleft (8–10 residues) or transport larger fragments that are subjected to further cleavage upon arrival in the lumen of the ER is not known. While it must be stressed that this pathway is still hypothetical, the most plausible hypothesis for the function of the TAP1/TAP2 heterodimer is that it transports peptides from the cytoplasm into the lumen of the ER.

Allelic polymorphism at the transporter genes could have functional consequences. For the rat, allelic differences in the products of the TAP1 and TAP2 genes do alter the sets of peptides which associate with a given class I protein (Powis *et al.*, 1992) and this suggests that, at least in the rat, the transporter actually plays a role in determining which peptides will be available for binding to class I. Interestingly, human transporter genes are polymorphic as well, although to a much lesser extent than that found in the rat. An antigen-processing polymorphism, similar to that found in the rat, has been described for a human family (Pazmany *et al.*, 1992), although in this case the gene(s) involved have not been identified. This raises the possibility that other genes could be involved in these processes and thereby contribute to the functional polymorphism of antigen processing.

### 1.2.5 Class I peptide loading

All of the available genetic and biochemical data suggest that class I antigens bind their peptides in the ER and/or cis-Golgi. Firstly, in TAP-deficient cells, class I antigens are often retained in the ER (DeMars *et al.*, 1985; Salter & Cresswell, 1986; Townsend *et al.*, 1989). Therefore peptide loading not only appears to occur in the ER but also seems to be required for egress from the ER and cis-Golgi and entry into medial-Golgi. Secondly, immunocytochemical experiments with one of the transporter proteins also provide evidence that peptide loading occurs in the ER and/or cis-Golgi, since by immunoelectron microscopy TAP1 immunoreactivity has been localized to the ER and cis-Golgi (Kleijmeer *et al.*, 1992). Finally, recent experiments by Lapham *et al.* (1993) provide direct evidence that naturally processed peptides associate with class I in a pre-Golgi complex compartment. In these experiments, class I proteins were retained in the ER either by the use of chemical agents which block their egress from this compartment or by using genetically engineered class I proteins that are retained in the ER by virtue of a specific sequence in their cytoplasmic tail. Specific peptides were then shown to associate with these retained molecules.

### 1.2.6 Allele specificity of peptide binding

Cells containing different class I alleles present different peptide epitopes to CTLs, a phenomenon known as MHC restriction. Sequencing of self peptide mixtures eluted from class I antigens has revealed allele-specific sequence motifs for antigenic peptides (Falk *et al.*, 1991; Rammensee, Falk & Rötzschke, 1993) (Table 1.1). Each of the motifs described so far contains one or two 'anchor' positions, which are occupied by either a fixed residue or one of a few related residues. Those residues not at anchor positions are variable, although some positions seem to be preferentially occupied. It is now clear that the motifs reflect differences in the fine structure of the groove. Many of the pockets situated along the length of the binding groove accommodate side chains of the peptide. However, certain pockets, in particular those responsible for accommodating the side chain of the anchors, are quite limited as to the nature of the side chain that they can accommodate: gener-

Table 1.1. *Class I allele-specific peptide motifs as revealed by pooled peptide sequencing or analysis of epitopes*

| | Position | | | | | | | | | |
|---|---|---|---|---|---|---|---|---|---|---|
| | 1 | 2 | 3 | 4 | 5 | 6 | 7 | 8 | 9 | References |
| H-2K$^d$ | | Y | | | | | | | L<br>I | Falk *et al.* (1991) |
| H-2K$^b$ | | | | | F<br>Y | | | L | | Falk *et al.* (1991) |
| H-2D$^b$ | | | | | N | | | | M<br>I<br>L | Falk *et al.* (1991) |
| H2-K$^k$ | | E | | | | | | I | | Rammensee *et al.* (1993) |
| H2-L$^d$ | | P | | | | | | | L | Lurquin *et al.* (1989)<br>Falk *et al.* (1991)<br>Schultz *et al.* (1991) |
| HLA-A2.5 | | | | | | | | | L | Rammensee *et al.* (1993) |
| HLA-A2.1 | | L | | | | | | | V | Falk *et al.* (1991)<br>Hunt *et al.* (1992a) |
| HLA-B35 | | P | | | | | | | Y | Hill *et al.* (1992) |
| HLA-B53 | | P | | | | | | | F<br>W | Hill *et al.* (1992) |
| HLA-B8 | | | K | | K | | | | L | Sutton *et al.* (1993) |
| HLA-A68 | | T | | | | | | | R | A. J. McMichael, unpublished data |
| HLA-A3.1 | | I<br>L | | | | | | | K | A. J. McMichael, unpublished data |
| HLA-A11 | | I<br>L | | | | | | | K | A. J. McMichael, unpublished data |
| HLA-B27 | | R | | | | | | | K<br>R | Jardetsky *et al.* (1991) |

The common features of peptide ligands for a given class I antigen are shown. Amino acids are represented by the one letter code. The side chains of these amino acids contact the allele-specific binding pockets in the class I molecule. For clarity, auxiliary anchors have been omitted.

ally these pockets can accommodate only one or at most a few related side chains. The binding pockets, therefore, dictate specific sequence constraints for the peptide at certain positions and the motifs are a reflection of these sequence constraints. Sequence constraints at only a few positions of the peptide allows not only for allele specificity in binding but at the same time also permits a single class I molecule to present a vast number of chemically distinct peptides.

## 1.3 Class II antigens

### *1.3.1 Structure and function*

Class II antigens are heterodimers composed of a 33–35 kDa $\alpha$ chain and a 25–29 kDa $\beta$ chain. Both chains form two extracellular domains (the $\alpha$ chain containing $\alpha_1$ and $\alpha_2$ and the $\beta$ chain forming $\beta_1$ and $\beta_2$), a single transmembrane domain and a carboxy-terminal cytoplasmic tail. Alpha helices contributed by the $\alpha_1$ and $\beta_1$ domains form a peptide-binding site that is similar to that formed by the $\alpha_1$ and $\alpha_2$ domains of class I antigens (Fig 1.2*b*). The three-dimensional structure of the extracellular portion of the HLA-DR1 proteins has been determined (Brown *et al.*, 1993), and, although the class I and class II peptide-binding grooves are similar in structure, there are certain differences. The most notable difference is that the ends of the class II groove are open and allow for overhanging of both amino- and carboxy-termini of bound peptides (Brown *et al.*, 1993).

### *1.3.2 Assembly and transport*

MHC class II molecules are assembled in the ER from an $\alpha$ and $\beta$ chain and the invariant chain. Although all naturally occurring class II positive cells express invariant chain, this chain is not an absolute prerequisite for the formation of class II $\alpha/\beta$ heterodimers (Miller & Germain, 1986; Sekaly *et al.*, 1986). Formation of $\alpha/\beta$ dimers in the ER could, in principle, result in the binding of peptides from the same pool of peptides that bind to class I and, since *in vitro* studies have shown that class II proteins only bind peptide once invariant chain is removed (Roche & Cresswell, 1990; Teyton *et al.*, 1990), it has been suggested that one of the functions of invariant chain is to prevent premature peptide binding in the ER.

Invariant chains can form trimers and these will acquire three sets of $\alpha$ and $\beta$ chains during co-expression, if they maintain this structure. Crosslinking studies suggest that class II molecules are released from the ER as a nine subunit complex consisting of three $\alpha$, three $\beta$ and three invariant chain complexes (Roche, Marks & Cresswell, 1991). After leaving the ER, these $\alpha/\beta$-invariant chain nonamers are transported through the Golgi apparatus and it is at this point that the cellular transport pathways for class I and class II diverge (Fig 1.3*b*). Whereas class I molecules appear to proceed directly to the cell surface, the class II molecules are sorted to the endocytic pathway where they will lose invariant chain and encounter peptides derived from internalized, degraded proteins (Neefjes *et al.*, 1990). The signal for sorting to the endocytic pathway is located in the cytoplasmic tail of the invariant chain (Bakke & Dobberstein, 1990; Lamb *et al.*, 1991; Lotteau *et al.*, 1990). A second function of the invariant chain is to target class II–invariant chain complexes to the endosomal compartments.

Where class II molecules actually enter the endocytic compartment and bind peptide is still a matter of debate. These molecules have been observed in early and/or late endosomes (Pieters *et al.*, 1991) as well as in lysosomes (Peters *et al.*, 1991). There are studies to suggest that lysosomes are the site of class II peptide generation (Harding *et al.*, 1991). Further, biochemical and electron microscopy data suggest that class II can bind peptides in lysosomes – which implies that lysosomes are a site for the generation and loading of class II peptides, although other sites for peptide generation and loading have not been excluded. Class II molecules are retained in the endocytic pathway for 1–3 hours before appearing on the cell surface (Neefjes *et al.*, 1990). During this time, invariant chain is degraded by proteases (Blum & Cresswell, 1988) and the peptide-binding capacity of the molecule is restored (Neefjes & Ploegh, 1992a).

### *1.3.3 Peptide generation*

Most exogenous antigens are processed by the endocytic pathway (Pierce *et al.*, 1988; Lanzavecchia, 1990), although some can be processed by cell surface proteases (Delovitch *et al.*, 1988). Antigens can be internalized by a number of different routes. For example, antigen can be internalized after binding surface immunoglobulin on B cells via clathrin-coated pits, or antigen–antibody complexes can bind Fc receptors and enter by endocytosis. Following internalization, the endocytic pathway encountered consists of a series of increasingly degradative and increasingly acidified compartments that terminate with the lysosomes (Steinman *et al.*, 1983). It is still not clear whether these compartments actually represent fixed compartments or are formed during some sort of maturation process of the vesicle. Nevertheless, internalized antigens do pass through these compartments in a temporal fashion and during their transit are subjected to increasing proteolytic activity. The early endosome, pH 6.0–6.5, is the first compartment encountered and appears to contain the proteolytic enzymes cathepsin B and cathepsin D. These proteases have been implicated in the production of antigenic peptides (Chain, Kaye & Shaw, 1988; Takahashi, Cease & Berzofsky, 1989; Diment, 1990; van Noort *et al.*, 1991) as well as in invariant chain dissociation from class II antigens (Blum & Cresswell, 1988; Nguyen, Knapp & Humphreys, 1989; Roche & Cresswell, 1991). Cathepsin B, in particular, appears to be essential for the degradation of the invariant chain. In the late endosome, pH 5.5, antigen is further subjected to cleavage by cathepsin B and cathepsin D and lysosomal enzymes that are targeted to the lysosome from this compartment (although the lysosomal enzymes are not fully active in this compartment because of the non-optimal pH). The final endocytic compartment is the highly degradative lysosome (pH 4.5–5.0), and here proteins will eventually be reduced to single amino acids or dipeptides that are transported into the cytoplasm. Exactly where in the endocytic pathway antigenic peptides are generated and associate with class II is not clear. While some antigenic peptides might be generated in the early stages of the endocytic pathway, complete degradation of antigen and the release of most antigenic peptides probably does not occur until the pre-lysosomal or lysosomal stages. This is illustrated by studies of the degradation of iodinated tetanus toxin (TT). In these experiments, iodinated TT was internalized following binding to surface immunoglobulin specific for the toxin. TT fragments were found in cell fractions containing early and late endosomes as well as those fractions containing lysosomes. Kinetics of the association of TT fragments with class II then suggested that, although TT degradation began early in the pathway, the actual fragments which bound class II were generated late in the pathway (Davidson, West & Watts, 1990). For some antigens though, antigenic peptides may be released in the initial degradation steps. Therefore, it appears that there is no exclusive role for an individual protease in the generation of peptides by the endocytic route, and it is more likely that many proteases are involved in the process so that the stage at which a peptide is produced and bound will be a function of both when it is produced in the degradative pathway and when it encounters an appropriate class II molecule with an exposed binding site.

### *1.3.4 Peptide delivery*

There appears to be at least one gene located in the MHC class II region (in the vicinity of the TAP genes) that is needed for the intracellular production and/or delivery of class II peptides. Human B cell lines that have homozygous deletions in this region are defective in the class II-restricted antigen presentation. Although levels of cell surface expression for class II appear normal in these mutant cell lines, the molecules are no longer recognized by certain, conformationally sensitive antibodies (Mellins *et al.*, 1991; Ceman *et al.*, 1992). Further, the molecules appear to be unstable, as indicated by the atypical dissociation of the $\alpha$ and $\beta$ chains in the presence of SDS detergent. These molecules also appear to be functionally impaired, as most class II-specific T cell clones do not recognize them and introduction into the culture medium of proteins that would normally be internalized, processed and presented does not result in T cell reaction (Mellins *et al.*, 1990; Ceman *et al.*, 1992). However, this can be reversed by addition of the appropriate peptides to the culture medium (Mellins *et al.*, 1990), which suggests a defect in either peptide generation or loading. Although the identity of the gene required of class II antigen processing is still unknown, it has been localized to a region of approximately 230 kb between the human DMB and DQB1 genes (Ceman *et al.*, 1992). This region does include the TAP1 and TAP2 genes, which are important for class I peptide transport, as well as the LMP2 and LMP7 genes, which may play a role in the generation of class I peptides. However, because antigen processing and presentation with class II molecules is normal in a mutant in which none of these four genes are expressed, at least one undiscovered gene in this region is required for class II antigen presentation (DeMars & Spies, 1992).

### *1.3.5 Allele specificity in peptide binding*

In contrast to class I-associated peptides, which are mostly nonamers, the length of class II peptides varies between 12 and 25 residues, with the majority being 15 amino acid residues long (Yu Rudensky *et al.*, 1991; Chicz *et al.*, 1992; Hunt *et al.*, 1992a,b). Unlike class I peptides, sequencing of the first peptides eluted from class II failed to reveal any allele-specific sequence motif for the peptides (Yu Rudensky *et al.*, 1991). Subsequent studies, however, established that there were, in fact, motifs but that they were obscured by a heterogeneity in length exhibited by the class II-associated peptides (Yu Rudensky *et al.*, 1992). That is, peptides containing a given core sequence were found in various lengths, with extensions at both the amino- and carboxy-termini (Table 1.2). These 'ragged' ended peptides are common to all studies analysing class II peptides and this difference in peptide structure as compared with that of class I may reflect differences in the mechanisms of antigen processing. Electron density measurements from peptides copurifying with DR1 indicates that the peptides are bound in an extended conformation with about 15 peptide residues contacting the DR molecule (Brown *et al.*, 1993). This density is not weak in the middle as observed for the class I molecules HLA-A2 and HLA-Aw68, where peptides of slightly different lengths are thought to be accommodated by bulging out of the binding site in the middle. Instead, the, density extends out both ends of the site and is consistent with the observation that class II-associated peptides vary from 12–25 residues in length and have ragged ends. One prominent pocket,

Table 1.2. *Sequence motifs in naturally processed class II associated peptides*

| | | | | | | | | | | | | | | | | | | | |
|---|---|---|---|---|---|---|---|---|---|---|---|---|---|---|---|---|---|---|---|
| I-E$^b$ consensus | | | X | X | **Y** | **L** | **Y** | X | X | X | X | **R** | **R** | X | X | **Y** | X | | |
| | | S | P | S | **Y** | **V** | **Y** | H | Q | F | E | **R** | **R** | A | K | **Y** | K | | |
| | | | G | K | **Y** | **L** | **Y** | E | I | A | R | **R** | **H** | P | Y | **F** | y | a | p |
| | X | P | Q | S | **Y** | **L** | **I** | H | E | X | X | X | I | S | | | | | |
| I-A$^s$ consensus | | X | X | X | X | **I** | **T** | X | X | X | X | **H** | X | X | X | | | | |
| | | I | L | R | K | **I** | **T** | D | S | G | P | **R** | V | P | I | G | p | n | |
| | W | P | S | Q | S | **I** | **T** | C | N | V | A | **H** | P | A | S | S | T | | |
| | N | V | E | V | H | **T** | **A** | Q | T | Q | T | **H** | R | E | D | Y | | | |
| | | K | P | T | E | **V** | **S** | G | K | L | V | **H** | A | N | F | G | T | | |
| | | X | P | Y | M | **F** | **A** | D | K | V | V | **H** | L | P | G | S | Q | | |

Possible anchor residues are shown in bold. Amino acids are represented by the one letter code; X stands for any residue. Data is taken from Yu Rudensky *et al.* (1991; 1992).

accommodating a peptide side chain at approximately the third position of a 15-mer, was identified near one end of the DR1-binding groove. Other pockets were inferred from the shape of the electron density profile and the location of class II polymorphic residues.

### *1.3.6 Recycling of class II and peptide exchange*

Class II molecules are transported to the cell surface from the endocytic route by an unknown pathway. A small percentage of empty (devoid of peptide) molecules do appear to reach the cell surface where they are able to bind peptide. It has been suggested that the recycling of cell surface class II provides a second opportunity for class II to encounter peptide via peptide exchange, although this is still subject to debate. Recent experiments have shown that the half-life of class II antigens and associated peptide is identical (Lanzavecchia, Reid & Watts, 1992) and that only newly synthesized class II molecules bind degraded antigen (Davidson *et al.*, 1991), implying that the exchange of peptides by recycling is unlikely. However, the possibility that a fraction of class II antigens are reloaded with peptides following their internalization cannot be excluded, since Adorini *et al.* (1991) have demonstrated exchange of endogenous by exogenous peptides loaded on class II and these data are consistent with the exchange of peptides by recycling.

## 1.4 Adhesion molecules and signalling events

Antigen-specific responses have been found to require not only the TCR but also a set of adhesion molecules on the T cells and their complementary structures on the target cell or antigen-presenting cell (APC). The primary adhesion molecules that contribute to the binding avidity between a T cell and any cell are CD2 and CD11a/CD18 (also known as LFA-1, lymphocyte function associated antigen), shown schematically in Fig. 1.4. Their respective ligands, CD58 (LFA-3) and CD54 (intercellular adhesion molecule-1 ICAM-1) are expressed on all potential APC. In addition to these molecules,

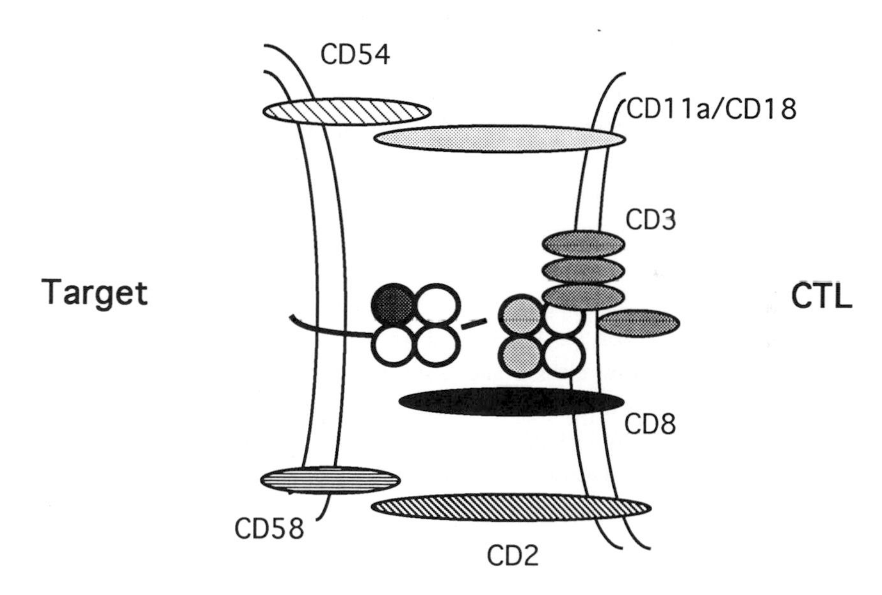


Fig. 1.4. Molecules that are important for T cell activation. In addition to the TCR/MHC interaction, interactions between T cell adhesion receptors and counter receptors are also important for T cell activation. CD2 and CD11a/Cd18 are adhesion molecules found on the T cell and they contribute significantly to the binding avidity between a T cell and any cell. Their respective counter receptors, CD58 and CD54, are expressed on all potential target cells. The TCR and CD8 (or CD4, see text) appear to be co-receptors that are brought together by co-recognition of the same peptide–MHC complex. In contrast to CD2 and CD11a/CD18, CD8 and CD4 appear to be more important for signalling than for adhesion.

the CD4 and CD8 glycoproteins, found on helper and cytotoxic T cells, respectively, also play an important role in T cell recognition. Monoclonal antibodies to any of these surface molecules can inhibit T cell responses, showing that T cell responses are highly complex processes that require the co-operation of a number of different surface molecules (Springer *et al.*, 1987; Blanchard *et al.*, 1988; Dougherty, Dransfield & Hogg, 1988; Bierer & Burakoff, 1989; Bierer *et al.*, 1989; Dustin, Olive & Springer, 1989). Subsequent studies have shown that all of these molecules are involved in cell–cell adhesion and many are also involved in the signalling events that result from the specific recognition of antigen.

T cell activation proceeds in two stages. The first involves a non-specific adhesion step that is independent of antigen recognition and the second involves the specific interaction between the TCR and peptide/MHC (Spits *et al.*, 1986; Bierer *et al.*, 1989; Matsui *et al.*, 1991). The cell–cell adhesion step allows T cells to screen all potential antigen-presenting cells and adhesion is aborted in the absence of specific antigen recognition by the TCR but intensified when the correct TCR–peptide/MHC contact is made. The antigen-independent adhesion results from binding of the CD2 and LFA-1 molecules on the T cell to the LFA-3 and ICAM molecules on the target cell (Spits *et al.*, 1986; Shaw & Luce, 1987; Springer *et al.*, 1987). CD2 and CD11a/CD18 may also do more than just promote cellular contact. Pairs of anti-CD2 monoclonal antibodies have been shown to stimulate T cell proliferation (Meuer *et al.*, 1984), while immobilized monoclonal antibodies specific for CD11a/CD18 induce co-mitogenic signals (Bierer & Burakoff, 1989).

The strong correlation between the type of MHC molecule recognized and CD8 or CD4 expression on T cells led to the proposal that CD8 and CD4 bind to determinants on class I and class II antigens, respectively. Indeed, CD8 has been shown to directly contact the $\alpha_3$ domain of class I (Salter *et al.*, 1989). At physiological densities, however, CD8 and CD4 seem to be not as important in adhesion as in signalling (Shaw *et al.*, 1986; Spits *et al.*, 1986). When this signalling contribution is blocked, as with monoclonal antibodies to either CD4 or CD8, the T cells require 100-fold higher concentrations of antigen in order to induce responsiveness (Janeway, 1988; von Boehmer, 1988; Parnes, 1989). CD4 and CD8 may also play an important role in the differentiation of immature thymocytes in the thymus. Early in T cell ontogeny, immature T cells express both CD4 and CD8. It is thought that co-association of one or the other of these molecules with the T cell receptor may help signal whether that receptor recognizes class I or class II and may then regulate the subsequent diffentiation of these cells into either $CD8^+$ cytotoxic or $CD4^+$ cells.

## References

Adorini, L., Moreno, J., Momburg, F., Hammerling, G.J., Guery, J.C., Valli, A. & Fuchs, S. (1991). Exogenous peptides compete for the presentation of endogenous antigens to major histocompatibility complex class II-restricted T cells. *Journal of Experimental Medicine*, 174, 945–948.

Arnold, D., Driscoll, J., Androlewicz, M., Hughes, E., Cresswell, P. & Spies, T. (1992). Proteasome subunits encoded in the MHC are not generally required for the processing of peptides bound by MHC class I molecules. *Nature (London)*, 360, 171–174.

Baas, E.J., van Santen, H.M., Kleijmeer, M.J., Geuze, H.J., Peters, P.J. & Ploegh, H.L. (1992). Peptide-induced stabilization and intracellular localization of empty HLA class I complexes. *Journal of Experimental Medicine*, 176, 147–156.

Babbitt, B.P., Allen, P.M., Matsueda, G., Haber, E. & Unanue, E.R. (1985). Binding of immunogenic peptides to Ia histocompatibility molecules. *Nature (London)*, 317, 359–361.

Bakke, O. & Dobberstein, B. (1990). MHC class II-associated invariant chain contains a sorting signal for endosomal compartments. *Cell*, 63, 707–716.

Benacerraf, B., Paul, W.E. & Green, I. (1967). The immune response of guinea pigs to hapten-poly-L-lysine conjugates as example of the genetic control of the recognition of antigenicity. *Cold Spring Harbor Symposium on Quantitative Biology*, 32, 569–574.

Bierer, B.E. & Burakoff, S.J. (1989). T-lymphocyte activation: the biology and function of CD2 and CD4. *Immunological Reviews*, 111, 267–294.

Bierer, B.E., Sleckman, B.P., Ratnofsky, S.E. & Burakoff, S.J. (1989). The biologic roles of CD2, CD4, and CD8 in T-cell activation. *Annual Review of Immunology*, 7, 579–599.

Bird, A. (1987). CpG islands as gene markers in the vertebrate nucleus. *Trends in Genetics*, 3, 342–347.

Blanchard, D., van Els, C., Aubry, J.P., de Vries, J.E. & Spits, H. (1988). CD4 is involved in a post-binding event in the cytolytic reaction mediated by human $CD4^+$ cytotoxic T lymphocyte clones. *Journal of Immunology*, 140, 1745–1752.

Blum, J.S. & Cresswell, P. (1988). Role for intracellular proteases in the processing and transport of class II HLA antigens. *Proceedings of the National Academy of Sciences of the USA*, 85, 3975–3979.

Brown, J.H., Jardetsky, T.S., Gorga, J.C., Stern, L.J., Urban, R.G., Strominger, J.L. & Wiley, D.C. (1993). Three-dimensional structure of the human class II histocompatibility antigen HLA-DR1. *Nature (London)*, 364, 33–39.

Brown, M.G., Driscoll, J. & Monaco, J.J. (1991). Structural and serological similarity of MHC-linked LMP and proteasome (multicatalytic proteinase) complexes. *Nature (London)*, 353, 355–357.

Buus, S., Sette, A., Colon, S.M., Miles, C. & Grey, H.M. (1987). The relation between major histocompatibility complex (MHC) restriction and the capacity of Ia to bind immunogenic peptides. *Science*, 235, 1353–1358.

Campbell, R.D. & Trowsdale, J. (1993). Map of the human MHC. *Immunology Today*, 14, 349–352.

Ceman, S., Rudersdorf, R., Long, E.O. & Demars, R. (1992). MHC class II deletion mutant expresses normal levels of transgene encoded class II molecules that have abnormal conformation and impaired antigen presentation ability. *Journal of Immunology*, 149, 754–761.

Cerundolo, V., Alexander, J., Anderson, K., Lamb, C., Cresswell, P., McMichael, A., Gotch, F. & Townsend, A. (1990). Presentation of viral antigen controlled by a gene in the major histocompatibility complex. *Nature (London)*, 345, 449–452.

Chain, B.M., Kaye, P.M. & Shaw, M.A. (1988). The biochemistry and cell biology of antigen processing. *Immunological Reviews*, 106, 33–58.

Chicz, R.M., Urban, R.G., Lane, W.S., Gorga, J.C., Stern, L.J., Vignali, D.A. & Strominger, J.L. (1992). Predominant naturally processed peptides bound to HLA-DR1 are derived from MHC-related molecules and are heterogeneous in size. *Nature (London)*, 358, 764–768.

Chothia, C., Boswe, D.R. & Lesk, A.M. (1988). The outline structure of the T-cell alpha/beta receptor. *EMBO Journal*, 7, 3745–3755.

Davidson, H.W., Reid, P.A., Lanzavecchia, A. & Watts, C. (1991). Processed antigen binds to newly synthesized MHC class II molecules in antigen-specific B lymphocytes. *Cell*, 67, 105–116.

Davidson, H.W., West, M.A. & Watts, C. (1990). Endocytosis, intracellular trafficking, and processing of membrane IgG and monovalent antigen/membrane IgG complexes in B lymphocytes. *Journal of Immunology*, 144, 4101–4109.

Davis, M.M. & Bjorkman, P.J. (1988). T-cell antigen receptor genes and T-cell recognition. *Nature (London)*, 334, 395–402.

Degen, E., Cohen Doyle, M.F. & Williams, D.B. (1992). Efficient dissociation of the p88 chaperone from major histocompatibility complex class I molecules requires both β2-microglobulin and peptide. *Journal of Experimental Medicine*, 175, 1653–1661.

Delovitch, T.L., Semple, J.W., Naquet, P., Bernard, N.F., Ellis, J., Champagne, P. & Phillips, M.L. (1988). Pathways of processing of insulin by antigen-presenting cells. *Immunological Reviews*, 106, 195–222.

DeMars, R., Rudersdorf, R., Chang, C., Petersen, J., Strandtmann, J., Korn, N., Sidwell, B. & Orr, H.T. (1985). Mutations that impair a post-transcriptional step in expression of HLA-A and -B antigens. *Proceedings of the National Academy of Sciences of the USA*, 82, 8183–8187.

DeMars, R. & Spies, T. (1992). New genes in the MHC that encode proteins for antigen processing. *Trends in Cell Biology*, 2, 81–86.

Deverson, E.V., Gow, I.R., Coadwell, W.J., Monaco, J.J., Butcher, G.W. & Howard, J.C. (1990). MHC class II region encoding proteins related to the multidrug resistance family of transmembrane transporters. *Nature (London)*, 348, 738–741.

Diment, S. (1990). Different roles for thiol and aspartyl proteases in antigen presentation of ovalbumin. *Journal of Immunology*, 145, 417–422.

Dougherty, G.J., Dransfield, I. & Hogg, N. (1988). Identification of a novel monocyte cell surface molecule involved in the generation of antigen-induced proliferative responses. *European Journal of Immunology*, 18, 2067–2071.

Dustin, M.L., Olive, D. & Springer, T.A. (1989). Correlation of CD2 binding and functional properties of multimeric and monomeric lymphocyte function-associated antigen 3. *Journal of Experimental Medicine*, 169, 503–517.

Falk, K., Rotzschke, O., Stevanovic, S., Jung, G. & Rammensee, H.G. (1991). Allele-specific motifs revealed by sequencing of self-peptides eluted from MHC molecules. *Nature (London)*, 351, 290–296.

Fischer Lindahl, K. & Wilson, D.B. (1977a). Histocompatibility antigen-activated cytotoxic T lymphocytes. I. Estimates of the absolute frequency of killer cells generated *in vitro*. *Journal of Experimental Medicine*, 145, 500–507.

Fischer Lindahl, K. & Wilson, D.B. (1977b). Histocompatibility antigen-activated cytotoxic T lymphocytes. II. Estimates of the frequency and specificity of precursors. *Journal of Experimental Medicine*, 145, 508–522.

Garrett, T.P., Saper, M.A., Bjorkman, P.J., Strominger, J.L. & Wiley, D.C. (1989). Specificity pockets for the side chains of peptide antigens in HLA-Aw68. *Nature (London)*, 342, 692–696.

Glynne, R., Powis, S.H., Beck, S., Kelly, A., Kerr, L.A. & Trowsdale, J. (1991). A proteasome-related gene between the two ABC transporter loci in the class II region of the human MHC. *Nature (London)*, 353, 357–360.

Goldberg, A.L. (1992). The mechanism and functions of ATP-dependent proteases in bacterial and animal cells. *European Journal of Biochemistry*, 203, 9–23.

Goldberg, A.L. & Rock, K.L. (1992). Proteolysis, proteasomes and antigen presentation. *Nature (London)*, 357, 375–379.

Gorer, P.A. (1936). The detection of antigenic differences in mouse erythrocytes by the employment of immune sera. *British Journal of Experimental Pathology*, 17, 42–50.

Harding, C.V., Collins, D.S., Slot, J.W., Geuze, H.J. & Unanue, E.R. (1991). Liposome-encapsulated antigens are processed in lysosomes, recycled, and presented to T cells. *Cell*, 64, 393–401.

Henderson, R.A., Michel, H., Sakaguchi, K., Shabanowitz, J., Appella, E., Hunt, D.F. & Engelhard, V.H. (1992). HLA-A2.1-associated peptides from a mutant cell line: a second pathway of antigen presentation. *Science*, 255, 1264–1266.

Hershko, A. & Ciechanover, A. (1992). The ubiquitin system for protein degradation. *Annual Review of Biochemistry*, 61, 761–807.

Hill, A.V.S., Elvin, J., Willis, A.C., Aidoo, M., Allsopp, C.E.M., Gotch, F.M., Gao, X.M., Takiguchi, M., Greenwood, B.M., Townsend, A.R.M., McMichael, A.J. & Whittle, H.C. (1992). Molecular analysis of the association of HLA-B53 and resistance to severe malaria. *Nature (London)*, 360, 434–439.

Hosken, N.A. & Bevan, M.J. (1990). Defective presentation of endogenous antigen by a cell line expressing class I molecules. *Science*, 248, 367–370.

Hunt, D.F., Henderson, R.A., Shabanowitz, J., Sakaguchi, K., Michel, H., Sevilir, N., Cox, A.L., Appella, E. & Engelhard, V.H. (1992a). Characterization of peptides bound to the class I MHC molecule HLA-A2.1, by mass spectrometry. *Science*, 255, 1261–1263.

Hunt, D.F., Michel, H., Dickinson, T.A., Shabanowitz, J., Cox, A.L., Sakaguchi, K., Appella, E., Grey, H.M. & Sette, A. (1992b). Peptides presented to the immune system by the murine class II major histocompatibility complex molecule I-Ad. *Science*, 256, 1817–1820.

Janeway, C.A. Jr (1988). T-cell development. Accessories or co-receptors? *Nature (London)*, 335, 208–210.

Jardetsky, T.S., Lane, W.S., Robinson, R.A., Madden, D.R. & Wiley, D.C. (1991). Identification of self peptides bound to purified HLA-B27. *Nature (London)*, 353, 326–329.

Kelly, A., Powis, S.H., Glynne, R., Radley, E., Beck, S. & Trowsdale, J. (1991). Second proteasome-related gene in the human MHC class II region. *Nature (London)*, 353, 667–668.

Kleijmeer, M.J., Kelly, A., Geuze, H.J., Slot, J.W., Townsend, A. & Trowsdale, J. (1992). Location of MHC-encoded transporters in the endoplasmic reticulum and cis-Golgi. *Nature (London)*, 357, 342–344.

Klein, J. (1975). *Biology of the Mouse Histocompatibility-2 Complex: Principles of Immunogenetics Applied to a Single System*. Berlin: Springer-Verlag.

Kronenberg, M., Siu, G., Hood, L.E. & Shastri, N. (1986). The molecular genetics of the T-cell antigen receptor and T-cell antigen recognition. *Annual Review of Immunology*, 4, 529–591.

Lamb, C.A., Yewdell, J.W., Bennink, J.R. & Cresswell, P. (1991). Invariant chain targets HLA class II molecules to acidic endosomes containing internalized influenza virus. *Proceedings of the National Academy of Sciences of the USA*, 88, 5998–6002.

Lanzavecchia, A. (1990). Receptor-mediated antigen uptake and its effect on antigen presentation to class II-restricted T lymphocytes. *Annual Review of Immunology*, 8, 773–793.

Lanzavecchia, A., Reid, P.A. & Watts, C. (1992). Irreversible association of peptides with class II MHC molecules in living cells. *Nature (London)*, 357, 249–252.

Lapham, C.K., Bacik, I., Yewdell, J.W., Kane, K.P. & Bennink, J.R. (1993). Class I molecules retained in the endoplasmic reticulum bind antigenic peptides. *Journal of Experimental Medicine*, 177, 1633–1641.

Ljunggren, G., Stam, N.J., Ohlen, C., Neefjes, J.J., Hoglund, P., Heemels, M.T., Bastin, J., Schumacher, T.N., Townsend, A., Karre, K. & Ploegh, H.L. (1990). Empty MHC class I molecules come out in the cold. *Nature (London)*, 346, 476–480.

Lotteau, V., Teyton, L., Peleraux, A., Nilsson, T., Karlsson, L., Schmid, S.L., Quaranta, V. & Peterson, P.A. (1990). Intracellular transport of class II MHC molecules directed by invariant chain. *Nature (London)*, 348, 600–605.

Loveland, B., Wang, C.R., Yonekawa, H., Hermel, E. & Lindahl, K.F. (1990). Maternally transmitted histocompatibility antigen of mice: a hydrophobic peptide of a mitochondrially encoded protein. *Cell*, 60, 971–980.

Lurquin, C., van Pel, A., Marianne, B., de Plaen, E., Szikora, J.P., Janssens, C., Reddehase, M.J., Lejeune, J. & Boon, T. (1989). Structure of the gene of tum$^-$ transplantation antigen P91A: the mutated exon encodes a peptide recognized with $L^d$ by cytolytic T cells. *Cell*, 58, 293–303.

Madden, D.R., Gorga, J.C., Strominger, J.L. & Wiley, D.C. (1991). The structure of HLA-B27 reveals nonamer self-peptides bound in an extended conformation. *Nature (London)*, 353, 321–325.

Martinez, C.K. & Monaco, J.J. (1991). Homology of proteasome subunits to a major histocompatibility complex-linked LMP gene. *Nature (London)*, 353, 664–667.

Matsui, K., Boniface, J.J., Reay, P.A., Schild, H., Fazekas de St Groth, B. & Davis, M.M. (1991). Low affinity interaction of peptide-MHC complexes with T cell receptors. *Science*, 254, 1788–1791.

McDevitt, H.O. & Chinitz, A. (1972). Genetic control of the antibody response: relationship between immune response and histocompatibility (H-2) type. *Science*, 175, 273–276.

Mellins, E., Kempin, S., Smith, L., Monji, T. & Pious, D. (1991). A gene required for class II-restricted antigen presentation maps to the major histocompatibility complex. *Journal of Experimental Medicine*, 174, 1607–1615.

Mellins, E., Smith, L., Arp, B., Cotner, T., Celis, E. & Pious, D. (1990). Defective processing and presentation of exogenous antigens in mutants with normal HLA class II genes. *Nature (London)*, 343, 71–74.

Meuer, S.C., Hussey, R.E., Fabbi, M., Fox, D., Acuto, O., Fitzgerald, K.A., Hodgdon, J.C., Protentis, J.P., Schlossman, S.F. & Reinherz, E.L. (1984). An alternative pathway of T-cell activation: a functional role for the 50 kd T11 sheep erythrocyte receptor protein. *Cell*, 36, 897–906.

Michalek, M.T., Grant, E.P., Gramm, C., Goldberg, A.L. & Rock, K.L. (1993). A role for the ubiquitin-dependent proteolytic pathway in MHC class I-restricted antigen presentation. *Nature(London)*, 363, 552–554.

Miller, J. & Germain, R.N. (1986). Efficient cell surface expression of class II MHC molecules in the absence of associated invariant chain. *Journal of Experimental Medicine*, 164, 1478–1489.

Momburg, F., Ortiz Navarrete, V., Neefjes, J., Goulmy, E., van de Wal, Y., Spits, H., Powis, S.J., Butcher, G.W., Howard, J.C., Walden, P. & Hammerling, G.J. (1992). Proteasome subunits encoded by the major histocompatibility complex are not essential for antigen presentation. *Nature (London)*, 360, 174–177.

Monaco, J.J., Cho, S. & Attaya, M. (1990). Transport protein genes in the murine MHC: possible implications for antigen processing. *Science*, 250, 1723–1726.

Monaco, J.J. & McDevitt, H.O. (1986). The LMP antigens: a stable MHC-controlled multisubunit protein complex. *Human Immunology*, 15, 416–426.

Neefjes, J.J. & Ploegh, H.L. (1992a). Inhibition of endosomal proteolytic activity by leupeptin blocks surface expression of MHC class II molecules and their conversion to SDS resistant α/β heterodimers in endosomes. *EMBO Journal*, 11, 411–416.

Neefjes, J.J. & Ploegh, H.L. (1992b). Intracellular transport of MHC class II molecules. *Immunology Today*, 13, 179–184.

Neefjes, J.J, Stollorz, V., Peters, P.J., Geuze, H.J. & Ploegh, H.L. (1990). The biosynthetic pathway of MHC class II but not class I molecules intersects the endocytic route. *Cell*, 61, 171–183.

Nguyen, Q.V., Knapp, W. & Humphreys, R.E. (1989). Inhibition by leupeptin and antipain of the intracellular proteolysis of Ii. *Human Immunology*, 24, 153–163.

Nuchtern, J.G., Bonifacino, J.S., Biddison, W.E. & Klausner, R.D. (1989). Brefeldin A implicates egress from endoplasmic reticulum in class I restricted antigen presentation. *Nature (London)*, 339, 223–226.

Ortiz Navarrete, V., Seelig, A., Gernold, M., Frentzel, S., Kloetzel, P.M. & Hammerling, G.J. (1991). Subunit of the '20S' proteasome (multicatalytic proteinase) encoded by the major histocompatibility complex. *Nature (London)*, 353, 662–664.

Parham, P., Lomen, C.E., Lawlor, D.A., Ways, J.P., Holmes, N., Coppin, H.L., Salter, R.D., Wan, A.M. & Ennis, P.D. (1988). Nature of polymorphism in HLA-A, -B, and -C molecules. *Proceedings of the National Academy of Sciences of the USA*, 85, 4005–4009.

Parnes, J.R. (1989). Molecular biology and function of CD4 and CD8. *Advances in Immunology*, 44, 265–311.

Pazmany, L., Rowland Jones, S., Huet, S., Hill, A., Sutton, J., Murray, R., Brooks, J. & McMichael, A. (1992). Genetic modulation of antigen presentation by HLA-B27 molecules. *Journal of Experimental Medicine*, 175, 361–369.

Peters, P.J., Neefjes, J.J., Oorschot, V., Ploegh, H.L. & Geuze, H.J. (1991). Segregation of MHC class II molecules from MHC class I molecules in the Golgi complex for transport to lysosomal compartments. *Nature (London)*, 349, 669–676.

Pierce, S.K., Morris, J.F., Grusby, M.J., Kaumaya, P., van Buskirk, A., Srinivasan, M., Crump, B. & Smolenski, L.A. (1988). Antigen-presenting function of B lymphocytes. *Immunological Reviews*, 106, 149–180.

Pieters, J., Horstmann, H., Bakke, O., Griffiths, G. & Lipp, J. (1991). Intracellular transport and localization of major histocompatibility complex class II molecules and associated invariant chain. *Journal of Cell Biology*, 115, 1213–1223.

Powis, J., Townsend, A.R.M., Deverson, E.V., Bastin, J., Butcher, G.W. & Howard, J.C. (1991). Restoration of antigen presentation to the mutant cell line RMA-S by an MHC-linked transporter. *Nature (London)*, 354, 528–531.

Powis, S.J., Deverson, E.V., Coadwell, W.J., Ciruela, A., Huskisson, N.S., Smith, H., Butcher, G.W. & Howard, J.C. (1992). Effect of polymorphism of an MHC-linked transporter on the peptides assembled in a class I molecule. *Nature (London)*, 357, 211–215.

Rammensee, H.G., Falk, K. & Rötzschke, O. (1993). Peptides naturally presented by MHC class I molecules. *Annual Review of Immunology*, 11, 213–244.

Rechsteiner, M. (1987). Ubiquitin-mediated pathways for intracellular proteolysis. *Annual Review of Cell Biology*, 3, 1–30.

Roche, P.A. & Cresswell, P. (1990). Invariant chain association with HLA-DR molecules inhibits immunogenic peptide binding. *Nature (London)*, 345, 615–618.

Roche, P.A. & Cresswell, P. (1991). Proteolysis of the class II-associated invariant chain generates a peptide binding site in intracellular HLA-DR molecules. *Proceedings of the National Academy of Sciences of the USA*, 88, 3150–3154.

Roche, P.A., Marks, M.S. & Cresswell, P. (1991). Formation of a nine-subunit complex by HLA class II glycoproteins and the invariant chain. *Nature (London)*, 354, 392–394.

Salter, R.D. & Cresswell, P. (1986). Impaired assembly and transport of HLA-A and -B antigens in a mutant TxB cell hybrid. *EMBO Journal*, 5, 943–949.

Salter, R.D., Norment, A.M., Chen, B.P., Clayberger, C., Krensky, A.M., Littman, D.R. & Parham, P. (1989). Polymorphism in the alpha 3 domain of HLA-A molecules affects binding to CD8. *Nature (London)*, 338, 345–347.

Saper, M.A., Bjorkman, P.J. & Wiley, D.C. (1991). Refined structure of the human histocompatibility antigen HLA-A2 at 2.6 Å resolution. *Journal of Molecular Biology*, 219, 277–319.

Sargent, C.A., Dunham, I. & Campbell, R.D. (1989). Identification of multiple HTF-island associated genes in the human major histocompatibility complex class III region. *EMBO Journal*, 8, 2305–2312.

Schultz, M., Aichele, P., Schneider, R., Hansen, T.H., Zinkernagel, R.M. & Hengartner, H. (1991). Major histocompatibility complex binding and T cell recognition of a viral nonapeptide containing a minimal tetrapeptide. *European Journal of Immunology*, 21, 1181–1185.

Schumacher, T.N., Heemels, M.T., Neefjes, J.J., Kast, W.M., Melief, C.J. & Ploegh, H.L. (1990). Direct binding of peptide to empty MHC class I molecules on intact cells and *in vitro*. *Cell*, 62, 563–567.

Sekaly, R.P., Tonnelle, C., Strubin, M., Mach, B. & Long, E.O. (1986). Cell surface expression of class II histocompatibility antigens occurs in the absence of the invariant chain. *Journal of Experimental Medicine*, 164, 1490–1504.

Shaw, S. & Luce, G.E. (1987). The lymphocyte function-associated antigen (LFA)-1 and CD2/LFA-3 pathways of antigen-independent human T cell adhesion. *Journal of Immunology*, 139, 1037–1045.

Shaw, S., Luce, G.E., Quinones, R., Gress, R.E., Springer, T.A. & Sanders, M.E. (1986). Two antigen-independent adhesion pathways used by human cytotoxic T-cell clones. *Nature (London)*, 323, 262–264.

Silver, M.L., Guo, H.C., Strominger, J.L. & Wiley, D.C. (1992). Atomic structure of a human MHC molecule presenting an influenza virus peptide. *Nature (London)*, 360, 367–369.

Snell, G.D. (1958). Histocompatibility genes of the mouse. II. Production and analysis of isogenic resistant lines. *Journal of the National Cancer Institute,* 21, 843–877.

Spies, T., Bresnahan, M., Bahram, S., Arnold, D., Blanck, G., Mellins, E., Pious, D. & DeMars, R. (1990). A gene in the human major histocompatibility complex class II region controlling the class I antigen presentation pathway. *Nature (London)*, 348, 744–747.

Spies, T., Bresnahan, M. & Strominger, J.L. (1989). Human major histocompatibility complex contains a minimum of 19 genes between the complement cluster and HLA-B. *Proceedings of the National Academy of Sciences of the USA*, 86, 8955–8958.

Spies, T., Cerundolo, V., Colonna, M., Cresswell, P., Townsend, A. & DeMars, R. (1992). Presentation of viral antigen by MHC class I molecules is dependent on a putative peptide transporter heterodimer. *Nature (London)*, 355, 644–646.

Spies, T. & DeMars, R. (1991). Restored expression of major histocompatibility class I molecules by gene transfer of a putative peptide transporter. *Nature (London)*, 351, 323–324.

Spits, H., van Schooten, W., Keizer, H., van Seventer, G., van de Rijn, M., Terhorst, C. & de Vries, J.E. (1986). Alloantigen recognition is preceded by nonspecific adhesion of cytotoxic T cells and target cells. *Science,* 232, 403–405.

Springer, T.A., Dustin, M.L., Kishimoto, T.K. & Marlin, S.D. (1987). The lymphocyte function-associated LFA-1, CD2, and LFA-3 molecules: cell adhesion receptors of the immune system. *Annual Review of Immunology*, 5, 223–252.

Steinman, R.M., Mellman, I.S., Muller, W.A. & Cohn, Z.A. (1983). Endocytosis and the recycling of plasma membrane. *Journal of Cell Biology*, 96, 1–27.

Sutton, J., Rowland-Jones, S., Rosenberg, W., Nixon, D., Gotch, F., Gao, X.M., Murray , N., Spoonas, A., Driscoll, P., Smith, M., Willis, A. & McMichael, A. (1993). A sequence pattern for peptides presented to cytotoxic T lymphotcytes by HLA-B8 revealed by analysis of epitopes and eluted peptides. *European Journal of Immunology*, 23, 447–453.

Takahashi, H., Cease, K.B. & Berzofsky, J.A. (1989). Identification of proteases that process distinct epitopes on the same protein. *Journal of Immunology*, 142, 2221–2229.

Teyton, L., O'Sullivan, D., Dickson, P.W., Lotteau, V., Sette, A., Fink, P. & Peterson, P.A. (1990). Invariant chain distinguishes between the exogenous and endogenous antigen presentation pathways. *Nature (London)*, 348, 39–44.

Townsend, A.R., Bastin, J., Gould, K. & Brownlee, G.G. (1986a). Cytotoxic T lymphocytes recognize influenza haemagglutinin that lacks a signal sequence. *Nature (London)*, 324, 575–577.

Townsend, A.R., Bastin, J., Gould, K., Brownlee, G., Andrew, M., Coupar, B., Boyle, D., Chan, S. & Smith, G. (1988). Defective presentation to class I-restricted cytotoxic T lymphocytes in vaccinia-infected cells is overcome by enhanced degradation of antigen. *Journal of Experimental Medicine*, 168, 1211–1224.

Townsend, A.R., Elliott, T., Cerundolo, V., Foster, L., Barber, B. & Tse, A. (1990). Assembly of MHC class I molecules analyzed *in vitro. Cell*, 62, 285–295.

Townsend, A.R., Gotch, F.M. & Davey, J. (1985). Cytotoxic T cells recognize fragments of the influenza nucleoprotein. *Cell*, 42, 457–467.

Townsend, A., Ohlen, C., Bastin, J., Ljunggren, H.G., Foster, L. & Karre, K. (1989). Association of class I major histocompatibility heavy and light chains induced by viral peptides. *Nature (London)*, 340, 443–448.

Townsend, A.R., Rothbard, J., Gotch, F.M., Bahadur, G., Wraith, D. & McMichael, A.J. (1986b). The epitopes of influenza nucleoprotein recognized by cytotoxic T lymphocytes can be defined with short synthetic peptides. *Cell*, 44, 959–968.

Trowsdale, J., Hanson, I., Mockridge, I., Beck, S., Townsend, A. & Kelly, A. (1990). Sequences encoded in the class II region of the MHC related to the 'ABC' superfamily of transporters. *Nature (London)*, 348, 741–744.

Trowsdale, J., Ragoussis, J. & Campbell, R.D. (1991). Map of the human MHC. *Immunology Today*, 12, 443–446.

van Noort, J.M., Boon, J., van der Drift, A.C., Wagenaar, J.P., Boots, A.M. & Boog, C.J. (1991). Antigen processing by endosomal proteases determines which sites of sperm-whale myoglobin are eventually recognized by T cells. *European Journal of Immunology*, 21, 1989–1996.

van Someren, H., Westerveld, A., Hagemeijer, A., Mees, J.R., Meera Khan, P. & Zaalberg, O.B. (1974). Human antigen and enzyme markers in man–chinese hamster somatic cell hybrids: evidence for synteny between the HLA-A, $PGM_3$, $ME_1$ and IPO-B loci. *Proceedings of the National Academy of Sciences of the USA*, 71, 962–965.

von Boehmer, H. (1988). The developmental biology of T lymphocytes. *Annual Review of Immunology*, 6, 309–326

Wei, M.L. & Cresswell, P. (1992). HLA-A2 molecules in an antigen-processing mutant cell contain signal sequence-derived peptides. *Nature (London)*, 356, 443–446.

Yap, K.L. & Ada, G.L. (1978). Transfer of specific cytotoxic T lymphocytes protects mice inoculated with influenza virus. *Nature (London)*, 273, 238–239.

Yewdell, J.W. & Bennink, J.R. (1989). Brefeldin A specifically inhibits presentation of protein antigens to cytotoxic T lymphocytes. *Science*, 244, 1072–1075.

Yu Rudensky, A., Preston-Hurlburt, P., Hong, S.-C., Barlow, A. & Janeway, C.A. Jr (1991). Sequence analysis of peptides bound to MHC class II molecules. *Nature (London)*, 353, 622–627.

Yu Rudensky, A., Preston-Hurlburt, P., al Ramadi, B.K., Rothbard, J. & Janeway, C.A. Jr (1992). Truncation variants of peptides isolated from MHC class II molecules suggest sequence motifs. *Nature (London)*, 359, 429–431.

Zinkernagel, R.M. & Althage, A. (1977). Antiviral protection by virus-immune cytotoxic T cells: infected target cells are lysed before infectious virus progeny is assembled. *Journal of Experimental Medicine*, 145, 644–651.

# 2

# Organization of the MHC

JIANNIS RAGOUSSIS
*Guy's Hospital, London*

## 2.1 Introduction

The MHC is localized on the short arm of chromosome 6 in band 21.3. It extends over 4 Mb of DNA, which have been intensively studied for several decades. Detailed physical maps have been derived and the entire complex has been cloned in cosmid and yeast artificial chromosome (YAC) vectors. About 80 genes have been identified so far, which include the genes encoding the class I or classical transplantation antigens, the class II immune response genes, the class III genes originally defined by complement components and a number of novel genes involved in antigen processing or of unknown function (Trowsdale, Ragoussis & Campbell, 1991). It is intriguing that so many genes involved in immune responses are closely linked to each other in the genome. The class I and class II sequences have been maintained together through evolution and can be found on the same chromosome in species like chicken, mouse and humans. The organization and functional relationships between MHC genes will be presented in this chapter along with a description of MHC transcripts.

## 2.2 The class I region

### *2.2.1 The human class I region*

The class I region is 2 Mb in length and contains the class I multigene family comprising about 20 non-allelic DNA sequences (Fig. 2.1). These can be divided into the polymorphic classical transplantation antigen genes HLA-A, HLA-B and HLA-C and the non-classical class I genes, the HLA–E, HLA-F and HLA-G genes and including a number of pseudogenes like HLA-J and HLA-H (Klein & Figueroa, 1986; Koller *et al.*, 1989). The classical class I

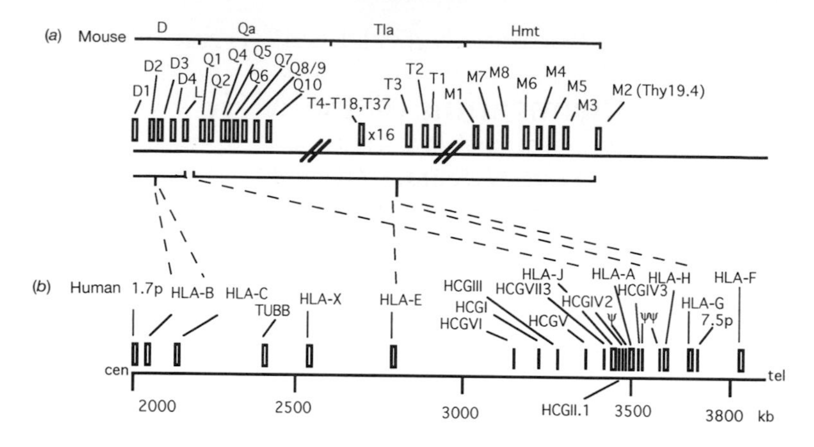


Fig. 2.1. Map of the human MHC class I region compared with the mouse. The human HLA-A, HLA-B and HLA-C loci are classical class I genes like the mouse D genes. The human HLA-E, HLA-H and HLA-F loci correspond to the mouse non-classical Qa/Tla genes. The solid bar indicates the region where deletions have been detected. The novel haemochromatosis candidate genes are designated HCG. cen: centromere; tel: telomere.

The mouse Tla region contains a haplotype-dependent number of genes, which are designated here as T4–T18 for reasons of simplicity. The mouse M3 gene encodes the molecule presenting the maternally transmitted antigen (see text). // bars indicate that these regions of the mouse genome have not been physically linked.

genes HLA-B, HLA-C and HLA-A present antigens to cytotoxic T cells since they can bind peptides from intracellularly degraded proteins and expose them at the cell surface. They are expressed on almost all nucleated cells in the body and their regulation and function are described in more detail in Chapter 1. HLA-B and HLA-C mark the centromeric end of the class I region. An isolated class I exon (1.7p) is located immediately centromeric of HLA-B (Bronson *et al.*, 1991). HLA-C lies 100 kb telomeric of HLA-B and HLA-A lies about 1400 kb further towards the telomere (Geraghty *et al.*, 1992). A number of class I pseudogenes have been mapped within 50 kb on either side of HLA-A, including HLA-J (Ragoussis *et al.*, 1989). An additional class I sequence, HLA-X, maps between HLA-C and HLA-E (Chimini *et al.*, 1990).

The non-classical genes can show a limited tissue or cell type expression, while their function is not well established. Some of them are presumed functionless and they may represent the remnants of gene duplication events. In particular the HLA-E gene, lying between HLA-C and HLA-A, is ubiquitously transcribed, but the protein is barely detectable on the cell surface. This has been shown to result from defective binding of HLA-E to available endogenous peptides (Ulbrecht *et al.*, 1992). HLA-F is located about 350 kb away from HLA-A (Schmidt & Orr, 1991; Gruen *et al.*, 1992) and marks the telomeric end of the class I region. It has a shorter cytoplasmic segment than all the other class I genes and its 3′ untranslated region is distinct from all the others of this family (Geraghty *et al.*, 1990). The region between HLA-F and HLA-A contains four pseudogenes, including the 7.5p fragment and the HLA-H (about 100 kb telomeric of HLA-A) and HLA-G (about 200 kb telomeric of HLA-A) genes (Shukla *et al.*, 1991; El Kahloun *et al.*, 1993). HLA-G shows a limited expression pattern (Chorney *et al.*, 1990; Kovats *et al.*, 1990). Polypeptides have been found in the extravillous cytotrophoblast and choriocarcinoma cell lines. The mRNA has been detected in foetal eye, thymus and choriocarcinoma (Shukla *et al.*, 1990). HLA-H is structurally related to HLA-A, suggesting that the two genes were formed by duplication of an ancestral gene. HLA-H shows limited diversity and all its alleles contain deleterious mutations with the effect that the class I heavy chain is not functional in antigen presentation (Zemmour *et al.*, 1990). This indicates that this gene has lost its function by force of natural selection and that the usage of MHC genes can change with time, as suggested by analysis of primate species (Watkins *et al.*, 1990). If HLA-H is the product of gene duplication by unequal chromosomal exchange, this would be reflected in gene number and structural variations in this region. Indeed deletions have been found encompassing 50 kb around HLA-H in particular haplotypes such as A24 and A23 (El Kahloun *et al.*, 1992; Venditti & Chorney, 1992). Although this degree of variation is limited by comparison with that found in the mouse and rat class I regions (Chimini *et al.*, 1988), it indicates that the MHC is an active genomic region where gene conversion and duplication or deletion are frequent events (Belich *et al.*, 1992; Watkins *et al.*, 1992). In the vicinity of HLA-E, a novel gene has been detected that belongs to the G-protein family (Denizot *et al.*, 1992). The region between HLA-E and HLA-F has been intensively investigated owing to the assignment of the gene responsible for haemochromatosis within 1 cM from HLA-A. Seven novel transcripts named HCGI–HCGVII have been isolated (El Kahloun *et al.*, 1993). HCGI encodes for a 1.2 kb mRNA expressed in duodenal mucosa. HCGII is expressed in the duodenal mucosa, placenta, lung and skeletal muscle as a 3.6 kb mRNA. HCGIV is expressed only in duodenal mucosa, where a 1.1 kb transcript has been detected. Messenger RNAs of HCGV (1.3 kb), HCGVI (0.9 kb) and HCGVII (1.7 kb) have been found in the duodenal mucosa and at higher levels in a number of other tissues. Current studies will reveal whether one of them is responsible for the disease or if they will bear any functional relationship with known MHC molecules.

As is the case with the class II and class III regions, the class I gene family is embedded around other, structurally unrelated genes. The first to be mapped is the β-tubulin gene (TUBB), between HLA-C and HLA-E (Volz *et al.*, 1994).

### *2.2.2 The mouse class I region*

The mouse class I region is divided into the D region, containing the classical polymorphic class I genes, and the Qa, Tla and Hmt regions, containing the less polymorphic 'non-classical' class I genes (Klein & Figueroa, 1986). (Nomenclature used for genes and protein products is that in use at 1992/93.) The D region extends over 250 kb and contains a haplotype-specific number of D (about four) and L genes, the human homologues of which are the HLA-A, HLA-B and HLA-C genes. In addition, two mouse class I genes, K and K2, are located at the other side of the complex at the class II region. The Qa region extends over 250 kb and contains (at present estimates) up to 10 haplotype-specific genes. It is not physically linked to the Tla region, which extends over 175 kb and contains a variable number of Tla genes (Richards *et al.*, 1989). The Hmt region is also not physically linked to the other two and contains eight non-classical class I M genes (Brorson *et al.*, 1989; Richards *et al.*, 1989; Fischer Lindahl *et al.*, 1991). The Qa, Tla and Hmt proteins are of largely unknown function but their lack of polymorphism suggests that they may present specific peptides derived either from self or from microbial proteins. For example, the Hmt M3 molecule presents peptides containing *N*-formyl-Met at the amino-terminus, indicating that they may play a role in bacterial antigen presentation (Wang, Loveland & Fischer Lindahl, 1991; Hedrick, 1992).

A comparison of the mouse and human class I regions shows overall conservation except that in humans the non-classical class I genes are dispersed among the classical class I genes and are not organized in distinct regions (Fig. 2.1).

## 2.3 The class II region

### *2.3.1 The human class II region*

The class II region consists of three subregions, DP, DQ and DR. Each of these subregions contains at least one functional A and B gene pair, which encode α and β chains that form a heterodimer on the cell surface (see

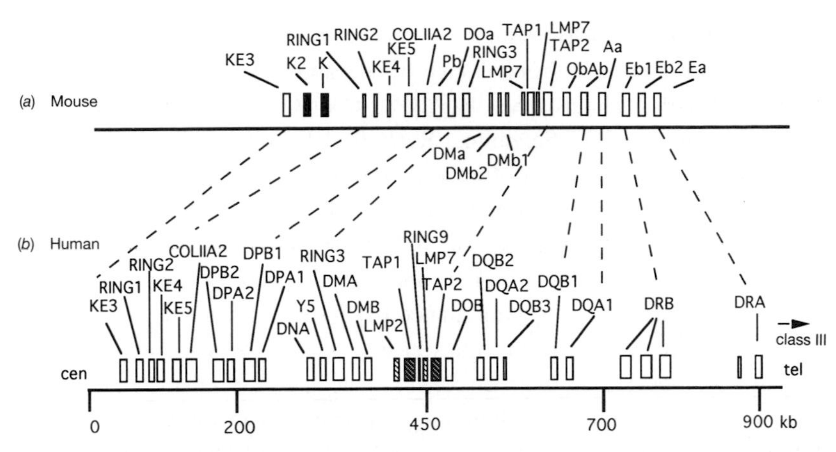


Fig. 2.2. The human class II region compared with that of the mouse. The genes encoding antigen-processing components are hatched. The mouse class I genes K and K2 (filled boxes) have been inserted in the mouse class II region and are not present at the same position in humans. cen: centromere; tel: telomere.

Fig. 2.2). These are the classical class II antigens. The class II region has been extensively mapped by pulsed-field gel electrophoresis (PFGE) resulting in a detailed map ranging over 1Mb of DNA (Dunham *et al.*, 1987; Kozono *et al.*, 1991; Ragoussis *et al.*, 1991). The distance of the most centromeric classical class II gene, DPB2, to the most telomeric DRA is 750 kb in most cases and can vary between haplotypes. A number of novel genes have been detected within and centromeric to the class II region and will be described.

The DR subregion contains one DRA and between one and four DRB genes and/or pseudogenes according to haplotype (illustrated in Fig. 2.2). The average length of this subregion is 240 kb in the DR3, DR5 and DR6 haplotypes. The DR2 haplotype contains 30 kb more DNA, while the DR4, DR7 and DR9 haplotypes contain 110 kb of additional DNA (Dunham *et al.*, 1989; Kendall, Todd & Campbell, 1991) The DR8 haplotype probably also contains more DNA than the DR3, DR5 and DR6 haplotypes, in the order of 150 kb (Ragoussis *et al.*, 1991). The DQ and DP subregions contain the same amount of DNA and number of genes in all the above investigated haplotypes. The DRA gene is the most telomeric classical class II gene. A DRB pseudogene has been mapped 20 kb centromeric of DRA (Meunier *et al.*, 1986), followed by one to four DRB genes or pseudogenes depending on the haplotype (Trowsdale *et al.*, 1991), as shown in Fig. 2.2.

The DQ subregion contains two functional genes: DQA and DQB. Three other DQ genes are present centromeric of DQA/B, termed DQA2, DQB2 and DQB3, for which no transcripts have yet been found (Ando *et al.*, 1989).

A separate B gene, DOB, encodes a distinct type of class II β chain. It features a pronounced hydrophobicity of the amino-terminal region and it may interact with α chains other than DP, DQ or DR, thus potentially fulfilling a different function (Servenius, Rask & Peterson, 1987). Immediately centromeric of DOB, a cluster of exciting novel genes has been found. These include the peptide-transporter genes TAP1 and TAP2, the proteasome components LMP2 and LMP7, as well as RING9, which encodes a transcript of currently unknown function. TAP1 and TAP2 (also called RING4 and RING11) have been identified simultaneously in the human and mouse MHC (Monaco, Cho & Attaya, 1990; Spies *et al.*, 1990; Trowsdale *et al.*, 1990). Two key findings led to their discovery. One was the characterization of human mutant cell lines with homozygous deletions in the class II region (Mellins *et al.*, 1991; Riberdy & Cresswell, 1992) and the genetic mapping of a locus called class I modifier (*cim*) affecting class I antigen presentation in the rat MHC class II region (Livingstone *et al.*, 1991; Powis, Howard & Butcher, 1991). TAP1 and TAP2 are members of the ABC (ATP-binding cassette) superfamily, the members of which are able to transport a wide spectrum of substrates through membranes. Defects in either TAP1 or TAP2 result in the formation of unstable class I molecules and the loss of presentation of intracellular antigens, suggesting that they transport peptides from the cytosol to the endoplasmic reticulum (Parham, 1990; Spies & DeMars, 1991; Powis *et al.*, 1992). It is particularly interesting that polymorphisms in the rat transporter genes affect the repertoire of peptides available for binding to class I antigens (Monaco, 1992). Such polymorphisms have been found also in the human TAP2 gene (Powis *et al.*, 1992). It has been further demonstrated that the TAP1 and TAP2 proteins form a complex and that ATP hydrolysis is important for their function *in vivo* (Kelly *et al.*, 1992). LMP2 and LMP7 encode proteins related to a large cytoplasmic protein complex, the proteasome, that plays a part in the degradation of intracellular proteins (Cameron *et al.*, 1990; Glynne *et al.*, 1991; Kelly *et al.*, 1991b). It is possible that LMP2 and LMP7 bind to this complex, leading to the production of peptides suitable for binding to class I or class II proteins (Robertson, 1991). However, transfection of LMP2 and LMP7 into mutant or deletion cell lines along with TAP1 and TAP2 has shown that the MHC proteasome components are not necessary for class I antigen presentation, leaving a question about their true function (Momburg *et al.*, 1992).

The DM locus lies immediately centromeric of LMP2 (Cho, Attaya & Monaco, 1991; Kelly *et al.*, 1991a). In humans, it contains the DMA and DMB genes, which code for two proteins similar to the known α and β chains, respectively. The primary sequence predicts that both proteins contain a transmembrane domain and two external domains. The membrane proximal domain is homologous to the immunoglobulin fold and both proteins contain residues diagnostic of class II $\alpha$ and $\beta$ chains, respectively. In contrast, the membrane distal domains are only marginally related to other class II genes. In addition, the immunoglobulin-like domains are as closely related to class I $\alpha_3$ domains as they are to class II $\alpha_2$ domains and to members of its own family. This suggests that the DM genes diverged from other MHC genes before any of the known class II genes were duplicated. The DM genes are transcribed in the same tissues as the class II genes and are inducible by interferon-$\gamma$ (IFN-$\gamma$).

RING3 is a ubiquitously expressed gene homologous to the *Drosophila melanogaster* female sterile homeotic (*fsh*) gene (Hanson, Poustka & Trowsdale, 1991; Okamoto *et al.*, 1991; Beck *et al.*, 1992). Fsh is implicated in early embryo segmentation and shows sequence homology to proteins involved in cell cycle control, cell division or cell growth regulation. The *fsh* gene is highly conserved and could play a role in human development. Close to RING3 is another expressed sequence, Y5 (Spies *et al.*, 1990).

The DNA gene (former DZ$\alpha$) encodes a typical class II $\alpha$ chain. It produces an unusually large 3 kb mRNA containing a typical polyadenylation signal. The function of the DNA gene is not clear since it was not possible until now to identify the encoded protein or its $\beta$ chain partner. The product of the mouse homologue of DNA, termed DO$\alpha$, can form a dimer with the DO$\beta$ molecule on the cell surface. Therefore, it is possible that the DNA and DOB products also form a functional dimer (Karlsson *et al.*, 1992). The DP subregion contains two pairs of A and B class II genes of which DPA1 and DPB1 are functional while DPA2 and DPB2 are pseudogenes.

The COL11A2 gene was mapped centromeric of DPB2 and the cloning of this region prompted the identification and mapping of a number of novel genes (Hanson *et al.*, 1991). These are KE5, KE3 and KE4, the last coding for a histidine-rich transmembrane protein first identified in the mouse (Abe *et al.*, 1988). RING2 and RING1 were first isolated in humans. RING1 is particularly interesting since it codes for a novel cysteine-rich protein that defines a new class of transcription factors (Trowsdale *et al.*, 1991).

### *2.3.2 The mouse class II region*

There is a striking similarity between the class II regions of mice and humans (Hanson & Trowsdale, 1991). Differences include the two unusually positioned class I genes H-2K and K2 in the mouse. These two genes have been translocated between KE3 and RING1 in the mouse. Other differences include the absence of any DPA1 or DPA2 homologue in the mouse and the lack of any DQA2, DQB2 or DQB3 homologues. Overall the 300 kb long mouse class II region is much more compact than the human class II region, which is at least three times longer.

## 2.4 The class III region

### *2.4.1 The human class III region*

The class III region spans approximately 1.1 Mb of DNA between the DRA and HLA-B genes. Initially it was identified by the presence of genes for the complement components C4, C2 and factor B. The 21-hydroxylase gene was mapped close to C4 and later the tumour necrosis factor cytokines TNF-α and TNF-β were mapped between C2 and HLA-B (Carroll *et al.*, 1987). The class III region has been analysed in detail and a large number of novel transcripts have been detected (Sargent, Dunham & Campbell, 1989a; Sargent *et al.*, 1989b; Spies *et al.*, 1989a; Spies, Bresnahan & Strominger, 1989b; Kendall, Sargent & Campbell, 1990; Milner & Campbell, 1992a). A diagrammatic representation of the class III region is shown in Fig. 2.3. The BAT1 (HLA-B associated transcript) gene encodes a 1.7 kb transcript that is ubiquitously expressed. The cytokines TNF-α and TNF-β are structurally related and produced by activated macrophages, natural killer (NK) cells and stimulated lymphocytes. Their pleiotropic activities include cytostatic and cytotoxic effects upon certain tumour cells. TNF-α (which is also known as

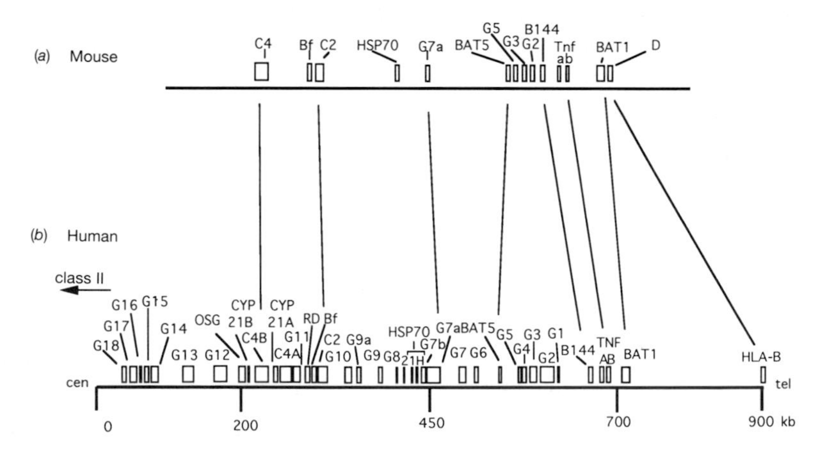


Fig. 2.3. The human class III region compared with that of the mouse. The mouse class III region extends over a shorter length of DNA but more is known about its gene content than in humans. cen: centromere; tel: telomere.

cachectin) can switch the metabolism of a cell from anabolic to catabolic processes and it has a number of actions that overlap with those of interleukin 1 (IL-1). One notable activity of TNF-α is that it enhances the transcription of the class I genes in a way similar to that of interferons (see Chapters 3 and 4). The B144 gene encodes a 0.8 kb transcript that is specific for B cells and macrophages and was first detected in the mouse (Tsuge *et al.*, 1987). G1 encodes a 0.6 kb transcript expressed in monocytes, macrophages and T lymphocytes but not in B lymphocytes. The 10 kDa G1 protein contains two potential calcium-binding domains in its amino-terminal portion and may play a role in T cell activation (Sargent *et al.*, 1989a). The G2 and G3 transcripts (or BAT2 and BAT3) are 6.0 kb and 3.8 kb, respectively, and are ubiquitously expressed. They encode proline-rich proteins that contain novel and RGD (Arg–Gly–Asp) repetitive elements (Banerji *et al.*, 1990). The RGD motif functions in cell adhesion by mediating the interaction of members of the integrin receptor superfamily with their ligands. These ligands are proteins like fibronectin, vitronectin, type I collagen and fibrinogen, which also contain one or two RGD sequences. G4 and G5 encode 2.5 kb and 1.6 kb transcripts, respectively, which are ubiquitously expressed and have no similarity to known proteins. The same is true for BAT-5 and G6, which encode for 2.0 kb and 1.4/1.5 kb transcripts, respectively. The 3.0 kb G7 transcript has been detected only in monocytes and macrophages but it does not show any homology to known sequences (Spies *et al.*, 1989b). In contrast, the 4.0 kb G7a transcript, which is also ubiquitously expressed, shows strong homology to the valyl tRNA synthetase of *Saccharomyces cerevisiae*. The G7a protein contains sequences characteristic of class I tRNA synthetases and its predicted molecular mass corresponds to that of other mammalian tRNA synthetases (Milner & Campbell, 1992a). The G7b gene spans approximately 10 kb between G7a and HSP70-Hom and encodes a 1 kb transcript that is ubiquitously expressed (Olavesen, Snoek & Campbell, 1993).

The HSP70-1, HSP70-2 and HSP70-Hom intron-less genes form a duplicated locus encoding the major heat shock proteins HSP70. The family of stress proteins of 70 kDa (HSP70) is a most abundant and highly conserved protein family. Members of this family are expressed in response to heat shock, free radicals, toxic metals and other stress stimuli. Indeed HSP70-1 and HSP70-2 are highly expressed as 2.4 kb mRNAs in cells subjected to heat shock at 42 °C. HSP70-1 is expressed constitutively at low levels, as is the HSP-70-Hom gene (which lacks the 5′ heat shock consensus sequence), and is expressed as a 3 kb mRNA irrespective of heat shock stimuli (Milner & Campbell, 1990; 1992b). The presence of the HSP70 genes in the MHC is most interesting. The HSP70s mediate protein–protein interactions and some are associated with translating ribosomes. They can bind to proteins via recognition of short amino acid sequences, thus imposing transient local conformation and interaction constraints on the substrate protein. It is also possible that the HSP70 molecule could chaperone peptides during antigen processing or display them on the cell surface (Rippman *et al.*, 1991).

The G8, G9 and G10 genes encode 1.0 kb, 1.9 kb and 2.6 kb transcripts, respectively, that are ubiquitously expressed and show no sequence homology to any known protein. G9a (BAT-8) shows the same expression pattern, with the difference that its 110 kDa protein product contains copies of a 33 amino acid residue motif known as the ankyrin repeat. Repeats of this type are presumed to play a role in intracellular protein–protein interactions involved in cell cycle control or tissue differentiation. In addition the carboxy-terminal region of G9a can bind zinc (Milner & Campbell, 1992a).

The C2, C4 and Bf glycoproteins are complement components. C2 and Bf have similar function and contain the catalytic centre of the classical and alternative complement pathways, respectively, while C4 is structurally similar to C3 and C5 and part of the classical pathway. The three genes are polymorphic, although not to the extent observed with classical class II genes. The 21-hydroxylase genes A and B have been mapped to the 3′ end of the C4 genes, although only 21-hydroxylase B is actively transcribed. Deletions or mutations in the active gene cause the disorder congenital adrenal hyperplasia (CAH), which has been shown to be linked to the HLA and in particular to the B47 allele. This demonstrates that a disease associated with the MHC can be caused by a linked non-classical class I or class II gene. The number of C4 and 21-hydroxylase genes varies between haplotypes causing expansion or contraction of the class III region (Dunham *et al.*, 1990). Two other genes have been identified between C2 and C4. They are the RD gene (encoding a 1.6 kb transcript) and G11 (encoding a 1.4 kb transcript), which are both ubiquitously expressed. The RD protein shows a characteristic dipeptide repeat that consists of alternating basic and acidic residues without any known function.

The OSG gene is a transcript encoded on the opposite strand of the 21-hydroxylase gene. Both the OSG and G12 genes may code for components of one tenascin-like gene product. The OSG product constitutes the carboxy-terminal part and G12 the amino-terminal part. Tenascin is a multidomain and multifunctional extracellular matrix protein similar to fibronectin and laminin and has neurite regulation properties (Milner & Campbell, 1992a). G13 encodes an abundant 2.7 kb mRNA showing sequence similarities to the leucine zipper family of transcription factors and could play a role in the regulation of transcription. The ubiquitously expressed genes G14, G15, G16,

G17 and G18 do not show any sequence similarity to known genes, although G16 and G17 may have copies on other chromosomes (Kendall *et al.*, 1990).

### *2.4.2 Comparison of the human and mouse class III regions*

The only apparent difference between mouse and human class III regions seems to be the fact that the BAT1 gene is very close to the class I D gene in the mouse (Gaskins *et al.*, 1990; Wroblewski *et al.*, 1990; Lafuse *et al.*, 1992). With the exception of a duplicated repeat (Leelayuwat *et al.*, 1992), no transcript has been found between BAT1 and HLA-B and there is very little space here in the mouse (approximately 10 kb). The distance between $E_\alpha$ and C4 in the mouse is 430 kb, which is very similar to that in humans. The position of the genes in the map is identical, as detected so far, although there may be some functional differences; for example, the non-haemolytic C4-like mouse protein Slp, which is not expressed in all mouse strains.

## 2.5 Conclusions

The way the MHC is organized shows several interesting characteristics. Many genes are duplicated, irrespective of their role in immune responses, for example 21-hydroxylase, HSP70s, DQA and DQB and many others. These have been maintained through evolution in the same part of the genome. There could be a number of reasons for the clustering of these genes. One possibility is that co-regulation plays an important role. For example, the transporter and proteasome-related genes are inducible by IFN-γ, like most of the class II genes around them. Another reason could be the co-evolution of functionally related genes. The genes responsible for antigen processing are linked to the ones responsible for antigen presentation and their tight linkage ensures that none is lost through segregation, while polymorphic genes can better adapt to each other. In addition, this linkage means that sequences between genes can be exchanged by non-homologous mechanisms, thus participating in the generation of polymorphisms (Belich *et al.*, 1992; Watkins *et al.*, 1992). Overall, these properties elevate the MHC into one of the most interesting and, therefore, most investigated regions of the human and murine genomes.

## References

Abe, K., Wei, J.F., Wei, F.S., Hsu, Y.C., Uehara, H., Artzt, K. & Bennett, D. (1988). Searching for coding sequences in the mammalian genome: the H-2K region of the mouse MHC is replete with genes expressed in embryos. *EMBO Journal*, 7, 3441–3449.

Ando, A., Kawai, J., Maeda, M., Tsuji, K., Trowsdale, J. & Inoko, H. (1989). Mapping and nucleotide sequence of a new HLA class II light chain gene, DQB3. *Immunogenetics*, 30, 243–249.

Banerji, J., Sands, J., Strominger, J.L. & Spies, T. (1990). A gene pair from the human major histocompatibility complex encodes large proline-rich proteins with mutiple repeated motifs and a single ubiquitin-like domain. *Proceedings of the National Academy of Sciences of the USA*, 87, 2374–2378.

Beck, S., Hanson, I., Kelly, A., Pappin, D.J.C. & Trowsdale, J. (1992). A homologue of the Drosophila female sterile homeotic (fsh) gene in the class II region of the human MHC. *DNA Sequence*, 2, 203–210.

Belich, M.P., Madrigal, J.A., Hildebrand, W.H., Zemmour, J., Williams, R.C., Luz, R., Petzl-Erler, M.L. & Parham, P. (1992). Unusual HLA-B alleles in two tribes of Brazilian Indians. *Nature (London)*, 357, 326–329.

Bronson, S.K., Pei, J., Taillon-Miller, P., Chorney, M.J., Geraghty, D.E. & Caplin, D.D. (1991). Isolation and characterization of yeast artificial chromosome clones linking the HLA-B and HLA-C loci. *Proceedings of the National Academy of Sciences of the USA*, 88, 1676–1680.

Brorson, K.A., Richards, S., Hunt III, S.W., Cheroutre, H., Fischer Lindahl, K. & Hood, L. (1989). Analysis of a new class I gene mapping to the Hmt region of the mouse. *Immunogenetics*, 30, 273–283.

Cameron, P.V., Tabarias, H.A., Pulendran, B., Robinson, W. & Dawkins, R.L. (1990). Conservation of the central MHC genome: PFGE mapping and RFLP analysis of complement, HSP70 and TNF genes in the goat. *Immunogenetics*, 31, 253–264.

Carroll, M.C., Katzman, P., Alicot, E.M., Koller, B.H., Geraghty, D.E., Orr, H.T., Strominger, J.L. & Spies, T. (1987). Linkage map of the human major histocompatibility complex including the tumor necrosis factor genes. *Proceedings of the National Academy of Sciences of the USA*, 84, 8535–8539.

Chimini, G., Boretto, J., Marguet, D., Lanau, F., Lauquin, G. & Pontarotti, P. (1990). Molecular analysis of the human MHC class I region using yeast artificial chromosome clones. *Immunogenetics*, 32, 419–426.

Chimini, G., Pontarotti, P., Nguyen, C., Toubert, A., Boretto, J. & Jordan, B.R. (1988). The chromosome region containing the highly polymorphic HLA class I genes displays limited large scale variability in the human population. *EMBO Journal*, 7, 395–400.

Cho, S., Attaya, M. & Monaco, J.J. (1991). New class II-like genes in the murine MHC. *Nature (London)*, 353, 573–576.

Chorney, M.J., Sawada, I., Gillespie, G.A., Srivastava, R., Pan, J. & Weissman, S.M. (1990). Transcription analysis, physical mapping, and molecular characterization of a non-classical human leukocyte antigen class I gene. *Molecular and Cellular Biology*, 10, 243–253.

Denizot, F., Mattei, M.G., Vernet, C., Pontarotti, P. & Chimini, G. (1992). YAC-assisted cloning of a putative G-protein mapping to the MHC class I region. *Genomics*, 14, 857–862.

Dunham, I., Sargent, C., Dawkins, R.L. & Campbell, R.D. (1989). An analysis of the variation in the long range genomic organisation of the MHC class II region by pulsed-field gel electrophoresis. *Genomics*, 5, 787–796.

Dunham, I., Sargent, C.A., Kendall, E. & Campbell, R.D. (1990). Characterization of the class III region in different MHC haplotypes by pulsed-field gel electrophoresis. *Immunogenetics*, 32, 175–182.

Dunham, I., Sargent, C.A., Trowsdale, J. & Campbell, R.D. (1987). Molecular map of the human major histocompatibility complex by pulsed-field gel electrophoresis. *Proceedings of the National Academy of Sciences of the USA*, 84, 7237–7241.

El Kahloun, A., Chauvel, B., Mauvieux, V., Dorval, I., Jouanolle, A., Gicquel, I., Le Gall, J.Y. & David, V. (1993). Localization of seven new genes around the HLA-A locus. *Human Molecular Genetics*, 2, 55–60.

El Kahloun, A., Vernet, C., Jouanolle, A.M., Boretto, J., Mauvieux, V., Le Gall, J.Y., David, V. & Pontarotti, P. (1992). A continuous restriction map from HLA-E to HLA-F: structural comparison between different HLA-A haplotypes. *Immunogenetics*, 35, 183–189.

Fischer Lindahl, K., Hermel, E., Loveland, B.E. & Wang, C.R. (1991). Maternally transmitted antigen of mice – a model transplantation antigen. *Annual Review of Immunology*, 9, 351–372.

Gaskins, H.R., Prochazka, M., Nadeau, J.H., Henson, V.W. & Leiter, E.H. (1990). Localization of a mouse heat shock Hsp70 gene within the H-2 complex. *Immunogenetics*, 32, 286–289.

Geraghty, D.E., Pei, J., Lipsky, B., Hansen, J.A., Taillon-Miller, P., Bronson, S.K. & Chaplin, D.D. (1992). Cloning and physical mapping of the HLA class I region spanning the HLA-E-to-HLA-F interval by using yeast artificial chromosomes. *Proceedings of the National Academy of Sciences of the USA*, 89, 2669–2673.

Geraghty, D.E., Wei, X., Orr, H.T. & Koller, B.H. (1990). Human leukocyte antigen F (HLA-F): an expressed HLA gene composed of a class I coding sequence linked to a novel transcribed repetitive element. *Journal of Experimental Medicine*, 171, 1–18.

Glynne, R., Powis, S.H., Beck, S., Kelly, A., Kerr, L., & Trowsdale, J. (1991). A proteasome-related gene between the two ABC transporter loci in the class II region of the human MHC. *Nature (London)*, 353, 357–360.

Gruen, J.R., Goei, V.L., Summers, K.M., Capossela, A., Powell, L., Halliday, J., Zoghbi, H., Shukla, H. & Weissman, S.M. (1992). Physical and genetic mapping of the telomeric major histocompatibility complex region in man and relevance to the primary hemochromatosis gene (HFE). *Genomics*, 14, 232–240.

Hanson, I.M., Poustka, A. & Trowsdale, J. (1991). New genes in the class II region of the human major histocompatibility complex. *Genomics*, 10, 417–424.

Hanson, I.M. & Trowsdale, J. (1991). Colinearity of novel genes in the class II regions of the MHC in mouse and human. *Immunogenetics*, 34, 5–11.

Hedrick, S.M. (1992). Dawn of the hunt for non-classical MHC function. *Cell*, 70, 177–180.

Karlsson, L., Surh, C.D., Sprent, J. & Peterson, P.A. (1992). An unusual class II molecule. *Immunology Today*, 13, 269–470.

Kelly, A.P., Monaco, J.J., Cho, S. & Trowsdale, J. (1991a). A new human HLA class II-related locus, DM. *Nature (London)*, 353, 571–573.

Kelly, A., Powis, S.H., Glynne, R., Radley, E., Beck, S. & Trowsdale, J. (1991b). Second proteasome-related gene in the human MHC class II region. *Nature (London)*, 353, 667–668.

Kelly, A., Powis, S.H., Kerr, L.A., Mockridge, I., Elliott, T., Bastin, J., Uchanska-Ziegler, B., Ziegler, A., Trowsdale, J. & Townsend, A. (1992). Assembly and function of the two ABC transporter proteins encoded in the human major histocompatibility complex. *Nature (London)*, 355, 641–644.

Kendall, E., Sargent, C.A. & Campbell, R.D. (1990). Human major histocompatibility complex contains a new cluster of genes between the HLA-D and complement C4 loci. *Nucleic Acids Research*, 18, 7251–7257.

Kendall, E., Todd, J.A. & Campbell, R.D. (1991). Molecular analysis of the MHC class II region in DR4, DR7, and DR9 haplotypes. *Immunogenetics*, 34, 349–357.

Klein, J. & Figueroa, F. (1986). Evolution of the major histocompatibility complex. *CRC Critical Reviews in Immunology*, 6, 295–386.

Koller, B.H., Geraghty, D.E., DeMars, R., Duvick, L., Rich, S.S. & Orr, H.T. (1989). Chromosomal organization of the human major histocompatibility complex class I gene family. *Journal of Experimental Medicine*, 169, 469–480.

Kovats, S., Main, E.K., Librach, C., Stubblebine, M., Fisher, J.S. & DeMars, R. (1990). A class I antigen, HLA-G, expressed in human trophoblasts. *Science*, 248, 220–223.

Kozono, H., Bronson, S.K., Taillon-Miller, P., Moorti, M.K., Jamry, I. & Chaplin, D.D. (1991). Molecular linkage of the HLA-DR, HLA-DQ, and HLA-DO genes in yeast artificial chromosomes. *Genomics*, 11, 577–586.

Lafuse, W.P., Lanning, D., Spies, T. & David, C.S. (1992). PFGE mapping and RFLP analysis of the S/D region of the H-2 complex. *Immunogenetics*, 36, 110–116.

Leelayuwat, C., Abraham, L.J., Tabarias, H., Christiansen, F.T. & Dawkins, R.L. (1992). Genomic organization of a polymorphic duplicated region centromeric of HLA-B. *Immunogenetics*, 36, 208–212.

Livingstone, A.M., Powis, S.J., Gunther, E., Cramer, D.V., Howard, J.C. & Butcher, G.W. (1991). Cim: an MHC class II-linked allelism affecting the antigenicity of a classical class I molecule for T lymphocytes. *Immunogenetics*, 34, 157–163.

Mellins, E., Kempin, S., Smith, L., Monji, T. & Pious, D. (1991). A gene required for class II-restricted antigen presentation maps to the major histocompatibility complex. *Journal of Experimental Medicine*, 174, 1607–1615.

Meunier, H.F., Carson, S., Bodmer, W.F. & Trowsdale, J. (1986). An isolated $\beta_1$ exon next to the DR$\alpha$ gene in the HLA-D region. *Immunogenetics*, 23, 172–180.

Milner, C.M. & Campbell, R.D. (1990). Structure and expression of the three MHC-linked HSP70 genes. *Immunogenetics*, 32, 242–251.

Milner, C.M. & Campbell, R.D. (1992a). Genes, genes and more genes in the human major histocompatibility complex. *BioEssays*, 14, 565–571.

Milner, C.M. & Campbell, R.D. (1992b). Polymorphic analysis of the three MHC-linked HSP70 genes. *Immunogenetics*, 36, 357–362.

Momburg, F., Ortiz-Navarette, V., Neefjes, J., Goulmy, E., van de Wal, Y., Spits, H., Powis, S.J., Butcher, G., Howard, J., Walden, P. & Hämmerling, G. (1992). Proteasome subunits encoded by the major histocompatibility complex are not essential for antigen processing. *Nature (London)*, 360, 174–177.

Monaco, J.J. (1992). Not so groovy after all? *Current Biology*, 2, 433–435.

Monaco, J.J., Cho, S. & Attaya, M. (1990). Transport protein genes in the murine MHC. *Science*, 250, 1723–1726.

Okamoto, N., Ando, A., Kawai, J., Yoshiwara, T., Tsuji, K. & Inoko, H. (1991). Orientation of HLA-DNA gene and identification of a CpG island-associated gene adjacent to DNA in human major histocompatibility complex class II region. *Human Immunology*, 32, 221–228.

Olavesen, M.G., Snoek, M. & Campbell, R.D. (1993). Localization of a new gene

adjacent to the HSP70 genes in human and mouse MHCs. *Immunogenetics*, 37, 394–396.
Parham, P. (1990). Antigen processing. Transporters of delight [news; comment]. *Nature (London)*, 348, 674–675.
Powis, S.H., Mockridge, I., Kelly, A., Kerr, L.A., Glynne, R., Gileadi, U., Beck, S. & Trowsdale, J. (1992). Polymorphism in a second ABC transporter gene located within the class II region of the human major histocompatibility complex. *Proceedings of the National Academy of Sciences of the USA*, 89, 1463–1467.
Powis, S.J., Howard, J.C. & Butcher, G.W. (1991). The major histocompatibility complex class II-linked cim locus controls the kinetics of intracellular transport of a classical class I molecule. *Journal of Experimental Medicine*, 173, 913–921.
Ragoussis, J., Bloemer, K., Pohla, H., Messer, G., Weiss, E.H. & Ziegler, A. (1989). A physical map including a new class I gene (cda12) of the human major histocompatibility complex (A2/B13 haplotype) derived from monosomy 6 mutant cell line. *Genomics*, 4, 301–308.
Ragoussis, J., Monaco, A., Mockridge, I., Kendall, E., Campbell, R.D. & Trowsdale, J. (1991). Cloning of the HLA class II region in yeast artificial chromosomes. *Proceedings of the National Academy of Sciences of the USA*, 88, 3753–3757.
Riberdy, J.M. & Cresswell, P. (1992). The antigen-processing mutant T2 suggests a role for MHC-linked genes in class II antigen presentation. *Journal of Immunology*, 148, 2586–2590.
Richards, S., Bucan, M., Brorson, K., Kiefer, M., Hunt, S.I., Lehreach, H. & Fischer Lindahl, K. (1989). Genetic and molecular mapping of the *Hmt* region of mouse. *EMBO Journal*, 8, 3749–3757.
Rippman, F., Taylor, W.F., Rothbard, J. & Green, N.M. (1991). A hypothetical model for the peptide binding domain of hsp70 based on the peptide binding domain of HLA. *EMBO Journal*, 10, 1053–1059.
Robertson, M. (1991). Proteasomes in the pathway. *Nature (London)*, 353, 300–301.
Sargent, C.A., Dunham, I. & Campbell, R.D. (1989a). Identification of multiple HTF-island associated genes in the human major histocompatibility complex class III region. *EMBO Journal*, 8, 2305–2312.
Sargent, C.A., Dunham, I., Trowsdale, J. & Campbell, R.D. (1989b). The human major histocompatibility complex contains genes for the major heat shock protein hsp70. *Proceedings of the National Academy of Sciences of the USA*, 86, 1968–1972.
Schmidt, C.M. & Orr, H.T. (1991). A physical linkage map of HLA-A, -G, –7.5p, and -F. *Human Immunology*, 31, 180–185.
Servenius, B., Rask, L. & Peterson, P.A. (1987). Class II genes of the human major histocompatibility complex. The DOβ gene is a divergent member of the class II β gene family. *Journal of Biological Chemistry*, 262, 8759–8766.
Shukla, H., Gillespie, G.A., Srivastava, R., Collins, F. & Chorney, M.J. (1991). A class I jumping clone places the HLA-G gene approximately 100 kilobases from HLA-H within the HLA-A subregion of the human MHC. *Genomics*, 10, 905–915.
Shukla, H., Swaroop, A., Srivastava, R. & Weissman, S.M. (1990). The mRNA of a human class I gene HLA G/HLA 6.0 exhibits a restricted pattern of expression. *Nucleic Acids Research*, 18, 2189–2197.
Spies, T., Blanck, G., Bresnahan, M., Sands, J. & Strominger, J.L. (1989a). A new

cluster of genes within the human major histocompatibility complex. *Science*, 243, 214–217.

Spies, T., Bresnahan, M. & Strominger, J.L. (1989b). Human major histocompatibility complex contains a minimum of 19 genes between the complement cluster and HLA-B. *Proceedings of the National Academy of Sciences of the USA*, 86, 8955–8958.

Spies, T., Bresnahan, M., Bahram, S., Arnold, D., Blanck, G., Mellins, E., Pious, D. & DeMars, R. (1990). A gene in the human major histocompatibility complex class II region controlling the class I antigen presentation pathway. *Nature(London)*, 348, 744–747.

Spies, T. & DeMars, R. (1991). Restored expression of major histocompatibility class I molecules by gene transfer of a putative peptide transporter. *Nature(London)*, 351, 323–324.

Trowsdale, J., Hanson, I., Mockridge, I., Beck, S., Townsend, A. & Kelly, A. (1990). Sequences encoded in the class II region of the MHC related to the 'ABC' superfamily of transporters. *Nature(London)*, 348, 741–744.

Trowsdale, J., Ragoussis, J. & Campbell, R.D. (1991). Map of the human MHC. *Immunology Today*, 12, 443–446.

Tsuge, I., Shen, F., Steinmetz, M. & Boyse, E.A. (1987). A gene in the H-2S:H-2D interval of the major histocompatibility complex which is transcribed in B cells and macrophages. *Immunogenetics*, 26, 378–380.

Ulbrecht, M., Kellermann, J., Johnson, J.P. & Weiss, E.H. (1992). Impaired intracellular transport and cell surface expression of non-polymorphic HLA-E: evidence for inefficient peptide binding. *Journal of Experimental Medicine*, 176, 1083–1090.

Venditti, C.P. & Chorney, M.J. (1992). Class I gene contraction within the HLA-A subregion of the human MHC. *Genomics*, 14, 1003–1009.

Volz, A., Weiss, E.H., Trowsdale, J. & Ziegler, A. (1994). Presence of an expressed β-tubulin gene (TUBB) in the HLA class I region may provide the genetic basis for HLA-linked microtubule dysfunction. *Human Genetics*, 93, 42–46.

Wang, C., Loveland, B.E. & Fischer Lindahl, K. (1991). H-2M3 encodes the MHC class I molecule presenting the maternally transmitted antigen of the mouse. *Cell*, 66, 335–345.

Watkins, D.I., Chen, Z.W., Hughes, A.L., Evans, M.G., Tedder, T.F. & Letvin, N.L. (1990). Evolution of the MHC class I genes of a New World primate from ancestral homologues of human non-classical genes. *Nature (London)*, 346, 60–63.

Watkins, D.I., McAdam, S.N., Liu, X., Strang, C.R., Milford, E.L., Levine, C.G., Garber, T.L., Dogon, A.L., Lord, C.I., Ghim, S.H., Troup, G.M., Hughes, A.L. & Letvin, N.L. (1992). New recombinant HLA-B alleles in a tribe of South American Amerindians indicate rapid evolution of MHC class I loci. *Nature (London)*, 357, 329–333.

Wroblewski, J.M., Kaminsky, S.G., Milisauskas, V.K., Pittman, A.M., Chaplin, D.D., Spies, T. & Nakamura, I. (1990). The B144-H-$2D^b$ interval and the location of a mouse homologue of the human D6S81E locus. *Immunogenetics*, 32, 200–204.

Zemmour, J., Koller, B.H., Ennis, P.D., Geraghty, D.E., Lawlor, D.A., Orr, H.T. & Parham, P. (1990). HLA-AR, an inactivated antigen-presenting locus related to HLA-A. *Journal of Immunology*, 144, 3619–3629.

# 3

# Interactions of cytokines in the regulation of MHC class I and class II antigen expression

IAN TODD
*University Hospital, Nottingham*

## 3.1 Introduction

T lymphocytes play a central role in antigen-specific immune responses through their interactions with other cell types. These interactions are guided by specific T cell recognition of antigen associated with cell surface molecules of the MHC. The two main types of MHC molecule (class I and class II) are both involved in antigen presentation, but to different types of T cell (expressing CD8 or CD4 antigens, respectively). In addition, they normally show very different patterns of expression within tissues: most cell types of the body express MHC class I antigens, whereas the expression of MHC class II antigens is restricted primarily to cells of the immune system – macrophages, dendritic cells, B lymphocytes and (in humans) activated T lymphocytes. Teleologically, this accords with the usual functions of the T cells, which interact with antigen associated with the different types of MHC protein. Thus, class I-restricted, $CD8^+$ cytotoxic T cells may be required to interact with antigen presented by almost any cell of the body (for example, following viral infection). By contrast, a major role of class II-restricted, $CD4^+$ helper T cells is to interact with, and help, other cells of the immune system.

## 3.2 Tissue distribution of MHC antigens

The distribution of MHC antigens is now known to be complex. The expression of class I antigens varies in intensity with, for example, a gradation of expression from strong to weak or negative through the following cell types: cells of the immune system, epithelial cells of the gastrointestinal and respiratory tracts, endocrine cells, striated muscle cells, hepatocytes and neurons (reviewed by Pujol-Borrell & Todd, 1987). In addition to their expression by cells of the immune system, class II antigens are also normally expressed by certain other cell types, including a proportion of capillary endothelial cells (Groenewegen, Buurman & van der Linden, 1985) and epithelial cells, e.g. in the gastrointestinal tract (Scott *et al.*, 1980).

Marked changes in MHC expression often occur in immunopathological situations, with increased expression of both class I and class II antigens and induction of *de novo* expression of class II in previously negative cell types. Such changes have been documented in various situations of infection, cell-mediated hypersensitivity, graft-versus-host reactions, graft rejection and cancer (Barclay & Mason, 1982; Milton & Fabre, 1985; Volc-Platzer *et al.*, 1984) and also in the affected tissues in a variety of autoimmune diseases. Examples of the last include thyroid follicular cells in autoimmune thyroid diseases (Hanafusa *et al.*, 1983; Aichinger, Fill & Wick, 1985; Jansson, Karlsson & Forsum, 1985), pancreatic β cells in type I diabetes mellitus (Bottazzo *et al.*, 1985; Foulis & Farquharson, 1986), bile duct epithelium in primary biliary cirrhosis (Ballardini *et al.*, 1984) and salivary ducts in Sjögren's syndrome (Lindhal *et al.*, 1985).

In view of the role of MHC molecules in antigen presentation, modulation of their expression could clearly have consequences for T cell activation. Indeed, it has been observed that susceptibility to killing by cytotoxic T cells can depend on the level of MHC class I expression by the target cells (Flyer, Burakoff & Faller, 1985) and that the level of class II expression by antigen-presenting cells contributes to the efficiency of T cell activation (Matis *et al.*, 1983; Lechler, Norcross & Germain, 1985).

## 3.3 General considerations in cytokine regulation of MHC expression

All of the situations described above, in which there is increased expression of MHC antigens within tissues, involve a component of cell-mediated immunity. Consistent with this, studies in the early 1980s showed that soluble factors secreted by activated T cells could induce MHC class II expression by macrophages (Steeg, Moore & Oppenheim, 1980; Calamai, Beller & Unanue, 1982) and this activity was subsequently shown to be caused by IFN-γ (see below). In the late 1980s and early 1990s, it has been found that a wide range of cytokines and other factors can modulate cellular expression of MHC antigens. This chapter is concerned primarily with these modulators of MHC expression in general and, where appropriate, reference is made to studies of the modulation of MHC expression by human thyroid epithelial cells (thyrocytes), in which the author has been involved, along with G. F. Bottazzo, R. Pujol-Borrell, M. Feldmann and their associates. These studies were undertaken in relation to the changes in MHC expression by thyrocytes observed in autoimmune thyroid diseases.

Much of the data on modulators of MHC expression are based on experimentation *in vitro*, making it dangerous to extrapolate directly to the 'real life' situation. It seems probable that the modulation of MHC expression *in vivo* is a complex phenomenon, with the phenotype observed being influenced by a number of interacting factors and processes. Furthermore, it should not be assumed that the effect of a mediator on MHC expression necessarily always involves direct action; in some instances the mediator may induce other factors which then modulate MHC expression. However, it is appropriate, at this point, to draw attention to some general principles, which will be exemplified when considering the individual modulators.

1. Some factors can induce or suppress MHC expression directly whereas others do so by synergizing with, or antagonizing, the action of other factors.
2. In some circumstances, class I antigens appear to be 'more readily' induced than class II in that some cytokines can directly modulate expression of class I but not class II antigens.
3. In considering the interactions between factors in the modulation of MHC expression, account must be taken of their relative concentrations, since this may influence the overall outcome.
4. It should be borne in mind that many of the cytokines and other factors considered will have a variety of other effects of immunological significance on cells in addition to effects on MHC expression; for example, they may affect expression of adhesion molecules such as integrins and selectins. Moreover, some factors may more generally influence the differentiation state of cells: in some circumstances, changes in MHC expression may actually be a consequence of these effects if MHC expression is linked to differentiation state.
5. To understand the effect on MHC expression of the interaction between a cytokine and a cell it is necessary to consider not only the cytokine involved, but also the cell type being acted upon, for the following reasons:

   - some cytokines (and other factors) influence MHC expression only in certain cell types
   - different cell types (or cells in particular differentiation states or stages of the cell cycle) may show different requirements for modulation of MHC expression; for example, they may or may not have to be acted upon by more than one factor for an effect to be observed

– the same cytokine may actually have opposite effects on MHC expression by different cell types: for example, inducing or inhibiting expression of MHC class II molecules.

### 3.3.1 Interferon-γ

Interferon-γ is produced exclusively by activated T cells and natural killer (NK) cells and is probably the cytokine that is most widely implicated in enhancing expression of MHC molecules. Like a variety of other cytokines, it stimulates increased expression of class I antigens by cells, but it is also a potent inducer of *de novo* class II expression in a wide range of cell types. In the early 1980s, it was shown that a factor secreted by activated T cells with the properties of IFN-γ was able to induce class II expression by, for example, macrophages (Steeg *et al.*, 1982b) and mast cell progenitors (Wong *et al.*, 1982). When, soon after this, recombinant IFN-γ became available, it was confirmed that this cytokine induced MHC class II expression by macrophage, lymphoid and myeloid cell lines (King & Jones, 1983; Wong *et al.*, 1983). It has further been shown that IFN-γ can induce class II expression by many other cells: for example epithelial cells (see below), endothelial cells and fibroblasts (Collins *et al.*, 1984; Lapierre, Fiers & Pober, 1988), brain cells (Wong *et al.*, 1984a) and muscle cells (Kalovidouris, 1992).

In terms of the signal transduction pathways stimulated by IFN-γ, a number of studies have implicated activation of protein kinase C in the IFN-γ-induced expression of both MHC class I (Seong *et al.*, 1991; Towata *et al.*, 1991) and class II (Benveniste *et al.*, 1991; Gumina *et al.*, 1991), although exceptions have been reported (Celada & Maki, 1991).

#### *3.3.1.1 Differential sensitivity of cells to IFN-γ*

When normal human thyrocytes were cultured with recombinant IFN-γ and then stained by indirect immunofluorescence with MHC-specific monoclonal antibodies (mAbs), both enhanced expression of class I and induction of class II HLA molecules were observed (Todd *et al.*, 1985). Optimal conditions for induction of cell surface HLA-DR expression were culture for at least six days with a concentration of IFN-γ of at least 100 U/ml. However, some class II induction could be observed within 24 hours of culture and a concentration of IFN-γ as low as 1 U/ml had an observable effect. Although many cell types respond to IFN-γ in this way, there are clear examples of differential sensitivity to the class II-inducing properties of this cytokine in different cell types. For example, human pancreatic β cells are essentially resistant to induction of class II expression when cultured with IFN-γ alone

(Pujol-Borrell *et al.*, 1986) but will express HLA class II antigens when cultured with a combination of IFN-γ and TNF-α or lymphotoxin (Pujol-Borrell *et al.*, 1987). Similarly, pancreatic β cells from normal rats are not induced to express class II when cultured with IFN-γ, whereas this cytokine does induce class II in β cells from diabetes-prone BB (Biobreeding) rats (Walker *et al.*, 1986): this could be the result of a genetic predisposition in this strain, or because of other effects on the cells *in vivo* related to the pathogenesis of diabetes.

There also appear to be certain species differences in susceptibility to class II induction by IFN-γ. Epidermal keratinocytes do not normally express class II molecules but do so when immune activity occurs in the epidermis in humans (Volc-Platzer *et al.*, 1984; Gottlieb *et al.*, 1986) and in mice (Stringer, Hicks & Botham, 1991). However, whereas human (Basham *et al.*, 1984) or, indeed, rat keratinocytes are induced to express MHC class II molecules when cultured with IFN-γ, mouse keratinocytes are resistant to such an effect when cultured either with IFN-γ alone or with IFN-γ and TNF-α, even though up-regulation of class I expression is induced (Yeoman, Anderton & Stanley, 1989). By contrast, class II expression can be induced in mouse keratinocytes *in vivo* by systemic administration of IFN-γ (Skoskiewicz *et al.*, 1985). The reasons for the resistance of mouse keratinocytes to class II induction by IFN-γ *in vitro* are unclear at present: for example, whether other inducers are required or the keratinocytes produce antagonistic factors.

There is much interest in MHC expression by trophoblast cells of foetus and placenta with respect to the immunological relationship between the mother and the semi-allogeneic foetus during pregnancy. These cells are devoid of MHC class II expression and are resistant to induction of class II by IFN-γ: *in vivo* treatment of mid-gestational pregnant mice with IFN-γ causes increased expression of MHC class I by spongiotrophoblasts, but no induction of class II (Mattsson *et al.*, 1991). However, placental class II expression has been induced by treatment of pregnant mice with 5-azacytidine (Athanassakis-Vassiliadis *et al.*, 1990), which is known to lead to decreased methylation of DNA. Inactivation of a number of genes has been shown to be associated with DNA methylation, including MHC class II genes (Reitz *et al.*, 1984); this may at least partly explain the lack of class II induction by IFN-γ in trophoblasts.

Different subpopulations of trophoblast cells also differ in expression of MHC class I antigens. In contrast to some trophoblast cells, syncytiotrophoblast cells (in humans) and labyrinthine trophoblast cells (in mice) are devoid of class I, and the latter have been shown to be resistant to induction of class

I expression by IFN-γ (Mattsson *et al.*, 1991). Again, DNA methylation may be involved here, since the Jar cell-line (a syncytiotrophoblast-like choriocarcinoma), which lacks HLA expression and has hypermethylated class I genes (Le Bouteiller *et al.*, 1991), is also resistant to inducers that have no effect on methylation, such as IFN (Hunt, Andrews & Wood, 1987). Villous cytotrophoblast cells, which contain mRNA for HLA-G but not the protein, are also refractory to IFN-γ (Hunt *et al.*, 1987); this may be partly because of a 13 bp deletion in the IFN consensus sequence/enhancer A region (see Chapter 4) of the HLA-G gene (Geraghty *et al.*, 1990). Another unusual feature of trophoblasts is that even those cells that are susceptible to induction of class I by IFN-γ appear to be resistant to TNF-α and IFN-α/β, which are usually considered as enhancers of class I expression (Hunt, Atherton & Pace, 1990; Chumbley *et al.*, 1991).

Another area of interest in differential MHC expression and regulation is in tumours. For example, many spontaneous tumours are class I deficient, which may influence their oncogenicity (see Chapters 13–15). This may, in part, relate to their responsiveness to inducers of MHC expression; in one study, it was found that two human haematopoietic tumour cell lines, class I-deficient K562 and class I positive Ramos, responded to IFN-γ with increased class I transcription, but this was sustained only in the Ramos cells (Radford *et al.*, 1991).

### *3.3.2 Alpha and beta interferons*

In contrast to IFN-γ, type I interferons (α and β) are structurally related to each other and can be produced by many different cell types. Initial studies on human thyrocytes showed that IFN-α and IFN-β could enhance expression of HLA class I molecules but, unlike IFN-γ, did not induce expression of class II molecules (Todd *et al.*, 1985). This stimulation of class I but not class II expression by IFN-α and IFN-β has also been observed with a variety of other cell types, including foetal monocytes and myeloid cell lines (Kelley, Fiers & Strom, 1984) and endothelial cells (Lapierre *et al.*, 1988). Furthermore, it has been found that IFN-α and IFN-β actually inhibit IFN-γ-mediated enhancement of class II expression in mouse macrophages (Ling, Warren & Vogel, 1985; Inaba *et al.*, 1986), fibroblasts and glial cells (Morris & Tomkins, 1989) as well as in human endothelial cells (Lapierre *et al.*, 1988; Manyak *et al.*, 1988) and thyrocytes (Guerin *et al.*, 1990). This down-regulation was shown to operate at the transcriptional level in mouse macrophages (Fertsch-Ruggio, Schoenberg & Vogel, 1988). By contrast, others have reported type I interferons to stimulate class II expression by melanoma and lymphoblastoid cell lines (Dolei, Capobianchi & Ameglio, 1983; Giacomini *et al.*, 1984), but this effect was slight, and certainly less than that of IFN-γ. IFN-α and IFN-β have also been reported to increase class II expression by human monocytes (Gerrard, Dyer & Mostowski, 1990; Rhodes & Stokes, 1982). Studies with different subspecies of IFN-α are of particular interest: Rhodes, Ivanyi & Cozens (1986) showed that whereas IFN-$\alpha_2$ had no effect on HLA class II expression by human monocytes, IFN-$\alpha_1$ increased class II expression by these cells. By contrast, we found that the same recombinant preparation of IFN-$\alpha_1$ not only did not induce HLA class II expression by human thyrocytes but inhibited class II induction by IFN-γ (Guerin *et al.*, 1990). Furthermore, Morris & Tomkins (1989) found that mouse IFN-$\alpha_2$ (unlike a natural mixture of IFN-α subspecies) weakly induced class II expression by murine fibroblasts or glial cells. All of these findings suggest that type I interferons can have different modulatory effects on class II expression, ranging from stimulatory to inhibitory, depending on the circumstances. Factors that might affect this could include the subspecies of IFN involved as well as the cell type, tissue source and animal species. A possible contribution to these differences might be whether or not a protein inhibitor of class II gene expression is synthesized by cells in response to type I interferons; Collins *et al.* (1986) reported the induction of such an inhibitor in human umbilical vein endothelial cells and that blocking its synthesis with cycloheximide led to induction of class II gene transcription by type I interferons.

In our studies on thyrocytes, we found that inhibition by IFN-α of class II induction by IFN-γ was dependent on the dose of IFN-α employed: in the presence of a relatively low dose of IFN-γ (5–10 U/ml), the degree of inhibition achieved increased up to the highest concentration of IFN-α employed, i.e. $10^4$ U/ml (Guerin *et al.*, 1990). However, it is conceivable that high concentrations of IFN-α may occur in pathological situations *in vivo*; for example, levels of type I interferons up to $10^3$ U/ml have been detected in brain tissue of mice infected with neurotropic Semliki Forest virus (Morris & Tomkins, 1989). It has been suggested by Morris & Tomkins (1989) that high levels of type I interferons induced during viral infections may inhibit inappropriate induction of class II by IFN-γ derived from T cells (or NK cells) responding to the virus (see Fig. 3.1): this could possibly contribute to inhibition of the development of autoimmune responses against the infected cell type.

### *3.3.3 Tumour necrosis factors*

The cytokine TNF-α has been shown to directly stimulate expression of MHC class I, but not class II, on, for example, endothelial cells and fibro-

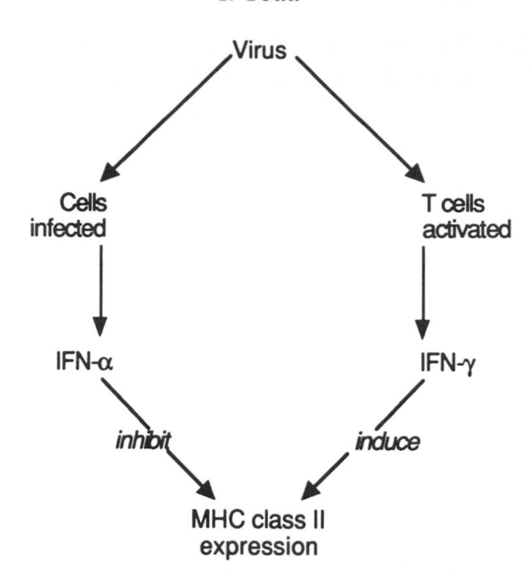


Fig. 3.1. Model for the opposing effects of IFN-α and IFN-γ on MHC class II expression as a result of the response to viral infection.

blasts (Collins *et al.*, 1986). At least in some cases, the effect on class I may be the result of stimulation of IFN-$\beta_2$ production by TNF, with the former cytokine then enhancing class I expression in an autocrine fashion (May, Helfgott & Sehgal, 1986). Both TNF-α and lymphotoxin (also known as TNF-β) can up-regulate (but not induce) HLA-DR expression on monocytes (but not lymphocytes) in a 3 hour incubation (Limb *et al.*, 1992).

In terms of interactions with IFN-γ, TNF-α has been reported both to augment (Chang & Lee, 1986; Pfizenmaier *et al.*, 1987; Arenzana-Seisdedos *et al.*, 1988) and to inhibit (Hoffman & Weinberg, 1987; Leeuwenberg *et al.*, 1988) induction of class II expression by IFN-γ. We observed that TNF-α synergized with IFN-γ in the induction of class II expression on human thyrocytes and that TNF-α caused increased binding of IFN-γ to these cells (Buscema *et al.*, 1989). This may be partly because of increased numbers of cell-surface receptors for IFN-γ, which have been shown to make cell lines more sensitive to IFN-γ for HLA-DR induction (Ucer *et al.*, 1985). It is interesting to note that, on tumour cell lines, the converse effect has also been observed, i.e. that IFN-γ increases expression of TNF-α receptors (Ruggiero *et al.*, 1986; Tsujimoto, Vip & Vilcek, 1986); however, this effect could be dissociated from the synergistic cytotoxic activity of these cytokines on the tumour cells, suggesting that the synergism was not explained by the receptor modulation (Tsujimoto, Feinman & Vilcek, 1986). Although TNF-α merely augments the induction of class II expression by IFN-γ on thyrocytes, it was noted in a previous section that the interaction of these cytokines is essentially obligatory for the induction of HLA class II expression by human pancreatic β cells (Pujol-Borrell *et al.*, 1987).

Recent studies suggest that whether TNF-α augments or inhibits IFN-γ-mediated class II induction may be related to the differentiation or maturation stage of the cells involved. Watanabe & Jacob (1991) found that, in the immature HL-60 and THP-1 cell lines, TNF-α enhanced IFN-γ-induced class II expression. However, when these cells were induced to differentiate by treatment with phorbol ester or IFN-γ, the additive effect of TNF-α on the IFN-γ-induced class II expression was eliminated. TNF-α was also found to down-regulate class II expression induced by IFN-γ in differentiated cells like skin fibroblasts and activated macrophages, and it decreased IFN-γ-induced class II expression by bone marrow cells in a maturation-dependent fashion. Similarly, Zimmer & Jones (1990) found that TNF-α augments the early induction of class II expression in thioglycollate-elicited peritoneal macrophages by IFN-γ but inhibits later expression.

### *3.3.4 Interleukin-3*

Interleukin-3 (IL-3) is one of the cytokines that can influence the survival, proliferation and differentiation of haematopoietic cells. As with the other cytokines discussed above, it can both positively and negatively modulate MHC class II expression. Some years ago, IL-3 (referred to then as P cell stimulating factor) was found to inhibit induction of class II on T-dependent mast cells by IFN-γ (Wong *et al.*, 1984b). However, whereas human cord blood haematopoietic cells lose class II expression upon exposure to IL-3 (Caux *et al.*, 1989), culture of murine macrophages with IL-3 leads to enhanced class II expression (Frendl & Beller, 1990). In a recent study, it was found that either IL-3 or IL-1α could induce HLA-DR expression by $CD34^+$ bone marrow cells and that a combination of the two was most effective (Srour *et al.*, 1992).

### *3.3.5 Interleukin-4*

Interleukin-4 (IL-4; previously known as B cell stimulatory factor-1) is able to induce MHC class II expression on certain cell types, but in a way which is distinct from that of IFN-γ. Thus, IL-4 induces an increase in class II expression by small, resting B cells, whereas IFN-γ will enhance class II expression on dividing, neoplastic B cells but not on resting B cells (Noelle *et al.*, 1984). The effect of IL-4 was relatively selective for class II expression, since levels of surface immunoglobulin and MHC class I molecules did not increase to the same extent. Other inducers of B cell activation and maturation can have a similar effect: antibodies to surface immunoglobulins (Ig) IgM and IgD were found to increase B cell surface expression of class II, but not other surface molecules like B1 and $\beta_2$-m (Godal *et al.*, 1985).

IL-4 and IFN-γ have also been found to have different quantitative and qualitative effects on class II expression by macrophages. Comparison of the effects of the two cytokines on mouse peritoneal macrophages showed IL-4 to induce class II expression on a smaller proportion of the cells, and with a shorter duration of expression, than was achieved with IFN-γ (Cao *et al.*, 1989). It has also been reported that IL-4 induces a 1.5- to 4-fold increase in both MHC class I and class II expression on mouse bone marrow-derived macrophages, whereas IFN-γ increases expression by 2- to 8-fold, with induction of class II mRNA occurring more rapidly than with IL-4 (Stuart, Zlotnik & Woodward, 1988). Furthermore, in this latter study, IL-4 differed from IFN-γ in being unable to induce MHC expression in thioglycollate-elicited peritoneal macrophages or in a myelomonocytic cell line. Finally, IL-4 does not influence class II expression on various other cell types in which IFN-γ can induce class II expression, such as thymic epithelial cells (Galy & Spits, 1991).

### *3.3.6 Interleukin-10*

Interleukin-10 (IL-10; previously called cytokine synthesis inhibitory factor) has effects on the immune system that are primarily immunosuppressive and anti-inflammatory (reviewed by de Waal Malefyt *et al.*, 1992). Various cell types produce IL-10 in both humans and mice, including T cells, B cells and macrophages. In addition, Epstein–Barr virus (EBV) produces a form of IL-10 (vIL-10; previously called BCRF1) that shares structural homology and biological functions with human IL-10 (de Waal Malefyt *et al.*, 1991b),

Interleukin-10 inhibits cytokine production by lymphocytes, including production of IFN-γ by T cells and NK cells: this appears to be an indirect effect via the action of IL-10 on monocytes and macrophages (see de Waal Malefyt *et al.*, 1992). This could down-regulate MHC class II expression through lack of stimulation by IFN-γ. However, human IL-10 and vIL-10 have also been reported to inhibit directly MHC class II expression by human monocytes – both constitutive expression and that induced by IFN-γ or by IL-4 (de Waal Malefyt *et al.*, 1991b). Activated monocytes are themselves a source of IL-10, which can down-regulate monocyte class II expression in an autocrine fashion (de Waal Malefyt *et al.*, 1991a). Human and viral IL-10 suppress antigen-specific T cell proliferation, possibly because of the down-regulation of monocyte class II expression limiting antigen presentation. Consistent with this notion are the observations that IL-10 does not affect class II expression by Epstein–Barr virus-transformed B lymphoblastoid cell lines and does not suppress antigen-specific T cell proliferative responses when these lymphoblastoid cells are used as the antigen-presenting cells (de Waal Malefyt *et al.*, 1991b).

In contrast to the above findings with human monocytes, mouse IL-10 has been reported to induce class II expression by small, resting B cells (Go *et al.*, 1990). In this respect, IL-10 produces the same effect as IL-4 but, unlike IL-4, does not induce CD-23 expression on B cells.

### *3.3.7 Colony-stimulating factors*

Colony-stimulating factors (CSFs) promote the proliferation and differentiation of bone marrow haematopoietic cells. Macrophage progenitors develop along the monocytic lineage under the influence of CSF-1 (also called macrophage-CSF), granulocyte-macrophage-CSF (GM-CSF) and, to some extent, IL-3 (already discussed above). Investigations into the effects of CSF-1 and GM-CSF on mouse bone marrow-derived macrophages indicate that GM-CSF has the more profound effect on class II expression by these cells. Fischer *et al.* (1988) reported that bone marrow-derived macrophages obtained by culture in CSF-1 were negative for expression of class II but could then be induced to express class II by culture with GM-CSF. Falk, Wahl & Vogel (1988) found both GM-CSF and CSF-1 to induce class II on bone marrow-derived macrophages, although GM-CSF was more effective. However, macrophages derived in the presence of CSF-1 showed enhanced susceptibility to class II induction by IFN-$\gamma$ (Warren & Vogel, 1985a), and CSF-1 appeared to be more effective in this respect than GM-CSF (Falk *et al.*, 1988).

GM-CSF has also been shown to influence MHC expression by other cell types; for example, it induces HLA-DR (but not usually HLA-DQ) expression by eosinophils (Lucey, Nicholson-Weller & Weller, 1989). It also induces increased translation of class I polypeptides (both heavy chain and $\beta_2$-m) by neutrophils, without increased transcription; there is also no increase in surface expression of class I by these cells because of increased turnover by shedding and internalization (Neuman *et al.*, 1992).

### *3.3.8 Transforming growth factor-β*

Transforming growth factor-β (TGF-β) was originally characterized in terms of its ability to bring about reversible transformation of fibroblasts. However, it can also inhibit the proliferation of certain cell types and has a range of effects on the immune system, most of which are suppressive (reviewed by Sasaki *et al.*, 1992). TGF-β is produced by transformed cells and also by some normal cells, including macrophages, lymphocytes and platelets. Furthermore, most cells possess high-affinity receptors for TGF-β. There are at least five isoforms of the molecule; the studies described below were undertaken mainly with TGF-$\beta_1$.

Similar to IL-10 (discussed above), TGF-β can inhibit IFN-γ production (Espevik *et al.*, 1988), which could, thus, be an indirect mechanism of down-regulating MHC class II, but can also directly suppress class II expression. For example, it reduces the induction of class II by IFN-γ in human peripheral blood mononuclear adherent cells and melanoma cell lines, this effect being evident at the transcriptional level (Czarniecki *et al.*, 1988). This suppression of IFN-γ-induced class II expression may be partly explained by the ability of TGF-β to down-regulate cell-surface expression and function of receptors for IFN-γ (Pinson *et al.*, 1992). However, TGF-β can also suppress class II expressed constitutively by some melanoma cells in the absence of IFN-γ (Czarniecki *et al.*, 1988).

TGF-β can have differential effects on different cells: comparing its effects on astrocytes from brainstem and forebrain, it inhibits proliferation and IFN-γ-induced class II expression by the former, but not the latter (Johns *et al.*, 1992).

### *3.3.9 Epidermal growth factor*

Epidermal growth factor (EGF) stimulates DNA synthesis by epidermal and epithelial cells; this includes thyrocytes, in which it also suppresses certain differentiated characteristics, such as iodide metabolism (Westermark, Karlsson & Westermark, 1983). EGF was found to partially suppress the induction of class II by IFN-γ in human thyrocytes (Todd *et al.*, 1990b). A similar suppressive effect was observed with TGF-α, which shows substantial amino acid sequence homology to EGF and binds to the same cell surface receptor. The ability of EGF to down-regulate class II expression is probably related to its effects on the thyrocytes in general rather than specifically affecting the action of IFN-γ. Therefore, we found that EGF inhibited class II induction not only when added to the cells at the same time as IFN-γ, but also when

Table 3.1. *Differential effects of mediators on HLA class II expression by different cell types*

| | HLA class II expression | |
|---|---|---|
| Mediators | Increased | Decreased |
| EGF, TGF-$\alpha$ | T cell-depleted peripheral blood mononuclear cells | Thyrocytes |
| IFN-$\alpha_1$ | Monocytes | Thyrocytes |

used to treat the thyrocytes either before or after IFN-$\gamma$. Furthermore, EGF also suppressed the enhancing effect of thyroid-stimulating hormone (TSH) on thyrocyte HLA class II expression (see below) either in the presence or absence of IFN-$\gamma$ (Todd *et al.*, 1990b).

In contrast to our findings with thyrocytes, Acres, Lamb & Feldmann (1985) found EGF (and also platelet-derived growth factor) to enhance class II expression by human blood-derived antigen-presenting cells. This is similar to the observations (discussed in Section 3.2.2) that human IFN-$\alpha_1$ suppresses IFN-$\gamma$-induced class II expression by thyrocytes (Guerin *et al.*, 1990) but enhances class II expression by monocytes (Rhodes *et al.*, 1986). It is interesting to note that these converse effects correlate with the normal tissue distribution of MHC class II, i.e. expression by conventional antigen-presenting cells of the immune system but not by, for example, most endocrine epithelial cells, like thyrocytes. The maintenance of this normal tissue distribution of class II expression may at least partly depend on a balance between factors which up-regulate and down-regulate its expression. Some factors, such as those discussed above, may play a dual role in this context by inducing class II in 'professional' antigen-presenting cells and inhibiting its expression by parenchymal cells (see Table 3.1).

### *3.3.10 Hormones and other factors*

In addition to cytokines and growth factors, a variety of other soluble mediators are known to influence the immune system, and this includes effects on MHC expression. Glucocorticoid hormones are potent anti-inflammatory and immunosuppressive agents that can modulate MHC expression. They inhibit IFN-$\gamma$-induced class II expression by mouse macrophages (Snyder & Unanue, 1982; Warren & Vogel, 1985b), acting at the transcriptional level (Fertsch-Ruggio *et al.*, 1988). Similar inhibitory effects of glucocorticoids on IFN-γ-induced class II expression have been reported on other cell types, e.g. T-dependent mast cells (Wong *et al.*, 1984b). Suppression of constitutive MHC class I expression by glucocorticoids has also been observed (von Knebel Doeberitz *et al.*, 1990).

In contrast to the above observations, glucocorticoids can also enhance MHC expression in some circumstances: whether up- or down-regulation of MHC occurs may depend, as noted for other factors, in part, on the cell type and species involved. Thus, dexamethasone slightly enhances IFN-γ-induced class II expression by human endothelial cells (Manyak *et al.*, 1988) and hydrocortisone has been reported to induce class II expression by human monocytes (Rhodes *et al.*, 1986).

Sex steroids have also been implicated in the modulation of MHC expression. Increased class II expression by guinea pig mammary gland epithelium has been noted during lactation. A similar effect on mammary class II expression could be produced by administering progesterone and oestrogen to guinea pigs for six weeks, and partial reversal of this expression could be achieved by administration of testosterone (Klareskog, Forsum & Peterson, 1980). In human fallopian tube epithelium, there is widespread pre-ovulatory HLA-DR expression lining the luminal surface, whereas the expression is withdrawn from the surface in the post-ovulatory phase. It was suggested that there may be hormonal regulation of this expression, with the changes reflecting a pre-ovulatory need for immunoreactivity against invading microbes and a post-ovulatory optimization of conditions for the semi-allogeneic pre-implantation embryo (Edelstam *et al.*, 1992).

TSH specifically regulates the activity of thyrocytes, stimulating thyroid hormone production. We examined its effects on MHC class II in human thyrocytes and found it to synergize with IFN-γ in inducing the expression of these molecules (Todd *et al.*, 1987b). TSH is known to stimulate the production of cyclic AMP (cAMP) in thyrocytes: we observed that addition of dibutyryl cAMP (dbcAMP) to thyrocyte cultures had similar effects to TSH on class II expression (except that the action of dbcAMP appeared to be additive rather than synergistic with IFN-γ), suggesting that the effect of TSH on class II is at least partly mediated via cAMP. This action probably represents part of the general effect of TSH on thyroid differentiation and metabolism, including stimulation of RNA and protein synthesis; TSH exerted its effect when used to stimulate the cells either before or after IFN-γ, as well as at the same time. We also observed a little stimulation of thyrocyte class II expression by TSH in the absence of IFN-γ; however, this appears to represent enhancement of pre-existing class II expression rather than *de novo* induction. The thyroid-stimulating antibodies characteristic of Graves' disease bind to the TSH receptor on thyrocytes and mimic the action of TSH in stimulating thyroid hormone production; they also appear to have the same effect as TSH on thyrocyte class II expression (Wenzel *et al.*, 1986).

Prostaglandins of the E series (PGE) inhibit IFN-γ-mediated induction of class II expression by mouse macrophages (Snyder, Beller & Unanue, 1982). PGE stimulates intracellular cAMP levels in macrophages, and dbcAMP was found to have similar inhibitory effects on macrophage class II expression, suggesting that cAMP is the second messenger for this effect of PGE. (Note that this is the opposite effect to that of dbcAMP on human thyrocytes, noted above.) Thromboxane was found to counteract the inhibitory effect of $PGE_2$ (Snyder *et al.*, 1982). Some other substances that down-regulate induction by IFN-γ of macrophage class II expression do so via stimulation of the production of prostaglandins and cAMP, e.g. bacterial endotoxin (Steeg, Johnson & Oppenheim, 1982a), and immune complexes binding via the Fcγ2b receptor, but not suppression of class II mediated via the Fcγ2a receptor (Hanaumi, Gray & Suzuki, 1984).

Serotonin, also known as 5-hydroxytryptamine (5-HT), is a major product of platelets and is released upon platelet aggregation at sites of inflammation. It suppresses class II induction by IFN-γ in mouse bone marrow macrophages (Sternberg, Trial & Parker, 1986). It may, thus, play a role in modulating macrophage class II expression at sites of inflammation.

The most active metabolite of vitamin $D_3$, 1, 25-dihydroxyvitamin $D_3$, can both positively and negatively influence the induction of class II expression by IFN-γ; an example of the former is on the human squamous cell carcinoma cell line HEp-2 (Tamaki & Nakamura, 1990), and an example of the latter is on the Pam 212 mouse keratinocyte cell line (Tani, Komura & Horikawa, 1992).

Glutamate and noradrenaline (norepinephrine) both act as neurotransmitters and both down-regulate class II induction on astrocytes by IFN-γ (Lee *et al.*, 1992). Interestingly, this inhibitory effect of glutamate was not seen on microglia. This may help to explain the observation that, although class II can be induced on both astrocytes and microglia *in vitro*, in central nervous system (CNS) tissue from animals with experimental allergic encephalomyelitis (EAE) (Hickey, Osborn & Kirby, 1985) or graft-versus-host disease (Hickey & Kimura, 1987), or in patients with active multiple sclerosis (Hayashi *et al.*, 1988; Lee *et al.*, 1990), class II expression is common in microglia but is rare in astrocytes.

Prothymosin α is a highly acidic polypeptide originally isolated from thymus, present in leucocytes and identified in various tissues. It has a number of immunopotentiating activities, including effects on MHC class II expression: it enhances class II expression by monocytes and HLA-DR-positive cell lines and induces class II in previously HLA-DR-negative tumour cells (Baxevanis *et al.*, 1992).

Certain plasma proteins are able to inhibit induction of class II expression by macrophages. One of these, $\alpha_2$-macroglobulin, is a large plasma glycoprotein that inhibits proteinases; it is converted from electrophoretically 'slow' to 'fast' forms by proteinases or by methylamine. The 'fast' forms bind to receptors on macrophages and have a number of immunosuppressive effects, including inhibition of antigen-induced T cell proliferation: they have been found to inhibit the induction of class II expression in macrophages by IFN-γ (Hoffman, Pizzo & Weinberg, 1987). Alpha-fetoprotein is a major protein in amnionic fluid and perinatal sera: it has immunosuppressive properties and also inhibits the induction of macrophage class II expression (Lu, Changelian & Unanue, 1984). This could play a role in facilitating the development of self tolerance in the foetus and neonate, as well as protecting the foetus from immunological rejection by the mother.

### 3.4 Effects of cytokines on MHC subregion expression

Several isotypes of MHC class I and class II antigens are expressed by each individual of a species: thus, in humans there are HLA-DP, HLA-DQ, and HLA-DR class II antigens and HLA-A, HLA-B and HLA-C class I antigens; mice express I-A and I-E class II antigens and H-2K, D and L class I antigens. There is no clear evidence for the different isotypic antigens having different functional roles (e.g. stimulating help versus suppression), but they do preferentially present different peptides. These different subregion products are also differentially susceptible to the effects of modulators of MHC expression.

In normal human tissues, HLA-DQ is expressed less widely and arises later in ontogeny than HLA-DR (Natali *et al.*, 1984). This correlates with the relative ease of induction of DR expression by IFN-γ compared with DQ, as observed on various cell types including endothelial cells (Collins *et al.*, 1984), thymic epithelium (Berrih *et al.*, 1985) and melanoma cell lines (Carrel, Schmidt-Kessen & Giuffre, 1985). In our investigations of ectopic HLA class II expression by thyrocytes in autoimmune thyroid diseases, we found that both the occurrence and intensity of expression in diseased thyroids followed the pattern DR > DP > DQ (Todd *et al.*, 1987a). This same hierarchy was observed when class II expression was induced in cultured thyrocytes derived from glands of non-autoimmune patients. Incubation with a relatively low concentration of IFN-γ (5–10 U/ml) resulted in moderate expression of DR, low expression of DP and little, if any, expression of DQ.

By contrast, incubation with a high dose of IFN-γ (500 U/ml), or a low dose plus TSH, resulted in higher expression of DR and DP and some induction of DQ.

The pattern of HLA-D subregion expression induced can vary from that described above, possibly depending on the cell type and/or particular cytokine(s) involved. For example, induction of HLA class II in human pancreatic β cells with IFN-γ plus TNF results in greater expression of DQ than DP, i.e. DR > DQ > DP (Soldevila *et al.*, 1990). In studies on human retinal pigment epithelial cells, Liversidge, Sewell & Forrester (1988) reported different effects of different cytokine preparations. Thus, moderate concentrations of IFN-γ (100–200 μ/ml) induced intermediate expression of DR and lesser expression of DP and DQ, whereas high concentrations of IFN-γ induced strong expression of all three subregion products. By contrast, a cytokine-containing supernatant derived from lymphocytes activated with concanavalin-A induced strong expression of DR and DQ but much weaker expression of DP. These authors also noted differences in the time course of induction by IFN-γ, with DR and DP being maximally induced within two to three days of culture whereas DQ expression increased gradually over eight days.

Other cytokines that have been studied in terms of their effects on HLA-D subregion expression include IL-4 and GM-CSF. Both of these were found to enhance expression of DR and DP, but not DQ, by monocytes, whereas IFN-γ enhanced expression of all three isotypes (Gerrard *et al.*, 1990). GM-CSF has also been shown to induce expression of DR in eosinophils, whereas DQ was not induced in most cases (Lucey *et al.*, 1989).

Different class I isotypes can also be differentially regulated by cytokines. In mouse tumour cell lines, it was found that IFN-γ enhances expression of $H\text{-}2D^k$ much more than $H\text{-}2K^k$ (Green & Phillips, 1986). Similarly, in humans, it has been found that IFN-α or IFN-γ enhances expression of HLA-B more than HLA-A: this is at least partly explained by two nucleotide differences in the promoters for these class I genes (Hakem *et al.*, 1991).

### 3.5 Evidence for cytokine modulation of MHC expression *in vivo*

As mentioned earlier in this chapter, caution should be exercised in relating effects of cytokines on MHC expression *in vitro* to what factors may determine the observed expression of MHC products *in vivo*. However, a number of approaches have indicated the possible relationships between cytokines and *in vivo* expression of MHC products.

There is indirect evidence that the maintenance of physiological MHC expression in various tissues is dependent on cytokine action and, in particular, IFN-γ. For example, thymic epithelial cells normally express MHC class II molecules *in vivo*, and this is important in the processes of thymic education of immature T cells. However, thymic epithelial cells gradually lose expression of class II molecules when cultured, indicating that this is not a constitutive property, but class II is maintained by the presence of IFN-γ in the cultures (Berrih *et al.*, 1985). Another example is provided by endothelial cells in which normal expression of class II was inhibited by *in vivo* treatment of dogs with cyclosporin A (Groenewegen *et al.*, 1985). This, again, is likely to relate to the action of IFN-γ and it is known that cyclosporin A inhibits IFN-γ synthesis by T cells (Reem, Cook & Vilcek, 1983). IFN-γ-secreting cells are present in normal humans (0.1–0.3% peripheral blood mononuclear cells) (Martinez-Maza *et al.*, 1984), presumably accounting for the physiological maintenance of non-constitutive class II expression.

One would anticipate that significant changes in the levels of MHC-modulating cytokines *in vivo* from those that occur physiologically would result in alterations in MHC expression. This has been demonstrated to be the case by administration of large doses of particular mediators. Systemic treatment of mice with IFN-γ was found to cause increases in MHC class I and class II expression in tissues throughout the body (Skoskiewicz *et al.*, 1985; Momburg *et al.*, 1986a,b). However, different cell types showed differential susceptibility to the treatment. For example, Skoskiewicz *et al.* (1985) reported that renal proximal tubules were induced to express large amounts of both class I and class II antigens, whereas distal tubules and collecting ducts were not. Momburg *et al.* (1986b) noted that, although class I expression was induced or enhanced in many cell types of the body, this was not the case for neurons, glial cells, gastric chief and parietal cells, and pancreatic cells. In addition, invariant chain (Ii) was induced more widely than were class II antigens, e.g. in bronchial epithelium (Momburg *et al.*, 1986a).

Evidence for synergy between cytokines *in vivo* is provided by the observation that the induction of keratinocyte class II expression by intradermal administration of IFN-γ to mice was enhanced by co-injection of TNF-α (Nakamura & Tamaki, 1990).

The use of cytokines in clinical trials is now providing information on their effects on MHC expression in humans. For example, administration of IFN-α to melanoma patients resulted in enhanced expression of class I antigens by peripheral blood mononuclear cells and raised levels of circulating class I antigens (Giacomini *et al.*, 1984).

The occurrence of enhanced MHC expression at sites of inflammation was discussed in the first section of this chapter. The involvement of cytokines in these processes is evidenced by studies demonstrating the occurrence of cytokines in inflamed tissues. In some investigations, mAbs against cytokines have been employed; for example, antibodies against IFN-α, IFN-β and IFN-γ were used to demonstrate these mediators in polymyositis at sites of muscle fibre damage and infiltration, where enhanced expression of MHC class I was also observed (Isenberg *et al.*, 1986). In muscular dystrophy, by contrast, enhanced class I expression was seen in the absence of IFNs, suggesting different regulation of the MHC expression in this disorder. Detection of cytokine expression at the transcriptional level using specific DNA probes has also proved to be informative, e.g. production of mRNA for IL-2 and IFN-γ by mononuclear cells from rheumatoid lesions (Buchan *et al.*, 1988).

Another approach to investigating the involvement of cytokines in regulation of MHC expression during inflammation is by blocking their activities during the inflammatory process. Thus, administration of antibodies against IFN-γ inhibited the induction of keratinocyte class II expression during hapten-induced delayed-type hypersensitivity in rats (Skoglund *et al.*, 1988) and prevented hyperexpression of class I in the pancreases of non-obese diabetic mice (Kay *et al.*, 1991). Halloran *et al.* (1992) found that localized inflammatory processes (rejection of an ascites tumour allograft or skin sensitization by oxazalone) induced increased MHC class I and class II expression in remote tissues such as liver, heart, pancreas and kidney. Anti-IFN-γ antibodies and cyclosporin A inhibited this, and transcripts for IFN-γ could be demonstrated in the uninvolved tissues. This suggests that IFN-γ released from the site of inflammation induces its own expression in remote tissues.

Inhibitory effects of cytokines on MHC expression *in vivo* have also been observed; for example, reduced class II expression by Küppfer cells induced by haemorrhagic shock can be inhibited by administration of antibodies to TNF (Ertel *et al.*, 1991).

Two final points of caution in interpreting cytokine effects on MHC expression, which apply particularly to the complex environment of interacting factors *in vivo*, are as follows: the first is that even if investigations implicate a particular cytokine in MHC modulation, its effect could be direct or could be via its influence on the expression of other mediators. Secondly, specific inhibitors of cytokine activity have been described; for some cytokines, soluble receptors and receptor antagonists have been identified, and neutralizing autoantibodies to interferons have been described in normal individuals (Ross *et al.*, 1990).

## 3.6 Interactions of mediators in determining MHC expression

Given the range of cytokines and other mediators, discussed in this chapter, which are known to modulate MHC expression, it seems likely that, in many

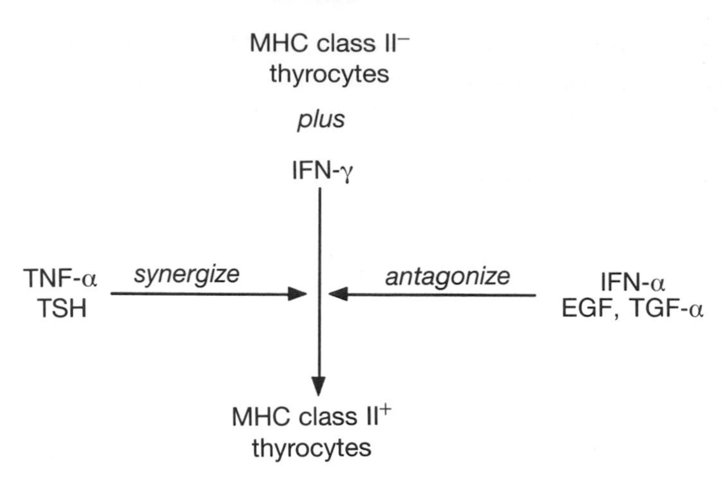


Fig. 3.2. Mediators that can stimulate or inhibit MHC class II expression by human thyrocytes.

instances, the MHC expression observed *in vivo* is the outcome of the effects of a number of these factors. However, one must then consider not only the presence but also the relative concentrations of different mediators in judging their relative efficacies. As a model, those factors may be considered that have been found to modulate HLA class II expression by human thyrocytes in terms of their possible involvement in determining the ectopic class II expression by thyrocytes in autoimmune thyroid diseases (Fig. 3.2).

In thyroid glands infiltrated by activated T cells, IFN-γ is likely to be the main inducer of thyrocyte class II expression (although the possible involvement of other factors, including viruses, remains a possibility). Indeed, IFN-γ has been demonstrated in the mononuclear cells infiltrating the thyroid in Graves' disease (Zheng *et al.*, 1992) and the IFN-γ-producing cells show strong spatial correlation with the thyrocytes expressing MHC class II antigens (Hamilton *et al.*, 1991). TNF-α and TSH can synergize with IFN-γ in thyrocyte class II induction *in vitro* (Todd *et al.*, 1987b; Buscema *et al.*, 1989). Classically, TNF-α was considered to be a macrophage product, but it can be produced by activated T cells (Turner, Londei & Feldmann, 1987) and is also produced by thyrocytes themselves in autoimmune glands (Zheng *et al.*, 1992). This latter observation raises the possibility that thyrocytes may play an autocrine role in enhancing their class II expression.

The concentrations of TSH which we found to be optimal in enhancing thyrocyte class II induction were of the order of 100–1000 mU/l. TSH concentrations are raised above normal in hypothyroid states like Hashimoto's thyroiditis, so it is possible that this hormone could play a role in the enhancement of thyrocyte class II expression in this disease. However, even in severe hypothyroidism, the maximum serum TSH concentrations achieved are about 100 mU/l, although local *in vivo* concentrations at the thyroid follicular surface may well be greater than those found in serum.

Immunoreactive EGF occurs in normal human serum at concentrations of about 0.2 ng/ml (Uchihashi *et al.*, 1984); we found that concentrations of EGF at least as low as this partially suppressed class II induction in thyrocytes *in vitro* by a low dose of IFN-γ (Todd *et al.*, 1990b). (It is also worth noting that TGF-α, which has a similar suppressive effect, is produced not only by neoplastic and early foetal cells, but also by certain normal adult cell types, as demonstrated in keratinocytes (Coffey *et al.*, 1987).) However, the circulating level of EGF *in vivo* appears to be affected by thyroid function, and its serum concentration has been reported to be raised in patients with hyperthyroidism and, to a lesser extent, in hypothyroid individuals (Uchihashi *et al.*, 1984). The former situation is consistent with the observation that the EGF content of submandibular glands was increased in mice injected with thyroid hormone (Gresik *et al.*, 1981). Other factors, including slow metabolism of EGF, could result in its raised level in hypothyroidism (Uchihashi *et al.*, 1984). Furthermore, TSH stimulated the expression of EGF receptors by thyroid cells (Westermark *et al.*, 1986). By comparison, we found that a relatively high dose of EGF (10 ng/ml) was unable to inhibit greatly thyrocyte class II induction by high concentrations of IFN-γ (1000 U/ml), as might be generated in an active autoimmune infiltrate (Todd *et al.*, 1990b). Therefore, any inhibitory activity of EGF that could be effective under normal circumstances might be overwhelmed in severe thyroiditis. Similar considerations could apply to the inhibition of IFN-γ-induced class II expression on thyrocytes by IFN-α, since only very high concentrations cause marked inhibition *in vitro* (Guerin *et al.*, 1990). However, it is known that high concentrations of IFN-α can be induced *in vivo* by viral infections (Morris & Tomkins, 1989).

Although the above considerations are based on *in vitro* experimentation and apply only to the limited range of factors investigated, they do indicate the complexity of signals influencing MHC expression to which cells may be subjected.

## 3.7 Conclusions: functional consequences of MHC class II modulation

Helper T cells, expressing CD4, play a central role in the induction of immune responses, and their activation is dependent on the expression of MHC class II molecules (McNicholas *et al.*, 1982; Matis *et al.*, 1982; 1983;

Lechler *et al.*, 1985). The modulation of class II expression can, therefore, have important consequences for immune responsiveness. Indeed, the ability of cytokines and other factors to modulate class II expression often correlates with their ability to influence the function of APCs (Walker, Lanier & Warner, 1982). This applies both to factors that can enhance class II expression and APC function, e.g. IFN-$\gamma$ and GM-CSF (Zlotnik *et al.*, 1983; Fischer *et al.*, 1988), and to factors that inhibit both of these, e.g. endotoxin and zymosan A (Yem & Parmely, 1981). However, antigen processing and presentation is a complex, multistep process that may be affected by cytokines in other ways in addition to their effects on MHC expression. It is, therefore, not surprising that, in some cases, factors can have opposite effects on class II expression and APC function: for example, IFN-$\alpha_1$ and hydrocortisone have been reported to enhance class II expression by human monocytes but to decrease their ability to stimulate antigen-specific T cell proliferation if present during antigen pulsing (Rhodes, Ivanyi & Cozens, 1986).

A further complexity is that T cell activation is dependent not only on the interaction between the T cell receptor and the antigen-MHC complex but also on what is termed 'co-stimulation' resulting from the interactions between various ligand/receptor pairs of adhesion molecules that can be expressed on the T cell and APC surfaces (reviewed by Liu & Linsley, 1992) (see Fig. 3.3). The expression of adhesion molecules is greater on activated or memory T cells than on unprimed T cells, and they can be induced on various other cell types by the action of cytokines (e.g. IFN-$\gamma$, IL-1, TNF-$\alpha$), or by other means of activation (Freedman *et al.*, 1987; Springer, 1990; Shimizu *et al.*, 1992). In the absence of these co-stimulatory signals, T cell activation may be suboptimal, or non-existent. Furthermore, without certain interactions, particularly those involving B7/CD28 or HSA, T cell interaction with an antigen–MHC class II complex results in the induction of unresponsiveness in the T cells (Liu & Linsley, 1992). Therefore induction of class II expression without the appropriate provision of co-stimulatory signals may actually lead to down-regulation, rather than enhancement, of specific immune responsiveness.

'Professional' APC, e.g. activated B cells and macrophages, can provide the full range of signals for antigen-specific T cell activation, including expression of MHC class II, integrins and B7, as well as production of cytokines like IL-1. By contrast, other cell types induced to express class II ectopically may thereby be endowed with the ability to present antigen to helper T cells without the provision for appropriate co-stimulation. Figure 3.4 presents a model of 'non-professional' APCs (e.g. tissue parenchymal cells, such as thyrocytes, expressing MHC class II), indicating their possible

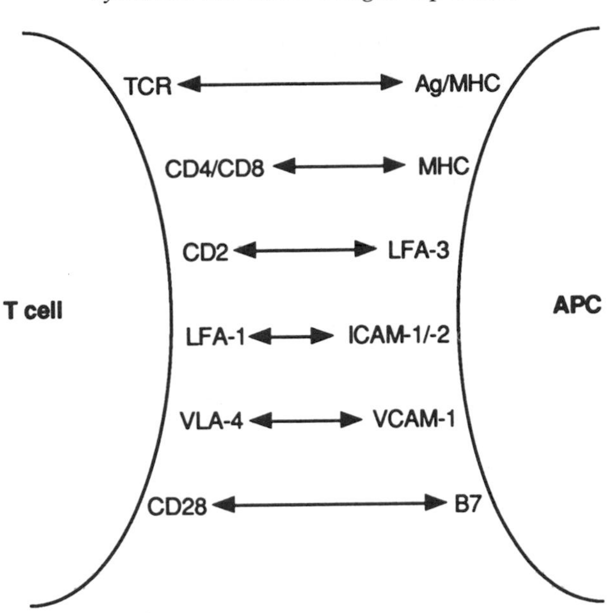


Fig. 3.3. Examples of the cell surface molecules involved in adhesion between T cells and antigen-presenting cells.

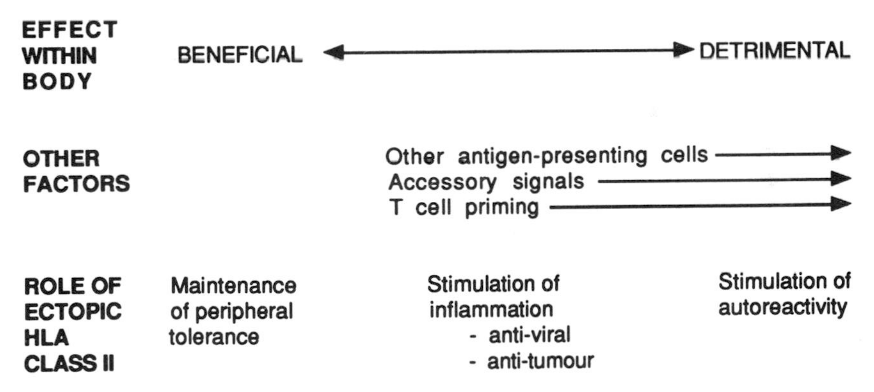


Fig. 3.4. Model for the possible spectrum of effects resulting from ectopic MHC (e.g. HLA) class II expression by 'non-professional' antigen-presenting cells.

spectrum of effects on T cells, depending on the contributions of additional factors; within this spectrum, the overall effects could be beneficial or detrimental. Some *in vitro* studies with class II-bearing thyrocytes have suggested that these cells can stimulate T cells (e.g. Londei, Bottazzo & Feldmann, 1985; Stein & Stadecker, 1987; Hirose *et al.*, 1988), whereas others have suggested that they cannot (Minami *et al.*, 1987). *In vivo* studies by Frohman,

Francfort & Cowing (1991) showed that when mouse thyroid isografts, exposed to IFN-γ to induce MHC class II expression, were implanted under the kidney capsule of recipients, the grafts were infiltrated and destroyed, but the recipients' own thyroid glands remained intact. In preliminary studies, we inserted isologous thyrocytes, pre-treated with IFN-γ to induce class II, directly into the thyroid lobes of rats. This treatment did not induce thyroid autoimmunity, but it should also be noted that the insertion of the cells (whether or not they were pre-treated with IFN-γ) appeared to cause specific suppression of thyroid autoantibody production induced by immunization with thyroid extract (Todd, Gibbon & Davenport, 1990a). The differences in results in these various studies are probably the result of the variable nature of the experimental design; however, they do indicate that class II$^+$ thyrocytes can have different antigen-presenting abilities, depending on circumstances.

In terms of the model shown in Fig. 3.4, the induction of MHC class II expression in parenchymal cells of normal tissues, without also expressing a sufficient range or amount of co-stimulatory molecules, may enable these cells to present their intrinsic self-antigens in a way that could induce non-responsiveness in potentially autoreactive T cells. This may be beneficial in helping to maintain peripheral self tolerance. It is unlikely that the numbers of professional APCs resident in normal tissues would be sufficient to provide co-stimulation, since this type of bystander effect is much less efficient than provision of T cell receptor ligand and B7 on the same cell surface (Liu & Janeway, 1992).

When tissues are subjected to some form of inflammation-inducing insult (e.g. infection or neoplasia), this is likely to induce changes that may facilitate T cell activation by the class II$^+$ parenchymal cells; if this involves presentation of, for example, microbial or tumour antigens, this could again be beneficial to the organism. In particular, the increased numbers and activation state of professional APCs could have two effects. Firstly, these APCs may then be able to provide a significant level of bystander co-stimulation to complement the antigen presentation by the parenchymal cells. Secondly, antigen presentation by the professional APCs themselves will generate primed T cells; these may be more readily activated than unprimed T cells and are more capable of receiving co-stimulation from bystander cells (Jenkins, Ashwell & Schwartz, 1988; Otten & Germain, 1991). Thus, they may be further stimulated by the parenchymal cells. Moreover, activation of the parenchymal cells by cytokines and other factors generated during the inflammatory process may improve their potential for T cell activation. Treatment of thyrocytes with IFN-γ enhances their ability to bind autologous T cells (Fukazawa *et al.*, 1991), and culturing thyrocytes with IFN-γ, IL-1β or

TNF-α induces expression of ICAM-1 (Weetman *et al.*, 1989; 1990; Tolosa *et al.*, 1992a,b). (It should be noted that thyrocytes themselves can synthesize IL-1 (Zheng *et al.*, 1991) as well as TNF-α (Zheng *et al.*, 1992).) Neural cell adhesion molecule (CD56) (Migita *et al.*, 1991) and LFA-3 (Tolosa *et al.*, 1992b) appear to be expressed constitutively by thyrocytes.

If the above inflammatory processes are too excessive or prolonged so that autoantigens are efficiently presented by parenchymal cells to autoreactive T cells that also receive adequate co-stimulatory signals, then the stimulation of autoimmunity could have detrimental consequences.

A better understanding of MHC expression in these processes, and its modulation by cytokines, may facilitate therapeutic approaches to enhance beneficial immune responses and to inhibit those which are detrimental.

## Acknowledgement

I am very grateful to Ms Claire Davenport for helpfully reviewing the manuscript.

## References

Acres, R.B., Lamb, J. & Feldmann, M. (1985). Effects of platelet derived growth factor and epidermal growth factor on antigen induced proliferation of human T cells lines. *Immunology*, 54, 9–16.

Aichinger, G., Fill, H. & Wick, G. (1985). *In situ* immune complexes: lymphocyte subpopulations and HLA-DR positive epithelial cells in Hashimoto thyroiditis. *Laboratory Investigation*, 52, 132–140.

Arenzana-Seisdedos, F., Mogensen, S.C., Vuillier, F., Fiers, W. & Virelizier, J.L. (1988). Autocrine secretion of tumor necrosis factor under the influence of interferon-gamma amplifies HLA-DR gene induction in human monocytes. *Proceedings of the National Academy of Sciences of the USA*, 85, 6087–6091.

Athanassakis-Vassiliadis, I., Galanopoulos, V.K., Grigoriou, M. & Papamatheakis, J. (1990). Induction of class II MHC antigen expression on the murine placenta by 5-azacytidine correlates with fetal abortion. *Cellular Immunology*, 128, 438–449.

Ballardini, G., Mirakian, R., Bianchi, F.B., Pisi, E., Doniach, D. & Bottazzo, G.F. (1984). Aberrant expression of HLA-DR antigens on bile duct epithelium in primary biliary cirrhosis: relevance to pathogenesis. *Lancet*, ii, 1009–1113.

Barclay, A.N. & Mason, D.W. (1982). Induction of Ia antigen in rat epidermal cells and gut epithelium by immunological stimuli. *Journal of Experimental Medicine*, 156, 1665–1676.

Basham, T.Y., Nickoloff, B.J., Merigan, T.C. & Morhenn U.B. (1984). Recombinant gamma interferon induces HLA DR expression on cultured human keratinocytes. *Journal of Investigative Dermatology*, 83, 88–92.

Baxevanis, C.N., Thanos, D., Reclos, G.J., Anastasopoulos, E., Tsokos, G.C., Papamatheakis, J. & Papamichail, M. (1992). Prothymosin α enhances human and murine MHC class II surface antigen expression and messenger RNA accumulation. *Journal of Immunology*, 148, 1979–1984.

Benveniste, E.N., Vidovic, M., Panek, R.B., Norris, J.G., Reddy, A.T. & Benos, D.J. (1991). Interferon-γ-induced astrocyte class II major histocompatibility complex gene expression is associated with both protein kinase C activation and $Na^+$ entry. *Journal of Biological Chemistry*, 266, 18119–18126.

Berrih, S., Arenzana-Seisdedos, F., Cohen, S., Devos, R., Charron, D. & Virelizier, J.-L. (1985). Interferon-γ modulates HLA class II antigen expression on cultured human thymic epithelial cells. *Journal of Immunology*, 135, 1165–1171.

Bottazzo, G.F., Dean, B.M., McNally, J.M., MacKay, E.H., Swift, P.G.F. & Gamble, D.R. (1985). *In situ* characterization of autoimmune phenomena and expression of HLA molecules in the pancreas in diabetic insulitis. *New England Journal of Medicine*, 313, 353–360.

Buchan, G., Barrett, K., Fujita, T., Taniguchi, T., Maini, R. & Feldmann, M. (1988). Detection of activated T cell products in the rheumatoid joint using cDNA probes to interleukin-2 (IL-2), IL-2 receptor and IFN-γ. *Clinical and Experimental Immunology*, 71, 295–301.

Buscema, M., Todd, I., Deuss, U., Hammond, L., Mirakian, R., Pujol-Borrell, R. & Bottazzo, G.F. (1989). Influence of tumour necrosis factor-α on the modulation by interferon-γ of HLA class II molecules in human thyroid cells and its effect on interferon-γ binding. *Journal of Clinical Endocrinology and Metabolism*, 69, 433–439.

Calamai, E.G., Beller, D.I. & Unanue, E.R. (1982). Regulation of macrophage populations. IV. Modulation of Ia expression in bone marrow-derived macrophages. *Journal of Immunology*, 128, 1692–1694.

Cao, H., Wolff, R.G., Meltzer, M.S. & Crawford, R.M. (1989). Differential regulation of class II MHC determinants on macrophages by IFN-γ and IL-4. *Journal of Immunology*, 143, 3524–3531.

Carrel, S., Schmidt-Kessen, A. & Giuffre, L. (1985). Recombinant interferon-γ can induce the expression of HLA-DR and -DC on DR-negative melanoma cells and enhance the expression of HLA-ABC and tumour-associated antigens. *European Journal of Immunology*, 15, 118–123.

Caux, C., Favre, C., Saeland, S., Duvert, V., Mannoni, P., Durand, I., Aubry, J.P. & deVries, J.E. (1989). Sequential loss of CD34 and class II MHC antigens on purified cord blood hematopoietic progenitors cultured with IL-3: characterization of $CD34^-$, $HLA\text{-}DR^+$ cells. *Blood*, 74, 1287–1294.

Celada, A. & Maki, R.A. (1991). IFN-γ induces the expression of the genes for MHC class II $I\text{-}A_\beta$ and tumor necrosis factor through a protein kinase C-independent pathway. *Journal of Immunology*, 146, 114–120.

Chang, R.J. & Lee, S.H. (1986). Effects of interferon-gamma and tumor necrosis factor-alpha on the expression of an Ia antigen on a murine macrophage cell line. *Journal of Immunology*, 137, 2853–2856.

Chumbley, G., Hawley, S., Carter, N.P. & Loke, Y.W. (1991). Human extravillous trophoblast MHC class I expression is resistant to regulation by interferon-α. *Journal of Reproductive Immunology*, 20, 289–296.

Coffey, R.J., Derynck, R., Wilcox, J.N., Bringman, T.S., Goustin, A.S., Moses, H.L. & Pittelkow, M.R. (1987). Production and auto-induction of transforming growth factor-α in human keratinocytes. *Nature (London)*, 328, 817–820.

Collins, T., Korman, A.J., Wake, C.T., Boss, J.M., Kappes, D.J., Fiers, W., Ault, K.A., Gimbrone, M.A., Strominger, J.L. & Pober, J.S. (1984). Immune interferon activates multiple class II major histocompatibility complex genes and the associated invariant chain gene in human endothelial cells and dermal fibroblasts. *Proceedings of the National Academy of Sciences of the USA*, 81, 4917–4921.

Collins, T., Lapierre, L.A., Fiers, W., Strominger, J.L. & Pober, J.S. (1986). Recombinant human tumor necrosis factor increases mRNA levels and surface expression of HLA-A,B antigens in vascular endothelial cells and dermal fibroblasts *in vitro*. *Proceedings of the National Academy of Sciences of the USA*, 83, 446–450.

Czarniecki, C.W., Chiu, H.H., Wong, G.H.W., McCabe, S.M. & Palladino, M.A. (1988). Transforming growth factor-$\beta_1$ modulates the expression of class II histocompatibility antigens on human cells. *Journal of Immunology*, 140, 4217–4223.

de Waal Malefyt, R., Abrams, J., Bennett, B., Figdor, C.G. & de Vries, J.E. (1991a). Interleukin-10 (IL-10) inhibits cytokine synthesis by human monocytes: an autoregulatory role of IL-10 produced by monocytes. *Journal of Experimental Medicine*, 174, 1209–1220.

de Waal Malefyt, R., Haanen, J., Spits, H., Roncarolo, M.-G., te Velde, A., Figdor, C., Johnson, K., Kastelein, R., Yssel, H. & de Vries, J.E. (1991b). Interleukin-10 (IL-10) and viral IL-10 strongly reduce antigen-specific human T cell proliferation by diminishing the antigen-presenting capacity of monocytes via down-regulation of class II major histocompatibility complex expression. *Journal of Experimental Medicine*, 174, 915–924.

de Waal Malefyt, R., Yssel, H., Roncarolo, M.-G., Spits, H. & de Vries, J.E. (1992). Interleukin-10. *Current Opinion in Immunology*, 4, 314–320.

Dolei, A., Capobianchi, M.R. & Ameglio, F. (1983). Human interferon-gamma enhances the expression of class I and class II major histocompatibility complex products in neoplastic cells more effectively than interferon-alpha or interferon-beta. *Infection and Immunity*, 40, 172–176.

Edelstam, G.A.B., Lundkvist, O.E., Klareskog, L. & Karlsson-Parra, A. (1992). Cyclic variation of major histocompatibility complex class II antigen expression in the human fallopian tube epithelium. *Fertility and Sterility*, 57, 1225–1229.

Ertel, W., Morrison, M.H., Ayala, A., Perrin, M.M. & Chaudry, I.H. (1991) Anti-TNF monoclonal antibodies prevent haemorrhage-induced suppression of Kupffer cell antigen presentation and MHC class II antigen expression. *Immunology*, 74, 290–297.

Espevik, T., Figari, I.S., Ranges, G.E. & Palladino, M.A. Jr (1988). Transforming growth factor-beta1 (TGF-β1) and recombinant human tumor necrosis factor-alpha reciprocally regulate the generation of lymphokine activated killer cell activity. Comparison between natural porcine platelet-derived TGF-β1 and TGF-β2, and recombinant human TGF-β1. *Journal of Immunology*, 140, 2312–2316.

Falk, L.A., Wahl, L.M. & Vogel, S.N. (1988). Analysis of Ia antigen expression in macrophages derived from bone marrow cells cultured in granulocyte-macrophage colony-stimulating factor or macrophage colony-stimulating factor. *Journal of Immunology*, 140, 2652–2660.

Fertsch-Ruggio, D., Schoenberg, D.R. & Vogel, S.N. (1988). Induction of macrophage Ia antigen expression by rIFN-γ and down-regulation by IFN-α/β and dexamethasone are regulated transcriptionally. *Journal of Immunology*, 141, 1582–1589.

Fischer, H.-G., Frosch, S., Reske, K. & Reske-Kunz, A.B. (1988). Granulocyte-macrophage colony-stimulating factor activates macrophages derived from bone marrow cultures to synthesis of MHC class II molecules and to augmented antigen presentation function. *Journal of Immunology*, 141, 3882–3888.

Flyer, D.C., Burakoff, S. & Faller, D. (1985). Retrovirus-induced changes in major histocompatibility complex antigen expression influence susceptibility to lysis by cytotoxic T lymphocytes. *Journal of Immunology*, 135, 2287–2292.

Foulis, A.K. & Farquharson, M.A. (1986). Aberrant expression of HLA-DR antigens by insulin-containing β cells in recent-onset type I diabetes mellitus. *Diabetes*, 35, 1215–1224.

Freedman, A.S., Freeman, G., Horowitz, J.C., Daley, J. & Nadler, L.M. (1987). B7, a B cell-restricted antigen that identifies preactivated B cells. *Journal of Immunology*, 139, 3260–3266.

Frendl, G. & Beller, D.I. (1990). Regulation of macrophage activation by IL-3. I. IL-3 functions as a macrophage activating factor with unique properties, inducing Ia and lymphocyte function-associated antigen-1 but not cytotoxicity. *Journal of Immunology*, 144, 3392–3399.

Frohman, M., Francfort, J.W. & Cowing, C. (1991). T-dependent destruction of thyroid isografts exposed to IFN-γ. *Journal of Immunology*, 146, 2227–2234.

Fukazawa, H., Hiromatsu, Y., Bernard, N., Salvi, M. & Wall, J.R. (1991). Binding of peripheral blood and thyroidal T lymphocytes to thyroid cell monolayers: possible role of 'homing-like' receptors in the pathogenesis of thyroid autoimmunity. *Autoimmunity*, 10, 181–188.

Galy, A.H.M. & Spits, H. (1991). IL-1, IL-4 and IFN-γ differentially regulate cytokine production and cell surface molecule expression in cultured human thymic epithelial cells. *Journal of Immunology*, 147, 3823–3830.

Geraghty, D.E., Wei, X., Orr, H.T. & Koller, B.H. (1990). Human leucocyte antigen F (HLA-F). An expressed HLA gene composed of a class I coding sequence linked to a novel transcribed repetitive element. *Journal of Experimental Medicine*, 171, 1–18.

Gerrard, T.L., Dyer, D.R. & Mostowski, H.S. (1990). IL-4 and granulocyte-macrophage colony-stimulating factor selectively increase HLA-DR and HLA-DP antigens but not HLA-DQ antigens on human monocytes. *Journal of Immunology*, 144, 4670–4674.

Giacomini, P., Aguzzi, A., Pestka, S., Fisher, P.B. & Ferrone, S. (1984). Modulation by recombinant DNA leukocyte (α) and fibroblast (β) interferons of the expression and shedding of HLA- and tumor-associated antigens by human melanoma cells. *Journal of Immunology*, 133, 1649–1655.

Go, N.F., Castle, B.E., Barrett, R., Kastelein, R., Dang, W., Mosmann, T.R., Moore, K.W. & Howard, M. (1990). Interleukin-10, a novel B cell stimulatory factor: unresponsiveness of X chromosome-linked immunodeficiency B cells. *Journal of Experimental Medicine*, 172, 1625–1631.

Godal, T., Davies, C., Smeland, E.B., Heikkila, R., Funderud, S., Steen, H.B. & Hilrum, K. (1985). Antibodies to surface IgM and IgD increase the expression of various class II antigens on human B cells. *European Journal of Immunology*, 15, 173–177.

Gottlieb, A.B., Lifshitz, B., Fu, S.M., Staiano-Coico, L., Wang, C.Y. & Carter, D.M. (1986) Expression of HLA-DR molecules by keratinocytes and presence of Langerhans cells in the dermal infiltrate of active psoriatic plaques. *Journal of Experimental Medicine*, 164, 1013–1028.

Green, W.R. & Phillips, J.D. (1986). Differential induction of H-2K vs H-2D class I major histocompatibility complex antigen expression by murine recombinant interferon-γ. *Journal of Immunology*, 137, 814–818.

Gresik, E.W., Schenkein, I., Van der Noen, H. & Barka, T. (1981). Hormonal regulation of epidermal growth factor and protease in the submandibular gland of the adult mouse. *Endocrinology*, 109, 924–929.

Groenewegen, G., Buurman, W.A. & van der Linden, C.J. (1985). Lymphokine dependence of *in vivo* expression of MHC class II antigens by endothelium. *Nature (London)*, 316, 361–363.

Guerin, V., Todd, I., Hammond, L.J. & Bottazzo, G.F. (1990). Suppression of HLA class II expression on thyrocytes by interferon-alpha 1. *Clinical and Experimental Immunology*, 79, 341–345.

Gumina, R.J., Freire-Moar, J., DeYoung, L., Webb, D.R. & Devens, B.H. (1991). Transduction of the IFN-γ signal for HLA-DR expression in the promonocytic line THP-1 involves a late-acting PKC activity. *Cellular Immunology*, 138, 265–279.

Hakem, R., Le Bouteiller, P., Jezo-Bremond, A., Harper, K., Campese, D. & Lemonnier, F.A. (1991). Differential regulation of HLA-A3 and HLA-B7 MHC class I genes by IFN is due to two nucleotide differences in their IFN response sequences. *Journal of Immunology*, 147, 2384–2390.

Halloran, P.F., Autenried, P., Ramassar, V., Urmson, J. & Cockfield, S. (1992). Local T cell responses induce widespread MHC expression. Evidence that IFN-γ induces its own expression in remote sites. *Journal of Immunology*, 148, 3837–3846.

Hamilton, F., Black, M., Farquharson, M.A., Stewart, C. & Foulis, A.K. (1991). Spatial correlation between thyroid epithelial cells expressing class II MHC molecules and interferon-gamma-containing lymphocytes in human thyroid autoimmune disease. *Clinical and Experimental Immunology*, 83, 64–68.

Hanafusa, T., Pujol-Borrell, R., Chiovato, L., Russell, R.C.G., Doniach, D. & Bottazzo, G.F. (1983). Aberrant expression of HLA-DR antigen on thyrocytes in Graves' disease: relevance for autoimmunity. *Lancet*, ii, 1111–1115.

Hanaumi, K., Gray, P. & Suzuki, T. (1984). Fcγ receptor-mediated suppression of γ-interferon-induced Ia antigen expression on a murine macrophage-like cell line ($P338D_1$). *Journal of Immunology*, 133, 2852–2856.

Hayashi, T., Morimoto, C.L., Burks, J.S., Kerr, C. & Hauser, S.L. (1988). Dual-label immunocytochemistry of the active multiple sclerosis lesion: major histocompatibility complex and activation antigens. *Annals of Neurology*, 24, 523–531.

Hickey, W.F. & Kimura, H. (1987). Graft-vs.-host disease elicits expression of class I and class II histocompatibility antigens and the presence of scattered T lymphocytes in rat central nervous system. *Proceedings of the National Academy of Sciences of the USA*, 84, 2082–2086.

Hickey, W.F., Osborn, J.P. & Kirby, W.M. (1985). Expression of Ia molecules by astrocytes during acute experimental allergic encephalomyelitis in the Lewis rat. *Cellular Immunology*, 91, 528–535.

Hirose, W., Lahat, N., Platzer, M., Schmitt, S. & Davies, T.F. (1988). Activation of MHC-restricted rat T cells by cloned syngeneic thyrocytes. *Journal of Immunology*, 141, 1098–1102.

Hoffman, M.R., Pizzo, S.V. & Weinberg, J.B. (1987). Modulation of mouse peritoneal macrophage Ia and human peritoneal macrophage HLA-DR expression by $\alpha_2$-macroglobulin 'fast' forms. *Journal of Immunology*, 139, 885–890.

Hoffman, M. & Weinberg, J.B. (1987). Tumor necrosis factor-alpha induces increased hydrogen peroxide production and Fc receptor expression, but not increased Ia antigen expression by peritoneal macrophages. *Journal of Leukocyte Biology*, 42, 704–707.

Hunt, J.S., Andrews, G.K. & Wood, G.W. (1987). Normal trophoblasts resist induction of class I HLA. *Journal of Immunology*, 138, 2481–2487.

Hunt, J.S., Atherton, R.A. & Pace, J.L. (1990). Differential responses of rat trophoblast cells and embryonic fibroblasts to cytokines that regulate proliferation and class I MHC antigen expression. *Journal of Immunology*, 145, 184–189.

Inaba, K., Kitaura, M., Kato, T., Watanabe, Y., Kawade, Y. & Muramatsu, S. (1986). Contrasting effects of $\alpha/\beta$- and $\gamma$-interferons on expression of macrophage Ia antigens. *Journal of Experimental Medicine*, 163, 1030–1035.

Isenberg, D.A., Rowe, D., Shearer, M., Novick, D. & Beverley, P.C.L. (1986). Localization of interferons and interleukin 2 in polymyositis and muscular dystrophy. *Clinical and Experimental Immunology*, 63, 450–458.

Jansson, R., Karlsson, A. & Forsum, U. (1985). Intrathyroid HLA-DR expression and T lymphocyte phenotypes in Graves' thyrotoxicosis, Hashimoto's thyroiditis and nodular colloid goitre. *Clinical and Experimental Immunology*, 58, 264–272.

Jenkins, M.K., Ashwell, J.D. & Schwartz, R.H. (1988). Allogeneic non-T spleen cells restore the responsiveness of normal T cell clones stimulated with antigen and chemically modified antigen-presenting cells. *Journal of Immunology*, 140, 3324–3330.

Johns, L.D., Babcock, G., Green, D., Freedman, M., Sriram, S. & Ransohoff, R.M. (1992). Transforming growth factor-$\beta_1$ differentially regulates proliferation and MHC class-II antigen expression in forebrain and brainstem astrocyte primary cultures. *Brain Research*, 585, 229–236.

Kalovidouris, A.E. (1992). The role of cytokines in polymyositis: interferon-$\gamma$ induces class II and enhances class I major histocompatibility complex antigen expression on cultured human muscle cells. *Journal of Laboratory and Clinical Medicine*, 120, 244–251.

Kay, T.W.H., Campbell, I.L., Oxbrow, L. & Harrison, L.C. (1991). Overexpression of class I major histocompatibility complex accompanies insulitis in the non-obese diabetic mouse and is prevented by anti-interferon-$\gamma$ antibody. *Diabetologia*, 34, 779–785.

Kelley, V.E., Fiers, W. & Strom, T.B. (1984). Cloned human interferon-$\gamma$, but not interferon-$\beta$ or -$\alpha$, induces expression of HLA-DR determinants by fetal monocytes and myeloid leukemic cell lines. *Journal of Immunology*, 132, 240–245.

King, D.P. & Jones, P.P. (1983). Induction of Ia and H-2 antigens on a macrophage cell line by immune interferon. *Journal of Immunology*, 131, 315–318.

Klareskog, L., Forsum, U. & Peterson, P.A. (1980). Hormonal regulation of the expression of Ia antigens on mammary gland epithelium. *European Journal of Immunology*, 10, 958–963.

Lapierre, L.A., Fiers, W. & Pober, J.S. (1988). Three distinct classes of regulatory cytokines control endothelial cell MHC antigen expression. Interactions with immune $\gamma$ interferon differentiate the effects of tumor necrosis factor and lymphotoxin from those of leukocyte $\alpha$ and fibroblast $\beta$ interferons. *Journal of Experimental Medicine*, 167, 794–804.

Le Bouteiller, P., Boucraut, J., Chimini, G., Vernet, C., Fauchet, R. & Pontarotti, P. (1991). Methylation status of CpG islands on the MHC class I chromosomal region of the trophoblast-derived human cell line, Jar. In *HLA, 2*, ed. T. Sasazuki. London: Oxford University Press.

Lechler, R.I., Norcross, M.A. & Germain, R.N. (1985). Qualitative and quantitative studies of antigen-presenting cell function by using I-A-expressing L cells. *Journal of Immunology*, 135, 2914–2922.

Lee, S.C., Collins, M., Vanguri, P. & Shin, M.L. (1992). Glutamate differentially inhibits the expression of class II MHC antigens on astrocytes and microglia. *Journal of Immunology*, 148, 3391–3397.

Lee, S.C., Moore, G.R.W., Golenwsky, G. & Raine, C.S. (1990). Multiple sclerosis: a role for astroglia in active demyelination suggested by class II MHC expression and ultrastructural study. *Journal of Neuropathology and Experimental Neurology*, 49, 122–136.

Leeuwenberg, J.F., Van Damme, J., Meager, T., Jeunhomme, T.M., & Buurman, W.A. (1988). Effects of tumour necrosis factor on the interferon-gamma-induced major histocompatibility complex class II antigen expression by human endothelial cells. *European Journal of Immunology*, 18, 1469–1472.

Limb, G.A., Hamblin, A.S., Wolstencroft, R.A. & Dumonde, D.C. (1992). Rapid cytokine up-regulation of integrins, complement receptor 1 and HLA-DR on monocytes but not on lymphocytes. *Immunology*, 77, 88–94.

Lindhal, G., Hedfors, E., Klareskog, L. & Forsum, U. (1985). Epithelial HLA-DR expression and T lymphocyte subsets in salivary glands in Sjögren's syndrome. *Clinical and Experimental Immunology*, 61, 475–482.

Ling, P.D., Warren, M.K. & Vogel, S.N. (1985). Antagonistic effect of interferon-β on the interferon-γ-induced expression of Ia antigen in murine macrophages. *Journal of Immunology*, 135, 1857–1863.

Liu, Y. & Janeway, C.A. Jr (1992). Cells that present both T cell receptor ligand and the costimulator are the most efficient inducers of clonal expansion of normal CD4 T cells. *Proceedings of the National Academy of Sciences of the USA*, 89, 3845–3849.

Liu, Y. & Linsley, P.S. (1992). Costimulation of T cell growth. *Current Opinion in Immunology*, 4, 265–270.

Liversidge, J.M., Sewell, H.F. & Forrester, J.V. (1988). Human retinal pigment epithelial cells differentially express MHC class II (HLA DP, DR and DQ) antigens in response to *in vitro* stimulation with lymphokine or purified IFN-γ. *Clinical and Experimental Immunology*, 73, 489–494.

Londei, M., Bottazzo, G.F. & Feldmann, M. (1985). Human T cell clones from autoimmune thyroid glands: specific recognition of autologous thyroid cells. *Science*, 228, 85–89.

Lu, C.Y., Changelian, P.S. & Unanue, E.R. (1984). α-Fetoprotein inhibits macrophage expression of Ia antigens. *Journal of Immunology*, 132, 1722–1727.

Lucey, D.R., Nicholson-Weller, A. & Weller, P.F. (1989). Mature human eosinophils have the capacity to express HLA-DR. *Proceedings of the National Academy of Sciences of the USA*, 86, 1348–1351.

Manyak, C.L., Tse, H., Fischer, P., Coker, L., Sigal, N.H. & Koo, G.C. (1988). Regulation of class II MHC molecules on human endothelial cells. Effects of IFN and dexamethasone. *Journal of Immunology*, 140, 3817–3821.

Martinez-Maza, O., Andersson, U., Andersson, J., Britton, S. & De Ley, M. (1984). Response of histocompatibility-restricted T cell clones is a function of the product of the concentrations of antigen and Ia molecules. *Proceedings of the National Academy of Sciences of the USA*, 80, 6019–6023.

Matis, L.A., Glimcher, L.H., Paul, W.F. & Schwartz, R.H. (1983). Magnitude of response of histocompatibility-restricted T cell clones is a function of the product of the concentration of antigen and Ia molecules. *Proceedings of the National Academy of Sciences of the USA*, 80, 6019–6023.

Matis, L.A., Jones, P.P., Murphy, D.B., Hedrick, S.M., Lerner, E.A., Janeway, C.A. Jr, McNicholas, J.M., & Schwartz, R.H. (1982). Immune response gene function correlates with the expression of an Ia antigen. II. A quantitative deficiency in $A_e$:$E_\alpha$ complex expression causes a corresponding defect in antigen-presenting cell function. *Journal of Experimental Medicine*, 155, 508–523.

Mattsson, R., Holmdahl, R., Scheynius, A., Bernadotte, F., Mattsson, A. & Van der Meide, P.H. (1991). Placental MHC class I antigen expression is induced in mice following *in vivo* treatment with recombinant interferon-γ. *Journal of Reproductive Immunology*, 19, 115–129.

May, L.T., Helfgott, D.C. & Sehgal, P.B. (1986). Anti-β-interferon antibodies inhibit the increased expression of HLA-B7 mRNA in tumour necrosis factor-treated human fibroblasts: structural studies of the $\beta_2$ interferon involved. *Proceedings of the National Academy of Sciences of the USA*, 83, 8957–8961.

McNicholas, J.M., Murphy, D.B., Matis, L.A., Schwartz, R.H., Lerner, E.A., Janeway, C.A. Jr & Jones, P.P. (1982). Immune response gene function correlates with the expression of an Ia antigen. I. Preferential association of certain $A_e$ and $E_\alpha$ chains results in a quantitative deficiency in expression of an $A_e$:$E_\alpha$ complex. *Journal of Experimental Medicine*, 155, 490–507.

Migita, K., Eguchi, K., Kawakami, A., Ida, H., Fukuda, T., Kurata, A., Ishikawa, N., Ito, K. & Nagataki, S. (1991). Detection of Leu-19 (CD-56) antigen on human thyroid epithelial cells by an immunohistochemical method. *Immunology*, 72, 246–249.

Milton, A.D. & Fabre, J.W. (1985). Massive induction of donor-type class I and class II major histocompatibility complex antigens in rejecting cardiac allografts in the rat. *Journal of Experimental Medicine*, 161, 98–112.

Minami, M., Ebner, S.A., Stadecker, M.J. & Dorf, M.E. (1987). The effects of phorbol ester on alloantigen presentation. *Journal of Immunology*, 138, 393–400.

Momburg, F., Koch, N., Moller, P., Moldenhauer, G., Butcher, G.W. & Hammerling, G.J. (1986a). Differential expression of Ia and Ia-associated invariant chain in mouse tissues after in vivo treatment with IFN-γ. *Journal of Immunology*, 136, 940–948.

Momburg, F., Koch, N., Moller, P., Moldenhauer, G. & Hammerling, G.J. (1986b). *In vivo* induction of H-2K/D antigens by recombinant interferon-γ. *European Journal of Immunology*, 16, 551–557.

Morris, A.G. & Tomkins, P.T. (1989). Interactions of interferons in the induction of histocompatibility antigens in mouse fibroblasts and glial cells. *Immunology*, 67, 537–539.

Nakamura, K. & Tamaki, K. (1990). Tumour necrosis factor enhances the interferon-γ-induced class II MHC antigen expression on murine keratinocytes *in vivo*. *Archives of Dermatological Research*, 282, 415–417.

Natali, P.G., Segatto, O., Ferrone, S., Tosi, R. & Corte, G. (1984). Differential tissue distribution and ontogeny of DC-1 and HLA-DR antigens. *Immunogenetics*, 19, 109–116.

Neuman, E., Huleatt, J.W., Vargas, H., Rupp, E.E. & Jack R.M. (1992). Regulation

of MHC class I synthesis and expression by human neutrophils. *Journal of Immunology*, 148, 3520–3527.
Noelle, R., Krammer, P.H., Ohara, J., Uhr, J.W. & Vitetta, E.S. (1984). Increased expression of Ia antigens on resting B cells: an additional role for B-cell growth factor. *Proceedings of the National Academy of Sciences of the USA*, 81, 6149–6153.
Otten, G.R. & Germain, R.N. (1991). Split anergy in $CD8^+$ T cells: receptor-dependent cytolysis in the absence of interleukin-2 production. *Science*, 251, 1228–1231.
Pfizenmaier, K., Scheurich, P., Schluter, C. & Kronke, M. (1987). Tumor necrosis factor enhances HLA-A,B,C and HLA-DR gene expression in human tumor cells. *Journal of Immunology*, 138, 975–980.
Pinson, D.M., LeClaire, R.D., Lorsbach, R.B., Parmely, M.J. & Russell, S.W. (1992). Regulation by transforming growth factor-$\beta_1$ of expression and function of the receptor for IFN-$\gamma$ on mouse macrophages. *Journal of Immunology*, 149, 2028–2034.
Pujol-Borrell, R. & Todd, I. (1987). Inappropriate HLA Class II in autoimmunity: is it the primary event? *Bailliere's Clinical Immunology and Allergy*, 1, 1–27.
Pujol-Borrell, R., Todd, I., Doshi, M., Bottazzo, G.F., Sutton, R., Gray, D., Adolf, G.R. & Feldmann, M. (1987). HLA class II induction in human islet cells by interferon-$\gamma$ plus tumour necrosis factor or lymphotoxin. *Nature (London)*, 326, 304–306.
Pujol-Borrell, R., Todd, I., Doshi, M., Gray, D., Feldmann, M. & Bottazzo, G.F. (1986). Differential expression and regulation of MHC products in the endocrine and exocrine cells of the human pancreas. *Clinical and Experimental Immunology*, 65, 128–139.
Radford, J.E. Jr, Chen, E., Hromas, R. & Ginder, G.D. (1991). Cell-type specificity of interferon-$\gamma$-mediated HLA class I gene transcription in human hematopoietic tumor cells. *Blood*, 77, 2008–2020.
Reem, G.H., Cook, L.A. & Vilcek, J. (1983). Gamma interferon synthesis by human thymocytes and T lymphocytes is inhibited by cyclosporin A. *Science*, 221, 63–65.
Reitz, M.S. Jr, Mann, D.L., Eiden, M., Trainor, C.D. & Clarke, M.F. (1984). DNA methylation and expression of HLA-DR$\alpha$. *Molecular and Cellular Biology*, 4, 890–897.
Rhodes, J., Ivanyi, J. & Cozens, P. (1986). Antigen presentation by human monocytes: effects of modifying major histocompatibility complex class II antigen expression and interleukin 1 production by using recombinant interferons and corticosteroids. *European Journal of Immunology*, 16, 370–375.
Rhodes, J. & Stokes, P. (1982). Interferon-induced changes in the monocyte membrane: inhibition by retinol and retinoic acid. *Immunology*, 45, 531–536.
Ross, C., Hansen, M.B., Schyberg, T. & Berg, K. (1990). Autoantibodies to crude human leucocyte interferon (IFN), native human IFN, recombinant human IFN-alpha 2b and human IFN-gamma in healthy blood donors. *Clinical and Experimental Immunology*, 82, 57–62.
Ruggiero, V., Tavernier, J., Fiers, W. & Baglioni, C. (1986). Induction of the synthesis of tumor necrosis factor receptors by interferon-$\gamma$. *Journal of Immunology*, 136, 2445–2450.
Sasaki, H., Pollard, R.B., Schmitt, D. & Suzuki, F. (1992). Transforming growth factor-$\beta$ in the regulation of the immune response. *Clinical Immunology and Immunopathology*, 65, 1–9.

Scott, H., Solheim, B.G., Brandtzaeg, P. & Thorsby, E. (1980). HLA-DR-like antigens in the epithelium of the human small intestine. *Scandinavian Journal of Immunology*, 12, 77–82.

Seong, D., Sims, S., Johnson, E., Lyding, J., Lopez, A., Garovoy, M., Talpaz, M., Kantarjian, H., Lopez-Berestein, G., Reading, C. & Deisseroth, A. (1991). Activation of class I HLA expression by TNF-alpha and gamma-interferon is mediated through protein kinase C-dependent pathway in CML cell lines. *British Journal of Haematology*, 78, 359–367.

Shimizu, Y., Newman, W., Tanaka, Y. & Shaw, S. (1992). Lymphocyte interactions with endothelial cells. *Immunology Today*, 13, 106–112.

Skoglund, C., Scheynius, A., Holmdahl, R. & Van der Meide, P.H. (1988). Enhancement of DTH reaction and inhibition of the expression of class II transplantation antigens by *in vivo* treatment with antibodies against γ-interferon. *Clinical and Experimental Immunology*, 71, 428–432.

Skoskiewicz, M.J., Colvin, R.B., Schneeberger, E.E. & Russell, P.S. (1985). Widespread and selective induction of major histocompatibility complex-determined antigens in vivo by γ interferon. *Journal of Experimental Medicine*, 162, 1645–1664.

Snyder, D.S., Beller, D.I. & Unanue, E.R. (1982). Prostaglandins modulate macrophage Ia expression. *Nature (London)*, 299, 163–165.

Snyder, D.S. & Unanue, E.R. (1982). Corticosteroids inhibit murine macrophage Ia expression and interleukin 1 production. *Journal of Immunology*, 129, 1803–1805.

Soldevila, G., Toshi, M., James, R., Lake, S.D., Sutton, R., Gray, D., Bottazzo, G.F. & Pujol-Borrell, R. (1990). HLA-DR, DP and DQ induction in human islet β cells by the combination IFN-γ and TNF-α. *Autoimmunity*, 6, 307–317.

Springer, T.A. (1990). Adhesion receptors of the immune system. *Nature (London)*, 346, 425–434.

Srour, E.F., Brandt, J.E., Leemhuis, T., Ballas, C.B. & Hoffman, R. (1992). Relationship between cytokine-dependent cell cycle progression and MHC class II antigen expression by human CD34+ HLA-DR$^-$ bone marrow cells. *Journal of Immunology*, 148, 815–820.

Steeg, P.S., Johnson, H.M. & Oppenheim, J.J. (1982a). Regulation of murine macrophage Ia antigen expression by an immune interferon-like lymphokine: inhibitory effect of endotoxin. *Journal of Immunology*, 129, 2402–2406.

Steeg, P.S., Moore, R.N., Johnson, H.M. & Oppenheim, J.J. (1982b). Regulation of murine macrophage Ia antigen expression by a lymphokine with immune interferon activity. *Journal of Experimental Medicine*, 156, 1780–1793.

Steeg, P.S., Moore, R.N. & Oppenheim, J.J. (1980). Regulation of murine macrophage Ia antigen expression by products of activated spleen cells. *Journal of Experimental Medicine*, 152, 1734–1744.

Stein, M.E. & Stadecker, M.J. (1987). Characterization and antigen presenting function of a murine thyroid-derived epithelial cell line. *Journal of Immunology*, 139, 1786–1791.

Sternberg, E.M., Trial, J. & Parker, C.W. (1986). Effect of serotonin on murine macrophages: suppression of Ia expression by serotonin and its reversal by 5-$HT_2$ serotonergic receptor antagonists. *Journal of Immunology*, 137, 276–282.

Stringer, C.P., Hicks, R. & Botham, P.A. (1991). The expression of MHC class II (Ia) antigens on mouse keratinocytes following epicutaneous application of contact sensitizers and irritants. *British Journal of Dermatology*, 125, 521–528.

Stuart, P.M., Zlotnik, A. & Woodward, J.G. (1988). Induction of class I and class II

MHC antigen expression on murine bone marrow-derived macrophages by IL-4 (B cell stimulatory factor 1). *Journal of Immunology*, 140, 1542–1547.
Tamaki, K. & Nakamura, K. (1990). Differential enhancement of interferon-γ-induced MHC class II expression of HEp-2 cells by 1,25-dihydroxyvitamin $D_3$. *British Journal of Dermatology*, 123, 333–338.
Tani, M., Komura, A. & Horikawa, T. (1992). 1α,25-Dihydroxyvitamin $D_3$ modulates Ia antigen expression induced by interferon-γ and prostaglandin $E_2$ production in Pam 212 cells. *British Journal of Dermatology*, 126, 266–274.
Todd, I., Gibbon, L. & Davenport, C. (1990a). The influence of thyroid follicular cells on the induction of thyroid autoimmunity. *Journal of Pathology*, 161, 339A.
Todd, I., Hammond, L.J., James, R.F.L., Feldmann, M. & Bottazzo, G.F. (1990b). Epidermal growth factor and transforming growth factor-alpha suppress HLA class II induction in human thyroid epithelial cells. *Immunology*, 69, 91–96.
Todd, I., Pujol-Borrell, R., Abdul-Karim, B.A.S., Hammond, L.J., Feldmann, M. & Bottazzo, G.F. (1987a). HLA-D subregion expression by thyroid epithelium in autoimmune thyroid diseases and induced *in vitro*. *Clinical and Experimental Immunology*, 69, 532–542.
Todd, I., Pujol-Borrell, R., Hammond, L.J., Bottazzo, G.F. & Feldmann, M. (1985). Interferon-γ induces HLA-DR expression by thyroid epithelium. *Clinical and Experimental Immunology*, 61, 265–273.
Todd, I., Pujol-Borrell, R., Hammond, L.J., McNally, J.M., Feldmann, M. & Bottazzo, G.F. (1987b). Enhancement of thyrocyte HLA class II expression by thyroid stimulating hormone. *Clinical and Experimental Immunology*, 69, 524–531.
Tolosa, E., Roura, C., Catalfamo, M., Marti, M., Lucas-Martin A., Sanmarti, A., Salinas, I., Obiols, G., Foz-Sala, M. & Pujol-Borrell, R. (1992a). Expression of intercellular adhesion molecule-1 in thyroid follicular cells in autoimmune, non-autoimmune, and neoplastic diseases of the thyroid gland: discordance with HLA. *Journal of Autoimmunity*, 5, 107–118.
Tolosa, E., Roura, C., Marti, M., Belfiore, A. & Pujol-Borrell, R. (1992b). Induction of intercellular adhesion molecule-1 but not of lymphocyte function-associated antigen-3 in thyroid follicular cells. *Journal of Autoimmunity*, 5, 119–135.
Towata, T., Hayashi, N., Katayama, K., Takehara, T., Sasaki, Y., Kasahara, A., Fusamoto, H. & Kamada, T. (1991). Signal transduction pathways in the induction of HLA class I antigen expression on HUH 6 cells by interferon-gamma. *Biochemical and Biophysical Research Communications*, 177, 610–618.
Tsujimoto, M., Feinman, R. & Vilcek, J. (1986). Differential effects of type I IFN and IFN-γ on the binding of tumour necrosis factor to receptors in two human cell lines. *Journal of Immunology*, 137, 2272–2276.
Tsujimoto, M., Vip, Y.K. & Vilcek, J. (1986). Interferon-γ enhances expression of cellular receptors for tumor necrosis factor. *Journal of Immunology*, 136, 2441–2444.
Turner, M., Londei, M. & Feldmann, M. (1987). Human T cells from autoimmune and normal individuals can produce tumor necrosis factor. *European Journal of Immunology*, 17, 1807–1814.
Ucer, U., Bartsch, H., Scheurich, P. & Pfizenmaier, K. (1985). Biological effects of γ-interferon on human tumor cells: quantity and affinity of cell membrane receptors for γ-IFN in relation to growth inhibition and induction of HLA-DR expression. *International Journal of Cancer*, 36, 103–108.
Uchihashi, M., Hirata, Y., Tomita, M., Nakajima, H., Fujita, T. & Kuma, K. (1984).

Concentrations of serum human epidermal growth factor (hEGF) in patients with hyper and hypothyroidism. *Hormonal and Metabolic Research*, 16, 676.

Volc-Platzer, B., Majdic, O., Knapp, W., Wolff, K., Hinterberger, W., Lechner, K. & Stingl, G. (1984). Evidence of HLA-DR antigen biosynthesis by human keratinocytes in disease. *Journal of Experimental Medicine*, 159, 1784–1789.

von Knebel Doeberitz, M., Koch, S., Drzonek, H. & zur Hausen, H. (1990). Glucocorticoid hormones reduce the expression of major histocompatibility class I antigens on human epithelial cells. *European Journal of Immunology*, 20, 35–40.

Walker, E.B., Lanier, L.L. & Warner, N.L. (1982). Concomitant induction of the cell surface expression of Ia determinants and accessory cell function by a murine macrophage tumor cell line. *Journal of Experimental Medicine*, 155, 629–634.

Walker, R., Cooke, A., Bone, A.J., Dean, B.M., van der Meide, P. & Baird, J.D. (1986). Induction of class II MHC antigens in vitro on pancreatic B cells isolated from BB/E rats. *Diabetologia*, 29, 249–251.

Warren, M.K. & Vogel, S.N. (1985a). Bone marrow-derived macrophages: development and regulation of differentiation markers by colony-stimulating factor and interferons. *Journal of Immunology*, 134, 982–989.

Warren, M.K. & Vogel, S.N. (1985b). Opposing effects of glucocorticoids on interferon-γ-induced macrophage Fc receptor and Ia antigen expression. *Journal of Immunology*, 134, 2462–2469.

Watanabe, Y. & Jacob, C.O. (1991). Regulation of MHC class II antigen expression. Opposing effects of tumor necrosis factor-α on IFN-β-induced HLA-DR and Ia expression depends on the maturation and differentiation stage of the cell. *Journal of Immunology*, 146, 899–905.

Weetman, A.P., Cohen, S.B., Makgoba, M.W. & Borysiewicz, L.K. (1989). Expression of an intercellular adhesion molecule, ICAM-1, by human thyroid cells. *Journal of Endocrinology*, 122, 185–191.

Weetman, A.P., Freeman, M.A., Borysiewicz, L.K. & Makgoba, M.W. (1990). Functional analysis of intercellular adhesion molecule-1-expressing human thyroid cells. *European Journal of Immunology*, 20, 271–275.

Wenzel, B.E., Gutekunst, R., Mansky, T., Schultek, Th. & Scriba, P.C. (1986). Thyrotropin and IgG from patients with Graves' disease induce class-II antigen on human thyroid cells. In *Thyroid and Autoimmunity*, ed. H.A. Drexhage & W.M. Wiersinga, pp. 141–144. Amsterdam: Excerpta Medica, Elsevier.

Westermark, K., Karlsson, F.A. & Westermark, B. (1983). Epidermal growth factor modulates thyroid growth and function in culture. *Endocrinology*, 112, 1680–1686.

Westermark, K., Westermark, B., Karlsson, A. & Ericson, L.E. (1986). Location of EGF receptors on porcine thyroid follicle cells and receptor regulation by thyrotropin. *Endocrinology*, 118, 1040–1046.

Wong, G.H.W., Bartlett, P.F., Clark-Lewis, I., Battye, F. & Schrader, J.W. (1984a). Inducible expression of H-2 and Ia antigens on brain cells. *Nature (London)*, 310, 688–691.

Wong, G.H.W., Clark-Lewis, I., Hamilton, J.A. & Schrader, J.W. (1984b). P cell stimulating factor and glucocorticoids oppose the action of interferon-γ in inducing Ia antigens on T-dependent mast cells (P cells). *Journal of Immunology*, 133, 2043–2050.

Wong, G.H.W., Clark-Lewis, I., McKimm-Breschkin, J.L., Harris, A.W. & Schrader, J.W. (1983). Interferon-γ induces enhanced expression of Ia and H-2 antigens on B lymphoid, macrophage, and myeloid cell lines. *Journal of Immunology*, 131, 788–793.

Wong, G.H.W., Clark-Lewis, I., McKimm-Breschkin, J.L. & Schrader, J.W. (1982). Interferon-γ-like molecule induces Ia antigens on cultured mast cell progenitors. *Proceedings of the National Academy of Sciences of the USA*, 79, 6989–6993.

Yem, A.W. & Parmely, M.J. (1981). Modulation of Ia-like antigen expression and antigen-presenting activity of human monocytes by endotoxin and zymosan A. *Journal of Immunology*, 127, 2245–2251.

Yeoman, H., Anderton, J.G. & Stanley, M.A. (1989). MHC class II antigen expression is not induced on murine epidermal keratinocytes by interferon-gamma alone or in combination with tumour necrosis factor-alpha. *Immunology*, 66, 100–105.

Zheng, R.Q.H., Abney, E., Chu, C.Q., Field, M., Grubeck-Loebenstein, B., Maini, R.N. & Feldmann, M. (1991). Detection of IL-6 and IL-1 production in human thyroid epithelial cells by non-radioactive *in situ* hybridization and immunohistochemical methods. *Clinical and Experimental Immunology*, 83, 314–319.

Zheng, R.Q.H., Abney, E.R., Chu, C.Q., Field, M., Maini, R.N., Lamb, J.R. & Feldmann, M. (1992). Detection of *in vivo* production of tumour necrosis factor-alpha by human thyroid epithelial cells. *Immunology*, 75, 456–462.

Zimmer, T. & Jones, P.P. (1990). Combined effects of tumor necrosis factor-α, prostaglandin $E_2$, and corticosterone on induced Ia expression on murine macrophages. *Journal of Immunology*, 145, 1167–1175.

Zlotnik, A., Shimonkevitz, R.P., Gefter, M.L., Kappler, J. & Marrack, P. (1983). Characterization of the γ-interferon-mediated induction of antigen-presenting ability in P388D1 cells. *Journal of Immunology*, 131, 2814–2820.

# 4

# Control of MHC class I gene expression

KEIKO OZATO
*National Institute of Child Health and Human Development, Bethesda*

## 4.1 Introduction

This chapter deals with transcriptional regulation of MHC class I genes in normal physiology and development. The main focus is transcription factors involved in regulating class I gene expression and the mechanism of action of these factors.

Class I genes are widely expressed in many adult tissues. However, levels of class I expression vary among different tissues; some tissues do not express class I genes at a significant level (Klein, 1975). While lymphoid tissues express class I genes at high levels, there is virtually no expression in the CNS. The lack of class I gene expression in the CNS constitutes a unique immunological environment, which accounts for persistent infection by some viruses (Joly, Mucke & Oldstone, 1991). Other tissues, such as pancreas and muscle, appear to be low in class I gene expression (David-Watine, Israel & Kourilsky, 1990a). Class I gene expression is developmentally controlled: class I mRNAs are not found until mouse embryos reach mid-somite stage (Ozato, Wan & Orrison, 1985). Levels of MHC class I expression in embryonic tissues vary but are higher in haemato/lymphopoietic tissues than in other tissues (Hedley *et al.*, 1989). Class I mRNA levels rise sharply during the first week of neonatal development in the mouse (Kasik, Wan & Ozato, 1987). So far, the molecule responsible for the primary onset of class I expression during embryogenesis, or that which triggers the second, post-natal upsurge of class I expression, has not been identified. Recent evidence from nuclear run-on experiments indicates that tissue-specific expression and the developmental regulation of class I genes is largely controlled at the level of transcription (D. Morello, personal communication).

Class I gene expression is induced by several cytokines at the level of transcription, the most potent being IFNs (Fellous *et al.*, 1982) and TNF-α (Collins *et al.*, 1986). All types of IFNs induce MHC class I transcription in a cell type-dependent fashion; for example, IFN-$\gamma$ induction of MHC class I transcription is very prominent in embryonal carcinoma cells and neuroblastoma cells but is weaker in fibroblasts and other cells (Sugita *et al.*, 1987; Drew *et al.*, 1993). In various embryonal carcinoma cells, class I gene transcription is induced by retinoic acid (Croce *et al.*, 1981; Daniel *et al.*, 1983), which appears to mimic developmental regulation of MHC class I genes *in vivo*.

## 4.2 The regulatory elements of MHC class I genes

Regulatory elements involved in transcriptional control of class I genes have been identified by a series of transfection experiments. They are mostly located in a relatively short stretch of the conserved upstream region (Israel *et al.*, 1986; 1989; Kimura *et al.*, 1986; Miyazaki, Appella & Ozato, 1986; Baldwin & Sharp, 1987; Shirayoshi *et al.*, 1987) (Fig. 4.1). Although many studies identifying class I *cis*-acting elements have been carried out for mouse H-2 genes, these elements are conserved in a number of human HLA-A, HLA-B, and HLA-C genes. Figure 4.1 shows a comparison of mouse and human MHC class I genes in this region. In the mouse, there are two enhancer elements called region I and region II, both of which function as moderately active enhancer elements in various cells (Baldwin & Sharp, 1987; Shirayoshi *et al.*, 1987; Burke *et al.*, 1989; Israel *et al.*, 1989). Region I is also called enhancer A (Israel *et al.*, 1987) and is homologous to the NF-$\kappa$B binding site ($\kappa$B). The $\kappa$B site was originally described for promoter and enhancer regions of immunoglobulin genes (Sen & Baltimore, 1986). Later, related motifs were found in many other genes including the IL-2 receptor, IL-6, HIV, MHC class II E$\alpha$, IFN-$\beta$ and $\beta_2$-m genes and have been proposed to be involved in regulating transcription of many genes in response to inflammatory and immune stimulation (Lenardo & Baltimore, 1989). Region II has the GGTCA motif, a prototype hormone-responsive element found in many genes controlled by steroid/thyroid hormones and retinoids (Martinez, Givel & Wahli, 1991). Region II has been shown to be involved in retinoic acid induction of class I transcription in embryonal carcinoma (EC) cells (Nagata *et al.*, 1992). The IFN consensus sequence (ICS) present downstream from region I (enhancer A) confers IFN inducibility (Israel *et al.*, 1986; Korber *et al.*, 1988; Shirayoshi *et al.*, 1988; Blanar *et al.*, 1989). This element conforms to the minimum IFN-responsive motif GAAANN (MacDonald *et al.*, 1990) and is homologous to the IFN (stimulated) response element (ISRE or IRE) of many genes that are induced by IFNs (see reviews by Levy & Darnell, 1990; Williams, 1991). Further downstream there is another conserved element, site

```
I. UPSTREAM

MOUSE
       -203 REGION II                      REGION I                ICS          -136
H-2L^d GGTGAGGTCAGGGGTGGGGAAGCCCAGGGCTGGGGATTCCCCATCTCCTCAGTTTCACTTCTGCACCT
H-2K^b A-----------------------------------------------A-------------------
H-2K^k --------------------------------------------------------------------
H-2D^d --------------------------------------------------------------------
H-2K^d ------------------------------------------------A---------------T---
H-2K^k ------------------------------------------------A---------------T---

HUMAN
                 -196                       REGION I                ICS   -146
HLA A2           ATGGATTGGGGAGTCCAGCCTTGGGGATTCCCCAACTCCGC**AGTTTCTTTCTC
HLA A3           -A--------------------*------------------**------------
HLA B7           C--C---------G-G---G--------------*----C-TG------AC----
HLA B27          C--C---------G-G------------------*----CACG------AC----
HLA Cw1          C--C-C-------G-G-C--G---A------T---*----C-TG------AC----

Qa and TLa
                                                                 ICS
     Q10    GGTGAGGTCAGGGGTTGGGAAGCCCAGGGCTGAGGATTCCCCATCTCCCCAGTTTCACTTC
     Q7     A---------A----A----------T--------------*-------------------
     TL     AGGGTCTGAGCCTGAAGGCGGGGGATTAGGTTGGGCTGTACCAGAACTAT----------T

                                               -194            -176
B-FIV Chicken class I                           CCTTTCGCTTTCGCTTCA


II. SITEα

      (MOUSE)                                              (HUMAN)

              -108              -91                        -118              -100
H-2L^d        CACTGATGACGCGCTGGC               HLA-B7     TACTCGTGACGCG*TCCC
H-2D^b        ------------------               HLA-B27    -------------*----
H-2D^d        ------------------               HLA-cw1    -----A-------*----
H-2K^b        -----T--------A-T-               HLA-cw5    -----A-------*----
H-2K^k        -----T--------A-T-               HLA-A2     -----AC------G----
H-2K^d        -----T--------A-T-               HLA-A3     -----AG------G----

Q7             CACTGATGACGTGTAGA
Q10            CACTGATTGAGGCTAGT
```

Fig. 4.1. Upstream *cis*-regulatory elements of MHC class I genes. Region II, region I, ICS and site α (core sequence underlined) are conserved in the classical class I genes. Qa, Tla genes show conservation in the ICS.

α, that also constitutively increases MHC class I transcription (Dey *et al.*, 1992) and which also acts as a cAMP-responsive element (Israel *et al.*, 1989). Site α (containing a core sequence TGACGC) is related to the cAMP-responsive motif (GTGACG/TA) found in other genes (Montminy *et al.*, 1986), although this element may serve multiple functions (Dey *et al.*, 1992). From Fig. 4.1, it is clear that region I, ICS and site α are highly conserved among many of the classical MHC class I genes. In all mouse and human classical class I genes examined so far, region I and the ICS occur in juxtaposition. Region II is invariable in mouse class I genes, although a homologous sequence is not found in the comparable position in human class I genes. However, region II-like motifs are present in multiple copies farther upstream of many HLA-A, HLA-B and HLA-C genes. Consistent with the functionality of these elements, each of the above elements has been shown to bind to a specific transcription/*trans*-acting factor. It is noteworthy that these *cis*-acting elements are not unique to class I genes but are shared among a wide variety of other genes. Clearly the MHC class I regulatory region is a composite of multiple independent *cis*-acting elements, a combination of which most likely confers a characteristic class I gene transcription pattern in a particular cell type.

The functional significance of these elements for class I gene expression *in vivo* is evident from transgenic studies carried out using various lengths of class I upstream region. Chamberlain *et al.* (1991) showed that a 660 bp upstream fragment of the HLA-B7 gene conferred tissue-specific and copy-number-dependent expression of the transgene in the mouse that closely paralleled expression of the endogenous H-2 gene. These authors further demonstrated that the upstream region between –660 and –100 was critical for conferring tissue-specific expression and IFN responsiveness. Furthermore, sequences in introns or in the 3′ regions did not play a significant regulatory role. Jones-Youngblood *et al.* (1990) showed that expression of the Q10/L chimaeric transgene was expressed in essentially the same pattern as the endogenous Q10 gene: this transgene has a 400 bp upstream region of the Q10 gene and an approximately 500 bp upstream region of the Qa7 gene.

The following are additional regulatory elements that have been reported, but whose functions are less well characterized since specific factors which bind to the elements have not yet been isolated.

1. A negative regulatory element that maps between region I and the ICS has been reported to repress transcription of a class I gene in undifferentiated EC cells (Flanagan *et al.*, 1991).
2. Another negative regulatory element that represses class I transcription in cells (other than EC cells) has been reported for a miniature swine MHC class I gene (–676 to –771) (Ehrlich, Maguire & Singer, 1988; Weissman & Singer, 1991).
3. Katoh *et al.* (1990) reported yet another negative regulatory element further upstream (from –1.5 to –1.8 kb from the initiation site) that is implicated in adenovirus E1A-mediated repression in a mouse class I gene (see Chapter 8).
4. Kralova, Jansa & Forejt (1992) have reported a positive element residing in the intron between the $\alpha_1$ and $\alpha_2$ domain of the H-2K$^b$ gene that enhances the level of gene expression.
5. A sequence surrounding the conserved palindromic motif (CCCATTGGG) present near the CAAT box binds either NF1 or an NF1-like factor and enhances transcription *in vitro* (Driggers *et al.*, 1992).

The less polymorphic non-classical class I genes Qa and Tla, encoded by a series of genes telomeric to the MHC class I genes (see Section 2.2.2), show a lower level of conservation in these elements (David-Watine *et al.*, 1990b). The region I sequence of both the Q10 and Q7 genes has nucleotide substitutions that reduce the affinity of nuclear factor binding (Handy *et al.*, 1989). Site α in Qa genes also shows divergence from that in the classical class I genes, although the ICS and region II in these genes are conserved. The regulatory region of Tla genes seems to be even more divergent, since there is no recognizable region I- and region II-like sequences in the Tla genes. Interestingly the Tla genes have the ICS motif, which appears to be functional in responding to IFNs (I.-M. Wang and R. Cook, personal communication). The notion that the ICS has sustained evolutionary conservation better than region I and region II is supported by the presence of the minimum ISRE in the upstream region of the chicken class I gene B-IV, which however does not have a well-conserved region I- (NF-κB-like) or region II-like motif. This chicken ICS has been shown to confer IFN induction of reporter genes not only in chicken but also in mammalian cells (Zöller *et al.*, 1992). Regulation of class I gene expression is being studied not only in mammalian species but also in avian and amphibian species (J. Flajnik, personal communication). Phylogenetic studies of the *cis*-acting elements of class I genes may eventually shed light on the evolution of class I gene regulation.

### 4.3 Co-ordinated regulation of the MHC class I and the $\beta_2$-m genes

The MHC class I and $\beta_2$-m genes are almost always co-expressed. Tissue specificity, developmental regulation and cytokine induction of the $\beta_2$-m gene are very similar to those of MHC class I genes (Chamberlain *et al.*, 1988; Morello *et al.*, 1982, 1985; Jaffe *et al.*, 1990; Drew *et al.*, 1993). Similar to class I genes, retinoic acid induces expression of the $\beta_2$-m gene in EC cells (Croce *et al.*, 1981; Daniel *et al.*, 1983). This co-ordinated regulation may be explained partly (but not entirely, see below) by the fact that the two genes have two *cis*-acting elements, the κB site and the ICS, in common (Kimura *et al.*, 1986; Israel *et al.*, 1987), although these two elements are oriented in an opposite fashion in the two genes (Fig. 4.2). Interestingly, the same juxtaposition of the κB and the ICS motifs occurs in the IFN-β gene with the same orientation as that in the $\beta_2$-m gene (Keller & Maniatis, 1988). The ICS-like motif in the IFN-β gene, called PRDI, serves as a virus-inducible element, responsible for rapid induction of the gene following viral infection. The virus-inducible element of the IFN-α gene contains the same motif (MacDonald *et al.*, 1990). Therefore, this shared element confers tran-

(*a*)

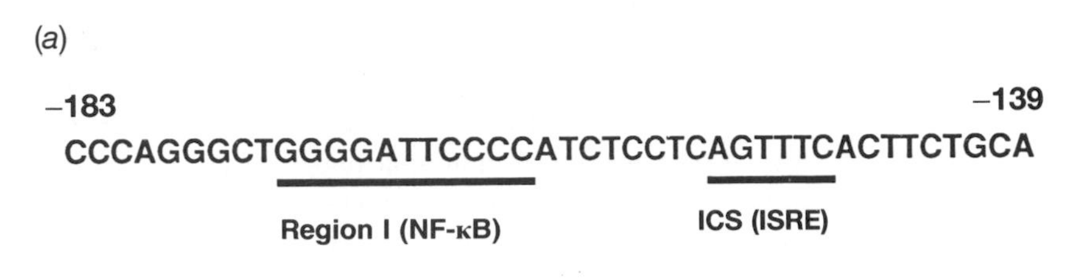


(*b*)

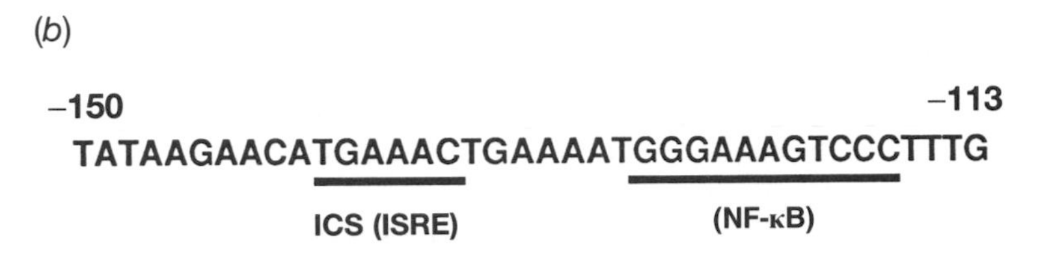


(*c*)

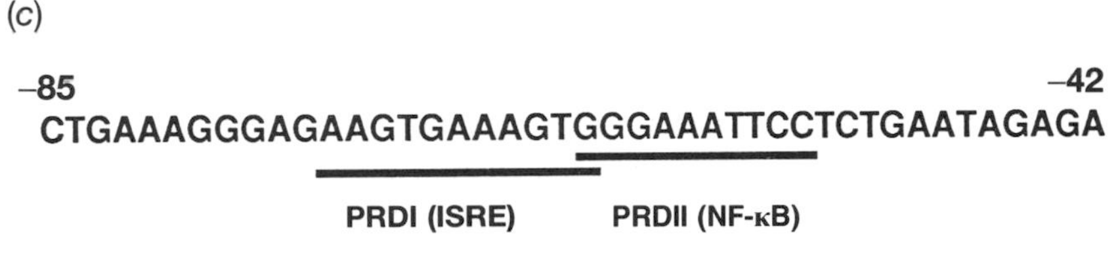


Fig. 4.2. Common *cis*-regulatory elements in (*a*) MHC class I, (*b*) $\beta_2$-m and (c) IFN-β genes. The NF-κB motif and the ISRE occur in juxtaposition in the three genes. The orientation of these elements in class I genes is opposite to that in the other two genes.

scriptional induction by both IFN and viruses, highlighting a common transcriptional pathway and the shared evolutionary origin for virus and IFN-mediated gene regulation. The fact that MHC class I, $\beta_2$-m and IFN-β genes have conserved regulatory elements in close proximity indicates that the three genes have an evolutionarily common origin in terms of their regulatory characteristics. This view is supported by the finding that the IFN-β and MHC class I genes have a cAMP-responsive element (Israel *et al.*, 1989; Du & Maniatis 1992).

### 4.4 *Trans*-acting factors that bind to the *cis* elements *in vitro*

The gel retardation assay (Fried & Crothers, 1981) has been used to detect factor-binding activities specific for class I regulatory elements present in various cells and tissues. It is now clear that the presence of nuclear factors in extracts does not necessarily correlate with the expression of MHC class I genes. For example, factors that bind to region II and site α are ubiquitous and are expressed even in the CNS and EC cells (Burke *et al.*, 1989; Drew *et al.*, 1993). However, it is interesting to note that region I-binding activity actually correlates well with tissue and developmental specificity of MHC class I gene expression: region I-binding activity is detected at high levels in lymphoid organs, while it is low or absent in brain and in embryonic tissues (Burke *et al.*, 1989). Neuroblastoma cells expressing a very low level of MHC class I and $\beta_2$-m genes have either undetectable or low levels of binding activity for region I (Lenardo *et al.*, 1988; Drew *et al.*, 1993). Region I-binding activity is similarly absent in some (but not all) EC cell lines (J. H. Segars *et al.*, unpublished results). Region I-binding activity can be induced by TNF-α or by retinoic acid treatment in these cells (Drew *et al.*, 1993; J. H. Segars *et al.*, unpublished results). ICS-binding activity has constitutive and inducible components. There are multiple IFN-inducible ICS-binding activities as distinguished by cycloheximide sensitivity (Shirayoshi *et al.*, 1988; Levy & Darnell, 1990). Factors capable of binding to class I regulatory elements are also capable of binding to related regulatory elements present in other genes; factor binding to region I, region II, ICS and site α in gel mobility shift analyses are competed by the related motifs, i.e. κB, thyroid/retinoic acid responsive elements, ISREs of other IFN-inducible genes and a cAMP-responsive motif, respectively. As described below, some of these factors have been cloned and their functional activities verified.

### 4.5 Factor binding to the MHC class I regulatory elements *in vivo*

Defining regulatory elements by reporter gene transfection assays is problematical, since it does not provide conclusive proof that elements identified are indeed functional in an endogenous gene *in vivo*. Genomic footprinting largely circumvents this problem, since it permits identification of factor binding *in vivo*. The introduction of the polymerase chain reaction (PCR) dramatically improved the sensitivity and resolution of the *in vivo* footprinting technique (Mueller & Wold, 1989; Garrity & Wold, 1992). To study *in vivo* factor binding to MHC class I genes, *in vivo* footprinting of the transgenic HLA-B7 gene and the endogenous H-2K$^b$ gene expressed in adult mouse tissues was performed (Dey *et al.*, 1992). The transgene HLA-B7, in both low- and high-copy numbers (approximately 1 and 30 copies, respectively), had a 660 bp upstream sequence and was expressed in a tissue-specific manner that paralleled expression of the endogenous H-2 genes. As shown schematically in

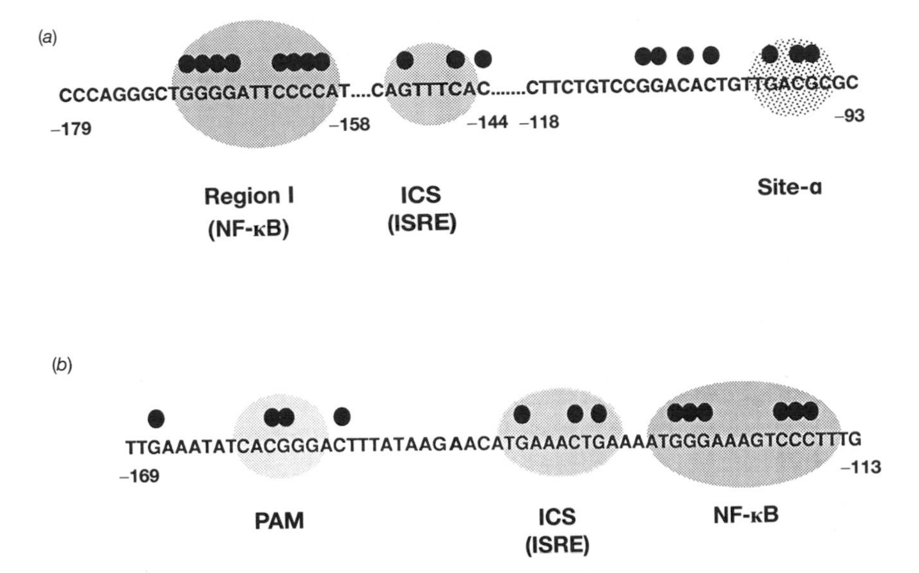


Fig. 4.3. *In vivo* footprinting of (*a*) class I (H-2K$^b$) and (*b*) $\beta_2$-m genes. *In vivo* occupancy is detected at region I (κB site), ICS and site α (PAM) of class I ($\beta_2$-m) in spleen, but not in brain, paralleling the tissue-specific expression of the two genes. Protected G residues are marked above with closed circles and the core sequence for each element is encircled by shaded oval (Dey *et al.*, 1992; Lonergan *et al.*, 1993).

Fig. 4.3, sequences corresponding to region I (enhancer A), ICS and site α in both transgene and endogenous gene were strongly protected in spleen. In these sites, identical G residues were protected in the transgene and the endogenous gene. In addition, in the spleen, a strong hypermethylation site was detected further upstream, close to region II. However, neither protection nor hypermethylation was observed in brain for either the transgene or the endogenous gene. Liver also showed *in vivo* footprinting of the same set of sites, although more weakly than spleen. An identical footprint was seen in the T cell line EL4, but not in F9 EC cells. Therefore the *in vivo* footprint of the MHC class I gene regulatory region correlates well with tissue-specific expression of MHC class I genes *in vivo*. Furthermore, sites exhibiting *in vivo* footprints correspond to the regulatory elements that had been identified by transfection assays, supporting their functional roles *in vivo*.

*In vivo* footprinting was also performed for the $\beta_2$-m gene (Lonergan *et al.*, 1993). It was found that the κB and ICS of the $\beta_2$-m gene, elements shared with MHC class I genes, and an additional upstream element termed PAM is protected in mouse spleen but not in brain. In this study,

PAM was identified as a new positive regulatory element in the $\beta_2$-m gene. Thus, *in vivo* protection of the $\beta_2$-m gene, like the MHC class I genes, correlates with the expression of the gene, suggesting that *in vivo* occupancy of the two genes is co-ordinately controlled. From these studies, it has become clear that *in vivo* factor occupancy does not correlate with the *in vitro* factor-binding activity. Gel retardation assays have revealed that there are abundant factor-binding activities for site $\alpha$, PAM and ICS in brain and F9 EC cells although there is no *in vivo* footprint for these elements. The lack of *in vivo* protection, despite the presence of *in vitro* factor-binding activities, has been noted for other genes (Mueller & Wold, 1989), including MHC class II genes (Kara & Glimcher, 1991). Since G residues showing *in vivo* protection were found to coincide with those showing contact *in vitro*, as judged by methylation interference assays, the factors detected in *in vitro* assays are likely to be involved in binding *in vivo*. Another significant aspect is that *in vivo* occupancy of both MHC class I and $\beta_2$-m genes occurs in an 'all or none' fashion: only those cases were found where either all the sites were concurrently occupied or none was occupied. No examples have been found where a single isolated site was protected. Therefore, it is likely that the presence of sequence-specific factors is not sufficient for generating *in vivo* factor binding to the endogenous class I genes, but it requires a 'higher order chromatin configuration' for regulatory elements to have access to nuclear factors. Unlike *in vitro* protein-binding conditions, DNA in the nucleus is tightly folded by nucleosomes and is organized by the nuclear scaffold matrices (Gross & Garrad, 1988; Adachi, Käs & Laemmli, 1989; Grunstein, 1990; Wolffe, 1990). The accessibility of regulatory elements to specific transcription factors is believed to be controlled by nucleosome unfolding and is affected by nuclear proteins such as topoisomerase, histone H1 and HMG. At present, very little is known as to what actually causes the MHC class I upstream region to gain access to the factors. In this respect, it is interesting to note that the transgene HLA-B7 exhibited the authentic *in vivo* occupancy pattern, even though it is integrated outside of the endogenous MHC and had only a 660 bp upstream region, suggesting that the genetic information dictating *in vivo* factor occupancy is contained within this short upstream segment of the gene. Maschek, Pulm & Hammerling (1989) have shown that upstream regions of H-2$D^b$ become hypersensitive to DNase I treatment in F9 cells after retinoic acid treatment, supporting the notion that induction of MHC class I genes expression coincides with a change in chromatin structure.

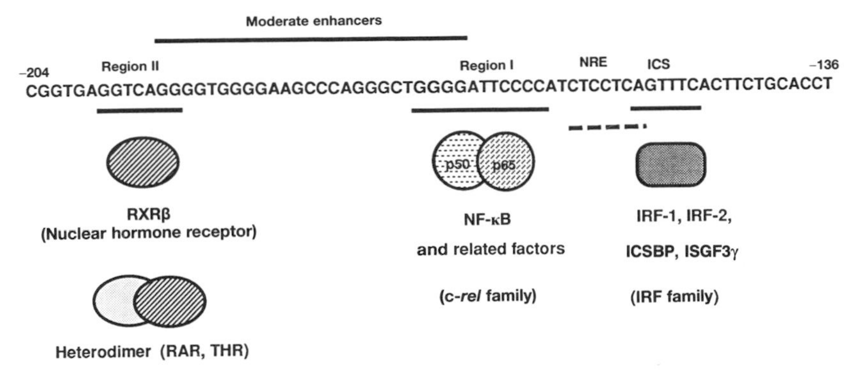


Fig. 4.4. Cloned transcription factors that bind to regulatory elements of class I genes. The sequence and nucleotide numbers are from the H-2L$^d$ gene.

## 4.6 Cloning of transcription factors that regulate MHC class I gene transcription

Transcription factors that bind to the ICS, region I and region II have been cloned, thereby opening new avenues of research addressing the mechanism(s) of MHC class I gene regulation (summarized in Fig. 4.4). These factors belong to families of regulatory proteins that are conserved across invertebrate and vertebrate species. These proteins comprise discrete domains that have discrete functions, in general the DNA-binding domain is the most conserved. As may have been expected, these transcription factors are not specific only for class I regulatory elements but are capable of binding to many related sequence motifs, although with varying affinities. The following is a summary of the families of proteins involved in transcription of MHC class I genes.

### *4.6.1 IFN regulatory factor (IRF)*

This family includes four transcription factors, IRF-1, IRF-2, ICS-binding protein (ICSBP) and ISGF3γ (Miyamoto *et al.*, 1988; Harada *et al.*, 1989; Driggers *et al.*, 1990; Pine *et al.*, 1990; Veals *et al.*, 1992). These proteins bind to the class I and $\beta_2$-m ICS, the ISRE of other IFN-inducible genes and the virus-inducible element of IFN genes (Fig. 4.5). The amino-terminal regions (about 115 amino acid residues) are homologous among the four proteins and represent the DNA-binding domain. This domain has a characteristic tryptophan repeat that is also found in the *myb* and *ets* proteins (Veals *et al.*, 1992). The carboxy-terminal domain is variable in size and sequence, although ICSBP and ISGF3γ share some similarity in this domain, as do

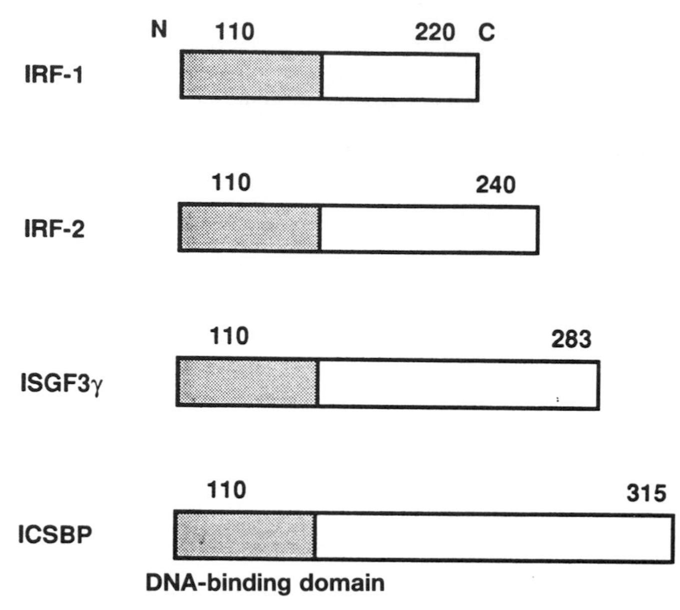


Fig. 4.5. The IRF family. IRF-1 and IRF-2 were isolated by Miyamoto *et al.* (1988) and Harada *et al.* (1990). ICSBP was described by Driggers *et al.* (1990) and ISGF3γ by Veals *et al.* (1992). All of these factors bind to the MHC class I ICS and ISRE of many other IFN-regulated genes as well as the virus-inducible element of IFN genes. IRF-1 has been shown to activate transcription, while IRF-2 and ICSBP repress transcription of class I genes (Harada *et al.*, 1990; Nelson *et al.*, 1993).

IRF-1 and IRF-2, suggesting that these pairs are functionally more related than the others. With respect to their expression pattern, IRF-1 and IRF-2 appear to be expressed broadly in many types of cell. By comparison, ICSBP expression is limited to lymphoid organs. The tissue and cell-type specificity of ISGF3 is not fully known (this factor has been analysed largely in HeLa and a few cultured cell lines). Not surprisingly, expression of these factors is induced by IFNs (IFN-α, IFN-β and IFN-γ) and viruses. The functional roles of IRF-1, IRF-2 and ICSBP have been studied by transfecting expression plasmids and measuring the activities of co-transfected reporter genes. Harada *et al.* (1990) have shown that IRF-1 activates several IFN- or virus-regulated genes, including MHC class I and the IFN-β gene, without requiring IFNs, indicating that IRF-1 is an activator of IFN-regulated genes. However, because IRF-1 activity is induced following *de novo* protein synthesis, early responses that do not require new protein synthesis are perhaps controlled by another factor(s) (see below; Pine *et al.*, 1990). However, IRF-2 has been found to repress transcription of IFN-induced genes, again including the MHC class I genes (Harada *et al.*, 1989; 1990). Several lines of evidence also indicate that IRF-1 and IRF-2 are involved in regulating IFN genes *in vivo* (Fujita *et al.*, 1992; Reis *et al.*, 1992). In common with IRF-2, ICSBP also represses IFN induction of class I, 2–5A synthetase, GBP (an IFN-γ-inducible gene) and ISG15 gene transcription (Nelson *et al.*, 1993). ICSBP is also capable of repressing constitutive transcription of several genes (Weisz *et al.*, 1992). The mechanism by which IRF-2 and ICSBP repress transcription is not fully understood. However, since both IRF-2 and ICSBP suppress IRF-1 induction of MHC class I and other genes, they almost certainly interfere with the activator function.

ISGF3γ is a 48 kDa protein subunit which functions in a unique fashion in that it requires to be assembled with three additional subunits, collectively termed ISGF3α, in the cytoplasm following IFN treatment, leading to nuclear translocation and binding to DNA target sites (Levy *et al.*, 1989; Fu *et al.*, 1992; Schindler *et al.*, 1992a). ISGF3γ appears to be the sole subunit conferring target DNA binding (Veals *et al.*, 1992). ISGF3α subunits are composed of 113, 91 and 84 kDa proteins that contain SH (*src* homology) domains (Fu, 1992; Schindler *et al.*, 1992b) which are phosphorylated on tyrosine following IFN treatment. *Tyk*2, the cytosolic non-receptor kinase, is likely to be involved in phosphorylation of these subunits (Velazquez *et al.*, 1992). This assembly and nuclear translocation occurs rapidly, without new protein synthesis. Other IRF members appear to be localized predominantly in the nucleus (J. Bovolenta, unpublished results). ISGF3 is believed to be responsible for early transcriptional induction triggered by IFNs and viruses, since it is the only binding activity that is induced immediately following IFN treatment. However, experimental verification of this claim remains to be made. MHC class I genes may be one of the target genes activated by ISGF3, since the IFN induction of MHC class I mRNA is rapid and occurs independently of protein synthesis in many cell types (Silverman *et al.*, 1988). Therefore, proteins belonging to the IRF family play a central role in both positive and negative transcriptional regulation mediated by IFNs and viruses. It is interesting that IFN- or virus-mediated regulation involves active repressors, which may be a necessary step to establish the desensitization process whereby many cells become refractory to further stimulation by IFNs after IFN stimulation (Larner, Chaudhuri & Darnell, 1986). This process may be part of the mechanism allowing the maintainance of homeostasis in gene regulation.

#### *4.6.2 NF-κB*

DNA clones encoding proteins that bind to the κB and related sites, including region I, have been described (Bours *et al.*, 1990; 1992; Ghosh *et al.*, 1990;

Kieran *et al.*, 1990; Nolan *et al.*, 1991, Schmid *et al.*, 1991; Ryseck *et al.*, 1992). These proteins belong to the conserved *rel* gene family that include the oncogene c-*rel*, p50, p65, p49 (50B) and *relB* (Ballard *et al.*, 1990). These proteins, either as homodimers or heterodimers, bind to κB and κB-like motifs present in many genes (Baeuerle, 1991). These proteins have a conserved DNA-binding domain of approximately 300 amino acid residues in the amino-terminal region. This domain is homologous to the *Drosophila* developmental gene *dorsal*. The carboxy-terminal domain of these proteins is highly variable in size and sequence. Some (but not all) members (e.g. p50 and p49) are produced as precursors that contain a large carboxy-terminal domain (Ghosh *et al.*, 1990; Kieran *et al.*, 1990; Schmid *et al.*, 1991) bearing the ankyrin repeat, part of which is removed upon cellular activation. While p50 and p49 are capable of binding to a target κB site by themselves as homodimers, p65 alone does not readily bind to target DNA (Baeuerle, 1991). Instead, the heterodimer composed of p50 and p65 (also p49 and p65) binds to target DNA with high affinity and activates transcription efficiently (Bours *et al.*, 1990; Ryseck *et al.*, 1992). Evidently p50 alone is not capable of *trans*-activating target genes in many cell types, since the *trans*-activating function is conferred by the carboxy-terminal domain of p65. The p50 homodimer, however, is apparently capable of activating a class I promoter in an *in vitro* transcription system (Fujita *et al.*, 1992), although this may not be relevant *in vivo*, since much of p50 in the cell appears to exist as a heterodimer with p65 (Baeuerle, 1991). It is noteworthy that cells have multiple *rel* members that are differentially involved in regulating various κB-bearing genes, creating large cell type- and target-dependent heterogeneity in transcription (Ballard *et al.*, 1990).

The functional activity of NF-κB is dictated by a unique cytoplasmic partitioning mechanism (Baeuerle, 1991). In resting cells, a p50–p65 heterodimer is associated with an inhibitory subunit (I-κB) that retains the complex in the cytoplasm (Baeuerle & Baltimore, 1988; Haskill *et al.*, 1991). Following external stimulation, I-κB is dissociated from the complex, initiating nuclear translocation of the heterodimer. A number of agents such as cytokines (TNF-α, IL-1), viruses, phorbol esters and protein synthesis inhibitors have been shown to initiate this process, which presumably involves phosphorylation of I-κB. Some (but not all) κB sequences have been shown to bind to high-mobility group (HMG) proteins that help enhance DNA-binding affinity by NF-κB proteins (Thanos & Maniatis, 1992). It will be of importance to determine which of the *rel* proteins bind to region I in particular cell types, whether they interact with HMG proteins and how region I binding differs from that of other κB motifs.

#### *4.6.3 The nuclear hormone receptor superfamily, RXR subgroup*

This superfamily represents the longest studied group of mammalian transcription factors. A wide variety of steroid hormone receptors, thyroid hormone receptors (THR), retinoic acid receptors (RAR), as well as vitamin D receptor (VDR) belong to this superfamily (see reviews by Beato, 1989; Evans, 1988). Like those of other families, proteins of this superfamily have a discrete domain composition that can also be functionally distinguished. The most conserved region is the DNA-binding domain of the $C_4C_5$ zinc finger that is located in the central part of the protein. The carboxy-terminal domain contains a region responsible for ligand binding and dimerization and exhibits some level of structural conservation. The amino-terminal domain, presumably involved in transcriptional activation/repression, is variable and least conserved. RXR is a new subgroup in this superfamily, composed of three members RXRα, RXRβ and RXRγ (Mangelsdorf *et al.*, 1992) (Fig. 4.6). RXRβ has also been isolated on the basis of its binding to region II of murine class I genes (Hamada *et al.*, 1989). The RXR members have a unique ability to form heterodimers with other nuclear hormone receptors and confer increased ability to bind DNA and to activate target genes. Receptors that heterodimerize with RXRs include THR, RAR, VDR and the more recently isolated peroxisome proliferator activator receptors (PPAR) (Kliewer

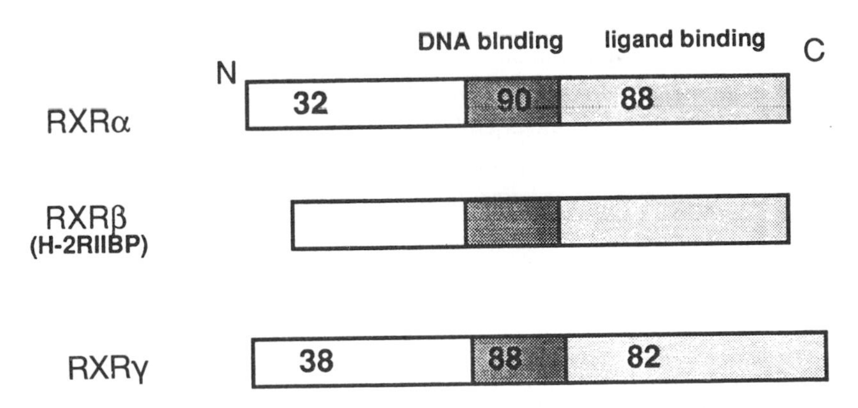


Fig. 4.6. The RXR subgroup of the nuclear hormone receptor superfamily. RXRβ was isolated on the basis of binding to region II (Hamada *et al.*, 1989) and is involved in retinoic acid induction of MHC class I transcription in EC cells (Nagata *et al.*, 1992). The central domain is the conserved $C_4C_5$ zinc finger. The carboxy-terminal domain contains nine repeats of hydrophobic amino acids, necessary for heterodimerization with other nuclear hormone receptors (Yu *et al.*, 1991; Marks *et al.*, 1992; Leid *et al.*, 1992). This domain is also responsible for binding of 9-*cis*-retinoic acid, a ligand for RXR.

*et al.*, 1992; Leid *et al.*, 1992; Marks *et al.*, 1992; Zhang *et al.*, 1992). Interestingly, RXR heterodimers have been shown to be kinetically more stable than RXR homodimers and bind to many target elements at higher affinity, at least in the absence of ligand (Marks *et al.*, 1992). By virtue of the ability to heterodimerize with multiple receptors and, thus, acquiring broad binding specificity, RXR confers a large combinatorial diversity upon nuclear hormone receptors. More recently, RXR has been shown to bind to 9-*cis*-, an isomer of retinoic acid that is derived from vitamin A (Heyman *et al.*, 1992; Levin *et al.*, 1992). Transcription of a murine class I gene is activated by RXRβ in a retinoic acid-dependent fashion in N-Tera2 human EC cells, which was dependent on region II (Nagata *et al.*, 1992). This activation most likely results from heterodimerization of RXRβ with RARs expressed in the EC cells. Region II binding activity could be synergistically enhanced by RXRβ when mixed with receptors that heterodimerize with RXRβ. Later studies have shown that a MHC class I reporter gene could be co-operatively activated by co-transfection of RXRβ and heterodimer partners (see below, Marks *et al.*, 1992). These findings lend credence to the hypothesis that MHC class I transcription is controlled by vitamins and hormones through this transcription factor superfamily. It will be of importance to determine whether these hormones (particularly the thyroid hormone) and vitamins (particularly vitamin A and derivatives) affect transcription of class I genes *in vivo*.

The availability of transcription factors facilitates mechanistic analyses of class I gene transcription in an *in vitro* transcription system (Driggers *et al.*, 1992). Studies on the effects of recombinant RXRβ on *in vitro* transcription from a murine class I promoter and other promoters with region II-like motifs showed that RXRβ in combination with 9-*cis*-retinoic acid drives template-specific, ligand-dependent transcription *in vitro* (I. J. Lee, P. H. Driggers, J. A. Medin and K. Ozato, unpublished results). An *in vitro* transcription system such as this may permit an analysis of how RXRβ (either as a homodimer or a heterodimer) interacts with the basal transcription machinery.

#### *4.6.4 RA induction of MHC class I transcription in N-Tera2 EC cells involves heterodimers of RXRβ–RARβ and NF-κB p50–p65*

The co-transfection assay is the most common way of addressing the functional role of a transcription factor. This approach, however, often involves the use of artificial reporters and host cells irrelevant to physiological regulation of the target genes in question. Although useful as an initial step, this approach does not provide definitive information about the physiological role of a transcription factor in gene regulation. To study the physiological role of NF-κB and RXR families, we have analysed RA induction of MHC class I genes in human N-Tera2 EC cells (J. H. Segars, T. Nagata, K. G. Becker, V. Bours, B. Neel, U. Siebenlist and K. Ozato, unpublished results). Expression of class I and $\beta_2$-m surface antigens was induced following retinoic acid treatment of the human EC cell line N-Tera2. This induction coincided with increased levels of class I mRNA, which occurred as a result of activation of MHC class I promoter activity. Analysis of mutant class I reporters led to the conclusion that this promoter activation was attributable largely to the activation of the region I and region II enhancers. Moreover, this activation coincided with the induction of nuclear factor-binding activities specific for region I and region II. Region I-binding activity was not present in undifferentiated N-Tera 2 cells but was induced following retinoic acid treatment. By testing the effect of antibodies specific for various *rel* proteins on region I binding, we concluded that retinoic acid-induced region I binding represented binding of a retinoic acid-induced NF-κB factor, the p50–p65 heterodimer. The p50–p65 heterodimer was produced as a result of *de novo* induction of p50 and p65 mRNAs, rather than liberation of cytoplasmic NF-κB. Region II-binding activity was present in undifferentiated cells at low levels but was greatly augmented by retinoic acid treatment, as the result of the binding of the nuclear hormone receptor heterodimer composed of RXRβ and RARβ. The RXRβ–RARβ heterodimer was also shown to bind to the retinoic acid-responsive elements of other genes, indicating that this heterodimer is involved in retinoic acid triggering of the regulatory cascade in these EC cells. Consistent with the functional role of these factors, class I reporter gene activities could be synergistically enhanced by co-transfection of RXRβ and RARα or β, or co-transfection of p50 and p65 in untreated Tera2 cells. Taken together, following retinoic acid treatment, the heterodimers of two transcription factor families are induced to bind to the class I enhancers, serving as the primary mechanism leading to retinoic acid induction of MHC class I gene expression in N-Tera2 EC cells. Because of multiple binding specificities, these heterodimers are likely to play fundamental roles in regulatory processes in EC cells and in mammalian embryos *in vivo*.

## 4.7 Conclusions

There is now an increasingly detailed understanding of the regulation of MHC class I transcription. Conserved *cis*-acting elements that control various aspects of gene regulation have been identified. A number of transcription factors have been characterized that affect transcription of MHC class I genes. 'Chromatin configuration' is likely to be an important factor governing access by transcription factors to endogenous class I genes *in vivo*. Based on this progress, fundamental mechanisms of class I gene transcription are being analysed *in vivo* and *in vitro*. It will be of great interest to investigate how various disease processes, including cancers, viral infections and autoimmunity affect class I regulatory mechanisms.

## References

Adachi, Y., Käs, E. & Laemmli, U.K. (1989). Preferential, cooperative binding of DNA topoisomerase II to scaffold-associated regions. *EMBO Journal*, 8, 3997–4006.

Baeuerle, P. & Baltimore, D. (1988). Activation of DNA-binding activity in an apparently cytoplasmic precursor of the NF-κB transcription factor. *Cell*, 53, 211–217.

Baeuerle, P. (1991). The inducible transcription activator NF-κB: regulation by distinct protein subunits. *Biochimica et Biophysica Acta*, 1072, 63–80.

Baldwin, A.S., Jr & Sharp, P.A. (1987). Binding of a nuclear factor to a regulatory sequence in the promoter of the mouse H-2K$^b$ class I major histocompatibility gene. *Molecular and Cellular Biology*, 7, 305–313.

Ballard, D.W., Walker, W.H., Doerre, S., Sista, P., Molitor, J.A., Dixon, E.P., Peffer, N.J., Hannink, M. & Greene, W.C. (1990). The v-*rel* oncogene encodes a κB enhancer binding protein that inhibits NF-κB function. *Cell*, 63, 803–814.

Beato, M. (1989). Gene regulation by steroid hormones. *Cell*, 56, 335–344.

Blanar, M.A., Baldwin, A.S., Jr, Flavell, R.A. & Sharp, P.A. (1989). A gamma-interferon-induced factor that binds the interferon response sequence of the MHC class I gene, H-2K$^b$. *EMBO Journal*, 8, 1139–1144.

Bours, V., Burd, P.R., Brown, K. Villalobos, J., Park, S., Ryseck, R.-P., Bravo, R., Kelly, K. & Siebenlist, U. (1992). A novel mitogen-inducible gene product related to p-50/p105-NFκB participates in *trans*-activation through a κB site. *Molecular and Cellular Biology*, 12, 685–695.

Bours, V., Villalobos, J., Burd, P.R., Kelly, K. & Siebenlist, U. (1990). Cloning of a mitogen-inducible gene encoding a κB DNA-binding protein with homology to the *rel* oncogene and to cell cycle motifs. *Nature (London)*, 348, 76–80.

Burke, P.A., Hirschfeld, S., Shirayoshi, Y., Kasik, J.W., Hamada, K., Appella, E. & Ozato, K. (1989). Developmental and tissue-specific expression of nuclear proteins that bind the regulatory element of the major histocompatibility complex class I gene. *Journal of Experimental Medicine*, 169, 1309–1321.

Chamberlain, J.W., Nolan, J.A., Conrad, H.A., Vasavada, H.A., Vasavada, H.H., Ploegh, H., Ganguly, S., Janeway, C.A., Jr & Weissman S.M. (1988). Tissue-specific and cell surface expression of major histocompatibility complex class I heavy (HLA-B7) and light ($\beta_2$-m) chain genes in transgenic mice. *Proceedings of the National Academy of Sciences of the USA*, 85, 7690–7694.

Chamberlain, J.W., Vasavada, H.A., Ganguly, S. & Weissman, S.M. (1991). Identification of cis-sequences controlling efficient position-independent tissue-specific expression of human major histocompatibility complex class I genes in transgenic mice. *Molecular and Cellular Biology*, 11, 3564-3572.

Collins, T., Lapierre, L.A., Fiers, W., Strominger, J.L., & Pober, J.S. (1986). Recombinant human tumor necrosis factor increases mRNA levels and surface expression of HLA-A,B antigens in vascular endothelial cells and dermal fibroblasts *in vitro*. *Proceedings of the National Academy of Sciences of the USA*, 83, 446–450.

Croce, C.M., Linnenbach, A., Huebner, K., Parnes, J.R., Margulies, D.H., Appella, E. & Seidman, J.G. (1981). Control of expression of histocompatibility antigens (H-2) and $\beta_2$-microglobulin in F9 teratocarcinoma stem cells. *Proceedings of the National Academy of Sciences of the USA*, 78, 5754–5758.

Daniel, F., Morello, D., Le Bail, O., Chambon, P., Cayre, Y. & Kourilsky, P. (1983). Structure and expression of the mouse $\beta_2$-microglobulin gene isolated from somatic and non-expressing teratocarcinoma cells. *EMBO Journal*, 2, 1061–1065.

David-Watine, B., Israel, A. & Kourilsky, P. (1990a). The regulation and expression of MHC class I genes. *Immunology Today*, 11, 286–292.

David-Watine, B., Logeat, F., Israel, A. & Kourilsky, P. (1990b). Regulatory elements involved in the liver specific expression of the mouse MHC class I Q10 gene: characterization of a new TATA binding factor. *International Immunology*, 2, 981–993.

Dey, A., Thornton, A.M., Lonergan, M., Weissman, S.M., Chamberlain, J.W. & Ozato, K. (1992). Occupancy of upstream regulatory sites *in vivo* coincides with major histocompatibility complex class I gene expression in mouse tissues. *Molecular and Cellular Biology*, 12, 3590–3599.

Drew, P.D., Lonergan, M., Goldstein, M.E., Lampson, L.A., Ozato, K. & McFarlin, D.E. (1993). Regulation of MHC class I and $\beta_2$-microglobulin gene expression in human neuronal cells: factor binding to conserved *cis*-acting regulatory sequences correlates with expression of the genes. *Journal of Immunology*, 150, 3300–3310.

Driggers, P.H., Elenbaas, B.A., An, J.B., Lee, I.J. & Ozato, K. (1992). Two upstream elements activate transcription of a major histocompatibility complex class I gene *in vitro*. *Nucleic Acids Research*, 20, 2533–2240.

Driggers, P.H., Ennist, D.L., Gleason, S.L., Mak, W., Marks, M.S., Levi, B., Flanagan, J.R., Appella, E. & Ozato, K. (1990). An interferon gamma-regulated protein that binds the interferon-inducible enhancer element of major histocompatibility complex class I genes. *Proceedings of the National Academy of Sciences of the USA*, 87, 3743–3747.

Du, W. & Maniatis, T. (1992). An ATF-CREB binding site protein is required for virus induction of the human interferon-β gene. *Proceedings of the National Academy of Sciences of the USA*, 89, 2150–2154.

Ehrlich, R., Maguire, J.E. & Singer, D.S. (1988). Identification of negative and positive regulatory elements associated with a class I major histocompatibility complex gene. *Molecular and Cellular Biology*, 8, 695–703.

Evans, R.M. (1988). The steroid and thyroid hormone receptor superfamily. *Science*, 240, 889–895.

Fellous, M., Nir, U., Wallach, D., Merlin, G., Rubinstein, M. & Revel, M. (1982). Interferon-dependent induction of mRNA for the major histocompatibility antigens in human fibroblasts and lymphoblastoid cells. *Proceedings of the National Academy of Sciences of the USA*, 79, 3082–3086.

Flanagan, J.R., Murata, M., Burke, P.A., Shirayoshi, Y., Appella, E., Sharp, P.A. & Ozato, K. (1991). Negative regulation of the major histocompatibility complex class I promoter in embryonal carcinoma cells. *Proceedings of the National Academy of Sciences of the USA*, 88, 8555–8559.

Fried, M. & Crothers, D.M. (1981). Equilibria and kinetics of lac repressor – operator interactions by polyacrylamide gel electrophoresis. *Nucleic Acids Research*, 9, 3047–3060.

Fu, X.-Y. (1992). A transcription factor with SH2 and SH3 domains is directly activated by an interferon α-induced cytoplasmic protein tyrosine kinase(s). *Cell*, 70, 323–335.

Fu, X.-Y., Schindler, C., Improta, T., Aebersold, R. & Darnell, J. E. Jr (1992). The proteins of ISGF-3, the interferon α-induced transcriptional activator, define a gene family involved in signal transduction. *Proceedings of the National Academy of Sciences of the USA*, 89, 7840–7843.

Fujita, T., Nolan, G.P., Ghosh, S. & Baltimore, D. (1992). Independent mode of transcription activation by the p50 and p65 subunits of NF-κB. *Genes and Development*, 6, 775–787.

Garrity, P.A. & Wold, B.J. (1992). Effects of different DNA polymerases in ligation-mediated PCR: enhanced genomic sequencing and *in vivo* footprinting. *Proceedings of the National Academy of Sciences of the USA*, 89, 1021–1025.

Ghosh, S., Gifford, A.M., Riviere, L., Tempst, P., Nolan, G.P. & Baltimore, D. (1990). Cloning of the p50 DNA binding subunit of NF-κB: homology to rel and dorsal. *Cell*, 62, 1019–1029.

Gross, D.S. & Garrad, W.T. (1988). Nuclease hypersensitive sites in chromatin. *Annual Review of Biochemistry*, 57, 159–197.

Grunstein, M. (1990). Nucleosomes: regulators of transcription. *Trends in Genetics*, 6, 395–400.

Hamada, K., Gleason, S.L., Levi, B.-Z., Hirschfeld, S., Appella, E. & Ozato, K. (1989). H-2RIIBP, a member of the nuclear hormone receptor superfamily that binds to both the regulatory element of major histocompatibility class I gene and the estrogen response element. *Proceedings of the National Academy of Sciences of the USA*, 86, 8289–8293.

Handy, D.E., Burke, P.A., Ozato, K., & Coligan, J.E. (1989). Site-specific mutagenesis of the class I regulatory element of the Q10 gene allows expression in non-liver tissues. *Journal of Immunology*, 142, 1015–1021.

Harada, H., Fujita, T., Miyamoto, M., Kimura, Y., Maruyama, M., Furia, A., Miyata, T. & Taniguchi, T. (1989). Structurally similar but functionally distinct factors, IRF-1 and IRF-2, bind to the same regulatory elements of IFN and IFN-inducible genes. *Cell*, 58, 729–739.

Harada, H., Willson, K., Sakakihara, J., Miyamoto, M., Fujita, T. & Taniguchi, T. (1990). Absence of the type I IFN system in EC cells: transcriptional activator (IRF-1) and repressor (IRF-1–2) are developmentally regulated. *Cell*, 63, 303–312.

Haskill, S., Beg, A.A., Tompkins, S.M., Morris, J.S., Yurochiko, A.D., Sampson-Johannes, A., Mondal, K., Ralph, P. & Baldwin, A.S. Jr (1991). Characterization of an immediate-early gene induced in adherent monocytes that encodes I-κB-like activity. *Cell*, 65, 1281–1289.

Hedley, M.L., Drake, B.L., Head, J.R., Tucker, P.W. & Forman, J. (1989). Differential expression of the class I MHC genes in the embryo and placenta during midgestational development in the mouse. *Journal of Immunology*, 142, 4046–4055.

Heyman, R.A., Mangelsdorf, D.J., Dyck, J.A., Stein, R.B., Eichele, G., Evans,

R.M. & Thaller, C. (1992). 9-cis Retinoic acid is a high affinity ligand for the retinoid X receptor. *Cell*, 68, 397–406.

Israel, A., Le Bail, O., Hatat, D., Piette, J., Kieran, M., Logeat, F., Wallach, D., Fellous, M. & Kourilsky, P. (1989). TNF stimulates expression of mouse MHC class I genes by inducing an NF-κB-like enhancer binding activity which displaces constitutive factors. *EMBO Journal*, 8, 3793–3800.

Israel, A., Kimura, A., Kieran, M., Yano, O., Kanellopoulos, J., Le Bail, O. & Kourilsky, P. (1987). A common positive *trans*-acting factor binds to enhancer sequences in the promoters of mouse H-2 and $\beta_2$-microglobulin genes. *Proceedings of the National Academy of Sciences of the USA*, 84, 2653–2657.

Israel, A., Kimura, A., Fournier, A., Fellous, M. & Kourilsky, P. (1986). Interferon response sequence potentiates activity of an enhancer in the promoter region of a mouse H-2 gene. *Nature (London)*, 322, 743–746.

Jaffe, L., Jeannotte, L., Bikoff, E.K. & Robertson, E.J. (1990). Analysis of $\beta_2$-microglobulin gene expression in the developing mouse embryo and placenta. *Journal of Immunology*, 145, 3474–3482.

Joly, E., Mucke, L. & Oldstone, M.B.A. (1991). Viral persistence in neurons explained by lack of major histocompatibility class I expression. *Science*, 253, 1283–1285.

Jones-Youngblood, S.L., Wieties, K., Forman, J. & Hammer, R.E. (1990). Effect of the expression of a hepatocyte-specific MHC molecule in transgenic mice on T cell tolerance. *Journal of Immunology*, 144, 1187–1195.

Kara, C.J., & Glimcher, L.H. (1991). *In vivo* footprinting of MHC class II genes: bare promoters in the bare lymphocyte syndrome. *Science*, 252, 709–712.

Kasik, J.W., Wan, Y.-J.Y. & Ozato, K. (1987). A burst of c-*fos* gene expression in the mouse occurs at birth. *Molecular and Cellular Biology*, 7, 3349–3352.

Katoh, S., Ozawa, K., Kondoh, S., Soeda, E., Israel, A., Shiroki, K., Fujinaga, K., Itakura, K., Gachelin, G. & Yokoyama, K. (1990). Identification of sequences responsible for positive and negative regulation by E1A in the promoter of the H-$2K^{bm1}$ class I MHC gene. *EMBO Journal*, 9, 127–135.

Keller, A.D. & Maniatis, T. (1988). Identification of an inducible factor that binds to a positive regulatory element of the human β-interferon gene. *Proceedings of the National Academy of Sciences of the USA*, 85, 3309–3313.

Kieran, M., Blank, V., Logeat, F., Vandekerckhove, J., Lottspeich, F., Le Bail, O., Urban, M.B., Kourilsky, P., Baeuerle, P.A. & Israel, A. (1990). The DNA binding subunit of NF-κB is identical to factor KBF1 and homologous to the *rel* oncogene product. *Cell*, 62, 1007–1018.

Kimura, A., Israel, A., LeBail, O. & Kourilsky, P. (1986). Detailed analysis of the mouse H-$2K^b$ promoter: enhancer-like sequences and their role in the regulation of class I gene expression. *Cell*, 44, 261–272.

Klein, J. (1975). *Biology of the Mouse Histocompatibility-2 Complex: Principles of Immunogenetics as Applied to a Single System*. New York: Springer-Verlag.

Kliewer, S.A., Umesono, K., Noonan, D.J., Heyman, R.A. & Evans, R.M. (1992). Convergence of 9-cis retinoic acid and peroxisome proliferator signalling pathways through heterodimer formation of their receptors. *Nature (London)*, 358, 771–774.

Korber, B., Mermod, N., Hood, L. & Stroynowski, I. (1988). Regulation of gene expression by interferon: control of H-2 promoter responses. *Science*, 239, 1302–1306.

Kralova, J., Jansa, P. & Forejt, J. (1992). A novel downstream regulatory element of the mouse H-$2K^b$ class I major histocompatibility gene. *EMBO Journal*, 11, 4591–4600.

Larner, A.C., Chaudhuri, A. & Darnell, J.E. Jr. (1986). Transcriptional induction by interferon: new protein(s) determine the extent and length of the induction. *Journal of Biological Chemistry*, 261, 453–459.

Leid, M., Kastner, P., Lyons, R., Nakshatri, H., Saunders, M., Zacharewski, T., Chen., J.-Y., Staub, A., Garnier, J.-M., Mader, S. & Chambon, P. (1992). Purification, cloning and RXR identity of the HeLa cell factor with which RAR or TR heterodimerizes to bind target sequence efficiently. *Cell*, 68, 377–395.

Lenardo, M.J. & Baltimore, D. (1989). NF-κB: a pleiotropic mediator of inducible and tissue specific gene control. *Cell*, 58, 227–229.

Lenardo, M.J., Rustgi, A.K., Schivella, A.R. & Bernards, R. (1988). Suppression of MHC class I gene expression by N-*myc* through enhancer inativation. *EMBO Journal*, 8, 3351–3356.

Levin, A.A., Sturzenbecker, L.J., Kazmer, S., Bosakowski, T., Huselton, C., Allenby, G. Speck, J., Krazeisen, C., Rosenberger, M., Lovey, A. & Grippo, J. (1992). 9-cis Retinoic acid stereoisomer binds and activates the nuclear receptor RXRα. *Nature (London)*, 355, 359–361.

Levy, D. & Darnell, J.E. Jr (1990). Interferon-dependent transcriptional activation: signal transduction without second messenger involvement? *New Biologist*, 2, 923–928.

Levy, D.E., Kessler, D.S., Pine, R. & Darnell, J.E. Jr (1989). Cytoplasmic activation of ISGF3, the positive regulator of interferon-β-stimulated transcription, reconstituted *in vitro. Genes and Development*, 3, 1362–1371.

Lonergan, M., Dey, A., Becker, K.G., Drew, P.D. & Ozato, K. (1993). A regulatory element in the $\beta_2$-microglobulin promoter identified by *in vivo* footprinting. *Molecular and Cellular Biology*, 13, 6629–6639.

MacDonald, N.J., Kuhl, D., Maguire, D., Näf, D., Gallant, P., Goswamy, A., Hug, H., Büeler, H., Chatuvedi, M., de la Fuente, J., Ruffner, H., Meyer, F. & Weismann, C. (1990). Different pathways mediate virus inducibility of the human IFN-$\alpha_1$ and IFN-β genes. *Cell*, 60, 767–779.

Mangelsdorf, D.J., Borgmeyer, U., Heyman, R.A., Zhou, J.Y., Ong, E.S., Oro, A.E., Kakizuka, A. & Evans, R.M. (1992). Characterization of three RXR genes that mediate the action of 9-*cis* retinoic acid. *Genes and Development*, 6, 329–344.

Marks, M.S., Hallenback, P.L., Nagata, T., Segars, J.H., Appella, E., Nikodem, V.V. & Ozato, K. (1992). H-2RIIBP (RXRβ) heterodimerizes with other nuclear hormone receptors and provides a combinatorial mechanism of gene regulation. *EMBO Journal*, 11, 1419–1435.

Martinez, E., Givel, F. & Wahli, W. (1991). A common ancestor DNA motif for invertebrate and vertebrate hormone response elements. *EMBO Journal*, 10, 263–268.

Maschek, V., Pulm, W. & Hammerling, G. (1989). Altered regulation of MHC class I genes in different tumour cell lines is reflected by distinct sets of DNAse I hypersensitive sites. *EMBO Journal*, 8, 2297–2304.

Miyamoto, M., Fujita, T., Kimura, Y., Maruyama, M., Harada, H., Sudo, Y., Miyata, T. & Taniguchi, T. (1988). Regulated expression of a gene encoding a nuclear factor, IRF-1, that specifically binds to IFN-β gene regulatory elements. *Cell*, 54, 903–913.

Miyazaki, J., Appella, E. & Ozato, K. (1986). Negative regulation of the major histocompatibility class I gene in undifferentiated embryonal carcinoma cells. *Proceedings of the National Academy of Sciences of the USA*, 83, 9537–9541.

Montminy, M.R., Sevarino, K.A., Wagner, J.A., Mandel, G. & Goodman, R.H.

(1986). Identification of a cyclic AMP responsive element within the rat somatostatin gene. *Proceedings of the National Academy of Sciences of the USA*, 83, 6682–6686.

Morello, D., Daniel, F., Baldacci, P., Cayre, Y., Gachelin, G. & Kourilsky, P. (1982). Absence of significant H-2 and $\beta_2$-microglobulin mRNA expression by mouse embryonal carcinoma cells. *Nature (London)*, 296, 260–262.

Morello, D., Duprey, P., Israel, A. & Babinet, C. (1985). Asynchronous regulation of mouse H-2D and $\beta_2$-microglobulin RNA transcripts. *Immunogenetics*, 22, 441–452.

Mueller, P.R. & Wold, B. (1989). *In vivo* footprinting of a muscle specific enhancer by ligation-mediated PCR. *Science*, 246, 780–786.

Nagata, T., Segars, J.-H., Levi, B.-Z. & Ozato, K. (1992). Retinoic acid dependent trans-activation of major histocompatibility complex class I promoters by a nuclear hormone receptor H-2RIIBP in undifferentiated embryonal carcinoma cells. *Proceedings of the National Academy of Sciences of the USA*, 89, 937–941.

Nelson, N., Marks, M.S., Driggers, P.H. & Ozato, K. (1993). Interferon consensus sequence-binding protein, a member of the interferon regulatory factor family, suppresses interferon-induced gene transcription. *Molecular and Cellular Biology*, 13, 588–599.

Nolan, G.P., Ghosh, S., Liou, H.-C., Tempst, P. & Baltimore, D. (1991). DNA binding and I-*k*B inhibition of the cloned p65 subunit of NF-κB, a *rel*-related polypeptide. *Cell*, 64, 961–969.

Ozato, K., Wan, Y.-J. & Orrison, B. (1985). Mouse major histocompatibility class I gene expression begins at midsomite stage and is inducible in earlier-stage embryos by interferon. *Proceedings of the National Academy of Sciences of the USA*, 82, 2427–2431.

Pine, R., Decker, T., Kessler, D.S., Levy, D.E. & Darnell, J.E. Jr (1990). Purification and cloning of interferon-stimulated gene factor 2 (ISGF2): ISGF2 (IRF-1) can bind to the promoters of both β-interferon- and interferon-stimulated genes but is not a primary transcriptional activator of either. *Molecular and Cellular Biology*, 10, 2448–2457.

Reis, L.F.L., Harada, H., Wolchok, J.D., Taniguchi, T. & Vilcek, J. (1992). Critical role of a common transcription factor, IRF-1, in the regulation of IFN-β and IFN-inducible genes. *EMBO Journal*, 11, 185–193.

Ryseck, R.-P., Bull, P., Takamiya, M., Bours, V., Siebenlist, U., Dobzanski, P. & Bravo, R. (1992). *rel*B, a new *rel* family transcription activator that can interact with p-50 NF-κB. *Molecular and Cellular Biology*, 12, 674–684.

Schindler, C., Fu, X.-Y., Improta, T., Aebersold, R. & Darnell, J.E. Jr (1992a). Proteins of transcription factor ISGF-3: one gene encodes the 91 and 84 kDa ISGF-3 proteins that are activated by interferon-α. *Proceedings of the National Academy of Sciences of the USA*, 89, 7836–7839.

Schindler, C., Shuai, K., Prezioso, V.R. & Darnell, J.E. Jr (1992b). Interferon-dependent tyrosine phosphorylation of a latent cytoplasmic transcription factor. *Science*, 257, 809–812.

Schmid, R.M., Perkins, N.D., Duckett, C.S., Andrews, P.C. & Nabel, G.J. (1991). Cloning of a NF-κB subunit which stimulates HIV transcription in synergy with p65. *Nature (London)*, 352, 733–736.

Sen, R. & Baltimore, D. (1986). Multiple nuclear factors interact with the immunoglobulin enhancer sequences. *Cell*, 46, 705–716.

Shirayoshi, Y., Burke, P.A., Appella, E. & Ozato, K. (1988). Interferon-induced transcription of a major histocompatibility class I gene accompanies binding of inducible nuclear factors to the interferon consensus sequence. *Proceedings of the National Academy of Sciences of the USA*, 85, 5884–5888.

Shirayoshi, Y., Miyazaki, J., Burke, P.A., Hamada, K., Appella, E. & Ozato, K. (1987). Binding of multiple nuclear factors to the 5′ upstream regulatory element of the major histocompatibility class I gene. *Molecular and Cellular Biology*, 7, 4542–4548.

Silverman, T., Rein, A., Orrison, B., Langlass, J., Bratthauser, G., Miyazaki, J. & Ozato, K. (1988). Establishment of cell lines from somite stage embryos and expression of major histocompatibility class I genes in these cells. *Journal of Immunology*. 14, 4378–4387.

Sugita, K., Miyazaki, J., Appella, E. & Ozato, K. (1987). Interferon regulates major histocompatibility class I gene expression through a 5′ flanking region. In *Proceedings of the 1986 Congress of the Society for International Interferon Research*, ed. K. Cantell & M. Schellenkens, pp. 265–272. Finland: Nijhoff.

Thanos, D. & Maniatis, T. (1992). The high mobility group protein HMG I(Y) is required for NF-κB dependent virus induction of the human IFN-β gene. *Cell*, 71, 777–789.

Veals, S.A., Schindler, C., Leonard, D., Fu, X.-Y., Aebersold, R., Darnell, J.E. Jr & Levy, D.E. (1992). Subunit of an α-interferon-responsive transcription factor is related to interferon regulatory factor and *myb* families of DNA-binding proteins. *Molecular and Cellular Biology*, 12, 3315–3324.

Velazquez, L., Fellous, M., Stark, G.R. & Pelligrini, S. (1992). A protein tyrosine kinase in the interferon α/β signalling pathway. *Cell*, 70, 313–322.

Weissman, J.D. & Singer, D.S. (1991). A complex regulatory DNA element associated with a major histocompatibility complex class I gene consists of both a silencer and enhancer. *Molecular and Cellular Biology*, 11, 4217–4225.

Weisz, A., Marx, P., Sharf, R., Appella, E., Ozato, K. & Levi, B. (1992). Human interferon consensus sequence binding protein is a negative regulator of enhancer elements common to interferon-inducible genes. *Journal of Biological Chemistry*, 267, 25589–25596.

Williams, B.R. (1991). Transcriptional regulation of interferon-stimulated genes. *European Journal of Biochemistry*, 200, 1–11.

Wolffe, A.P. (1990). New approaches to chromatin function. *New Biologist*, 2, 211–218.

Yu, V.C., Delsert, C., Anderson, B., Holloway, J.M., Devary, O., Naar, A.M., Kim, S.Y., Boutin, J.-M., Glass, C.K. & Rosenfeld, M.G. (1991). RXRβ: a co-regulator that enhances binding of retinoic acid, thyroid hormone and vitamin D receptors to their cognate response elements. *Cell*, 67, 1251–1266.

Zhang, X.-K., Lehmann, J., Hoffmann, B., Dawson, J. I., Cameron, J., Graupner, G., Hermann, T., Tran, P. & Pfahl, M. (1992). Homodimer formation of retinoid X receptor induced by 9-cis retinoic acid. *Nature (London)*, 358, 587–591.

Zöller, B., Ozato, K., Kroemer, G., Auffray, C. & Jungwirth, C. (1992). Interferon induction of chicken MHC class I gene transcription: phylogenetic conservation of the interferon responsive element. *Virology*, 191, 141–149.

# 5

# Control of MHC class II gene expression

JENNY PAN-YUN TING and BARBARA J. VILEN
*University of North Carolina*

## 5.1 Introduction

The MHC class II (also known as immune response-associated antigens, or Ia) genes are a multigene family encoded on chromosome 6 in humans (Fig. 5.1) and chromosome 17 in mice (see Chapter 2). The functional role of the MHC class II antigens in the immune response is described extensively in Chapters 1 and 3. This includes their function in antigen presentation, thymic education and T cell recognition/activation, their contribution to transplant acceptance or rejection, their genetic association with diseases, their aberrant up-regulation in a number of autoimmune and inflammatory diseases as well as their absence in specific immunodeficiencies. The involvement of MHC class II gene products in a large number of clinical conditions confers practical importance to the study of their regulation. Furthermore, the tissue-, cell-specific, differentiation-dependent and cytokine-regulated expression of MHC class II genes represent features that have made these genes a model system to study gene regulation. In this chapter, the expression and regulation of MHC class II genes in normal and disease states will be discussed, as well as the molecular basis for the constitutive and inducible modes of class II gene regulation.

## 5.2 Role of the MHC class II antigens in the immune response

The specificity of the immune system is conferred by the binding of foreign antigens through receptors on the surfaces of B and T lymphocytes, the immunoglobulin (Ig) molecule and the T cell receptor (TCR), respectively. While the Ig receptor on B cells can recognize soluble antigen, T cells must recognize antigen bound to MHC gene products. For $CD8^+$ cytotoxic T cells, antigenic peptides are primarily recognized when bound to the MHC class I

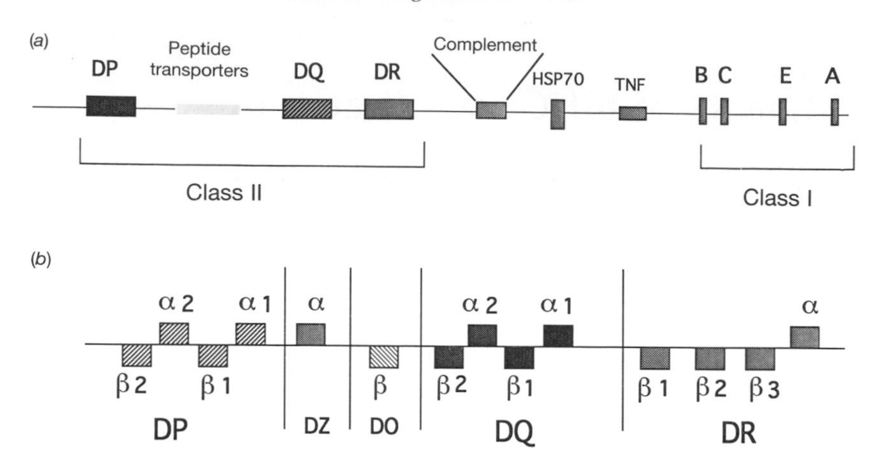


Fig. 5.1. The human HLA region (*a*) and class II MHC region (*b*). The class II MHC genes are clustered and located on the short arm of human chromosome 6. In the midst of the class II genes are the TAP and LMP genes (see Chapters 1 and 2). The class II MHC multigene family includes a total of 14 genes.

antigens, while $CD4^+$ $T_H$ cells recognize MHC class II–peptide complexes (Germain, 1986). The specificity of the TCR recognition is influenced both by the polymorphic residues in the antigen-binding groove of the MHC molecule and by the amino acid sequence of the bound, antigen-derived peptide. This dual recognition requirement is known as MHC restriction.

MHC class II proteins complexed with foreign antigen are mostly recognized by $CD4^+$ $T_H$ cells. This recognition event results in activation of $T_H$ cells, followed by autocrine growth in which the stimulated cells proliferate in response to up-regulation of their IL-2 and IL-2 receptor (IL-2R) genes (Smith, 1984; Greene & Leonard, 1986; Waldmann, 1989). In addition, the cells differentiate and acquire the ability to produce other lymphokines, such as IL-4 and IFN-γ. These lymphokines serve to activate other cells within the immune system (Swain *et al.*, 1988; Gajewski *et al.*, 1989; Salmon, Kitas & Bacon, 1989). Therefore, it is clear that the MHC class II proteins play a central role in eliciting an immune response to foreign antigen. This chapter will focus on how the MHC class II genes are regulated and how specific patterns of gene expression correlate with disease.

## 5.3 Normal tissue-specific and inducible patterns of MHC class II expression

Selective expression of the MHC class II antigens is required to maintain normal immune function. Class II antigens are constitutively expressed on immunocompetent B cells and dendritic cells. In addition, class II MHC antigens can be induced on a number of cell types by a large number of molecules. Constitutive expression has been best studied in B cells where both transcriptional and post-transcriptional mechanisms control cell surface class II expression. Post-transcriptional increases in cell surface expression appear to involve a factor that stabilizes class II mRNA 8–10-fold although the molecular mechanism has yet to be elucidated (Maffei *et al.*, 1989). Transcriptional control of class II gene expression remains a predominant mode of class II gene regulation (Basta, Sherman & Ting, 1988; Blanar, Bottger & Flavell, 1988; Boettger, Blanar & Flavell, 1988; Rosa & Fellous, 1988; Basta *et al.*, 1989).

The expression pattern of cell surface class II antigens during B cell maturation is tightly regulated. In mice, class II mRNA and protein are absent in pre-B cells but appear in mature B cells (McKearn & Rosenburg, 1985; Lala *et al.*, 1979; Kincade *et al.*, 1981). In contrast, human pre-B cells are class II positive. Expression of class II on the cell surface of resting B cells is relatively low compared with activated B cells or transformed B cells where class II mRNA and protein levels are maintained at constitutively high levels. Terminal differentiation of mature B cells into plasma cells results in down-regulated MHC class II expression (Latron *et al.*, 1988; Dellabona *et al.*, 1989) through a dominant suppression mechanism. This has been demonstrated by somatic cell fusion of plasma cells and mature B cells (Halper *et al.*, 1978; Mond *et al.*, 1980; Venkitaraman, Culbert & Feldmann, 1987).

Expression of class II antigens on the surface of B cells can be modulated by crosslinking of a number of B cell surface molecules such as IgM, IgD, B220 and Lyb2 (Noelle *et al.*, 1986; McMillan *et al.*, 1988). Viral or mycoplasma infection and bacterial lipopolysaccharide enhance class II expression (Long, Mach & Accolla, 1984; Ephrussi *et al.*, 1985; Yamamura *et al.*, 1985; Massa, Dorries & ter Meulen, 1986; Maniatis, Goodbourn & Fischer, 1987; Gaulton *et al.*, 1989; Stuart, Cassell & Woodward, 1989; Mirkovitch & Darnell, 1991). These may all be linked to the evolutionary role of MHC molecules in fighting infection.

Up-regulation of MHC class II expression on both B lymphocytes and macrophages also occurs following IL-4 stimulation (Mond *et al.*, 1981; Monroe & Cambier, 1983; Roehm *et al.*, 1984; Noelle *et al.*, 1985; 1986; Rousset *et al.*, 1988). This induction by IL-4 can be down-regulated by IFN-γ (Mond *et al.*, 1980). Evidence exists for both transcriptional and post-transcriptional mechanisms of IL-4 up-regulation of class II expression (Noelle *et al.*, 1986; Polla *et al.*, 1986). The former is associated with an IL-4-inducible DNA-binding protein with specificity for two DNA sequences within 1000 bp of the Aα promoter (Boothby *et al.*, 1988; Finn *et al.*, 1990c).

An IL-4-responsive motif was deduced from these two sites and a promoter fragment that contained this motif was demonstrated to be IL-4 responsive. These observations strongly suggest that the site defined is regulated by IL-4 (Finn *et al.*, 1990b).

IFN-γ is a potent regulator of class II MHC genes in a wide variety of both lymphoid and non-lymphoid cell types. These include fibroblasts, Kuppfer cells, Langerhans cells, astrocytes, microglia, skin keratinocytes, thymic epithelium, vascular endothelium and renal tubular cells (reviewed in Benoist & Mathis, 1990; Cogswell, Zeleznik-Le & Ting, 1991; Glimcher & Kara, 1992). The mechanism of induction by IFN-γ appears to be the direct result of an increased rate of transcription, although in some cell types post-transcriptional events may also be involved (Blanar *et al.*, 1988; Rosa & Fellous, 1988; Amaldi *et al.*, 1989; Kern *et al.*, 1989). The time course of class II induction by IFN-γ is rather slow, with a rise in the mRNA level beginning at 2–4 hours and becoming maximal by 48 hours (Rosa *et al.*, 1983; Collins *et al.*, 1984). The requirement for new class II protein synthesis during induction by IFN-γ appears to vary among cell types (Basta *et al.*, 1988; Blanar *et al.*, 1988; Bottger *et al.*, 1988; Amaldi *et al.*, 1989; Woodward, Omer & Stuart, 1989). Studies of second messenger involvement have implicated protein kinase C activation (Fan, Goldberg & Bloom, 1988; Mattila, Hayry & Renkonen, 1989; Benveniste *et al.*, 1991), changes in intracellular cAMP levels (Frohman *et al.*, 1988; Basta *et al.*, 1989; Sasaki, Levison & Ting, 1990), activation of the $Na^+/H^+$ antiport (Prpic *et al.*, 1989; Benveniste *et al.*, 1991) and calcium-dependent cellular pathways (Koide *et al.*, 1988; Nezu *et al.*, 1990). Although most of the studies are in agreement, a few of the findings are not. This may be the result of differences in the cell lines and cell types studied. An equally likely cause for conflicting findings is that these studies rely on a pharmacological approach and the specificity, quality, and purity of the pharmacological reagent can greatly influence the outcome of the experiments.

At the molecular level, MHC class II gene regulation by IFN-γ is mediated by three DNA elements present in the proximal promoter of all MHC class II genes, W/Z/S, X and Y elements (see Section 5.7 below). The requirement for three elements is in striking contrast to the IFN-γ induced expression of many other genes, among these the MHC class I genes (see Chapter 4). IFN-γ induction of the latter group of genes is mediated by the presence of a canonical interferon consensus sequence (ICS, see Section 4.2) distinct from W/Z/S, X or Y (Israel *et al.*, 1986; Sugita *et al.*, 1987; Blanar *et al.*, 1989). Interestingly, the promoters of several other IFN-γ inducible genes such as the Ii chain and the FcRγ also contain W/Z/S, X and/or Y elements (Brown, Barr &

Ting, 1990; Doyle *et al.*, 1990; Zhu & Jones, 1990; Pearse, Feinman & Ravetch, 1991).

The existence of multiple IFN-γ regulated pathways is also supported by genetic analyses, which have provided an important insight into the mechanism of class II induction by IFN-γ. Using genetic selection, two groups (Loh *et al.*, 1992; Mao *et al.*, 1993) have isolated mutants that are defective in the induction of MHC class II by IFN-γ. These studies show, firstly, that the induction of MHC class II genes requires unique signal transduction components, although some of the pathways are shared with MHC class I gene induction and, secondly, that there are at least five genetic complementation groups, suggesting the involvement of multiple genetic components in the induction of MHC class II. Complementation of these mutant lines with cDNA clones from expression libraries is likely to lead to the eventual identification of these genes.

In addition to IL-4 and IFN-γ, other lymphokines and cellular factors such as IL-10, GM-CSF, glutamate, IFN-α/β, TNF-α, glucocorticoids, prostaglandins and α-fetoprotein have also been shown to alter class II expression (Snyder, Beller & Unanue, 1982; Snyder & Unanue, 1982; Lu, Changelian & Unanue, 1984; Ling, Warren & Vogel, 1985; Dennis & Mond, 1986; Inaba *et al.*, 1986; Fertsch *et al.*, 1987; Koerner, Hamilton & Adams, 1987; Fertsch-Ruggio, Schoenberg & Vogel, 1988; Manyak *et al.*, 1988; Willman *et al.*, 1989; Go *et al.*, 1990; Zimmer & Jones, 1990; Lee *et al.*, 1992). Among these, IL-10, IFN-α/β, prostaglandins and α-foetoprotein all act as negative regulators. However, opposite effects have been observed for some of these factors (Pfizenmaier *et al.*, 1987; Arenzana-Seisdedos *et al.*, 1988; Alvaro-Gracia, Zvaifler & Firestein, 1989; Go *et al.*, 1990). For example, TNF-α can either enhance or decrease class II MHC expression (Chang & Lee, 1986; Zimmer & Jones, 1990; Benveniste, Sparacio & Bethea, 1989; Watanabe & Jacob, 1991; Pfizenmaier *et al.*, 1987; Arenzana-Seisdedos *et al.*, 1988). Noradrenaline also has opposing effects depending on the transformation state of the cell (Frohman *et al.*, 1988; Basta *et al.*, 1989; Sasaki *et al.*, 1990). A logical explanation, verified by experimental evidence, is that differences in lineages and differentiative states of the cells affect their responses to cytokines (Watanabe & Jacob, 1991).

## 5.4 Aberrant MHC class II expression and disease correlation

There are a large number of diseases that have association with specific MHC class II alleles. More current applications of PCR, which was originally used to elucidate MHC gene polymorphisms, have permitted the finer dissection of allelic DNA sequences and diseases. Type I insulin-dependent diabetes mellitus, IDDM (Lo *et al.*, 1988; Sarvetnick *et al.*, 1988), is polygenic in inheritance, although greater than 50% is contributed by the MHC class II D region (Rotter & Landlaw, 1984; Thomson, 1984). Several HLA serotypes (DR4, DR3 and DR1) are positively associated with this disease, with susceptibility mapping to the HLA-DQB gene (Svejgaard, Platz & Ryder, 1983; Todd *et al.*, 1987). In endocrine autoimmune diseases, such as Addison's and Graves' disease, there is a preponderance of HLA-DR3 (Hanafusa *et al.*, 1983). This particular serotype has also been associated with higher incidence of dermatitis herpetiformis, systemic lupus erythematosus (SLE) and myasthenia gravis (Svejgaard *et al.*, 1983). Likewise, MHC class II serotypes DR2 and DR4 are positively associated with multiple sclerosis (MS) and rheumatoid arthritis, respectively (Nepom & Erlich, 1991). Juvenile dermatomyositis is linked to the HLA-DQA1 allele (Reed, Pachman & Ober, 1991). This HLA class II-associated susceptibility to autoimmune disease may partially reflect aberrant expression of MHC class II genes in these individuals.

Although expression of class II antigens is critical for normal immune function, local aberrant expression is frequently observed in autoimmunity (see Chapters 17 and 18). Heightened or aberrrant expression of MHC class II antigens has been associated with autoimmune or autoimmune-like states in pancreatic islet β cells (Bottazzo *et al.*, 1983; 1985; Pujol-Borrell *et al.*, 1987; Campbell *et al.*, 1988), macrophages (Kelley & Roths, 1982), brain glial cells (Traugott, Scheinberg & Raine, 1985; Matsumoto *et al.*, 1986), synovial lining cells (Teyton *et al.*, 1987), thyroid epithelial cells (Londei *et al.*, 1984) and activated T cells (Yu *et al.*, 1980; Smith & Roberts-Thomson, 1990). A well-studied model is murine experimental autoimmune encephalomyelitis (EAE), where presentation of self peptides from myelin basic protein (MBP) by glial cells to T lymphocytes results in inflammatory brain lesions associated with demyelination (Fontana *et al.*, 1987). *In vitro* studies have shown that IFN-α induction of MHC class II antigens on cultured astrocytes and microglia allow these cells to present MBP, implicating aberrant MHC class II gene expression in this CNS disease (Fierz *et al.*, 1985; Fontana, Fierz & Wekerle, 1984). In addition, anti-Ia antibodies injected into the animals can alleviate many of the symptoms. Similar effects were observed when autoimmune (NZB × NZW) F1 mice and BB rats with thyroid and pancreatic autoimmunity were treated with anti-Ia antibodies (Adelman, Watling & McDevitt, 1983; Boitard *et al.*, 1985), again supporting a role for Ia in autoimmunity.

### 5.5 Lack of expression of MHC class II antigens leads to disease

In contrast to the hyperexpression of MHC class II antigens in a number of autoimmune diseases, patients suffering from the congenital bare lymphocyte syndrome (BLS) lack MHC class II antigen and mRNA expression and are moderately to severely immunodeficient (de Preval, Hadam & Mach, 1988; Hume *et al.*, 1989). BLS patients have normal numbers of circulating B and T lymphocytes yet they exhibit recurring bacterial or viral infections because of their inability to mount an immune response against foreign antigen (Rijkers *et al.*, 1987; Plaeger-Marshall *et al.*, 1988). These infections result directly from the failure to present processed foreign antigen because of a lack of MHC class I (type I BLS), MHC class II (type II BLS) or both class I and class II antigens (type III BLS) (Touraine *et al.*, 1978; Marcadet *et al.*, 1985; de Preval *et al.*, 1985; Rijkers *et al.*, 1987; Griscelli, Lisowska-Grospierre & Mach, 1989). Type II BLS is characterized by a lack of MHC class II mRNA and protein, although the structural genes are intact and can be activated upon fusion with normal class II positive cells (de Preval *et al.*, 1985; Hume & Lee, 1989). The MHC locus and the disease locus segregate independently, supporting the idea that the defect is caused by a *trans*-acting factor(s) (de Preval *et al.*, 1985).

To determine if the mutations causing different forms of type II BLS result from the same or different genes, complementation analyses have been performed. In addition, $DR^-$ variant cell lines generated by *in vitro* mutagenesis or irradiation have been extremely useful because they generally propagate better in culture than BLS cell lines (Gladstone & Pious, 1978; Accolla, 1983; Calman & Peterlin, 1987). Fusion between different defective cell lines results in the reappearance of class II protein, indicating the existence of complementary genes (Yang *et al.*, 1988; Hume & Lee, 1989). Complementation analysis from naturally occurring, patient-derived class II negative cell lines and *in vitro*-derived mutants has demonstrated at least four different genetic mutations that are capable of producing the class $II^-$ phenotype, revealing that multiple *trans*-acting factors are involved in regulating MHC class II gene expression (Hume & Lee, 1989; Benichou & Strominger, 1991). Some of the *in vitro* generated mutant lines have been used to produce human–mouse hybrids. These studies identified a mouse locus, aIr-1, located on murine chromosome 16 that activates class II expression in $DR^-$ RJ 2.2.5 variant cells (Accolla *et al.*, 1986; Guardiola *et al.*, 1986). The molecular identification of this locus will be of importance.

The use of genomic *in vivo* footprinting has been useful in analysing BLS cell lines. *In vivo* footprinting is a procedure that delineates protein–DNA

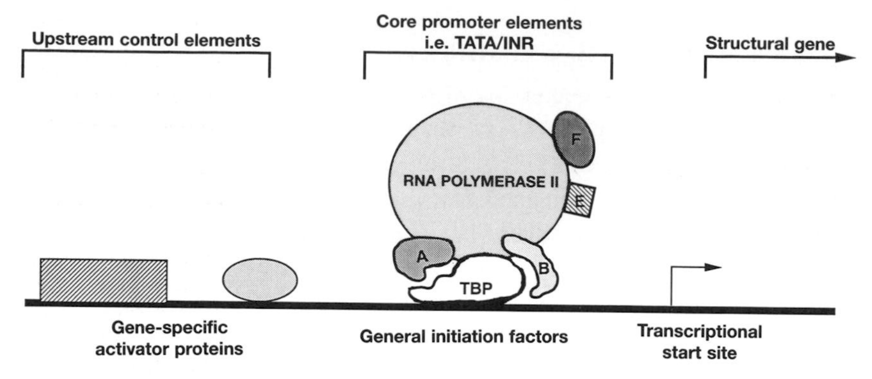


Fig. 5.2. A generic TATA-containing promoter. The promoter shown contains core promoter elements that consist of a TATA element or an initiator sequence (such as INR). TBP binds TATA, leading to the subsequent binding of a cascade of proteins and RNA polymerase II. The promoter also contains upstream control elements, which are the targets of gene-specific activator proteins. These elements are found among many promoters and combinations of such elements most likely define the fine specificity of gene expression.

interaction in the nuclei of intact cells (Mueller & Wold, 1989). Application of this procedure reveals two phenotypes for the different types of BLS (Kara & Glimcher, 1991). The first involves normal protein interactions with the promoter, suggesting a defect in gene activation since the gene is not expressed despite normal protein–DNA interaction. The second phenotype reveals a lack of protein interactions on the promoter, suggesting a defect in promoter accessibility. This may involve a protein that is responsible for opening a normally closed chromatin structure.

## 5.6 Transcriptional control of eukaryotic genes

Because transcriptional regulation is a major pathway that regulates MHC class II genes, a brief review of these mechanisms will be presented (reviewed in Greenblatt, 1991; Roeder, 1991). Studies of the mechanisms controlling transcription of genes encoding mRNAs show requirements for general initiation factors that bind core promoter elements in addition to gene-specific factors that bind to distal control elements. Transcription of a particular gene requires the co-ordinated binding and subsequent protein complex formation of the basal transcription factor including RNA polymerase II and upstream activators. Figure 5.2 is a diagram of a generic TATA-containing promoter.

Intense biochemical study of basal initiation factors has led to the elucidation of a multisubunit complex, the pre-initiation complex, which acts through the minimal promoter TATA element. The formation of a functional pre-initiation complex involves ordered interactions of TFIIA, TFIIB, TFIID (TATA-binding protein; TBP), TFIIE, TFIIF and RNA polymerase II. TFIID is the only general factor with an intrinsic capability for site-specific DNA binding. Many eukaryotic genes contain TATA elements, albeit a large number also lack this core promoter element. In promoters lacking a TATA element, an initiator sequence (INR) may be responsible for directing the start site of transcription (Smale *et al.*, 1990; Sawadogo & Sentenac, 1990). The exact location of these INR elements varies, as does the sequence, although recently a factor, designated TFII-I, has been identified that binds specifically to an INR with the consensus YAYTCYYY (Y = pyrimidine) in several promoters (Roy *et al.*, 1991).

Although pre-initiation complex formation can be sufficient for accurate transcription initiation, transcriptional activity is greatly stimulated by promoter-specific activator proteins which bind a vast array of *cis*-acting elements (Ptashne, 1988). Thus, individual genes possess unique arrangements of contiguous or overlapping response elements where distinct factors bind (Dynan, 1989).

The ability of transcription factors not only to bind DNA but also to interact with the basal transcription machinery requires that they have a number of functional domains including *trans*-activation and DNA-binding domains. DNA-binding domains such as the helix–turn–helix (HTH) motif (Harrison & Aggarwal, 1990; Harrison, 1991), the zinc-binding finger domains (Miller, McLachlan & Klug, 1985; Freedman *et al.*, 1988; Pan & Coleman, 1990) and the leucine zipper motif have been well characterized.

Unlike the highly ordered conformations assumed by DNA-binding domains, transcriptional activation domains appear to be less structured in that a well-defined amino acid sequence is not required to elicit activation function. Acidic activation domains (Ma & Ptashne, 1987; Lin & Green, 1991), glutamine-rich and proline-rich activation domains have all been identified. The latter two domains have been identified in NF-Y, which controls class II MHC gene expression (Laughton & Scott, 1984; Kumar *et al.*, 1987; Williams *et al.*, 1988).

In addition to the primary event of DNA–protein interaction, protein–protein interactions are fundamental to transcriptional regulation in eukaryotic cells. Such interactions are involved both in formation of the pre-initiation complex between RNA polymerase II and the general transcription factors and in contacts between sequence-specific DNA-binding proteins and the promoter proximal complex (Ptashne, 1988; Pugh & Tjian, 1990; Conaway *et al.*, 1991; Dynlacht, Hoey & Tjian, 1991; Tanese, Pugh & Tjian, 1990).

## 5.7 Molecular regulation of the MHC class II genes: DNA elements and transcription factors

Transcriptional regulation of the human and murine class II genes requires a number of 5′ regulatory sequences located within the promoter region of the genes. Extensive studies have been done on many of the murine and human class II genes. These studies reveal common regulatory elements that are unique in their arrangement in class II MHC promoters, as well as gene-specific and allelic-specific variations. These common elements are important for the co-ordinate regulation of MHC class II genes, while the gene- and allelic-specific differences most probably account for unco-ordinated gene expression. In this section, emphasis will be placed on DNA elements common to all MHC class II promoters. Gene-specific and allele-specific regulatory elements will be discussed briefly.

Among MHC class II genes studied to date, all have a trimeric regulatory component consisting of three DNA motifs, the W/Z/S, X and Y elements (reviewed in Benoist & Mathis, 1990; Cogswell *et al.*, 1991; Glimcher & Kara, 1992) (see Fig. 5.3). These are not only conserved in sequence but also in their spatial arrangement. Interestingly, the promoter region of the class II MHC-associated invariant chain gene also contains these same elements (Brown *et al.*, 1990; Doyle *et al.*, 1990; Zhu & Jones, 1990). These elements most likely control the co-ordinate expression of the MHC class II genes together with the invariant chain gene.

Among the three elements, the Y element is the most well characterized. This 10 bp element is conserved among all the murine and human class II alleles and contains an inverted CCAAT sequence. Substitution mutagenesis and *in vitro* transcription analyses have identified it as a positive regulator of constitutive and IFN-γ-induced DRA expression (Sherman *et al.*, 1989b; Finn *et al.*, 1990a,b; Hume & Lee, 1990; Tsang, Nakanishi & Peterlin, 1990; Zeleznik-Le, Azizkhan & Ting, 1991; Moses *et al.*, 1992). In addition, this element may play a role in determining the start site of transcription, as mutation of the element introduces an aberrant transcript in *in vitro* run-off transcription assays (Zeleznik-Le *et al.*, 1991).

A predominant factor binds to the Y element (Fig. 5.4). This factor was initially identified by gel retardation analysis and is known by various names, including NF-Y, YEBP and CBP/CP1 (Dorn *et al.*, 1987a; Vuorio, Maity & de Crombrugghe, 1990; Zeleznik-Le *et al.*, 1991). NF-Y was initially ident-

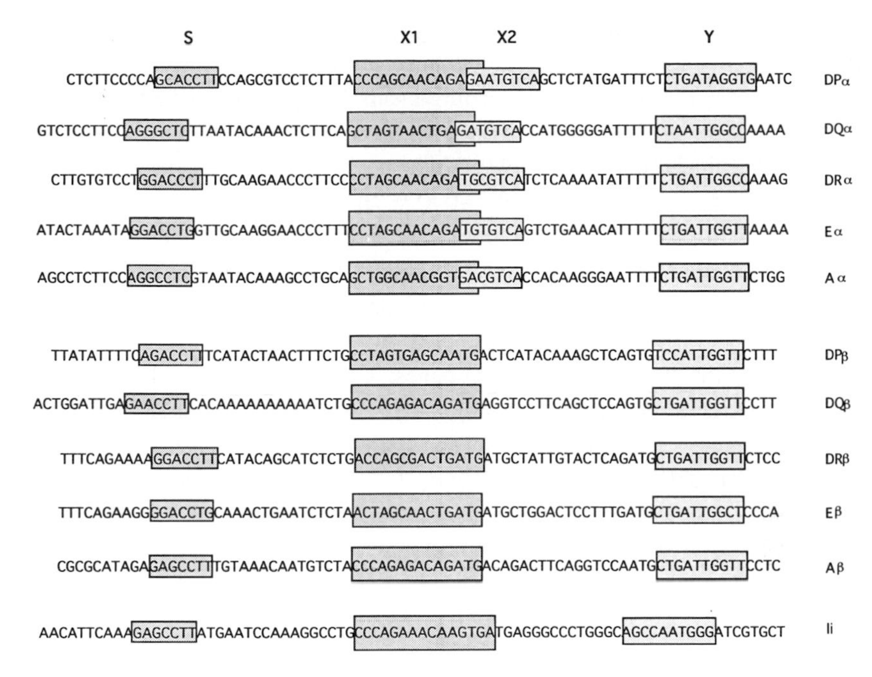


Fig. 5.3. Conserved regions of class II MHC promoters: the 5′ regulatory regions. The conserved S, $X_1$, $X_2$ and Y elements of class II MHC promoters and the Ii promoter are shown here.

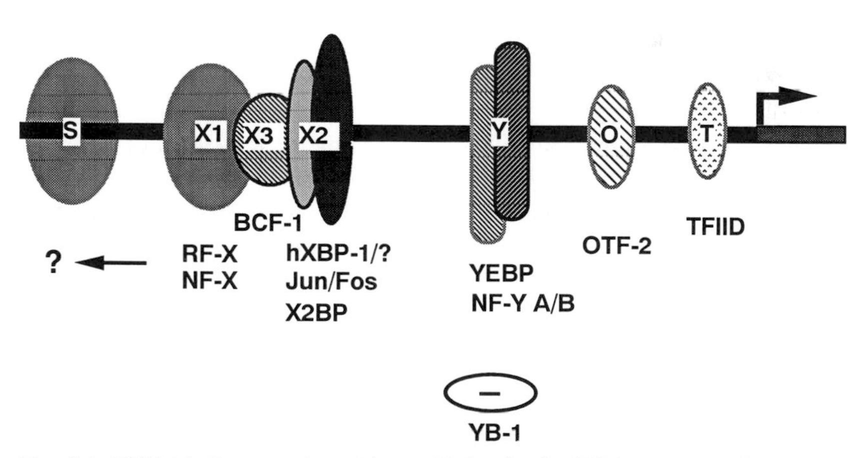


Fig. 5.4. DNA-binding proteins with specificity for the DRA promoter. A prototype class II MHC promoter is shown here because many of the proteins are known to bind sequences in this promoter. Proteins which interact with other class II promoters are described in the text. Some of the elements are unique to DRA, such as the TATA (T) and octamer (O) motifs. Others shown are conserved in all class II MHC promoters, as shown in Fig. 5.3.

ified as a complex specific for the Eα promoter (Dorn *et al.*, 1987a). The molecule is heteromeric, consisting of three proteins required for DNA binding. Two of the three individual components have been purified and their corresponding genes identified (Hatamochi *et al.*, 1988; Celada & Maki, 1989; van Huijsduijnen *et al.*, 1990; Vuorio *et al.*, 1990; Maity *et al.*, 1992). The A monomer is approximately 40 kDa and the B monomer is 30 kDa. They are highly conserved throughout evolution, from yeast, plants, amphibians to mammals (van Huijsduijnen *et al.*, 1990). YEBP is an affinity-enriched fraction that contains DRA Y-binding activity and can up-regulate the DRA as well as a viral thymidine kinase promoter (Zeleznik-Le *et al.*, 1991). Since the cloning of the NF-Y genes, it was found that YEBP contains NF-Y proteins, as determined serologically; therefore, NF-Y and YEBP are similar, if not identical (B. J. Vilen and J. P.-Y. Ting, unpublished observations). CBF, the rat homologue of NF-Y, binds to the rat collagen CCAAT sequence (Maity, Vuorio & de Crombrugghe, 1990; Vuorio *et al.*, 1990). NF-Y/YEBP can recognize some, but not all, CCAAT motifs in non-class II MHC genes such as thymidine kinase (Chodosh *et al.*, 1988), collagen (Hatamochi *et al.*, 1988; Maity *et al.*, 1990; Vuorio *et al.*, 1990), albumin (Wuarin, Mueller & Schibler, 1990; Tronche *et al.*, 1991) and the adenovirus major late promoter (Chodosh *et al.*, 1988), suggesting it could play a more general role in transcriptional regulation than simply regulating class II expression (Dorn *et al.*, 1987a).

YB-1 is another Y element-binding protein cloned from a λgt11 library through its ability to bind a region of the DRA promoter including the spacer and Y element (Didier *et al.*, 1988). Subsequent studies have revealed that the binding specificity of YB-1 is relatively non-stringent, as it is capable of binding a number of unrelated sequences including the X element of the class II box. Furthermore, it can bind to single-stranded DNA in a relatively specific fashion. Under some circumstances, the expression of YB-1 mRNA is inversely correlated with cell surface class II expression, implying that it may play a role in the transcriptional repression of class II MHC genes (Didier *et al.*, 1988). Indeed, evidence in our laboratory shows that the constitutive expression of YB-1 in an expression vector can inhibit IFN-γ-induced expression of DRA (J. P.-Y. Ting, N. Zeleznik-Le, A. Painter, T. L. Moore, A. M. Brown & B. D. Schwartz, unpublished observations), supporting its role as a negative regulator.

The X element is located approximately two helical turns upstream of the Y element. Mutational analysis and *in vitro* transcription competitions of X-binding proteins has shown that the X box is required for positive function in B cells and for IFN-γ-mediated induction (Dorn *et al.*, 1987b; Sherman *et al.*, 1987, 1989b; Koch *et al.*, 1988; Dedrick & Jones, 1990; Finn *et al.*, 1990a; Hume & Lee, 1990; Tsang *et al.*, 1990; Viville *et al.*, 1991). Mutation of this region has further delineated the X box into three elements, X1, X2 and X3, shown in Fig. 5.4 (Sloan *et al.*, 1992; Voliva *et al.*, 1992). While the X1 and X2 boxes are needed for constitutive expression of MHC class II genes in B cells, in some genes only the X1 element is required for IFN-γ-mediated induction (Moses *et al.*, 1992). The X element may also play a role in directing the start site of transcription, as site-specific competitions for X box-binding proteins in an *in vitro* transcription assay cause aberrantly initiated transcripts (Zeleznik-Le *et al.*, 1991). This result was also seen when the Y element was mutated, suggesting a role for the class II X/Y element in determining the start site of transcription (Dorn *et al.*, 1987b). Recently, a third X element has been delineated between the X1 and X2 elements. The X3 element represents a consensus motif that complexes with HTH domains and binds B cell factor 1, BCF-1 (Voliva *et al.*, 1992).

There are a number of proteins that show specificity for the X1 element. These may be necessary to accommodate the many regulatory events required for constitutive, inducible and differential expression of the MHC class II genes. RF-X is a DNA–protein complex that is missing in some BLS-derived cell lines (Reith *et al.*, 1988, 1989; Stimac *et al.*, 1991; Herrero Sanchez *et al.*, 1992; Hasegawa *et al.*, 1993). Interestingly, an X-binding protein(s) is also absent in an *in vitro*-mutagenized, DR-negative variant cell line, although its relationship to RF-X remains undetermined (Stimac *et al.*, 1991). A recombinant protein, named rRF-X, was originally identified from a λgt11 expression library as having X1-binding activity (Reith *et al.*, 1988). Unfortunately, rRF-X turned out to be distinct from RF-X because rRF-X protein and mRNA are expressed normally in BLS cell lines and the rRF-X genes in normal and BLS lines are indistinguishable (Herrero Sanchez *et al.*, 1992). Identification of the deficient X box-binding protein in BLS is clearly of great interest.

NF-X protein binds to the X1 box of the murine Eα gene (Dorn *et al.*, 1987b). It contacts the X element in a similar fashion to RF-X, although it is distinguished from RF-X by differential sensitivity to protease and to non-specific competitors (Dorn *et al.*, 1987b; Reith *et al.*, 1988; Hasegawa *et al.*, 1993). It has been suggested that NF-X is the murine homologue of rRF-X.

IFNEX, IFN-γ-enhanced X-binding protein, forms a protein–DNA complex primarily involving the X2 and some X1 nucleotides (Moses *et al.*, 1992). This protein is present at low levels in uninduced primary rat astrocytes, but is enhanced upon IFN-γ induction (Moses *et al.*, 1992.)

Serologically, IFNEX does not appear to be rRF-X. Using *in vivo* genomic footprinting, a procedure that reveals protein contact points on DNA in intact cells, it was revealed that IFN-γ enhances protein interactions over the X region (Wright & Ting, 1992). This finding is consistent with the *in vitro* existence of IFNEX, whose activity or expression is enhanced by interferon treatment (Wright & Ting, 1992).

The X2 element encompasses the 3′ end of the X box and extends into the spacer region (Fig. 5.4). The X2 sequence of some class II MHC promoters is homologous to a TRE (TPA response element) or AP-1 site and of others is more homologous to a consensus CRE (cAMP response element) element. An intact X2 site is necessary for constitutive DRA promoter activity in B cells, although others have reported no effect of X2 mutations (Liou *et al.*, 1990; Tsang *et al.*, 1990; Sloan, Hasegawa & Boss, 1992). Similar discrepancies have been found in the role of X2 in IFN-γ mediated induction (Tsang *et al.*, 1990; Moses *et al.*, 1992; Sloan *et al.*, 1992).

The hXBP-1 gene codes for an X2-binding protein isolated by its ability to bind the X2 site of Aα (Liou *et al.*, 1990). This protein is a member of the leucine zipper class of transcription factors and has been shown to interact with the c-*fos* protein but not *c-jun*, although *c-jun* homodimers have been reported to bind the X2 site of DRA (Andersson & Peterlin, 1990; Ono *et al.*, 1991). This finding is supported by *in vitro* experiments demonstrating that antibodies to c-*fos* block DRA gene expression (Ono *et al.*, 1991). Expression of anti-sense hXBP decreases DR and DP cell surface expression but not DQ in a B cell line constitutively expressing class II molecules (Ono *et al.*, 1991). In addition, anti-sense hXBP-1 reduces IFN-γ-induced expression of DRA but not DQB. The X2-binding protein mXBP was cloned based on its ability to bind the murine Aα X2 box and has a leucine zipper motif (Liou, Boothby & Glimcher, 1988). Another X2-binding protein, NF-S, has been described by its strong affinity to the X2 site of DQA and weaker binding to DRA and DPA (Kobr *et al.*, 1990).

Studies of MHC class II promoter function in transgenic mice are of particular biological interest and provide novel insights into the roles of X and Y elements in the tissue-specific and cytokine-inducible expression of these genes in the whole organism (Lemeur *et al.*, 1985; Pinkert *et al.*, 1985; Yamamura *et al.*, 1985). These experiments were performed by introducing the Eα gene under its own promoter into mouse strains that do not express this gene (Lemeur *et al.*, 1985). A 2 kb fragment of upstream sequence was sufficient to confer faithful regulation of the gene. Mutation of the Y element led to little Eα expression in the thymic medulla and had no effect on expression in thymic cortex (Pinkert *et al.*, 1985; Widera *et al.*, 1987; van

Ewijk *et al.*, 1988). In contrast, mutation of the X element produced the opposite pattern of expression. Therefore, X and Y elements are not functional in all cells. It would be of interest to determine if differential usage of X and Y contributes to different levels of class II MHC antigen expression and eventually to the process of thymic education and restriction.

W/Z/S is a positive regulatory element located upstream of the X box (Boss & Strominger, 1986; Polla *et al.*, 1986; Servenius, Rask & Peterson, 1987; Thanos, Mavrothalassitis & Papamatheakis, 1988; Tsang, Nakanishi & Peterlin, 1988; Cogswell *et al.*, 1990; Dedrick & Jones, 1990). At present, no proteins have been cloned that bind to this region. Genomic *in vivo* footprinting has revealed no detectable DNA–protein interactions over this region. The molecular mechanism that explains the requirement for this region in IFN-γ-mediated and constitutive expression remains unknown (Kara & Glimcher, 1991; Wright & Ting, 1992). In the DPA gene, an S element has been located but it is adjacent to the X element.

In addition to common elements shared among MHC class II promoters, several gene-specific, or allele-specific differences are noteworthy. A functional TATA element has only been found in the DRA promoter but not in the Eα or Aα promoters (Dedrick & Jones, 1990; Viville *et al.*, 1991). In DRA, mutation of this element greatly diminished gene expression in a transient transfection assay (Matsushima, Itoh-Lindstrom & Ting, 1992) and produced aberrantly initiated transcripts by *in vitro* transcription analysis. This element plays an important role in high DR gene expression in activated human T cells as mutation of the DRA TATA element to another TATA sequence, which is apparently functional but distinct, abrogated the expression of the DRA promoter activity in human T cells (Matsushima *et al.*, 1992). One possible explanation is that specific TBP-associated molecules may be activated upon T cell activation. The existence of a functional versus non-functional TATA box in human DRA versus murine MHC class II promoters may be one reason for Ia expression on human T cells but not murine T cells.

Another gene-specific element is the octamer sequence found in the DRA promoters of all primates but not in their murine homologues. The presence of the octamer motif in DRA may account for its high expression in human cells. A number of proteins can bind to this element; however, it appears that the B cell-specific Oct-2 but not the ubiquitous Oct-1 regulates the DRA gene (Zeleznik-Le *et al.*, 1992). This is consistent with mutagenesis analysis, which showed that the octamer sequence is essential for DRA expression in $DR^+$ lymphoid cell types but not non-lymphoid $DR^+$ cell types (Sherman *et al.*, 1989a; Tsang *et al.*, 1990). Genomic *in vivo* footprint analysis confirmed the lymphoid-specific function of the octamer element where the octamer site is occupied in $DR^+$ lymphoid cell lines but not in a $DR^+$, IFN-γ-induced glioblastoma line (Wright & Ting, 1992).

A positive regulatory element (J) has been delineated in the DPA and DQB promoters and is involved in both basal and IFN-γ-inducible gene expression (Sugawara *et al.*, 1991). At present, it is not entirely clear if this is a gene-specific element. In the murine Eα gene, a distal B cell enhancer has been mapped between −1960 and −1180 in transgenic mice (Widera *et al.*, 1987; Dorn *et al.*, 1988; van Ewijk *et al.*, 1988). Within this region, an inverted X–Y motif has been found. In addition, a similar X–Y motif has been found in the distal upstream region of Aβ and Eβ. The distal sequences for the promoters of most other MHC class II genes are not available; therefore, it is difficult to conclude if these are gene-specific sequences.

Allele- and locus-specific variations have been observed in the promoter of DQB that produced alterations in promoter strength (Andersen *et al.*, 1991; Shewey *et al.*, 1992). The allele- and locus-specific polymorphism within the promoter regions of these genes is relevant in studying disease association with MHC class II phenotypes. In considering the genetic linkage between the HLA phenotype and diseases, one has to be aware that polymorphism in the regulatory motif may result in varying levels in the expression of specific MHC class II genes that could directly affect antigen-presentation efficiency.

In addition to the mere interaction of *cis*-acting elements and corresponding *trans*-acting factors, protein–protein interactions among the DNA-binding proteins are probably occurring over MHC class II promoters. We have shown that the trimeric S, X and Y have to be spaced in a very specific fashion for gene induction. In both constitutive and IFN-γ-induced expression of DRA, the spacing between the S and X elements cannot be altered, while spacing between the X and Y elements tolerates addition or deletion of whole helical turns, but not half-helical turns of the DNA (Vilen, Cogswell & Ting, 1991; Vilen, Penta & Ting, 1992). These data suggest that the proteins binding to X and Y may have to bind to the same phase of the DNA, or, alternatively, that they are part of a larger complex that requires stereospecificity for function or assembly. Recent *in vivo* footprinting studies have also revealed that mutation of the Y element abrogated *in vivo* binding over the X elements, although the reverse is not true (K. L. Wright and J. P.-Y. Ting, unpublished observations). This would suggest that binding to the Y element is an early event in protein–DNA interaction over this promoter. Interestingly, studies of *in vitro* transcription of the Eα and albumin genes showed that antibodies to NF-Y inhibited re-initiation of transcription (Mantovani *et al.*, 1992). In total, these studies suggest that NF-Y is distinct from many other upstream-binding proteins and serves an early function in initiation of transcription.

## 5.8 Conclusions

MHC class II gene regulation is of importance in understanding the role of aberrant MHC expression in autoimmunity and immunodeficiencies, in antigen presentation and thymic education and in disease association. The importance of understanding MHC class II gene regulation is likely to increase in the future because of the interest in using foreign MHC genes in cancer immunotherapy and the potential problems that MHC genes could create in gene therapy. Much is known of some of the cytokines that induce these genes, and many of the *cis*-acting elements and *trans*-acting proteins have been well defined. Future directions are likely to involve the functional analysis of individual proteins that interact with the MHC class II promoter, higher-order protein–protein interactions between proteins binding upstream activating sequences and their subsequent interaction with the basal transcription machinery and application of this knowledge in modulating the immune response in a clinical setting.

## References

Accolla, R.S. (1983). Human B cell variants immunoselected against a single Ia antigen subset have lost expression of several Ia antigen subsets. *Journal of Experimental Medicine*, 157, 1053–1058.

Accolla, R.S., Jotterand-Bellomo, M., Scarpellino, L., Maffei, A., Carra, G. & Guardiola, J. (1986). aIr-1, a newly found locus on mouse chromosome 16 encoding a *trans*-acting activator factor for MHC class II gene expression. *Journal of Experimental Medicine*, 164, 369–374.

Adelman, N.E., Watling, D.L. & McDevitt, H.O. (1983). Treatment of (NZB × NZW)F1 disease with anti-I-A monoclonal antibodies. *Journal of Experimental Medicine*, 158, 1350–1355.

Alvaro-Gracia, J.M., Zvaifler, N.J. & Firestein, G.S. (1989). Cytokines in chronic inflammatory arthritis. Granulocyte/macrophage colony stimulating factor-mediated induction of class II MHC antigen on human monocytes: a possible role in rheumatoid arthritis. *Journal of Experimental Medicine*, 170, 865–875.

Amaldi, I., Reith, W., Berte, C. & Mach, B. (1989). Induction of HLA class II genes by IFN-γ is transcriptional and requires a *trans*-acting protein. *Journal of Immunology*, 142, 999–1004.

Andersen, L.C., Beaty, J.S., Nettles, J.W., Seyfried, C.E., Nepom, G.T. & Nepom, B.S. (1991). Allelic polymorphisms in transcriptional regulatory regions of HLA-DQB genes. *Journal of Experimental Medicine*, 173, 181–192.

Andersson, G. & Peterlin, B.M. (1990). NF-X2 that binds to the DRA X2-Box is activation protein I: expression cloning of c-*jun*. *Journal of Immunology*, 145, 3456–3462.

Arenzana-Seisdedos, F., Mogenson, S.C., Vuillier, F., Fiers, W. & Virelizier, J.L. (1988). Autocrine secretion of tumor necrosis factor under the influence of interferon-γ amplifies HLA-DR gene induction in human monocytes. *Proceedings of the National Academy of Sciences of the USA*, 85, 6087–6091.

Basta, P.V., Moore, T.L., Yokota, S. & Ting, J.P.-Y. (1989). A β-adrenergic agonist modulates DRα gene transcription via enhanced cAMP levels in a glioblastoma multiform cell line. *Journal of Immunology*, 142, 2895–2901.

Basta, P.V., Sherman, P.A. & Ting, J.P.-Y. (1988). Detailed delineation of an interferon-γ-responsive element important in human HLA-DRA gene expression in a glioblastoma multiform line. *Proceedings of the National Academy of Sciences of the USA*, 85, 8618–8622.

Benichou, B. & Strominger, J.L. (1991). Class II antigen-negative patient and mutant B cell lines represents at least three, and possibly four, distinct genetic defects by complementation analysis. *Proceedings of the National Academy of Sciences of the USA*, 88, 4285–4288.

Benoist, C. & Mathis, D. (1990). Regulation of major histocompatibility complex class-II genes: X, Y, and other letters of the alphabet. *Annual Review of Immunology*, 8, 681–715.

Benveniste, E.N., Sparacio, S.M. & Bethea, J.R. (1989). Tumor necrosis factor-α enhances interferon-γ-mediated class II antigen expression on astrocytes. *Journal of Neuroimmunology*, 25, 209–219.

Benveniste, E.N., Vidovic, M., Panek, R.B., Norris, J.G., Reddy, A.T. & Benos, D.J. (1991). Interferon-γ induced astrocyte class II major histocompatibility complex gene expression is associated with both protein kinase C activation and $Na^+$ entry. *Journal of Biological Chemistry*, 266, 18119–18126.

Blanar, M.A., Baldwin, A.S., Flavell, R.A. & Sharp, P.A. (1989). A γ-interferon-induced factor that binds the interferon response sequence of the MHC class I gene, H-$2K^b$. *EMBO Journal*, 8, 1139–1144.

Blanar, M.A., Boettger, E.C. & Flavell, R.A. (1988). Transcriptional activation of HLA-DRA by interferon-γ requires a *trans*-acting protein. *Proceedings of the National Academy of Sciences of the USA*, 85, 4672–4676.

Boitard, C., Michie, S., Serrurier, P., Butcher, G.W., Larkins, A.P. & McDevitt, H.O. (1985). *In vivo* prevention of thyroid and pancreatic autoimmunity in the BB rat by antibody to class II major histocompatibility complex gene products. *Proceedings of the National Academy of Sciences of the USA*, 82, 6627–6631.

Boothby, M., Gravallese, E., Liou, H.-C. & Glimcher, L.H. (1988). A DNA binding protein regulated by IL-4 and by differentiation in B cells. *Science*, 242, 1559–1562.

Boss, J.M. & Strominger, J.L. (1986). Regulation of a transfected human class II MHC gene in human fibroblasts. *Proceedings of the National Academy of Sciences of the USA*, 83, 9139–9143.

Bottazzo, G.F., Dean, B.M., McNally, J.M., Mackey, E.H., Swift, P.G. & Ganek, D.R. (1985). *In situ* characterization of autoimmune phenomena and expression of HLA molecules in the pancreas of diabetes insulitis. *New England Journal of Medicine*, 313, 353–360.

Bottazzo, G.F., Pujol-Borrell, R., Hanafusa, T. & Feldman, M. (1983). Role of aberrant HLA-DR expression and antigen presentation in induction of endocrine autoimmunity. *Lancet*, ii, 1115–1119.

Bottger, E.C., Blanar, M.A & Flavell, R.A. (1988). Cycloheximide, an inhibitor of protein synthesis, prevents γ-interferon-induced expression of class II mRNA in a macrophage cell line. *Immunogenetics*, 28, 215–220.

Brown, A.M., Barr, C.L. & Ting, J.P.-Y. (1990). Sequences homologous to class II MHC W, X, and Y elements mediate constitutive and interferon-γ induced expression of human class II-associated invariant chain gene. *Journal of Immunology*, 146, 3183–3189.

Calman, A.F. & Peterlin, B.M. (1987). Mutant human B cell lines deficient in class II major histocompatibility complex transcription. *Journal of Immunology*, 139, 2489–2495.

Campbell, I.L., Oxbrow, L., Koulmanda, M. & Harrison, L. (1988). IFN-γ induces islet cell MHC antigens and enhances autoimmune streptozotocin-induced diabetes in the mouse. *Journal of Immunology*, 140, 1111–1116.

Celada, A. & Maki, R.A. (1989). DNA binding of the mouse class II major histocompatibility CCAAT factor depends on two components. *Molecular and Cellular Biology*, 9, 3097–3100.

Chang, R.J. & Lee, S.H. (1986). Effects of interferon-γ and tumor necrosis factor-α on the expression of an Ia antigen on a murine macrophage cell line. *Journal of Immunology*, 137, 2853–2856.

Chodosh, L.A., Baldwin, A.S., Carthew, R.W. & Sharp, P.A. (1988). Human CCAAT-binding proteins have heterologous subunits. *Cell*, 53, 11–24.

Cogswell, J.P., Basta, P.V. & Ting, J.P.-Y. (1990). X-box binding proteins positively and negatively regulate transcription of the HLA-DRA gene through interaction with discrete upstream W and V elements. *Proceedings of the National Academy of Sciences of the USA*, 87, 7703–7707.

Cogswell, J.P., Zeleznik-Le, N. & Ting, J.P.-Y. (1991). Transcriptional regulation of the HLA-DRA gene. *Critical Reviews in Immunology*, 11, 87–112.

Collins, T., Korman, A.J., Wake, C.T., Boss, J.M., Kappes, D.J., Fiers, W., Ault, K.A., Gimbrone, M.A., Strominger, J.L. & Pober, J.S. (1984). Immune interferon activates multiple class II major histocompatibility complex genes and the associated invariant chain gene in human endothelial cells and dermal fibroblasts. *Proceedings of the National Academy of Sciences of the USA*, 81, 4917–4921.

Conaway, J.W., Hanley, J.P., Garrett, K.P. & Conaway, R.C. (1991). Transcription initiated by RNA polymerase II and transcription factors from liver. Structure and action of transcription factors epsilon and tau. *Journal of Biological Chemistry*, 266, 7804–7811.

Dedrick, R.L. & Jones, P.P (1990). Sequence elements required for activity of a murine MHC class II promoter bind common and cell-type specific nuclear factors. *Molecular and Cellular Biology*, 10, 593–604.

Dellabona, P., Latron, F., Maffei, A., Scarpellino, L. & Accolla, R.S. (1989). Transcriptional control of MHC class II gene expression during differentiation from B cells to plasma cells. *Journal of Immunology*, 142, 2902–2910.

Dennis, G.J. & Mond, J.J. (1986). Corticosteroid-induced suppression of murine B cell immune response antigens. *Journal of Immunology*, 136, 1600–1604.

de Preval, C., Lisowska-Grospierre, B., Loche, M., Griscelli, C. & Mach, B. (1985). A *trans*-acting class II regulatory gene unlinked to the MHC controls expression of HLA class II genes. *Nature (London)*, 318, 291–293.

de Preval, C., Hadam, M.R. & Mach, B. (1988). Regulation of genes for HLA class II antigens in cell lines from patients with severe combined immunodeficiency. *New England Journal of Medicine*, 318, 1295–1300.

Didier, D.K., Schiffenbauer, J., Woulfe, S.L., Zacheis, M. & Schwartz, B.D. (1988). Characterization of the cDNA encoding a protein binding to the major histocompatibility complex class II Y box. *Proceedings of the National Academy of Sciences of the USA*, 85, 7322–7326.

Dorn, A., Bollekens, J., Staub, A., Benoist, C. & Mathis, D. (1987a). A multiplicity of CCAAT box-binding proteins. *Cell*, 50, 863–872.

Dorn, A., Durand, B., Marfing, C., Le Meur, M., Benoist, C. & Mathis, D. (1987b). Conserved major histocompatibility complex class II boxes-X and -Y are transcriptional control elements and specifically bind nuclear proteins. *Proceedings of the National Academy of Sciences of the USA*, 84, 6249–6253.

Dorn, A., Fehling, H.J., Koch, W., Le Meur, M, Gerlinger, P., Benoist, C. & Mathis, D. (1988). B cell control region at the 5′ end of a major histocompatibility complex class II gene: sequence and factors. *Molecular and Cellular Biology*, 8, 3975–3987.

Doyle, C., Ford, P.J., Ponath, P.D., Spies, T. & Strominger, J.L. (1990). Regulation of the class II-associated invariant chain gene in normal and mutant B lymphocytes. *Proceedings of the National Academy of Sciences of the USA*, 87, 4590–4594.

Dynan, W.S. (1989). Modularity in promoters and enhancers. *Cell*, 58, 1–4.

Dynlacht, B.D., Hoey, T. & Tjian, R. (1991). Isolation of coactivators associated with the TATA-binding protein that mediate transcriptional activation. *Cell*, 66, 563–576.

Ephrussi, A., Church, G.M., Tonegawa, S. & Gilbert, W. (1985). B lineage-specific interactions of an immunoglobulin enhancer with cellular factors *in vivo*. *Science*, 227, 134–140.

Fan, X.D., Goldberg, M. & Bloom, B.R. (1988). Interferon-γ-induced transcriptional activation is mediated by protein kinase C. *Proceedings of the National Academy of Sciences of the USA*, 85, 5122–5125.

Fertsch, D., Schoenberg, D.R., Germain, R.N., Tou, J.Y. & Vogel, S.N. (1987). Induction of macrophage Ia antigen expression by IFN-γ and down-regulation by IFN-γ/β and dexamethasone are mediated by changes in steady state levels of Ia mRNA. *Journal of Immunology*, 139, 244–249.

Fertsch-Ruggio, D., Schoenberg, D.R. & Vogel, S.N. (1988). Induction of macrophage Ia antigen expression by rIFN-γ and down-regulation by IFN α/β and dexamethasone are regulated transcriptionally. *Journal of Immunology*, 141, 1582–1589.

Fierz, A., Endler, B., Reske, K., Wekerle, H. & Fontana, A. (1985). Astrocytes as antigen presenting cells. Induction of Ia antigen expression on astrocytes by T cells via immune interferon and its effects on antigen presentation. *Journal of Immunology*, 134, 3785–3793.

Finn, P.W., Kara, C.J., Douhan III, J. I., Van, T.T., Folsom, V. & Glimcher, L.H. (1990a). Interferon-γ regulates binding of two nuclear protein complexes in a macrophage cell line. *Proceedings of the National Academy of Sciences of the USA*, 87, 914–918.

Finn, P.W., Kara, C.J., Grusby, M.J., Folsom, V. & Glimcher, L.H. (1990b). Upstream elements of the MHC class II Eβ gene active in B cells. *Journal of Immunology*, 146, 4011–4015.

Finn, P.W., Kara, C.J., Van, T.T., Douhan III, J., Boothby, M.R. & Glimcher, L.H. (1990c). The presence of a DNA binding complex correlates with $E^{\beta}$ class II MHC gene expression. *EMBO Journal*, 9, 1543–1549.

Fontana, A., Fierz, W. & Wekerle, H. (1984). Astrocytes present myelin basic protein to encephalitogenic T cell lines. *Nature (London)*, 307, 273–276.

Fontana, A., Frei, K., Bodmer, S. & Hofer, E. (1987). Immune-mediated

encephalitis: on the role of antigen-presenting cells in brain tissue. *Immunological Reviews*, 100, 185–201.
Freedman, L.P., Luisi, B.F., Korszun, Z.R., Basavappa, R., Sigler, P.B. & Yamamoto, K.R. (1988). The function and structure of the metal coordination sites within the glucocorticoid receptor DNA binding domain. *Nature (London)*, 334, 543–546.
Frohman, E.M., Vayuvegula, B., Gupta, S. & van den Noort, S. (1988). Norepinephrine inhibits γ-interferon-induced major histocompatibility class II (Ia) antigen expression on cultured astrocytes via $\beta^2$-adrenergic signal transduction mechanisms. *Proceedings of the National Academy of Sciences of the USA*, 85, 1292–1296.
Gajewski, T.F., Schell, S.R., Nau, G. & Fitch, F.W. (1989). Regulation of T cell activation: differences among T cell subsets. *Immunological Reviews*, 111, 79–110.
Gaulton, G.N., Stein, M.E., Safko, B. & Stadecker, M.J. (1989). Direct induction of Ia antigen on murine thyroid-derived epithelial cells by reovirus. *Journal of Immunology*, 142, 3821–3825.
Germain, R.N. (1986). Immunology. The ins and outs of antigen processing and presentation. *Nature (London)*, 322, 687–689.
Gladstone, P. & Pious, D. (1978). Stable variants affecting B cell alloantigens in human lymphoid cells. *Nature (London)*, 271, 459–461.
Glimcher, L.H. & Kara, C.J. (1992). Sequences and factors: a guide to MHC class II transcription. *Annual Review of Immunology*, 10, 13–49.
Go, N.F., Castle, B.E., Barrett, R., Kastelein, R., Dang, W., Mosmann, T.R., Moore, K.W. & Howard, M. (1990). Interleukin 10 (IL-10), a novel B cell stimulatory factor: unresponsiveness of 'XID' B cells. *Journal of Experimental Medicine*, 172, 1625–1631.
Greenblatt, J. (1991). RNA polymerase-associated transcription factors. *Trends in Biochemical Sciences*, 16, 408–417.
Greene, W.C. & Leonard, W.J. (1986). The human interleukin-2 receptor. *Annual Review of Immunology*, 4, 69–95.
Griscelli, C., Lisowska-Grospierre, B. & Mach, B. (1989). Combined immunodeficiency with defective expression in MHC class II genes. *Immunodeficiency Reviews*, 1, 135–153.
Guardiola, J., Scarpellino, L., Carra, G. & Accolla, R.S. (1986). Stable integration of mouse DNA into Ia-negative human B-lymphoma cells causes re-expression of the human Ia positive phenotype. *Proceedings of the National Academy of Sciences of the USA*, 83, 7415–7418.
Halper, J., Fu, S.M., Wang, C.-Y., Winchester, R. & Kunkel, H.G. (1978). Patterns of expression of 'Ia like' antigens during the terminal stages of B cell development. *Journal of Immunology*, 120, 1480–1484.
Hanafusa, T., Pujol-Borrell, R., Chiovato, L., Russell, R.C.G., Doniach, D. & Bottazzo, G.F. (1983). Aberrant expression of HLA-DR antigen anthyrocytes in Graves' disease: relevance for autoimmunity. *Lancet*, ii, 1111–1119.
Harrison, S.C. (1991). A structural taxonomy of DNA-binding domains. *Nature (London)*, 353, 715–719.
Harrison, S.C. & Aggarwal, A.K. (1990). DNA recognition by proteins with the helix-turn-helix motif. *Annual Review of Biochemistry*, 59, 933–969.
Hasegawa, S.L., James, L.R., Sloan III, J.H. & Boss, J.M. (1993). Protease treatment of nuclear extracts distinguishes between class II MHC X1 box DNA-binding proteins in wild type and class II deficient B cells. *Journal of Immunology*, 150, 1781–1793.

Hatamochi, A., Golumbek, P.T., Van Schaftingen, E. & de Crombrugghe, B. (1988). A CCAAT DNA binding factor consisting of two different components that are both required for DNA binding. *Journal of Biological Chemistry*, 263, 5940–5947.

Herrero Sanchez, C., Reith, W., Silacci, P. & Mach, B. (1992). The DNA binding defect observed in major histocompatibility complex class II regulatory mutants concerns only one member of a family of complexes binding to the X boxes of class II promoters. *Molecular and Cellular Biology*, 12, 4076–4085.

Hume, C.R. & Lee, J.S. (1989). Congenital immunodeficiencies associated with absence of HLA class II antigens on lymphocytes results from distinct mutations in *trans*-acting factors. *Human Immunology*, 26, 288–309.

Hume, C.R. & Lee, J.S. (1990). Functional analysis of *cis*-linked regulatory sequences in the HLA-DRA promoter by transcription *in vitro*. *Tissue Antigens*, 36, 108–115.

Hume, C.R., Shookster, L.A., Collins, N., O'Reilly, R. & Lee, J.S. (1989). Bare lymphocyte syndrome: altered class II expression in two B cell lines. *Human Immunology*, 25, 1–11.

Inaba, K., Kitaura, M., Kato, T., Watanabe, Y., Kawade, Y. & Muramatsu, S. (1986). Contrasting effects of α/β- and γ-interferons on expression of macrophage Ia antigens. *Journal of Experimental Medicine*, 163, 1030–1035.

Israel, A., Kimura, A., Fournier, A., Fellows, M. & Kourilsky, P. (1986). Interferon response sequence potentiates activity of an enhancer in the promoter region of a mouse H-2 gene. *Nature* (*London*), 322, 743–746

Kara, C.J. & Glimcher, L.H. (1991). *In vivo* footprinting of MHC class II genes: bare promoter in the bare lymphocyte syndrome. *Science*, 252, 709–712.

Kelley, V.E. & Roths, J.B. (1982). Increase in macrophage Ia expression in autoimmune mice: role of the *lpr* gene. *Journal of Immunology*, 129, 923–928.

Kern, M.J., Stuart, P.M., Omer, K.W. & Woodward, J.G. (1989). Evidence that IFN-γ does not affect MHC class II gene expression at the post-transcriptional level in a mouse macrophage cell line. *Immunogenetics*, 30, 258–265.

Kincade, P.W., Lee, G., Watanabe, T., Sun, L. & Scheid, M.P. (1981). Antigens displayed on murine B-lymphocyte precursors. *Journal of Immunology*, 127, 2262–2266.

Kobr, M., Reith, W., Herrero Sanchez, C. & Mach, B. (1990). Two DNA binding proteins discriminate between the promoters of different members of the major histocompatibility complex class II multigene family. *Molecular and Cellular Biology*, 10, 965–971.

Koch, W., Candeias, S., Guardiola, J., Accolla, R., Benoist, C. & Mathis, D. (1988). An enhancer factor defect in a mutant Burkitt lymphoma cell line. *Journal of Experimental Medicine*, 167, 1781–1790.

Koerner, T.J., Hamilton, T.A. & Adams, D.O. (1987). Suppressed expression of surface Ia by lipopolysaccharide: evidence for regulation at the level of accumulation of mRNA. *Journal of Immunology*, 139, 239–243.

Koide, Y., Ina, Y., Nezu, N. & Yoshida, T.O. (1988). Calcium influx and the $Ca^{2+}$-calmodulin complex are involved in interferon-γ-induced expression of HLA class II molecules on HL-60 cells. *Proceedings of the National Academy of Sciences of the USA*, 85, 3120–3124.

Kumar, V., Green, S., Stack, G., Berry, M., Jin, J.R. & Chambon, P. (1987). Functional domains of the human estrogen receptor. *Cell*, 51, 941–951.

Lala, P.K., Johnson, G.T., Battye, F.C. & Nossal, G.T. (1979). Maturation of B-lymphocytes. Concurrent appearance of increasing Ig, Ia and mitogen responsiveness. *Journal of Immunology*, 122, 334–340.

Latron, F., Jotterand-Bellomo, M., Maffei, A., Scarpellino, L., Bernard, M., Strominger, J.L. & Accolla, R.S. (1988). Active suppression of major histocompatability complex class II gene expression during differentiation from B cells to plasma cells. *Proceedings of the National Academy of Sciences of the USA*, 85, 2229–2233.

Laughton, A. & Scott, M.P. (1984). Sequence of a *Drosophila* segmentation gene: protein structure homology with DNA binding proteins. *Nature (London)*, 310, 25–31.

Lee, S.C., Collins, M., Vanguri, B. & Shin, M. (1992). Glutamate differentially inhibits the expression of class II MHC antigens in astrocytes and microglia. *Journal of Immunology*, 148, 3391–3397.

Lemeur, M., Gerlinger, P., Benoist, C. & Mathis, D. (1985). Correcting an immune-response deficiency by creating Eα gene transgenic mice. *Nature (London)*, 316, 38–42.

Lin, Y.-S. & Green, M.R. (1991). Mechanism of action of an acidic transcriptional activator *in vitro*. *Cell*, 64, 971–981.

Ling, P.D., Warren, M.K. & Vogel, S.N. (1985). Antagonistic effect of interferon-β on the interferon-γ-induced expression of Ia antigen in murine macrophages. *Journal of Immunology*, 135, 1857–1863.

Liou, H.-C., Boothby, M.R., Finn, P.W., Davidon, R., Nabavi, N., Zeleznik-Le, N., Ting, J.P.-Y. & Glimcher, L.H. (1990). A new member of the leucine zipper class of proteins that binds to the HLA-DRA promoter. *Science*, 247, 1581–1584.

Liou, H.-C., Boothby, M.R. & Glimcher, L.H. (1988). Distinct cloned MHC class II MHC DNA binding proteins recognize the X box transcription element. *Science*, 242, 69–71.

Lo, D., Burkly, L.C., Widera, G., Cowing, C., Flavell, R.A., Palmiter, R.D. & Brinster, R.L. (1988). Diabetes and tolerance in transgenic mice expressing class II MHC molecules in pancreatic beta cells. *Cell*, 53, 159–168.

Loh, J.E., Chang, C.-H., Fodor, W.L. & Flavell, R.A. (1992). Dissection of the interferon-γ MHC class II signal transduction pathway reveals that type I and type II interferon systems share common signalling component(s). *EMBO Journal*, 11, 1351–1363.

Londei, M., Lamb, J.R., Bottazzo, G.F. & Feldman, M. (1984). Epithelial cells expressing aberrant MHC class II determinants can present antigen to cloned human T cells. *Nature (London)*, 312, 639–641.

Long, E.O., Mach, B. & Accolla, R.S. (1984). Ia-negative B cell variants reveal a coordinate regulation in transcription of the HLA class II gene family. *Immunogenetics*, 19, 349–353.

Lu, C.Y., Changelian, P.S. & Unanue, E.R. (1984). α-Fetoprotein inhibits macrophage expression of Ia antigens. *Journal of Immunology*, 132, 1722–1727.

Ma, J. & Ptashne, M. (1987). A new class of yeast transcriptional activators. *Cell*, 51, 113–119.

Maffei, A., Perfetto, C., Ombra, N., Del Pozzo, G. & Guardiola, J. (1989). Transcriptional and post-transcriptional regulation of human class II genes require the synthesis of short-lived proteins. *Journal of Immunology*, 142, 3657–3661.

Maity, S.N., Sinha, S., Ruteshouser, E.C. & de Crombrugghe, B. (1992). Three different polypeptides are necessary for DNA binding of mammalian heteromeric CCAAT binding factor. *Journal of Biological Chemistry*, 267, 16574–16580.

Maity, S.N., Vuorio, T. & de Crombrugghe, B. (1990). The B subunit of a rat heteromeric CCAAT-binding transcription factor shows a striking sequence identity with the yeast HAP2 transcription factor. *Proceedings of the National Academy of Sciences of the USA*, 87, 5378–5382.

Maniatis, T., Goodbourn, S. & Fischer, J.A. (1987). Regulation of inducible and tissue-specific gene expression. *Science*, 236, 1237–1245.

Mantovani, R., Pessara, U., Tronche, F., Li, X.Y., Knapp, A.M., Pasquali, J.L., Benoist, C. & Mathis, D. (1992). Monoclonal antibodies to NF-Y define its function in MHC class II and albumin gene transcription. *EMBO Journal*, 11, 3315–3322.

Manyak, C.L., Tse, H., Fischer, P., Coker, L., Sigal, N.H. & Koo, G.C. (1988). Regulation of class II MHC molecules on human endothelial cells. Effects of IFN and dexamethasone. *Journal of Immunology*, 140, 3817–3821.

Mao, C., Davies, D., Kerr, I.M. & Stark, G.R. (1993). Mutant human cells defective in induction of major histocompatibility complex class II genes by interferon-γ. *Proceedings of the National Academy of Sciences of the USA*, 90, 2880–2884.

Marcadet, A., Cohen, D., Dausset, J., Fisher, A., Durandy, A. & Griscelli, C. (1985). Genotyping with DNA probes in combined immunodeficiency syndrome with defective expression of HLA. *New England Journal of Medicine*, 312, 1287–1292.

Massa, P.T., Dorries, R. & ter Meulen, V. (1986). Viral particles induce Ia antigen expression on astrocytes. *Nature (London)*, 320, 543–546.

Matsumoto, Y., Hara, N., Tanaka, R. & Fujiwara, M. (1986). Immuno-histochemical analysis of the rat central nervous system during experimental allergic encephalomyelitis, with special reference to Ia-positive cells with dendritic morphology. *Journal of Immunology*, 136, 3668–3676.

Matsushima, G.K., Itoh-Lindstrom, Y. & Ting, J.P.-Y. (1992). Activation of HLA-DRA gene in primary human blood T-lymphocytes: novel usage of TATA and the X/Y promoter elements. *Molecular and Cellular Biology*, 12, 5610–5619.

Mattila, P., Hayry, P. & Renkonen, R. (1989). Protein kinase C is crucial in signal transduction during IFN-γ induction in endothelial cells. *FEBS Letters*, 250, 362–366.

McKearn, J.P. & Rosenburg, N. (1985). Mapping cell surface antigens on mouse pre-B cell lines. *European Journal of Immunology*, 15, 295–298.

McMillan, V.M., Dennis, G.J., Glimcher, L.H., Finkelman, F.D. & Mond, J.J. (1988). Corticosteroid induction of $Ig^+Ia^-$ B cells *in vitro* is mediated via interaction with the glucocorticoid cytoplasmic receptor. *Journal of Immunology*, 140, 2549–2555.

Miller, J., McLachlan, A.D. & Klug, A. (1985). Repetitive zinc-binding domains in the protein transcription factor IIIA from *Xenopus* oocytes. *EMBO Journal*, 4, 1609–1614.

Mirkovitch, J. & Darnell, J.E. Jr (1991). Rapid *in vivo* footprinting technique identifies protein bound to the TTR gene in the mouse liver. *Genes and Development*, 5, 83–93.

Mond, J.J., Kessler, S., Finkelman, F.D., Paul, W.E. & Scher, I. (1980). Heterogeneity of Ia expression on normal B cells, neonatal B cells and on cells from B cell-defective CBA/N mice. *Journal of Immunology*, 124, 1675–1682.

Mond, J.J., Seghal, E., Kung, J. & Finkelman, F.D. (1981). Increased expression of I-region-associated antigen(Ia) on B cells after cross-linking of surface immunoglobulin. *Journal of Immunology*, 127, 881–885.

Monroe, J.G. & Cambier, J.C. (1983). B cell activation. B cell plasma membrane depolarization and hyper-I-A antigen expression induced by receptor immunoglobulin crosslinking are coupled. *Journal of Experimental Medicine*, 158, 1589–1599.

Moses, H., Panek, R.B., Beveniste, E.N. & Ting, J.P.-Y. (1992). Usage of primary cells to delineate IFN-γ-responsive DNA elements in the HLA-DRA promoter and to identify a novel IFN-γ-enhanced nuclear factor. *Journal of Immunology*, 148, 3643–3651.

Mueller, P.R. & Wold, B. (1989). *In vivo* footprinting of a muscle specific enhancer by ligation mediated PCR. *Science*, 246, 780–786.

Nepom, G.T. & Erlich, H. (1991). MHC class II molecules and autoimmunity. *Annual Review of Immunology*, 9, 493–525.

Nezu, N., Ryu, K., Koide, Y. & Yoshida, T.O. (1990). Regulation of HLA class II molecule expression by IFN-γ. The signal transduction mechanism in glioblastoma cell lines. *Journal of Immunology*, 145, 3126–3135.

Noelle, R., Krammer, P.H., Ohara, J., Uhr, J.W. & Vitteta, E.S. (1985). Increased expression of Ia antigens on resting B cells: an additional role for B cell growth factor. *Proceedings of the National Academy of Sciences of the USA*, 81, 6149–6153.

Noelle, R.J., Kuziel, W.A., Maliszewski, C.R., McAdams, E., Vitteta, E.S. & Tucker, P.W. (1986). Regulation of the expression of multiple class II genes in murine B cells by B cell stimulatory factor-1 (BSF-1). *Journal of Immunology*, 137, 1718–1723.

Ono, S.J., Liou, H.-C., Davidon, R., Strominger, J.L. & Glimcher, L.H. (1991). Human X-box-binding protein 1 is required for the transcription of a subset of human class II major histocompatibility genes and forms a heterodimer with c-*fos*. *Proceedings of the National Academy of Sciences of the USA*, 88, 4309–4312.

Pan, T. & Coleman, J.E. (1990). GAL4 transcription factor is not a 'zinc finger' but forms a $Zn(II)_2Cys_6$ binuclear cluster. *Proceedings of the National Academy of Sciences of the USA*, 87, 2077–2081.

Pearse, R.N., Feinman, R. & Ravetch, J.V. (1991). Characterization of the promoter of the human gene encoding the high-affinity IgG receptor: transcriptional induction by γ-interferon is mediated through common DNA response elements. *Proceedings of the National Academy of Sciences of the USA*, 88, 11305–11309.

Pfizenmaier, K., Scheurich, P., Schluter, C. & Kronke, M. (1987). Tumor necrosis factor enhances HLA-A,B,C, and HLA-DR gene expression in human tumor cells. *Journal of Immunology*, 138, 975–980.

Pinkert, C.A., Widera, G., Cowing, C., Heber-Katz, E., Palmiter, R.D., Flavell, R.A. & Brinster, R.L. (1985). Tissue-specific, inducible and functional expression of the $E\alpha^d$ MHC class II gene in transgenic mice. *EMBO Journal*, 4, 2225–2230.

Plaeger-Marshall, S., Haas, A., Clement, L.T., Giorgi, J.V., Chen, I.S.Y., Quan, S.G., Gatti, R.A, & Stiehm, E.R. (1988). Interferon-induced expression of class II major histocompatibility antigens in the major histocompatibility complex class II deficiency syndrome. *Journal of Clinical Immunology*, 8, 285–295.

Polla, B.S., Poljak, A., Ohara, J., Paul, W.E. & Glimcher, L.H. (1986). Regulation of class II gene expression: analysis in B cell stimulatory factor 1-inducible murine pre-B cell lines. *Journal of Immunology*, 137, 3332.

Prpic, V., Yu, S.F., Figueiredo, F., Hollenbach, P.W., Gawdi, G., Herman, B., Uhing, R.J. & Adams, D.O. (1989). Role of $Na^+/H^+$ exchange by interferon-γ in enhanced expression of JE and $I\text{-}A^\beta$ genes. *Science*, 244, 469–471.

Ptashne, M. (1988). How eukaryotic transcriptional activators work. *Nature (London)*, 335, 683–689.

Pugh, B.F. & Tjian, R. (1990). Mechanism of transcriptional activation by Spl: evidence for co-activators. *Cell*, 61, 1187–1197.

Pujol-Borrell, R., Todd, I., Doshi, M., Bottazzo, G.F., Sutton, R., Gray, D., Adolf, G.R. & Feldman, M. (1987). HLA class II induction in human islet cells by interferon-γ plus tumor necrosis factor or lymphotoxin. *Nature (London)*, 326, 304–306.

Reed, A.M., Pachman, L. & Ober, C. (1991). Molecular genetic studies of major histocompatibility complex genes in children with juvenile dermatomyositis: increased risk associated with HLA-DQA1*0501. *Human Immunology*, 32, 235–240.

Reith, W., Barras, E., Satola, S., Kobr, M., Reinhart, D., Sanchez, C.H. & Mach, B. (1989). Cloning of the major histocompatibility complex class II promoter binding protein affected in a hereditary defect in class II gene regulation. *Proceedings of the National Academy of Sciences of the USA*, 86, 4200–4204.

Reith, W., Satola, S., Sanchez, C.H., Amaldi, I., Lisowska-Grospierre, B., Griscelli, C., Hadam, M.R. & Mach, B. (1988). Congenital immunodeficiency with a regulatory defect in MHC Class II gene expression lacks a specific HLA-DR promoter binding protein, RF-X. *Cell*, 53, 897–906.

Rijkers, G.T., Roord, J.J., Koning, F., Kuis, W. & Zegers, B.J.M. (1987). Phenotypical and functional analysis of B-lymphocytes of two siblings with combined immunodeficiency and defective expression of major histocompatibility antigens on mononuclear cells. *Journal of Clinical Immunology*, 7, 98–106.

Roeder, R.G. (1991). The complexities of eukaryotic transcription initiation: regulation of pre-initiation complex assembly. *Trends in Biochemical Sciences*, 16, 402–408.

Roehm, N., Leibson, J.L., Zlotnik, A., Kappler, J., Marrack, P. & Cambier, J.C. (1984). Interleukin induced increase in Ia expression by normal mouse B cells. *Journal of Experimental Medicine*, 160, 679–694.

Rosa, F.M. & Fellous, M. (1988). Regulation of HLA-DR gene by IFN-γ. *Journal of Immunology*, 140, 1660–1664.

Rosa, F.H., Hatat, D., Abadie, A., Wallach, D., Revel, M. & Fellous, M. (1983). Differential regulation of HLA-DR mRNAs and cell surface antigens by interferon. *EMBO Journal*, 2, 1585–1589.

Rotter, J.I. & Landlaw, E.M. (1984). Measuring the genetic contribution of a single locus to a multilocus disease. *Clinical Genetics*, 26, 529–542.

Rousset, F., de Waal-Malefijt, R.W., Slierendregt, B., Aubry, J.-P., Bonnefoy, J.-Y., Defrance, T., Banchereau, J. & de Vries, J.E. (1988). Regulation of Fc receptor of IgE (CD23) and class II MHC antigen expression on Burkitt's lymphoma cell lines by human IL-4 and IFN-γ. *Journal of Immunology*, 140, 2625–2632.

Roy, A.L., Meisterernst, M., Pognonec, P. & Roeder, R.G. (1991). Cooperative interaction of an initiator-binding transcription initiation factor and the helix-loop-helix activator USF. *Nature (London)*, 354, 245–248.

Salmon, M., Kitas, G.D. & Bacon, P.A. (1989). Production of lymphokine mRNA by $CD45R^+$ and $CD45R^-$ helper T cells from human peripheral blood and by human $CD4^+$ T cell clones. *Journal of Immunology*, 143, 907–912.

Sarvetnick, N., Liggitt, D., Pitts, S.L., Hansen, S.E. & Stewart, T.A. (1988). Insulin-dependent diabetes mellitus induced in transgenic mice by ectopic expression of class II MHC and interferon-gamma. *Cell*, 52, 773–782.

Sasaki, A., Levison, S. & Ting, J.P.-Y. (1990). Differential suppression of

interferon-γ-induced Ia antigen expression in cultured rat astroglia and microglia by second messengers. *Journal of Neuroimmunology*, 29, 213–222.
Sawadogo, M. & Sentenac, A. (1990). RNA polymerase B (II) and general transcription factors. *Annual Review of Biochemistry*, 59, 711–754.
Servenius, B., Rask, L. & Peterson, P.A. (1987). Class II genes of the human major histocompatibility complex. *Journal of Biological Chemistry*, 262, 8759–8766.
Sherman, P.A., Basta, P.V., Heguy, A., Wloch, M.K., Roeder, R.G. & Ting, J.P.-Y. (1989a). The octamer motif is a B-lymphocyte-specific regulatory element of the HLA-DRA gene promoter. *Proceedings of the National Academy of Sciences of the USA*, 86, 6739–6743.
Sherman, P.A., Basta, P.V., Moore, T.L., Brown, A.M. & Ting, J.P.-Y. (1989b). Class II box consensus sequences in the HLA-DRA gene: transcriptional function and interaction with nuclear proteins. *Molecular and Cellular Biology*, 9, 50–56.
Sherman, P.A., Basta, P.V. & Ting, J.P.-Y. (1987). Upstream DNA sequences required for tissue-specific expression of the HLA-DRA gene. *Proceedings of the National Academy of Sciences of the USA*, 84, 4254–4258.
Shewey, L.M., Beaty, J.S., Andersen, L.C. & Nepom, G.T. (1992). Differential expression of related HLA class II DQB genes caused by nucleotide variation in transcriptional regulatory elements. *Journal of Immunology*, 148, 1265–1273.
Sloan, J.H., Hasegawa, S.L. & Boss, J.M. (1992). Single base pair substitutions within the HLA-DRA gene promoter separate the functions of the X1 and X2 boxes. *Journal of Immunology*, 148, 2591–2599.
Smale, S.T., Schmidt, M.C., Berk, A.J. & Baltimore, D. (1990). Transcriptional activation by Sp1 as directed through TATA or initiator: specific requirement for mammalian transcription factor IID. *Proceedings of the National Academy of Sciences of the USA*, 87, 4509–4513.
Smith, K.A. (1984). Interleukin 2. *Annual Review of Immunology*, 2, 319–333.
Smith, M.D. & Roberts-Thomson, P.J. (1990). Lymphocyte surface marker expression in rheumatic diseases: evidence for prior activation of lymphocytes *in vivo*. *Annals of Rheumatic Diseases*, 49, 81–87.
Snyder, D.S., Beller, D.I. & Unanue, E.R. (1982). Prostaglandins modulate macrophage Ia expression. *Nature (London)*, 299, 163–165.
Snyder, D.S. & Unanue, E.R. (1982). Corticosteroids inhibit murine macrophage Ia expression and interleukin 1 production. *Journal of Immunology*, 129, 1803–1805.
Stimac, E., Urieli-Shoval, S., Kempin, S. & Pious, D. (1991). Defective HLA-DRA X box binding in the class II transactive transcription factor mutant 6.1.6 and in cell lines from class II immunodeficient patient. *Journal of Immunology*, 146, 4398–4405.
Stuart, P.M., Cassell, G.H. & Woodward, J.G. (1989). Induction of class II MHC antigen expression in macrophages by mycoplasma species. *Journal of Immunology*, 142, 3392–3399.
Sugawara, M., Ponath, P.D., Shin, J., Yang, Z. & Strominger, J.L. (1991). Delineation of a previously unrecognized *cis*-acting element required for HLA class II gene expression. *Proceedings of the National Academy of Sciences of the USA*, 88, 10347–10351.
Sugita, K., Miyazaki, J., Appella, E. & Ozato, K. (1987). Interferons increase transcription of a major histocompatibility class I gene via a 5′ interferon consensus sequence. *Molecular and Cellular Biology*, 7, 2625–2630.

Svejgaard, A., Platz, P. & Ryder, L.P. (1983). HLA and disease 1982 – a survey. *Immunological Reviews*, 70, 193–218.

Swain, S.L., McKenzie, D.T., Weinberg, A.D. & Hancock, W. (1988). Characterization of T-helper 1 and 2 cell subsets in normal mice. Helper T cells responsible for IL-4 and IL-5 production are presented as precursors that require priming before they develop into lymphokine-secreting cells. *Journal of Immunology*, 141, 3445–3455.

Tanese, N., Pugh, B.F. & Tjian, R. (1990). Co-activators for a proline-rich activator purified from the multi-subunit human TFIID complex. *Genes and Development*, 5, 2212–2224.

Teyton, L., Lotteau, V., Turmel, P., Arenzana-Seisdedos, F., Virelizier, J.-L., Pujol, J.-P., Loyau, G., Piatier-Tonneau, D., Auffray, C. & Charron, D.J. (1987). HLA DR, DQ and DP antigen expression in rheumatoid synovial cells: a biochemical and quantitative study. *Journal of Immunology*, 138, 1730–1738.

Thanos, D., Mavrothalassitis, G. & Papamatheakis, J. (1988). Multiple regulatory regions of the 5′ side of the mouse Eα gene. *Proceedings of the National Academy of Sciences of the USA*, 85, 3075–3079.

Thomson, G. (1984). HLA DR antigens and susceptibility to insulin-dependent diabetes mellitus. *American Journal of Human Genetics*, 36, 1309–1317.

Todd, J.A., Bell, J.I. & McDevitt, H.O. (1987). HLA-$DQ_{\beta}$ gene contributes to susceptibility and resistance to insulin-dependent diabetes mellitus. *Nature (London)*, 329, 599–604.

Touraine, J.L., Betvel, H., Souillet, G. & Jeune, M. (1978). Combined immunodeficiency disease associated with absence of cell surface HLA A and B antigens. *Journal of Pediatrics*, 93, 47–51.

Traugott, U., Scheinberg, L.C. & Raine, C.S. (1985). On the presence of Ia-positive endothelial cells and astrocytes in multiple sclerosis lesions and its relevance to antigen presentation. *Journal of Neuroimmunology*, 8, 1–14.

Tronche, F., Rollier, A., Sourdive, D., Cereghini, S. & Yaniv, M. (1991). NFY or a related CCAAT binding factor can be replaced by other transcriptional activators for co-operation with HNF1 in driving the rat albumin promoter *in vivo*. *Journal of Molecular Biology*, 222, 31–43.

Tsang, S.Y., Nakanishi, M. & Peterlin, B.M. (1988). B cell-specific and interferon-γ inducible regulation of the HLA-DRA gene. *Proceedings of the National Academy of Sciences of the USA*, 85, 8598–8602.

Tsang, S.Y., Nakanishi, M. & Peterlin, B.M. (1990). Mutational analysis of the DRA promoter: *cis*-acting sequences and *trans*-acting factors. *Molecular and Cellular Biology*, 10, 711–719.

van Ewijk, W., Ron, Y., Monaco, J., Kappler, J., Marrack, P., Lemeur, M., Gerlinger, P., Durand, B., Benoist, C. & Mathis, D. (1988). Compartmentalization of MHC class II gene expression in transgenic mice. *Cell*, 53, 357–370.

van Huijsduijnen, R., Li, X.Y., Black, D., Matthes, H., Benoist, C. & Mathis, D. (1990). Co-evolution from yeast to mouse: cDNA cloning of the two NF-Y (CP-1/CBF) subunits. *EMBO Journal*, 9, 3119–3127.

Venkitaraman, A.R., Culbert, E.J. & Feldman, M. (1987). A phenotypically dominant regulatory mechanism suppresses major histocompatability complex class II gene expression in a murine plasmacytoma. *European Journal of Immunology*, 17, 1411–1446.

Vilen, B.J., Cogswell, J.P. & Ting, J.P.-Y. (1991). Requirement for stereospecific alignment of the X and Y elements in MHC class II DRA function. *Molecular and Cellular Biology*, 11, 2406–2415.

Vilen, B.J., Penta, J.F. & Ting, J.P.-Y. (1992). Structural constraints within a trimeric transcriptional regulatory region: constitutive and interferon-γ inducible expression of the HLA-DRA gene. *Journal of Biological Chemistry*, 267, 23728–23734.

Viville, S., Jongeneel, V., Koch, W., Mantovani, R., Benoist, C. & Mathis, D. (1991). The Eα promoter: a linker scanning analysis. *Journal of Immunology*, 146, 3211–3217.

Voliva, C.F., Aronheim, A., Walker, M.D. & Peterlin, B.M. (1992). B cell factor 1 is required for optimal expression of the DRA promoter in B cells. *Molecular and Cellular Biology*, 12, 2383–2390.

Vuorio, T., Maity, S.N. & de Crombrugghe, B. (1990). Purification and molecular cloning of the 'A' chain of a rat heterodimeric CCAAT-binding protein. Sequence identity with the yeast HAP3 transcription factor. *Journal of Biological Chemistry*, 265, 22480–22486.

Waldmann, T.A. (1989). The multi-subunit interleukin-2 receptor. *Annual Review of Biochemistry*, 58, 875–911.

Watanabe, Y. & Jacob, C.O. (1991). Regulation of MHC class II antigen expression. Opposing effects of tumor necrosis factor-α on IFN-γ-induced HLA-DR and Ia expression depends on the maturation and differentiation state of the cell. *Journal of Immunology*, 146, 899–905.

Widera, G., Burkly, L.C., Pinkert, C.A., Bottger, E.C., Cowing, C., Palmiter, R.D., Brinster, R.L. & Flavell, R.A. (1987). Transgenic mice selectively lacking MHC class II (I–E) antigen expression on B cells: an *in vivo* approach to investigate Ia gene function. *Cell*, 51, 175–187.

Williams, T., Admon, A., Luscher, B. & Tjian, R. (1988). Cloning and expression of AP-2, a cell type-specific transcription factor that activates inducible enhancer elements. *Genes and Development*, 2, 1557–1569.

Willman, C.L., Stewart, C.C., Miller, V., Yi, T.-L. & Tomasi, T.B. (1989). Regulation of MHC class II gene expression in macrophages by hematopoietic colony-stimulating factors (CSF). Induction by granulocyte/macrophage CSF and inhibition by CSF-1. *Journal of Experimental Medicine*, 170, 1559–1567.

Woodward, J.G., Omer, K.W. & Stuart, P.M. (1989). MHC class II transcription in different mouse cell types. Differential requirement for protein synthesis between B cells and macrophages. *Journal of Immunology*, 142, 4062–4069.

Wright, K.L. & Ting, J.P.-Y. (1992). *In vivo* footprint analysis of the HLA-DRA gene promoter: cell-specific interaction at the octamer site and up-regulation of X box binding protein by interferon-γ. *Proceedings of the National Academy of Sciences of the USA*, 89, 7601–7605.

Wuarin, J., Mueller, C. & Schibler, U. (1990). A ubiquitous CCAAT factor required for efficient *in vitro* transcription from the mouse albumin promoter. *Journal of Molecular Biology*, 214, 865–874.

Yamamura, K., Kikutani, H., Folsom, V., Clayton, L.K., Kimoto, M., Akira, S., Kashiwamura, S., Tonegawa, S. & Kishimoto, T. (1985). Functional expression of a microinjected $E\alpha^d$ gene in C57BL/6 transgenic mice. *Nature (London)*, 316, 67–69.

Yang, Z., Accolla, R.S., Pious, D., Zegers, B.J. & Strominger, J.L. (1988). Two distinct genetic loci regulating class II gene expression are defective in human mutant and patient cell lines. *EMBO Journal*, 7, 1965–1972.

Yu, D.T.Y., McCune, J.M., Fu, S.M., Winchester, R.J. & Kunkel, H.G. (1980). Two types of Ia positive T cells. Synthesis and exchange of Ia antigens. *Journal of Experimental Medicine*, 152, 895-985.

Zeleznik-Le, N.J., Azizkhan, J.C. & Ting, J.P.-Y. (1991). An affinity-purified serum inducible CCAAT box-binding protein (YEBP) functionally regulates the expression of a human class II MHC gene and another serum inducible gene. *Proceedings of the National Academy of Sciences of the USA*, 88, 1873–1877.

Zeleznik-Le, N., Itoh-Lindstrom, Y., Clarke, J.B., Moore, T.L. & Ting, J.P.-Y. (1992). The B cell specific nuclear factor OTF-2 positively regulates transcription of the human class II transplantation gene DRA. *Journal of Biological Chemistry*, 267, 7677–7682.

Zhu, L. & Jones, P.P. (1990) Transcriptional control of the invariant chain gene involves promoter and enhancer elements common to and distinct from major histocompatibility complex class II genes. *Molecular and Cellular Biology*, 10, 3906–3916.

Zimmer, T. & Jones, P.P. (1990). Combined effects of TNF-α, prostaglandin E2, and corticosterone on induced Ia expression on murine macrophages. *Journal of Immunology*, 145, 1167–1175.

# 6

# Modulation of MHC antigen expression by viruses

D. JOHN MAUDSLEY
*University of Warwick*

## 6.1 Introduction

Many animal viruses have the property of being able to modulate the expression of the MHC antigens of infected host cells and sometimes of uninfected cells of the host animal. Since the MHC is central to the immune system, this may be beneficial to the virus in evading an immune response and, therefore, have consequences for disease. The significance of such modulation can be seen from how widespread it is, its effects on pathogenicity and its prevalence in common and important human (and animal) infections. It will be shown here and below (Chapters 7 to 12) that modulation of MHC by viruses (and, briefly, other pathogens) is indeed widespread, does affect pathogenicity and is present in common and important infections. It will, therefore, be possible to conclude that modulation of MHC antigens is a highly important characteristic of viruses in general.

This chapter describes the different viruses for which modulation of MHC has been observed. Different mechanisms involved in modulation, both potential and actual, are summarized with suitable examples where known. This information is related to the pathogenicity of the virus where data are available. In the overall context of infectious disease, a brief note is made of other infectious organisms (i.e. microorganisms other than viruses) that are known to affect expression of functional MHC antigens with immunological consequences. Conclusions are drawn from these data on the importance of the modulation of MHC expression and on suitable directions for further research.

## 6.2 Modulation by viruses of MHC antigen expression is widespread

Host modulation of MHC antigen expression during virus infection is well documented (Maudsley, Morris & Tomkins, 1989). This process can be summarized in several steps. Infected cells respond to virus infection by release of IFN-α/β. Virus-specific $T_H$ cells secrete IFN-γ in response to suitably presented viral antigen. IFN-α/β stimulates increased expression of MHC class I antigens and IFN-γ stimulates the induction of and increase in both MHC class I and class II antigen expression (see Chapter 3). Increased expression of MHC class I and class II antigens aids the immune system by increasing the probability of virus-specific T cells seeing sufficient MHC–peptide complexes to stimulate a response mediated by either cytotoxic or $T_H$ cells (see Chapters 1 and 3).

What has been described above is mainly a host response, although affected by viral characteristics. In this and the following six chapters, the modulation by viruses of host MHC expression will be described, that is modulation by mechanisms other than those resulting from normal host responses involving interferons and other cytokines. This includes mechanisms whereby the normal responses are subverted.

At one extreme, a few viruses (e.g. the coronavirus JHM and measles virus) stimulate inappropriate or excessive MHC antigen expression by mechanisms probably not involving IFNs (Massa, Brinkmann & ter Meulen, 1987a; Massa *et al.*, 1987b). When this occurs in tissues not normally expressing MHC antigens, or expressing either only low levels of MHC antigens or only selected MHC antigens, this may be beneficial to the virus by eliciting inappropriate autoimmune responses that may interfere with the host response to the virus (Sedgwick & ter Meulen, 1989; see also Chapter 16).

A large number of viruses have been found to down-regulate the expression of MHC antigens. It is easy to see how this could benefit the virus by reducing the chance of virus-specific T cells recognizing sufficient numbers of MHC–peptide complexes on target or accessory cells to stimulate a response, thus impairing the immune response to virus infection (see Chapters 1 and 3). Some of the earliest evidence of viruses down-regulating MHC antigen expression was obtained at about the same time as experiments showing 'MHC restriction' of T cell responses to viruses (Gardner, Bowen & Blanden, 1975; Koszinowski & Ertl, 1975). Although 'restriction' was recognized as a general phenomenon at the time, modulation of MHC antigens as a frequent characteristic of viruses was not recognized until much later (e.g. Maudsley *et al.*, 1989; Maudsley & Pound, 1991). Some of these early experiments showed both cytotoxic T cell responses to virus-infected cells and also down-regulation of MHC antigens on cells infected *in vitro*. Where allogeneic MHC antigen-specific T cells were used as 'controls', this down-regulation of MHC antigen expression resulted in reduced effectiveness of the T cell response. Evidence has also been obtained from studies *in vivo*, where clinical relevance of the down-regulation of MHC antigen expression has been shown (e.g. for hepatitis B virus (HBV), see Chapter 10). However, because of the opposing effects of IFN (and other cytokines) and the virus, modulation can often be more clearly seen in *in vitro* systems where the concentrations of cytokines can be controlled (e.g. for retroviruses (see Chapter 7); Maudsley, 1991; Maudsley & Pound, 1991). Table 6.1 summarizes available data for different viruses that modulate MHC expression.

Modulation of MHC antigens is clearly widespread across a range of virus families from poxviruses to adenoviruses (Ad) and from togaviruses to retroviruses. More examples of retroviruses modulating MHC antigen expression are known than of any other viral group. This may reflect the relative ease of studying persistent retrovirus infection of cells *in vitro* compared with some of the other families of viruses and strongly suggests that with further research more viruses (especially from other families less amenable to study) will be found to modulate MHC antigen expression.

## 6.3 Mechanisms involved in modulating MHC antigen expression

As noted above, IFN and other cytokines are naturally involved in regulating the expression of MHC antigens. Chapters 1 to 5 describe in greater detail the control of MHC expression by the host. Figure 6.1 briefly summarizes the stages involved in the expression of MHC antigens and identifies 12 possible sites of interference by viruses.

Viruses could potentially interfere with the control of MHC expression at any stage from the production of IFN-$\alpha/\beta$ by infected cells or IFN-$\gamma$ by $T_{H1}$ cells onwards. The 12 potential sites identified fall into five broad categories: those affecting transcription of MHC genes (sites 1 to 5); translation (site 6); those affecting association of MHC polypeptides and transport to the cell surface (sites 7, 9, 11 and 12); those affecting the function of the surface antigens (site 10) and those affecting availability of suitable virus peptides (site 8) (see Table 6.2).

All 12 sites are catalogued below with a brief description of each and examples where known.

### *Site 1 Inhibition of interferon production*

At the level of interfering with IFN-$\gamma$ production, two viruses are candidates, namely Epstein–Barr virus (EBV) and human immunodeficiency virus (HIV). The EBV-encoded IL-10 homologue BCRF1 inhibits (like IL-10) the $T_{H1}$ production of IFN-$\gamma$ but (unlike IL-10) does not itself induce MHC class II antigens (Go *et al.*, 1990; Vieira *et al.*, 1991). HIV, like a number of other

Table 6.1. *Summary of the modulation of MHC expression by viruses*

| Virus family | Viruses | Effects on MHC: Class I | Effects on MHC: Class II | References |
|---|---|---|---|---|
| Papovaviruses | HPV16 | ↓ | | Cromme *et al.* (1993); see Chapter 9 |
| | SV40 | ↓ | | Breau *et al.* (1992) |
| Adenoviruses | Ad12 virus | ↓ | | Schrier *et al.* (1983) |
| | Ad2 virus | ↓ | | Andersson *et al.* (1987); see Chapter 8 |
| Herpesviruses | HCMV | ↓ | ↓ | Brown *et al.* (1990); Buchmeier & Cooper (1989); see Chapter 12 |
| | HSV 1 and 2 | ↓ | | Jennings *et al.* (1985) |
| | EBV | | ↓ | de Waal Malefyt *et al.* (1991); see Chapter 11 |
| Poxviruses | Vaccinia virus | ↓ | | Kohonen-Corish, Blanden & King (1989) |
| | Ectromelia virus | ↓ | | Gardner *et al.* (1975) |
| | Myxoma virus | ↓ | | Boshkov, Macen & McFadden (1992) |
| | Malignant rabbit fibroma virus | ↓ | | Boshkov *et al.* (1992) |
| Hepadnaviruses | HBV | ↓ | | Foster *et al.* (1991)<br>Twu *et al.* (1988); see Chapter 10 |
| Togaviruses | West Nile virus | ↑ | ↑ | Liu *et al.* (1989) |
| Orthomyxoviruses | Influenza A virus | ↓ or ↑ | | Townsend *et al.* (1989); Parham (1989) |
| Paramyxoviruses | Measles virus | ↑ | ↑ | Massa *et al.* (1987b) |
| Coronaviruses | JHM virus | | ↑ | Massa *et al.* (1987a) |
| Rhabdoviruses | VSV | ↓ | | Hecht & Summers (1972) |
| Reoviruses | | ↑ | ↑ | Campbell *et al.* (1989); Gaulton *et al.* (1989) |

| | | | | |
|---|---|---|---|---|
| Retroviruses | HIV | ↓ | ↓, ↑ | Scheppler *et al.* (1989); Nong *et al.* (1991); Kannagi *et al.* (1987) |
| | SIV | | ↑ | Kannagi *et al.* (1987) |
| | Ki-MuSV | ↓ | ↓ | Maudsley & Morris (1989a,b) |
| | Ki-MuLV | ↓ | ↓ | Maudsley & Morris (1989a,b) |
| | Ha-MuSV | ↓ | ↓ | D. J. Maudsley, unpublished data |
| | MoMuLV | ↑ | | Flyer, Burakoff & Faller (1985) |
| | MoMuSV | ↓ | | Flyer *et al.* (1985) |
| | MoMuLV/MuSV | — | ↓ | D. J. Maudsley, unpublished data |
| | RSV | ↓ | ↓ | Gogusev *et al.* (1988); Powell, Mala & Wick (1987) |
| | RadLV | ↓, ↑ | | Meruelo *et al.* (1986) |
| | HTLV-I | ↑ | ↑ | Sawada *et al.* (1990); Lehky *et al.* (1994) |

↑ indicates induction of or increases in the expression of MHC antigens; ↓ indicates an inhibition of or a decrease in the expression of MHC antigens.

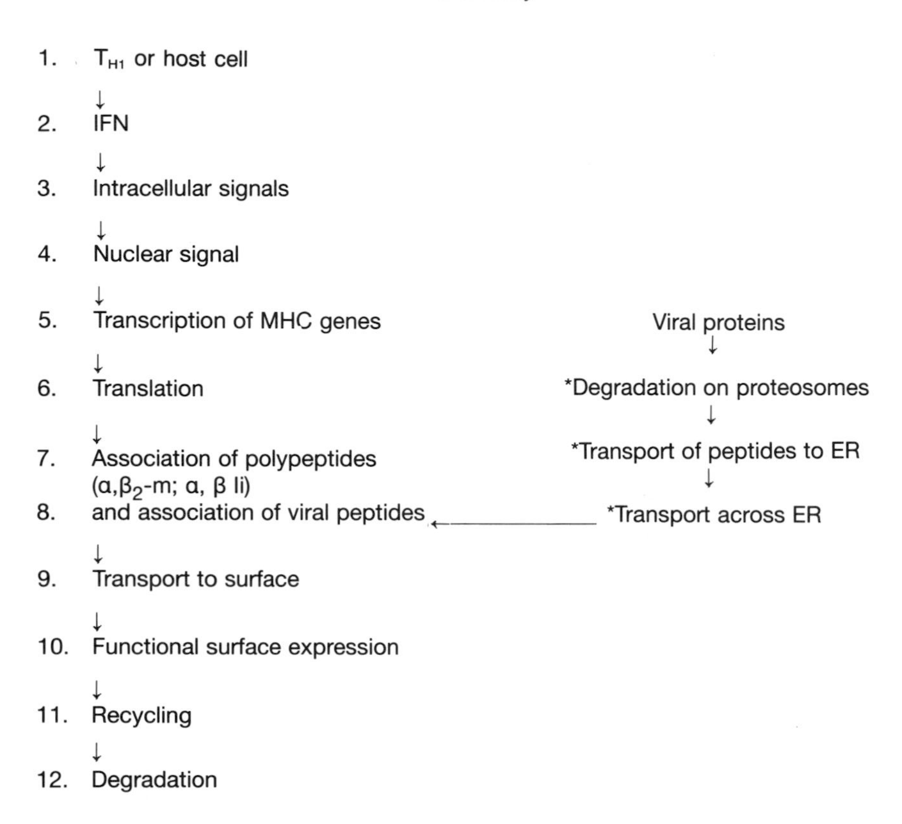


Fig. 6.1. Summary of potential sites of interference by viruses of cell surface expression of functional MHC molecules. *Transcription of genes for proteosome components, transporters and the peptide pump are also regulated by IFN-γ.

viruses, directly infects $T_H$ cells and, if by no other mechanism than depletion of these cells, probably interferes with IFN-γ production. At the level of interfering with IFN-α/β production, some retroviruses (the murine sarcoma viruses, MuSV) and HBV are candidates. MuSV has been reported to prevent the production of IFN-α/β by cells infected with murine leukaemia viruses (MuLV; Chapter 7) and Twu *et al.* (1988) showed that an HBV DNA fragment containing the core antigen gene suppressed the production of human IFN-β. This may have a role in preventing cells infected by HBV from producing IFN, possibly even if superinfected by another virus.

*Site 2 Neutralization of interferon*

Interferon itself could be neutralized by some virus product, either by its enzyme activity or by it binding to and neutralizing the interferon. The HIV

Table 6.2. *Examples of different viruses interfering with functional MHC antigen expression at different sites*

| Site of virus interference | Virus | References |
| --- | --- | --- |
| Site 1, secretion of IFN$\alpha/\beta$ or $\gamma$ | EBV | de Waal Malefyt *et al.* (1991) |
| | HBV | Twu *et al.* (1988) |
| Site 2, IFN effectiveness | HIV | Nong *et al.* (1991) |
| Site 3, intracellular signalling | Ki-MuSV and other retroviruses | Maudsley (1991) |
| Site 4, nuclear signalling | | See Chapter 7 |
| Site 5, transcription | Ad12 | See Chapter 8 |
| | HBV | Foster *et al.* (1991); see Chapter 10 |
| Site 6, translation | None known | |
| Site 7, MHC antigen–peptide complex formation | HCMV | See Chapter 12 |
| Site 8, availability of viral peptides | Possibly all in sites 1–5 above, possibly influenza | Townsend *et al.* (1989); Parham (1989) |
| Site 9, intracellular transport | Ad2 | Andersson *et al.* (1987); see Chapter 8 |
| Site 10, interference with the function of MHC antigens expressed at the cell surface | EBV | See Chapter 11 |
| Sites 11 and 12, recycling and degradation of MHC antigens | No clear examples | |

Potential sites of viral interference are shown in Fig. 6.1.

p24 product has been reported to prevent uninfected target cells from responding to IFN-γ by increasing class II expression; the mechanism underlying this process is unknown (Nong *et al.*, 1991).

*Site 3 Effects on signal transduction from membrane to nucleus*

A number of viruses interfere with the transduction of signals from the cell surface (IFN receptors etc.) to the nucleus and with other intracellular signalling involved in regulating MHC transcription. Several retroviruses containing oncogenes interfere here by virtue of the products of their oncogenes (Maudsley & Morris, 1989a,b; Maudsley, 1991). HIV may affect expression through this site.

*Site 4 Effects on signals in the nucleus and transcription factors*

Direct or indirect effects by virus products in the nucleus, particularly those affecting transcription factors, affect the transcription of MHC genes. Again some of the nuclear oncogene-carrying viruses could fall into this category (see Maudsley, 1991 and Chapters 7, 8 and 11).

It is also possible that virus-specific sequences that are homologous to ISREs or IREs could sequester ISRE- (or IRE-) specific transcription factors, thus inhibiting transcription, for example VSV, certain retroviruses (Maudsley & Morris, 1989b) and HBV (Onji *et al.*, 1987).

*Site 5 Inhibition of gene transcription*

Direct effects on MHC gene transcription may be possible by virally encoded inhibitory transcription factors (HBV, see Chapter 10; Ad12, see Chapter 8). In addition, viral RNA polymerase ‘taking over’ the host cell transcription machinery could effectively switch off certain host cell functions and reduce transcription of certain host cell genes, such as MHC genes. Poxviruses, for example, may fall in the latter category (Gardner *et al.*, 1975; Koszinowski & Ertl, 1975).

*Site 6 Post-transcriptional events and translation*

There are no known examples of a virus affecting MHC mRNA processing or translation, but any viral mechanism capable of switching off host macromolecular synthesis may non-specifically affect these processes.

*Site 7 Inhibition of MHC antigen assembly*

If a viral gene product could effectively interfere with the association of MHC chains (heavy and $\beta_2$-m chains for class I or α and β chains, with or without Ii, for class II) then this would prevent the formation of functional mature molecules and might prevent further processing of the MHC chains, including their transport to the cell surface (site 9). The human cytomegalovirus (HCMV) product H301 binds to $\beta_2$-m and this may interfere with its association with class I heavy chains and peptide, possibly contributing to the reduced expression at the cell surface (see Chapter 12).

*Site 8 Prevention of viral peptides associating with MHC antigens*

This is clearly very closely linked to site 7 and the availability of MHC polypeptides. Association of viral peptides with MHC antigens depends on two criteria: (a) availability of suitable viral peptides and (b) competition with other peptides. Availability of peptides depends on processing and transport of viral peptides, which in turn depends on proteases and, largely, on proteosome components and peptide transporters (see Chapters 1 and 2). Firstly, virus-encoded protease inhibitors could interfere with this step (Smith, Howard & Chan, 1989). Secondly, at least some proteosome components and peptide transporters are encoded in the MHC and are responsive to IFN-γ. Mechanisms that inhibit transcription of MHC genes (sites 1–5, above) may also inhibit transcription of genes of these components of peptide processing and hence may inhibit the supply of suitable peptides (see Fig. 6.1). Competition with inappropriate peptides could take two forms. Firstly, influenza viruses encode peptides that will bind to MHC class I antigens but do not permit transport to the cell surface (Townsend *et al.*, 1989). This has been proposed as being of potential benefit to the virus by competing for available MHC antigens and reducing expression of suitable MHC–peptide complexes at the cell surface (Parham, 1989). Secondly, viruses may encode peptides that, although they bind to MHC antigens (class I and class II) and permit expression at the cell surface, do not elicit an immune response. This could be the result of either some similarity to a host peptide or another reason for a deficit in the T cell repertoire. However, no examples of this latter phenomenon are known.

*Site 9 Inhibition of transport to the cell surface*

In the case of MHC class I antigens, anything that prevents correct association of the heavy chain with $\beta_2$-m and peptide will inhibit transport to the surface and thus prevent surface expression (sites 7 and 8). Interaction with a viral gene product might also inhibit transport. In the case of HCMV, it is not clear whether binding of the H301 gene product to $\beta_2$-m inhibits association with MHC class I heavy chains and, presumably, peptide, but transport to the cell surface is inhibited by some mechanism (see Chapter 12).

Ad2 also blocks transport to the cell surface because of the 19 kDa product of the E3 gene binding to class I MHC molecules and inhibiting terminal glycosylation (Burgert & Kvist, 1985; Andersson, McMichael & Peterson, 1987; see also Chapter 8).

*Site 10 Interference with the function of cell surface-expressed MHC molecules*

Reports of interference with the function of cell surface-expressed MHC molecules fall into two, or possibly three, categories. Firstly, expression of functional MHC antigens at the cell surface can be interfered with by virus-encoded molecules binding to MHC antigens. This may occur for HCMV to some extent (see Chapter 12). Clearer examples are seen for bacteria, where staphylococcal enterotoxins A and B (SEA, SEB) and toxic shock syndrome toxin-1 (TSST-1) bind to MHC class II antigens causing inappropriate stimulation, thus interfering with the immune system and causing severe disease (Fraser, 1989; Scholl, Diez & Geha, 1989; Uchiyama *et al.*, 1989).

Semliki Forest Virus (SFV) and simian virus 40 (SV40) use MHC antigens as receptors (Breau, Atwood & Norkin, 1992) and this may interfere with MHC function (although there is no evidence which bears upon this point).

Secondly, some viruses, for example HCMV (see Chapter 12) and EBV (Chapter 11), affect the expression of cell adhesion molecules and other accessory molecules required for effective antigen-specific stimulation of responding T cells. Where expression of these molecules is reduced (e.g. in the case of EBV in Burkitt's lymphoma), although the MHC–peptide complex may be recognized by antigen-specific T cells, it may be unable to stimulate a suitable response. In addition, HIV affects CD4 expression by two mechanisms, one involving gp120 and one involving the *nef* gene product (Guy *et al.*, 1987; Garcia & Miller, 1991). Binding of gp120 may affect CD4 expression and (since CD4 associates with MHC class II antigens) may indirectly affect MHC expression or function.

Where neither of the above mechanisms has been shown, there may be a third reason for non-functionality of MHC antigens. For example, MuSV induces differentiation of melanoma cells to form transformed cells that are known to express MHC class II antigens (Albino *et al.*, 1986). For reasons yet unknown, these Ia antigens are non-functional and this may be a result of some property of the virus. Nevertheless, the apparent induction of Ia antigens by the virus, which otherwise might stimulate an immune response to clear the virus, in this case does no harm to the virus because the Ia antigens are non-functional.

*Site 11 Effects on recycling MHC antigens*

The rate of recycling of cell surface molecules may be important for efficient expression of different peptides by target cells and antigen-presenting cells, particularly in association with MHC class II antigens on accessory cells (Chapters 1 and 3).

*Site 12 Effects on degradation*

Changes to the rate of degradation of MHC antigens would be expected to alter the level of their expression at the cell surface. No clear examples of this have been reported, but effects of excess gp 120 in HIV-infected cells may cause increased degradation of CD4 and with it MHC class II antigens in infected cells.

Twelve levels or sites where viruses may act to interfere with the expression of MHC molecules have been identified and described. This information is summarized in Table 6.2. It is worth noting that for a number of viruses the mechanisms involved in modulating MHC antigen expression are at present unknown and where they are known the information is often incomplete. Hence only after further research will it be possible to assign them to one or more of the 12 sites of interference.

## 6.4 Pathogenicity

It is clearly important to know whether modulation of MHC antigen expression affects the pathogenicity of the virus. Where there are increases in expression resulting from virus infection there are clear examples of the pathological consequences; for example, JHM virus-mediated MHC class II expression correlates with the induction of experimental allergic encephalitis (Massa *et al.*, 1987a).

Increased expression generally results in greater inflammation and autoimmune disease. The more widespread decrease in MHC antigen expression is expected to result in a reduction in the ability of the host's immune system to mount an effective response. Firstly, this may allow an infection to be initiated and, in particular, may increase the length of time taken for the host immune system to begin to control such an infection. This is most clear for HBV infection (see Chapter 10). Secondly, such inhibition of the host immune system may enable the establishment of a persistent infection even if a strong immune response is eventually mounted against the virus. Again HBV infection gives a clear example of the dependence of establishment of persistent infection on reduced MHC expression. Restoration of MHC antigen expression to the hepatocytes can result in eventual eradication of the virus (see Chapter 10). Persistent infection caused by reduced MHC expression and consequent immune evasion does have the benefit to the host of reduced tissue damage – at least until an effective immune response is mounted. Presumably the virus is able to spread further in the absence of an effective immune response, possibly resulting in more tissue damage in the long term than if an immune response had been mounted earlier. Other examples of persistent viral infection probably being assisted by reduced MHC expression include adenovirus (see Chapter 8), EBV (see Chapter 11) and HIV infection (see Chapter 7).

It can be concluded, therefore, that there is clear experimental (and clinical) evidence that the modulation of MHC antigens by viruses has consequences for the kinetics and pathology of infection.

## 6.5 Significance of viral traits

As stated in the Introduction, the importance of a particular characteristic of viruses can be taken as roughly proportional to (or at least related to) how widespread the characteristic is between virus families, whether it affects the pathogenicity of the virus and whether it occurs in important or common infections. It has been seen that the modulation of host MHC antigen expression occurs in a large number of virus families. Moreover, a variety of different mechanisms is used to modulate host MHC antigen expressions (and sometimes more than one mechanism for a single virus). Clear examples of modulation by viruses of MHC antigen expression having effects on viral pathogenicity, both experimentally and clinically, have been described. It was noted also that those viruses that have been described as modulating MHC expression include viruses that are among the most common viruses known (e.g. EBV), viruses that are both relatively common and cause severe disease (e.g. HBV, where there are over 200 million cases worldwide), and viruses that are probably, if given time, always fatal (e.g. HIV).

By the criteria adopted here – prevalence across virus families, effects on pathogenicity and prevalence in common or important infections – modulation of MHC expression is clearly a highly significant characteristic of any virus.

## 6.6 Evidence of modulation of MHC antigens in infections caused by non-viral pathogens

Given that the modulation of MHC antigen expression appears to be a very important characteristic of infection by a number of viruses, it might be expected that this may also be an important characteristic of infection by other pathogens. This is indeed the case, although the number of known examples is relatively limited at present. Three important examples are given below.

As already noted, certain strains of the bacterium *Staphylococcus aureus* produce TSST-1. In addition *S. aureus* may also produce other proteins or toxins including SEA and SEB. These molecules initially bind to MHC class II (or Ia) molecules, thus disrupting the host's immune system causing toxic shock syndrome, a serious and often lethal condition. In this case over-stimulation of the immune system results from production of the bacterial protein.

The protozoan *Leishmania* sp. can cause serious disease in humans. *Leishmania* can also infect mouse strains. The ability of *Leishmania* to interfere with MHC class II antigen cycling and MHC–peptide presentation appears to correlate with pathogenicity in certain strains of mice (Reiner, Winnie & McMaster, 1987; Reiner *et al.*, 1988; Kwan *et al.*, 1992). That is, when the pathogen can inhibit MHC class II expression, it causes disease, but where it cannot inhibit MHC expression it does not cause disease. This suggests that, at least in some situations, the ability of a particular pathogen to successfully infect a host is dependent on its ability to successfully modulate host MHC antigen expression.

*Mycobacterium microti*, which can cause disease in both humans and animals, has been reported to affect MHC class II (Ia) antigen expression (Kaye, Sims & Feldmann, 1986).

In the case of *S. aureus*, a protein, secreted by the bacterium, acts at a distance. In the cases of *Leishmania* and *Mycobacteria*, these organisms infect cells directly and, therefore, can benefit from modulation of MHC on the infected host cells. On this basis, it, therefore, seems reasonable to expect that modulation of host MHC antigens will be more important in general for those pathogens that are intracellular (as are all viruses) and possibly less important for extracellular pathogens.

## 6.7 Conclusions

It is clear that modulation of MHC antigen expression by viruses is important and widespread across virus families. It is also clear that it deserves further investigation, both in greater depth where viral modulation of MHC antigen expression has already been observed and in investigation of other important viruses where such modulation has not yet been looked for, or not yet observed.

In subsequent chapters (Chapters 7 to 12), the modulation of MHC antigen expression by several viruses will be described in greater depth. These chapters cover a number of clinically and experimentally important individual viruses or virus groups and illustrate the different mechanisms employed in modulating host MHC antigens by a number of different viruses.

## References


Albino, A.P., Houghton, A.N., Eisinger, M., Lee, J.S., Kantar, R.R.S., Oliff, A.I. & Old, L.J. (1986). Class II histocompatibility antigen expression in human melanocytes transformed by Harvey murine sarcoma virus and Kirsten murine sarcoma virus retroviruses. *Journal of Experimental Medicine*, 164, 1710–1722.

Andersson, M., McMichael, A. & Peterson, P.A. (1987). Reduced allorecognition of adenovirus 2 infected cells. *Journal of Immunology*, 138, 3960–3966.

Boshkov, L.K., Macen, J.L. & McFadden, G. (1992). Virus-induced loss of class I MHC antigens from the surface of cells infected with myxoma virus and malignant rabbit fibroma virus. *Journal of Immunology*, 148, 881–887.

Breau, W.C., Atwood, W.J. & Norkin, L.C. (1992). Class I major histocompatibility proteins are an essential component of the simian virus 40 receptor. *Journal of Virology*, 66, 2037–2045.

Brown, H., Smith, G., Beck, S. & Minson, T. (1990). A complex between the MHC class I homologue encoded by human cytomegalovirus and $\beta_2$-microglobulin. *Nature (London)*, 347, 770–772.

Buchmeier, N.A. & Cooper, N.R. (1989). Suppression of monocyte functions by human cytomegalovirus. *Immunology*, 66, 278–283.

Burgert, H.G. & Kvist, S. (1985). An adenovirus type 2 glycoprotein blocks cell surface expression of human histocompatibility class I antigens. *Cell*, 41, 987–997.

Campbell, I.L., Harrison, L.C., Aschcroft, R.G. & Jack, I. (1988). Reovirus infection enhances expression of class I MHC proteins on human beta-cell and rat RINmSF cell. *Diabetes*, 37, 362–365.

Cromme, F.V., Snijders, P.J.F., van den Brule, A.J.C., Kenemans, P., Meijer, C.J.L.M. & Walboomers, J.M.M. (1993). MHC class I expression in HPV 16 positive cervical carcinomas is post-transcriptionally controlled and independent from c-*myc* over-expression. *Oncogene*, 8, 2969–2975.

de Waal Malefyt, R., Haanen, J., Spits, H., Roncarolo, M.G., te Velde, A., Figdor, C., Johnson, K., Kastelein, R., Yssel, H. & de Vries, J.E. (1991). Interleukin 10 (IL-10) and viral IL-10 strongly reduce antigen specific human T cell proliferation by diminishing the antigen-presenting capacity of monocytes via down-regulation of class II major histocompatibility complex expression. *Journal of Experimental Medicine*, 174, 915–924.

Flyer, D.C., Burakoff, S.J. & Faller, D.V. (1985). Retrovirus-induced changes in major histocompatibility complex antigen expression influence susceptibility to lysis by cytotoxic T lymphocytes. *Journal of Immunology*, 135, 2287–2292.

Foster, G.R., Ackrill, A.M., Goldin, R.D., Kerr, I.M., Thomas, H.C. & Stark, G.R. (1991). Expression of the terminal protein region of hepatitis B virus inhibits cellular responses to interferons $\alpha$ and $\gamma$ and double-stranded RNA. *Proceedings of the National Academy of Sciences of the USA*, 88, 2888–2892.

Fraser, J.D. (1989). High-affinity binding of staphylococcal enterotoxins A and B to HLA-DR. *Nature (London)*, 339, 221–223.

Garcia, J.V. & Miller, A.D. (1991). Serine phosphorylation-independent down-regulation of cell-surface CD4 by *nef*. *Nature (London)*, 350, 508–511.

Gardner, I.D., Bowen, N.A. & Blanden, R.V. (1975). Cell-mediated cytotoxicity against ectromelin virus-infected target cells III. Role of the H-2 complex. *European Journal of Immunology*, 5, 122–127.

Gaulton, G.N., Stein, M.E., Safko, B. & Starecker, M.J. (1989). Direct induction of Ia antigen on murine thyroid-derived epithelial cells by reovirus. *Journal of Immunology*, 142, 3821–3825.

Go, N.F., Castle, B.E., Barrett, R., Kastelein, R., Dang, W., Mosmann, T.R., Moore, K.W. & Howard, M. (1990). Interleukin 10, a novel B cell stimulatory factor: unresponsiveness of X chromosome-linked immunodeficiency B cells. *Journal of Experimental Medicine*, 172, 1625–1631.

Gogusev, J., Teutsch, B., Morin, M.T., Mongial, F., Hagnenan, F., Suskind, G. & Rabotti, G.F. (1988). Inhibition of HLA class I antigen and mRNA expression induced by Rous sarcoma virus in transformed human fibroblasts. *Proceedings of the National Academy of Science of the USA*, 85, 203–207.

Guy, B., Kieny, M.P., Riviere, Y., Le Peuch, C., Dott, K., Girard, M., Montagnier, L. & Lecocq, J.-P. (1987). HIV F/3′ *orf* encodes a phosphorylated GTP-binding protein resembling an oncogene product. *Nature (London)*, 330, 266–269.

Hecht, T.T. & Summers, D.F. (1972). Effect of vesicular stomatitis virus infection on the histocompatibility antigens of L cells. *Journal of Virology*, 10, 578–585.

Jennings, S.R., Rice, P.L., Kloszewski, E.D., Anderson, R.W., Thompson, D.L. & Tevethia, S.S. (1985). Effect of herpes simplex types 1 and 2 on surface expression of class I major histocompatibility complex antigens on infected cells. *Journal of Virology*, 56, 757–766.

Kannagi, M., Kiyotaki, M., King, N.W., Lord, C.I. & Latvin, N.L. (1987). Simian immunodeficiency virus induces expression of class II major histocompatibility complex structures on infected target cells *in vitro*. *Journal of Virology*, 61, 1421–1476.

Kaye, P.M., Sims, M. & Feldmann, M. (1986). Regulation of macrophage accessory cell activity by mycobacteria II. *In vitro* inhibition of Ia expression by *Mycobacterium microti*. *Clinical and Experimental Immunology*, 64, 28–34.

Kohonen-Corish, M.R.J., Blanden, R.V. & King, N.J. (1989). Induction of cell surface expression of HLA antigens by human IFN-γ encoded by recombinant vaccinia virus. *Journal of Immunology*, 143, 623–627.

Koszinowski, U. & Ertl, H. (1975). Lysis mediated by T cells and restricted by H-2 antigen of target cells infected with vaccinia virus. *Nature (London)*, 255, 552–554.

Kwan, W.C., McMaster, W.R., Wang, N. & Reiner, N.E. (1992). Inhibition of expression of major histocompatibility complex class II molecules in macrophages infected with *Leishmania donovani* occurs at the level of gene transcription via a cyclic AMP-independent mechanism. *Infection and Immunity*, 60, 2115–2120.

Lehky, T.J., Cowan, E.P., Lampson, L.A. & Jacobson, S. (1994). Induction of HLA class I and class II expression in human T-lymphotropic virus type I-infected neuroblastoma cells. *Journal of Virology*, 68, 1854–1863.

Liu, Y., King, N., Kesson, A., Blanden, R.V. & Mullbacher, A. (1989). Flavivirus infection up-regulates the expression of class I and class II major histocompatibility antigens on and enhances T cell recognition of astrocytes *in vitro*. *Journal of Neuroimmunology*, 21, 157–168.

Massa, P.T., Brinkmann, R. & ter Meulen, V. (1987a). Inducibility of Ia antigen on astrocytes by murine coronavirus JHM is rat strain dependent. *Journal of Experimental Medicine*, 166, 259–264.

Massa, P.T., Schimpl, A., Wecker, E. & ter Meulen, V. (1987b). Tumor necrosis factor amplifies measles virus-mediated Ia induction on astrocytes. *Proceedings of the National Academy of Sciences of the USA*, 84, 7242–7245.

Maudsley, D.J. (1991). Role of oncogenes in the regulation of MHC antigen expression. *Biochemical Society Transactions*, 19, 291–296.

Maudsley, D.J. & Morris, A.G. (1989a). Regulation of IFN-γ induced host cell MHC antigen expression by Kirsten MSV and MLV. I. Effects on class I antigen expression. *Immunology*, 67, 21–25.

Maudsley, D.J. & Morris, A.G. (1989b). Regulation of IFN-γ induced host cell MHC antigen expression by Kirsten MSV and MLV. II. Effects on class II antigen expression. *Immunology*, 67, 26–31.

Maudsley, D.J., Morris, A.G. & Tomkins, P.T. (1989). Regulation by interferon of the immune response to viruses via the major histocompatibility complex antigens. In *Immune Responses, Virus Infections and Disease*, ed. N.J. Dimmock & P.D. Minor, pp. 15–33. Oxford: IRL Press.

Maudsley, D.J. & Pound, J.D. (1991). Modulation of MHC antigen expression by viruses and oncogenes. *Immunology Today*, 12, 429–431.

Meruelo, D., Kornreich, R., Rossomando, A., Pampeno, C., Boral, A., Silver, J.L., Buxbaum, J., Weiss, E.H. & Devlin, J.J. (1986). Lack of class I H-2 antigens in cells transformed by radiation leukaemia virus is associated with methylation and rearrangement of H-2 DNA. *Proceedings of the National Academy of Sciences of the USA*, 83, 4504–4508.

Nong, Y., Kandil, O., Tobin, E.H., Rose, R.M. & Remold, H.G. (1991). The HIV core protein p24 inhibits interferon-γ-induced increase of HLA-DR and cytochrome-b heavy chain messenger RNA levels on the human monocyte-like cell line THP1. *Cellular Immunology*, 132, 10–16.

Onji, M., Lever, A.M.L., Saito, I. & Thomas, H.C. (1987). Hepatitis B virus reduces the sensitivity of cells to interferons. *Journal of Interferon Research*, 7, 690.

Parham, P. (1989). MHC molecules: a profitable lesson in heresy. *Nature (London)*, 340, 426–428.

Powell, P.C., Mala, K. & Wick, G. (1987). Aberrant expression of Ia-like antigens on tumour cells of regressing but not progressing Rous sarcomas. *European Journal of Immunology*, 17, 723–726.

Reiner, N.E., Ng, W., Ma, T. & McMaster, W.R. (1988). Kinetics of interferon γ binding and induction of major histocompatibility complex class II mRNA in *Leishmania*-infected macrophages. *Proceedings of the National Academy of Sciences of the USA*, 85, 4330–4334.

Reiner, N.E., Winnie, N.G. & McMaster, W.R. (1987). Parasite–accessory cell interactions in murine leishmaniasis II. *Leishmania donovani* suppresses macrophage expression of class I and class II major histocompatibility complex gene products. *Journal of Immunology*, 138, 1926–1932.

Sawada, M., Suzumura, A., Yoshida, M. & Marunouchi, T. (1990). Human T-cell leukemia virus type I *trans* activation induces class I major histocompatibility complex antigen expression in glial cells. *Journal of Virology*, 64, 4002–4006.

Scheppler, J.A., Nicholson, J.K.A., Swan, D.C., Ahmed-Ansari, A. & McDougal, J.S. (1989). Down-modulation of MHC-I in a $CD4^+$ T cell line, CEM-E5, after HIV-1 infection. *Journal of Immunology*, 143, 2858–2866.

Scholl, P.R., Diez, A. & Geha, R.S. (1989). Staphylococcal enterotoxin B and toxic shock syndrome toxin-1 bind to distinct sites on HLA-DR and HLA-DQ molecules. *Journal of Immunology*, 143, 2583–2588.

Schrier, P.I., Bernards, R., Vaessen, R.T.M.J., Houweling, A. & van der Eb, A.J. (1983). Expression of class I major histocompatibility antigens is switched off by highly oncogenic adenovirus 12 in transformed rat cells. *Nature (London)*, 305, 771–775.

Sedgwick, J.D. & ter Meulen, V. (1989). Induction of autoimmunity following viral infection of the central nervous system. In *Immune Responses, Virus Infection and Disease*, ed. N.J. Dimmock & P.D. Minor, pp. 115–124. Oxford: IRL Press.

Smith, G.C., Howard, S.T. & Chan, Y.S. (1989). Vaccinia virus encodes a family of genes with homology to serine protease inhibitors. *Journal of General Virology*, 70, 2333–2343.

Townsend, A., Ohlen, C., Bastin, J., Ljunggren, H.-G., Foster, L. & Karre, K. (1989). Association of class I major histocompatibility heavy and light chains induced by viral peptides. *Nature (London)*, 340, 443–448.

Twu, J.-S., Lee, C.-H., Lin, P.-M. & Schloemer, R.H. (1988). Hepatitis B virus suppresses expression of human β-interferon. *Proceedings of the National Academy of Sciences of the USA*, 85, 252–256.

Uchiyama, T., Imanishi, K., Saito, S., Araake, M., Yan, X.-J., Fuijikawa H., Igarashi, H., Kato, H., Obata, F., Kashiwagi, N. & Inoki, H. (1989). Activation of human T cells by toxic shock syndrome toxin-1: the toxin-binding structures expressed on human lymphoid cells acting as accessory cells are class II molecules. *European Journal of Immunology*, 19, 1803–1809.

Vieira, P., de Waal-Malefyt, R., Dang, M.-N., Johnson, K.E., Kastelein, R., Fiorentino, D.F., de Vries, J.E., Roncarolo, M.-G., Mosmann, T.R. & Moore, K.W. (1991). Isolation and expression of human cytokine synthesis inhibitory factor cDNA clones: homology with Epstein–Barr virus open reading frame BCRF1. *Proceedings of the National Academy of Sciences of the USA*, 88, 1171–1176.

## Note added in proof

Two recent publications (see below) demonstrate a novel mechanism by which viruses modulate MHC antigen expression: HSV expresses a cytoplasmic protein, ICP47, which binds to the transporter associated with antigen processing (TAP) and prevents peptide translocation into the ER. Consequently, α chains do not associate stably with $\beta_2$-m, and transport to the cell surface is prevented. This is a clear example of a virus acting at site 8, as defined above.

Früh, K., Ahn, K., Djaballah, H. *et al.* (1995). A viral inhibitor of peptide transporters for antigen presentation. *Nature (London)*, 375, 415–418.

Hill, A., Jugovic, P., York, I. *et al.* (1995). Herpes simplex virus turns off the TAP to evade host immunity. *Nature (London)*, 375, 411–415.

# 7

# Modulation of MHC antigen expression by retroviruses

DOUGLAS V. FALLER
*Boston University School of Medicine*

## 7.1 Introduction

The growth cycles and oncogenic properties of the murine and human retroviruses are inextricably linked to the immune system. Those viruses that cause leukaemias, lymphomas or immunodeficiencies do so by infecting and often activating immune cells. Conversely, those retroviruses that induce solid tumours must evolve ways to aid the host cell in evading the cellular immune system. One major molecular mechanism by which these retroviruses can either activate or evade the immune system is by control of MHC class I antigen expression in the cells they infect. An effect of murine retrovirus infection on MHC antigen expression was first suspected in the late 1970s, when it was observed that thymocytes obtained from animals several weeks after infection with leukaemia viruses appeared to express higher levels of MHC class I antigens than thymic cells from control animals. Conversely, down-regulation of MHC expression on solid tumours induced by oncogene-containing (sarcoma) retroviruses had also been observed. Because of the experimental constraints of these *in vivo* systems, however, proof of a causal relationship between retrovirus infection and MHC regulation was lacking. More recent studies have demonstrated a direct action of retroviruses on MHC gene regulation and have begun to elucidate the ways in which these compact viruses, with only 6000–10 000 bases of coding sequence, regulate the histocompatibility antigen expression of their host cells.

## 7.2 Murine leukaemia viruses

The murine retroviruses can be broadly divided into two classes: the leukaemia viruses and the sarcoma viruses. The murine leukaemia viruses (MuLV) cause a persistent infection of cells. In certain cells, leukaemogenesis occurs after an extended latent period. MuLV are not, by themselves, acutely transforming or oncogenic. Closely related to the MuLV are the murine sarcoma viruses (MuSV), which contain oncogenes, induce fibrosarcomas acutely *in vivo* and transform fibroblasts *in vitro*. In the course of acquiring an oncogene in their genome by transduction of a cellular proto-oncogene, the murine sarcoma viruses lost much of their own viral coding sequences and, therefore, require MuLV as helper viruses for growth and replication.

### *7.2.1. Moloney murine leukaemia virus*

Our laboratory began investigation of MHC regulation by MuLV during analysis of the recognition of virus-induced tumours by the cellular immune system. We and others had found that CTLs directed against MuLV- or MuSV-induced tumours are H-2 restricted, in that recognition of the virus-specific target antigens must be in association with the appropriate syngeneic MHC gene products (Gomard *et al.*, 1976; Shearer, Rehn & Schmitt-Verhulst, 1976; Flyer, Burakoff & Faller, 1983; 1985a,b). The specificity of sarcoma virus-specific CTLs is directed against proteins of the MuLV helper virus (Enjuanes, Lee & Ihle, 1979; Collins, Britt & Chesebro, 1980), although other antigens may also be recognized (Peace *et al.*, 1991). Using viral gene transfection, we established that the recognition of MuSV-induced tumours by specific CTLs in the BALB/c system is mediated primarily by the viral glycoprotein gp70 (Flyer *et al.*, 1986). Clonal CTL lines, whose specificity for *env* (envelope) gene products has been demonstrated by their reactivity with *env* gene-transfected cells, are able to recognize and kill MuSV:MuLV-induced tumours. Significantly, we also found that tumours induced by MuLV and MuSV control their own recognition by the immune system, by direct regulation of MHC class I expression in the infected cells.

In experiments designed to quantify the susceptibility of BALB/c fibroblasts infected with Moloney murine leukaemia virus (MoMuLV) to lysis by allogeneic-reactive CTLs, we observed that MoMuLV-infected fibroblasts exhibited a markedly increased susceptibility to lysis by CTLs in comparison with uninfected fibroblasts. Analysis of the level of MHC class I antigen expression on the surface of MuLV-infected cells by cell surface antigen staining with anti-H-2 mAbs demonstrated that all three MHC class I antigens (H-2K, D and L) were expressed at much higher levels on leukaemia virus-infected fibroblasts in comparison with uninfected or sarcoma virus-infected fibroblasts (Fig. 7.1). Expression of all three types of MHC class I antigen was increased up to 10-fold after MoMuLV infection and such cells were

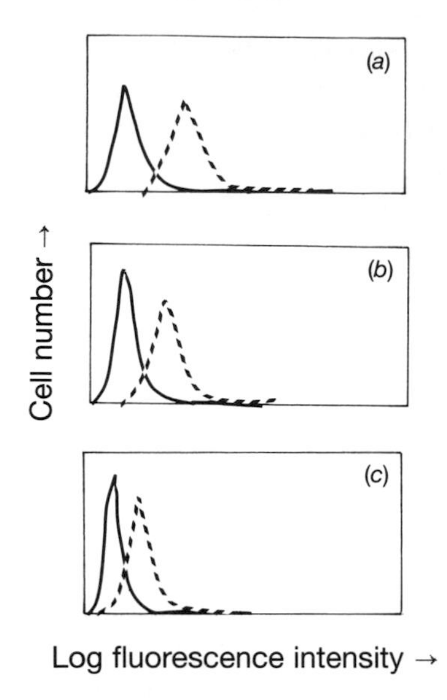


Fig. 7.1. Effect of MuLV infection of BALB/c-3T3 fibroblasts on cell surface class I MHC antigen expression. Binding of mAbs specific for H-2K$^d$ (*a*), H-2D$^d$ (*b*) and H-2L$^d$ (*c*) was measured by flow cytofluorometry after reacting infected or uninfected BALB/c cells with anti-H-2 mAbs, followed by FITC-conjugated goat anti-mouse IgG. Uninfected BALB/c-3T3, ———; MoMuLV-infected BALB/c-3T3, – – –.

now readily recognized by specific CTLs. However, co-infection of these cells with the v-*ras* or v-*mos* oncogene-containing sarcoma viruses caused a rapid and profound fall in the levels of MHC proteins on the surface of the transformed cell. Despite the large amount of viral antigens made by these tumour cells, they became invisible to the cellular immune system: they were no longer recognized by allo-specific, virus-specific or tumour-specific CTLs (Flyer *et al.*, 1985a,b; 1986). Therefore, the acutely transforming, oncogene-containing sarcoma viruses profoundly suppress MHC expression, while the non-transforming leukaemogenic viruses strongly induce MHC expression.

Having demonstrated the profound biological and immunological effects of control of MHC antigen expression by retroviruses, we began to characterize the regulation of MHC class I expression by retroviruses at the molecular level. Induction of MHC class I by MoMuLV occurs at the transcriptional level, as shown by analysis of steady-state mRNA transcripts and by nuclear run-on experiments, and involves all three class I genes (K, D and L) and $\beta_2$-m. This transcriptional regulation occurs via a *trans* effect, mediated by the virus. We have ruled out the possibility that induction of the MHC genes by MoMuLV is an indirect result of retroviral infection, mediated by elaboration of IFN, in a number of ways:

1. No soluble factors are elaborated from MoMuLV-infected cells that can increase MHC expression on other cells
2. There are no transcripts for IFN in MuLV-infected cells
3. Antibodies against IFN have no effect on the induction of MHC expression by MoMuLV (Flyer *et al.*, 1986; Faller, Flyer & Wilson, 1987; Wilson, Flyer & Faller, 1987)
4. *Trans*-activation by MuLV can be demonstrated using a transfected class I promoter–reporter gene construction in which the IFN-response element has been deleted (Koka *et al.*, 1991a).

Infection of other mouse fibroblast cell lines (NIH-3T3) with MoMuLV also induced MHC class I gene expression (Hassan *et al.*, 1990). MoMuLV infection rendered these cells sensitive to chemical transformation, and exposure of these MoMuLV-infected cells to 3-methylcholanthrene resulted in rapid cell transformation with further enhancement of H-2K class I MHC gene expression. Tumorigenicity studies suggested that the increased expression of the H-2K gene correlated with resistance to allogeneic rejection of these cells in BALB/c mice, allowing tumour growth.

The natural cellular targets for MuLV include lymphocytes as well as fibroblasts. Indirect evidence of MHC regulation by MuLV in lymphocytes has been generated in a number of studies in which increases in H-2 expression on thymocytes of mice inoculated intrathymically with radiation leukaemia virus or the AKR virus have been demonstrated (Chazan & Haran-Ghera, 1976; Meruelo *et al.*, 1978; Henley, Wise & Acton, 1984). More direct evidence for MHC class I control by retroviruses in murine and human lymphoid cells has been obtained using cultured cells. We have shown that the introduction of MoMuLV into primary, hapten-specific murine B lymphocytes, or murine T lymphocytes, resulted in increases in MHC class I antigen expression (see below).

We have reported that infection of human T lymphocyte lines with a murine leukaemia virus resulted in induction of a family of lymphocyte-specific cell surface antigens in addition to MHC antigens (Koka, van de Mark & Faller, 1991b). These proteins, which are induced by MoMuLV, include MHC class I and II antigens, CD2, CD3, CD4 and the TCR (Fig. 7.2). The expression of other cell surface proteins, such as LFA-3, was unaffected by the presence of the retrovirus. This up-regulation occurred at the level of the mRNA transcripts encoding these proteins and was the result of

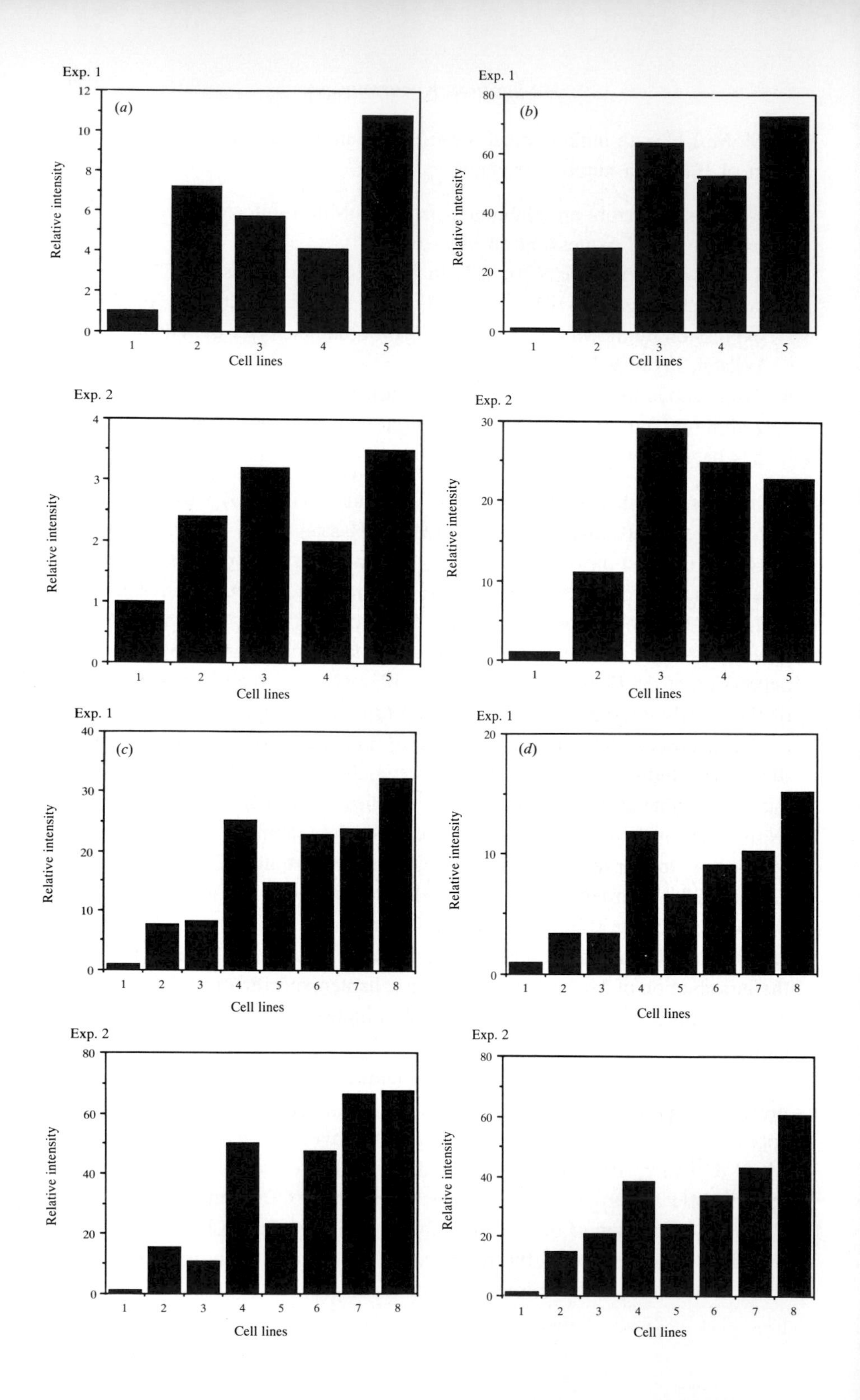

Exp. 1
(a)
Relative intensity
Cell lines
Exp. 1
(b)
Relative intensity
Cell lines
Exp. 2
Relative intensity
Cell lines
Exp. 2
Relative intensity
Cell lines
Exp. 1
(c)
Relative intensity
Cell lines
Exp. 1
(d)
Relative intensity
Cell lines
Exp. 2
Relative intensity
Cell lines
Exp. 2
Relative intensity
Cell lines

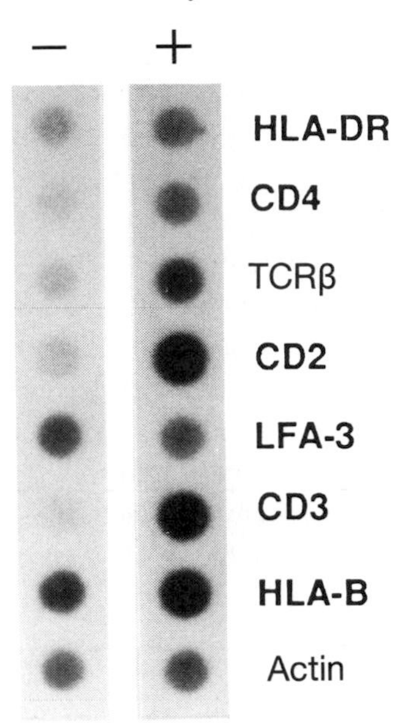


Fig. 7.3. Transcriptional activity of genes encoding T cell surface proteins in Jurkat cells in the presence of MoMuLV sequences. Nuclei were prepared from Jurkat cells, uninfected (–) or containing ZIP-NEO-SV(X) (+). Run-on transcription was allowed to proceed and the labelled RNA reaction products were exhaustively hybridized to a nitrocellulose membrane containing denatured DNA probes specific for HLA-DRβ, CD4, Jurkat TCR β chain (TCRβ), CD2, LFA-3, CD3, HLA-B7 and β-actin. An autoradiogram of the RNase-treated membrane is shown here.

increased transcription of the respective genes (Fig. 7.3). The increases in transcription were the result of a *trans*-activation process by the leukaemia virus. This pattern of host cell gene induction by MoMuLV is quite significant. Most of the molecules we found to be up-regulated, specifically the TCR, CD2, CD3, MHC class I, MHC class II and CD4 antigens, are members

---

Fig. 7.2. Immunofluorescence analysis of T cell surface proteins on a human T cell line (Jurkat) in the presence or absence of the MoMuLV vector ZIP-NEO-SV(X). Cells were reacted with mAbs specific for (*a*) MHC class I; (*b*) CD2; (*c*) CD3 or (*d*) TCR. (*a*) and (*b*) Cell lines used were: 1, Jurkat uninfected; 2, Jurkat ZIP-NEO-SV(X) clone 11; 3, Jurkat ZIP-NEO-SV(X) clone 14; 4, Jurkat ZIP-NEO-SV(X) clone 1; 5, Jurkat ZIP-NEO-SV(X) clone 2. (*c*) and (*d*) Cell lines used were: 1, Jurkat uninfected; 2, Jurkat ZIP-NEO-SV(X) clone 12; 3, Jurkat ZIP-NEO-SV(X) clone 13; 4, Jurkat ZIP-NEO-SV(X) clone 14; 5, Jurkat ZIP-NEO-SV(X) clone 17; 6, Jurkat ZIP-NEO-SV(X) clone 1; 7, Jurkat ZIP-NEO-SV(X) clone 2; 8, Jurkat ZIP-NEO-SV(X) clone 4. In each figure, the relative fluorescence intensity of the Jurkat uninfected cells was assigned an arbitrary value of 1. Exp. 1 and Exp. 2 represent two independent immunofluorescence analyses of the same cell lines at different times using mAbs of different origins as the primary antibodies.

of the immunoglobulin gene superfamily (Hood, Kronenberg & Hunkapillar, 1985; Sayre *et al.*, 1987) and all can serve as activation antigens for T cells.

MoMuLV induces increases in the transcription of genes in addition to those for T cell-specific antigens, including the lymphokine monocyte chemoattractant protein-1 (MCP-1) and type IV collagenase genes. The activation of all these genes is clearly the result of a *trans*-activation process by the leukaemia virus. The transient introduction of chimaeric genes, consisting of MHC class I gene promoter sequences, collagenase promoter sequences, or MCP gene promoter sequences driving the reporter gene chloramphenicol acetyl transferase (CAT), into human T cells or human or murine mesenchymal cells containing murine retrovirus stimulated transcription of the reporter gene. Up to this time, no *trans*-activating activities had been ascribed to the MuLVs. Subgenomic portions of the MuLV containing the long terminal

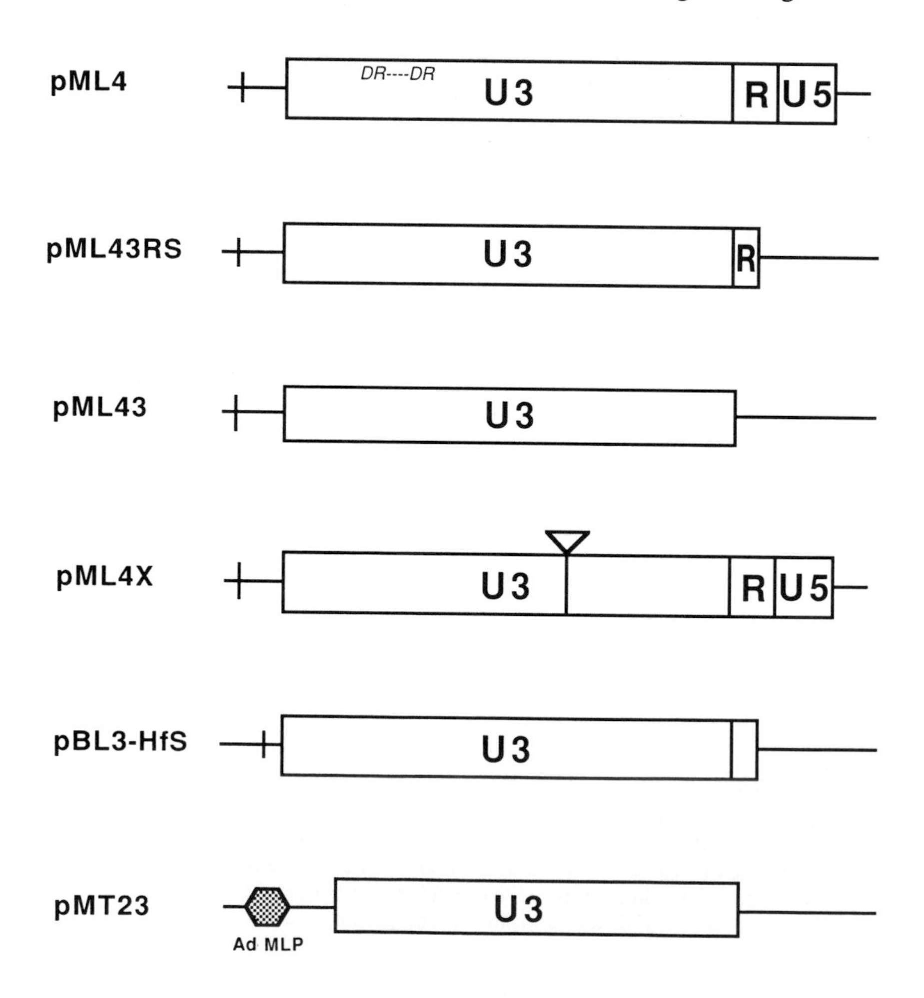

repeats (LTR) were sufficient to produce *trans*-activation of the same set of T cell or MHC genes as the whole leukaemia virus (Koka *et al.*, 1991a,b). The region of the murine retrovirus responsible for this *trans*-activating activity appeared to lie within the viral LTR. An entire proviral genome, as well as substantially deleted MuLV vectors, were capable of *trans*-activating MHC and $\beta_2$-m genes in fibroblasts and lymphocytes (Faller *et al.*, 1987; Wilson *et al.*, 1987; Koka *et al.*, 1991a,b). We developed a co-transfection assay to facilitate mapping of those retroviral sequences responsible for *trans*-activating activity (Koka *et al.*, 1991a; Choi *et al.*, 1991a,b; Choi & Faller, 1994; 1995). In these experiments, mutated or deleted retroviral constructs were co-transfected into cells along with MHC promoter–CAT constructs and the CAT activity was assayed after various times. Constructs containing only the 5′ and 3′ viral LTRs and small amounts of adjacent viral sequence possessed strong *trans*-activating activity (Fig. 7.4 and Table 7.1). Constructs containing only the 5′ LTR or a single chimaeric LTR also possessed full transcriptional activator activity. Insertion of a 9 bp linker into the U3 region of the LTR decreased *trans*-activating activity by 50-fold and deletion of four specific bases within the U3 region of the LTR abolished this activity completely. We have also been able to remove a 230 bp fragment from the

---

Fig. 7.4. Transient expression of MHC-promoter–CAT chimaeric genes after co-transfection with MuLV subgenomic elements or mutated LTR elements or LTR fragments.

Diagrams of the LTR constructions used in these *trans*-activation experiments. The deleted MoMuLV retroviral vectors were pML4 (whole LTR), pML43 (U3 region only) or pML43RS (U3 and R regions only). DR indicates location of direct repeats in the LTR. Other LTR constructions and mutations consisted of: ML4X (a deletion in the U3 region); BL3-HfS (deleted U5 and R regions); and pMT23 (296 bp of U3 region of MoMuLV LTR cloned into an expression vector, driven by the adenovirus major late promoter).

Uninfected BALB/c cells were co-transfected with a control promoterless plasmid (pSV0) or the deleted MoMuLV retroviral vector or the LTR constructions and mutations, along with chimaeric genes consisting of a CAT reporter gene driven by 800 bp of the $K^d$ MHC promoter ($MHC_{800}$–CAT) or by 64 bp of the $K^d$ MHC promoter ($MHC_{core}$–CAT). Positive (+) and negative (–) controls consisted of cells transfected with RSV–CAT or pSV0–CAT, respectively. After 48 hours, lysates of the transfected cells were assayed for CAT activity, and the autoradiogram of the reaction products was used as a template to excise the reaction products from the TLC plates for quantification. Results are shown in Table 7.1.

The retroviral vectors ML4, ML43, and ML43RS show strong *trans*-activation of both MHC–CAT constructions, as does the LTR construction BL3-HfS, in which the R, U5 and upstream *env* sequences have been deleted. Mutating the U3 region (ML4X) reduces the *trans*-activation. A fragment of the U3 region is also capable of strong *trans*-activation when driven by a heterologous promoter whether in the normal (pMT23) or inverted (pMT231) position.

Table 7.1. Trans-*activation of MHC-promoter–CAT reporter genes by mutated viral LTRs*

| LTR construction[a] | Relative CAT activity[b] | |
|---|---|---|
| | $MHC_{800}$–CAT | $MHC_{core}$–CAT |
| pSV0–CAT (alone) | 0 | 0 |
| pRSV–CAT (alone) | 100 | 100 |
| pSV0 plus MHC–CAT | 0 | 0 |
| pML4 plus MHC–CAT | 120 | ND |
| pML43 plus MHC–CAT | 130 | 100 |
| pML43RS plus MHC–CAT | 125 | ND |
| pML4X plus MHC–CAT | 25 | ND |
| pBL3-HfS plus MHC–CAT | 150 | ND |
| pMT23 plus MHC–CAT | 145 | 125 |
| pMT23I plus MHC–CAT | 95 | 75 |

ND, Not done.
[a] LTR constructions and experimental procedure are described in Fig. 7.4.
[b] Chloramphenicol acetyl transferase (CAT) activity relative to that found in cells transfected with pRSV–CAT (arbitrarily assigned a value of 100).

U3 region of the LTR, insert it into an expression vector containing a heterologous promoter and demonstrate that this fragment *trans*-activates MHC genes in a co-transfection assay. Interestingly, the fragment we have defined has been reported to contain a determinant of leukaemogenicity in MuLV strains (Hanecak, Pattengale & Fan, 1991). Mutations within the U3 region are a consistent finding in the highly leukaemogenic MCF (mink cell focus-inducing) viruses that were derived from lymphomas induced by weakly leukaemogenic viruses (Rommelaere, Faller & Hopkins, 1978). Therefore the *trans*-activating, MHC-inducing activity of the MuLV maps within the LTR – a region with known open reading frames but with no previously identified protein product. We have identified a transcript from the LTR region in cells transfected with constructs containing the MoMuLV LTR alone. There are a number of potential start sites for transcription within the LTR other than the classical start site for the viral genomic transcript and we have used RNase protection analysis and primer extension of LTR-containing probes to map all transcripts derived from the LTR. One major transcript and several smaller transcripts have been observed to arise from the LTR. The *trans*-activating 230 bp U3 fragment has demonstrable activity even when not driven by a heterologous promoter and transcripts from it are still detected in the absence of an identifiable RNA polymerase II promoter. (Choi & Faller, 1994; 1995). Studies with RNA polymerase II inhibitors

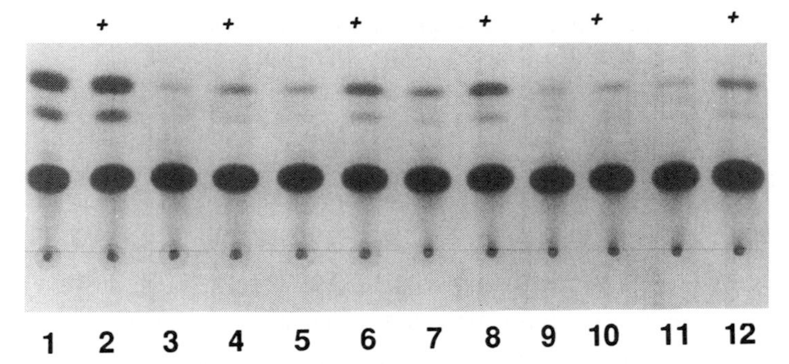


Fig. 7.5. Transient expression of MHC-promoter–CAT chimaeric genes after transfection into Jurkat cells. Uninfected Jurkat cells or cells infected with the ZIP-NEO-SV(X) vector (designated '+MoMuLV' in the figure) were transfected with chimaeric genes consisting of a CAT reporter gene driven by: the RSV-LTR (RSV–CAT) (lanes 1 and 2); 65 bp of $K^b$ promoter (p$K^b$core–CAT) (lanes 3 and 4); 650 bp of $K^b$ promoter (p$K^b$PsN–CAT) (lanes 5 and 6); 1.2 kb of $K^b$ promoter (p$K^b$PvN–CAT) (lanes 7 and 8); 2.1 kb of $K^b$ promoter (p$K^b$HN–CAT) (lanes 9 and 10); or 4.2 kb of $K^b$ promoter (p$K^b$EN–CAT) (lanes 11 and 12). Lysates of the transfected cells were assayed for CAT activity, and the autoradiograms of the reaction products are shown here. The autoradiogram was used as a template to excise the reaction products from the TLC plates for quantification.

and mutational analysis suggest that the MuLV *trans*-activator is an RNA polymerase III 'small RNA' transcript. This appears to be another example of the function of certain untranslated RNAs to act as regulators of cellular or viral genes (Choi & Faller, 1995).

Studies employing serial truncation of the murine class I MHC promoter–CAT chimaeric constructs to map the *cis*-responsive elements indicated that the response to *trans*-activation by MuLV is preserved even when only 65 bp of promoter sequence remain (Koka *et al.*, 1991a). Thus, serial deletion of the MHC H-2$K^b$ promoter from base pair –3000 to –65 does not abrogate its ability to be induced by MoMuLV (Fig. 7.5). Analogous truncations of the promoters of the other inducible genes behave in a similar fashion to the MHC class I genes. For example, the gene for the cytokine MCP-1 (encoded by the gene JE) is up-regulated by MoMuLV LTR and a JE promoter fragment containing only 74 bp upstream from the transcriptional start site is sufficient for *trans*-activation.

A potential consensus sequence for this viral *trans*-activating activity has been identified (Choi *et al.*, 1991a,b). In a number of MoMuLV-responsive cellular genes (including MHC H-2$K^b$ and JE), a 60–80 bp promoter region is sufficient for 20–30-fold activation by an MuLV provirus or MuLV-LTR in co-transfection experiments. In addition to their CAAT and TATA boxes, both promoters contain a homologous region around position –65. This area has not been previously implicated in the regulation of these genes and, therefore, represents a new molecular mechanism whereby tumour-causing viruses may regulate MHC expression in the cells they infect. Using this consensus sequence, we have predicted that promoter elements derived from certain other genes or viruses, including CMV and HIV, would be *trans*-activated if co-transfected with MuLV. Subsequent studies have demonstrated that this consensus sequence does confer inducibility by MuLV. The HIV promoter and the CMV promoter are *trans*-activated when co-transfected with an MuLV LTR. To prove that this consensus sequence is the *cis*-acting element responsive to MuLV *trans*-activation, we have established that, firstly, deletion of this region abrogated MuLV *trans*-activation, secondly, transfer of this region to a heterologous promoter resulted in the ability of that promoter to be *trans*-activated by MuLV, and, finally, introduction of single base substitutions within this region resulted in alterations in *trans*-activation. This 'MuLV-responsive' promoter element has been used to determine if there are differences in DNA-binding proteins in nuclear extracts from cells infected with MuLV or expressing the *trans*-activating LTR. Alterations in proteins binding to this consensus sequence, using extracts obtained from uninfected versus MuLV-infected cells, have been identified. MuLV LTR *trans*-activates the transcription factor complex AP-1 which regulates genes including collagenase IV, MCP-1/JE and c-*jun*. *Trans*-activation is dependent on an intact AP-1 consensus sequence in the promoter of these genes (Weng, Choi & Faller, 1995).

In contrast to the induction of MHC expression during acute infection by MoMuLV, analysis of lymphomas or thymomas induced by MoMuLV after a long latent period revealed different patterns of MHC antigen expression. One thymoma (BM5R) induced by MoMuLV expressed no detectable H-2 antigen or mRNA. When BM5R cells were fused with an H-2-expressing thymoma cell line, the H-2-negative phenotype was dominant (Baldacci *et al.*, 1986), suggesting the presence of a *trans*-acting repressor factor in these tumour cells. Other lymphoma cells induced by MoMuLV *in vivo* showed progressive loss of H-2 antigen expression when maintained in culture (Cikes, Friberg & Klein, 1973). The role of selection *in vivo* during the prolonged pre-leukaemic interval required for tumour induction cannot be ascertained in these studies.

### *7.2.2 Radiation leukaemia virus*

The first report that retroviral infection or virus-induced transformation could result in alterations of class I gene expression was by Meruelo and colleagues in 1978. An increase in H-2 expression on thymocytes occurred within 14 days of intrathymic radiation leukaemia virus (RadLV) inoculation of certain strains of mice. Resistance to RadLV-induced leukaemia is mediated by genes in the H-2D region of the murine MHC (Meruelo *et al.*, 1977). Susceptibility to RadLV-induced disease in various mouse strains was correlated with the changes in H-2 expression after virus inoculation and after tumorigenesis (Meruelo *et al.*, 1978; Meruelo, 1979). Thus H-2D antigen expression was elevated after virus inoculation of thymocytes from tumour-resistant mice, but not on thymocytes of tumour-susceptible mice. In addition, RadLV antigen expression was greater in tumour-susceptible animals, which showed little, if any, change in H-2 antigen expression. Finally, H-2 antigens disappeared from the surface of RadLV-transformed cells when overt leukaemia developed. Thus, resistance to RadLV-induced neoplasia was associated with increased H-2 antigen expression, whereas the onset of leukaemia was associated with the disappearance of these MHC antigens. The pattern of elevated H-2 antigen expression and tumour resistance was further correlated with the strength of the immune response of the host to virally infected cells. Cell-mediated immunity against RadLV-transformed or RadLV-infected cells could be detected in tumour-resistant mice. In addition, resistant mice were shown to generate more effector cells when infected with RadLV than did the tumour-susceptible mice. Further evidence of the importance of cellular immunosurveillance in controlling tumorigenesis by RadLV was obtained in studies in which thymocytes from resistant or susceptible mice, which had been infected with RadLV, were transferred into irradiated, adult, syngeneic $F_1$ mice. Cells from either strain of mouse caused tumours in a time period that was too short to be the result of *de novo* induction of the leukaemia in the recipient adults. Rather, it was more likely that pre-leukaemic cells generated by RadLV in the thymus of the resistant or susceptible donor mice were able to escape from immunosuppression in the recipient and to develop into a full leukaemia.

The likelihood of *in vivo* selection is always a significant problem in analyses of MHC expression several weeks after intrathymic injection of virus. *In vivo* selection has not been ruled out as the cause of loss of MHC class I expression after RadLV-induced transformation of thymocytes. The H-2D increases occurring after RadLV infection, however, do not appear to result from selected proliferation of a 'high-H-2' subpopulation of cells in the thymus. At least 85% of thymus cells showed increased cell surface H-2 antigen expression as little as 36 hours after virus inoculation. In addition, K and D determinants showed different time courses of increased expression both between susceptible and resistant mice and also on thymocytes of a single mouse strain (Meruelo *et al.*, 1978). RadLV appears to preferentially infect immature T cells. Less than 10% of all thymocytes in resistant mice were positive for viral antigen expression at 36 hours after virus inoculation, yet greater than 85% of all such cells displayed marked increases in H-2D antigen expression.

Similar experiments to those described above, using variants of the RadLV (called A-RadLV), have yielded somewhat contradictory results, reporting virus-induced elevations in H-2 antigen expression to be only transient (lasting six to seven weeks) in tumour-resistant animals, whereas the increase in MHC antigen expression was persistent in the tumour-sensitive animals, lasting until the development of leukaemia (Katz, Peled & Haran-Ghera, 1985). Furthermore, these studies showed that the majority of RadLV-induced leukaemias expressed more H-2D and H-2K gene products than did normal thymocytes. One possible explanation for these discrepancies between the various RadLV strains and their correlation with MHC expression and leukaemogenesis may be found in genetic studies which have shown that the susceptibility to RadLV of the Kaplan strain maps to the H-2D region (Meruelo *et al.*, 1977; Lonai & Haran-Ghera, 1980), whereas susceptibility to the A-RadLV has been mapped to the I subregions A, B and J of the MHC, suggesting that a humoral response may play a more important role than cellular immunity in the leukaemogenic properties of the A-RadLV strain.

The molecular mechanism by which RadLV alters MHC class I gene expression is not known. These studies are complicated by the need for intrathymic injection of virus and subsequent analysis of antigen expression levels in a mixed population of thymocytes many days later. There is some evidence for a direct effect of RadLV on MHC gene transcription and a requirement for viral integration. RadLV replication has been proposed as necessary for the increase in H-2 antigen expression. Both ultraviolet inactivation and X-irradiation of virus prevented the RadLV-induced increases in expression of H-2 antigens while not affecting virus adsorption or entry into cells (Decleve *et al.*, 1977; Meruelo & Kramer, 1981). These experiments, however, do not rule out effects of RadLV that are independent of integration. A requirement for viral replication and integration was also suggested by the finding that the *Fv-1* locus, which restricts both of these events, also regulated the viral increase in H-2 antigen expression (Mellor *et al.*, 1984). Blocking viral integration with FUdR inhibited H-2D antigen induction. Rearrangements in MHC class I DNA appear to accompany RadLV transformation, as do changes in the methylation state of the MHC class I genes (Meruelo *et al.*, 1986), but the role of these alterations in the observed changes in MHC class I antigen expression is not clear. The possibility of retroviral integration near MHC class I genes as a mechanism for their activation or inactivation has been proposed (Rossomando & Meruelo, 1986). Proviral insertion into the cellular DNA is not always random (King *et al.*, 1985). The 'preferential' retroviral integration site *Pim-1* has been linked to the H-2 complex (Brown *et al.*, 1988). However, since the *Pim-1* integrations were analysed only in tumours, the possibility exists that integrations were indeed random, but only those integrations that occurred by chance within the *Pim-1* locus were observed because of preferential proliferation of such cells.

In the RadLV system, as in the MoMuLV system, activation of H-2D$^d$ antigen expression appears to operate by *trans*-regulation of *cis*-acting elements in the H-2 gene. Inoculation of hybrid mice (resistant × susceptible $F_1$, whose cells express both the H-2D$^d$ and the H-2D$^q$ antigens) with RadLV leads to increased cellular expression of resistant haplotype antigens (H-2D$^d$) but not the other susceptible haplotype antigen (H-2D$^q$) (Meruelo & Kramer, 1981), suggesting that the *cis*-acting elements upstream of H-2D differ from those upstream of H-2D$^q$. C57BL/6 mice transgenic for an 8.0 kb DNA fragment containing the entire H-2D$^d$ gene, along with 5′ and 3′ flanking DNA, demonstrated induction of H-2D$^d$ antigen expression after infection with the RadLV (Brown *et al.*, 1988). Activation of MHC class I genes by RadLV does not appear to be a secondary effect mediated by IFN (Sonnenfeld *et al.*, 1981). There is, however, some evidence that a soluble mediator produced by virus-infected cells can act on neighbouring cells uninfected by RadLV (Meruelo *et al.*, 1978). Soluble extracts from RadLV-infected thymocytes were capable of selectively increasing H-2D antigen expression on thymocytes of tumour-resistant but not of tumour-susceptible mice (Meruelo & Kramer, 1981). A substantial portion of this activity of the soluble extracts could be removed when virus particles were removed from the extracts. The mechanisms underlying the induction of MHC class I in newly infected thymocytes by RadLV have been studied at the molecular level (Nobunaga *et al.*, 1992). Increases in H-2D antigen expression were observed as early as 12 days after inoculation. H-2D antigen expression continued to increase for up to four to five weeks and then began a steady decline. Nuclei prepared from thymocytes at day 17 after inoculation showed a 3–5-fold increase in transcription from the H-2D$^d$ gene, compared with uninfected thymocytes. A DNA-binding protein activity was found to be increased in RadLV-infected cells. This activity bound to an AP-1-like sequence upstream of the D$^d$ gene and also bound to an AP-1 consensus sequence. The protein(s) that constitutes this binding activity, however, does not appear to be the *fos* or *jun* protein which comprises the AP-1 transcription complex. It has not yet been demonstrated that the DNA-binding protein identified in these studies is indeed responsible for the *trans*-activation of MHC genes by the RadLV.

Modulation of MHC expression by RadLV may influence the natural history of the viral infection in ways other than altering susceptibility to immune surveillance. Clonotypic anti-TCR antibodies completely inhibit RadLV binding to a RadLV-induced thymoma. Anti-CD4 and anti-H-2K$^b$ antibodies but not anti-H-2D$^b$ antibodies were found to inhibit partially the ability of RadLV to infect these cells. Conversely, when free virus particles were adsorbed to receptors on the thymoma cells, the binding of several anti-H-2K$^b$ antibodies was specifically inhibited. Because the TCR has been shown to co-modulate with H-2K$^b$ antigens, it has been proposed that the TCR may be a receptor structure for RadLV and that the H-2K$^b$ molecule may be part of a complex that can form around the TCR (O'Neill, 1989).

The mechanism by which MHC class I expression is down-regulated in those cells eventually transformed after RadLV infection is unclear. Selection *in vivo* for MHC class I loss is likely. In those RadLV-induced tumours with no MHC class I antigen expression, all class I mRNA is similarly lacking. As in the case of sarcoma virus-transformed cells (Flyer *et al.*, 1986), IFN can reinduce expression of H-2 antigens in the RadLV-transformed cell lines (Sonnenfeld *et al.*, 1981).

### *7.2.3 AKR/Gross leukaemia virus*

MoMuLV and RadLV are leukaemogenic retroviruses that have undergone prolonged passage in the laboratory and selection for leukaemogenic potential. As such, they are deemed 'exogenous' leukaemia viruses and have no counterparts in the mouse genome. The endogenous viruses of the BALB/c or AKR strains do exist intact in the genomes of these strains, are inducible and can confer a propensity for leukaemia to the mouse strains in which they are found. The endogenous MuLV are not, by themselves, acutely transforming or oncogenic. They can undergo a recombinational event with other cellular retroviral sequences to generate a new leukaemogenic virus after infecting an animal. These new recombinant viruses (known as MCF, for mink cell focus-inducing) can now induce a lymphoma or leukaemia after a latency of several months (Hartley *et al.*, 1977). These endogenous viruses and their MCF counterparts have also been demonstrated to affect MHC expression in their host cells. Analysis of this effect is complicated by several factors. These studies almost invariably analyse MHC expression in tumours that have arisen spontaneously or in a latent fashion after virus inoculation. Therefore, the originally inoculated or reactivated virus may not be the proximal agent in the leukaemogenic process. These studies also suffer from the lack of an appropriate control cell for comparison of class I MHC antigen expression levels.

Two variants of a T lymphoblastoid cell line that differed by the presence or absence of productive infection with an AKR/Gross leukaemia virus were studied for H-2 antigen expression. Selective elevation of H-2 antigen and preferential association of the Thy-1 antigen with MuLV virus variants and a productively infected variant line suggested that retrovirus replication in T lymphoid cells can alter the quantitative expression of cell surface antigens and perhaps also the arrangement of the antigens on the surface of the cell. Age-related changes in the expression of cell surface antigens on pre-leukaemic AKR thymocytes include an increase in the expression of H-2 antigens and a decrease in the expression of Thy-1 (Kawashima *et al.*, 1976). The expression of H-2$K^b$ and H-2$D^d$ was markedly elevated in the Gross virus-infected cells by factors of 7- to 15-fold (Henley *et al.*, 1984).

MCF 1233 induces lymphoblastic T cell lymphomas and also follicle centre cell or lymphoblastic B cell lymphomas. Induction of T cell lymphomas by MCF 1233 is influenced by a helper T cell-dependent, H-2 Ia-restricted anti-viral immune response (Vasmel *et al.*, 1988). T cell lymphomas are frequently induced in the H-2 Ia-non-responder type mice and rarely induced by the virus in H-2 Ia-responder type mice. MCF 1233-infected thymocytes show moderate to high MHC class I expression without strain-specific features. Tumours isolated from non-responders expressed high levels of both *env* and *gag* viral proteins and 15 of the 17 tumours from non-responder mice expressed high levels of H-2 class I K and D antigens. Ten of eleven lymphomas from responder animals, however, lacked *env* and/or *gag* determinants. The only responder lymphoma that continued to express strong *env* and *gag* expression had lost expression of H-2K and D antigens. Notably, H-2-dependent clearance of MCF 1233-infected thymocytes was not correlated with the amount of class I antigens present at the cell surface. The H-2 Ia-regulated anti-viral immune response, therefore, not only reduced T cell lymphoma incidence but also selected for loss of expression of viral antigens among those lymphomas that did occur, allowing them to escape from immunosurveillance (Vasmel *et al.*, 1989).

Analysis of MHC expression on tumours induced by endogenous retroviruses suggests that regulation of MHC expression may be occurring, although direct proof is lacking in such studies. *In vitro* work from our laboratory has studied more directly the effects of AKV and Gross viruses on MHC expression. In experiments comparable to the ones described for the LTR of MoMuLV, the LTR from AKV has been shown to *trans*-activate the endogenous H-2K$^d$ gene in stable transfectants and to *trans*-activate chimaeric genes containing the H-2K$^d$ promoter linked to CAT in co-transfection assays. In all cases, however, the magnitude of *trans*-activation by the AKV LTR was considerably less than that of the MoMuLV LTR.

It is quite likely that strong immunoselection occurs during tumorigenesis. In contrast to RadLV immune recognition, the H-2K locus product is a major, if not the sole, restricting element for the CTLs directed against AKR/Gross MuLV-induced tumours or spontaneous AKR lymphomas. Modulation of this particular MHC-restricting element expression may constitute a particularly efficient tumour escape mechanism. The Gross virus-induced AKR T cell leukaemia line K36.16 expresses normal levels of H-2D antigens but very low levels of H-2K antigens (Schmidt & Festenstein, 1982). K36.16 tumour cells are resistant to H-2K-restricted killing by T cells. The H-2K antigen is necessary for CTL recognition of the Gross virus antigen that is expressed on these cells. Transfection of exogenous H-2K sequences and their expression at high levels results in rejection of transplanted K36.16 tumour cells. Thus, the absence of H-2K expression on these cells directly correlates with their ability to grow in immunocompetent hosts.

Five cultured tumour cell lines derived from BALB.K mice inoculated with Gross MuLV were examined for MHC class I expression (Klyczek, Murasko & Blank, 1987). All lines were Thy-1 positive, L3T4 negative and Lyt-2 negative. Two of the cell lines expressed no detectable cell surface or intracellular H-2K$^k$ or H-2D$^k$ class I proteins. Three other cell lines expressed both K$^k$ and D$^k$ proteins. Treatment with IFN-$\gamma$ increased K$^k$ and D$^k$ expression at 48 hours on the three cell lines that were originally positive. One of the class I-negative cell lines could be induced by IFN-$\gamma$ to express D$^k$ but not K$^k$. The remaining cell line was not inducible for any class I expression. IFN-$\alpha$/$\beta$ and TNF both enhanced cell surface K$^k$ and D$^k$ expression on the three positive cell lines. Neither cytokine was able to induce expression of class I on either of the two constitutively negative cell lines. The failure to induce class I protein did not result from lack of receptor for the cytokines or to any obvious structural rearrangement or deletion of the class I genes. Expression of $\beta_2$-m mRNA was enhanced by IFN-$\gamma$ in all five tumour cell lines. In similar studies examining the response of thymic lymphoma cell lines from mice expressing the H-2K haplotype induced by AKV/Gross MuLVs, IFN-$\gamma$ again induced expression of H-2D$^k$ but not H-2K$^k$ (Green & Phillips, 1986). Analysis of somatic cell hybrids from fusions of AKR-SL3 with an H-2K$^b$/H-2D$^b$ IFN-$\gamma$-inducible partner tumour cell showed full induction of K$^k$ in response to IFN-$\gamma$ in the hybrid populations. Therefore, the lack of K$^k$ augmentation by IFN-$\gamma$ is not likely to result from a K$^k$-specific repressor factor operating *in trans* (Rich *et al.*, 1990). These findings suggest that there may be a correlation (albeit indirect) between the lack of IFN responsiveness of H-2K$^k$ and the induction of tumours by the AKR/Gross MuLV.

When five spontaneous AKR/Gross virus leukaemic cell lines were analysed for H-2 class I antigen expression (H-2K), a novel class I antigen was detected on the surface of all five cell types, which was reactive with an H-2D$^d$-specific mAb (Labeta *et al.*, 1989). It was also recognized *in vitro* by anti-H-2D$^d$-specific, allo-reactive CTLs. An allo-antigen-like glycoprotein could be immunoprecipitated from the tumour cell lines. In some cases, this new antigen appeared to associate with the syngeneic class I molecules on the cell. Analysis of cellular mRNA demonstrated that a transcribed sequence similar to that of exon 5 of H-2D$^d$ was present in all of the tumours. Analysis of the glycoprotein showed that its amino-terminal and C2 domains appeared to be D$^d$-like. Exon 5 encodes the transmembrane region of the H-2D$^d$ gene product. The origin of this new antigen is as yet unknown, but it may be identical to the AKR Q5 or to other Qa-like genes. Activation of the Qa/Tla region-encoded products in the course of AKR leukaemogenesis may be common (Festenstein *et al.*, 1983). Whether this novel antigen is induced by the virus remains to be determined. One of these tumour cell lines that expressed minimal or undetectable levels of H-2K was transfected with a normal AKR H-2K gene. The transfected tumour cell lines were readily rejected by immunocompetent mice, whereas the non-transfected tumour cell line grew well in these mice. The efficiency of rejection was dependent on the amount of H-2K antigen expressed after transfection and rejection could be prevented by pre-incubation of the transfected tumour cell lines with specific anti-H-2K$^k$ mAbs. Furthermore, these transfected cell lines also could act as vaccines, in that when mice were rechallenged with the original non-transfected K36.16 tumour cell lines, they were also rejected (Hui, Grosveld & Festenstein, 1984). Another Gross MuLV-induced tumour cell line, which expressed no H-2K or H-2D class I antigens, appeared to express a different tumour-specific transplantation antigen, which induced tumour rejection *in vivo* and CTL generation *in vivo* without prior immunization. No transcripts hybridizing to a broadly reactive MHC class I DNA probe were found in the cells, but $\beta_2$-m and a $\beta_2$-m-associated protein were found on the surface of these cells. Therefore, although these tumour cells appeared to express a class I-like MHC antigen, this antigen did not appear to be encoded by known class I genes (Klyczek & Blank, 1989).

The majority of 15 primary tumours derived from AKR mice showed low levels of H-2K gene product expression (Festenstein, 1989). The H-2K antigen epitope on these tumours was concealed by increased attachment of sialic acid moieties and treatment with neuraminidase unmasked the H-2K epitopes (Labeta *et al.*, 1989). Analysis of H-2 antigen expression on the surface of three different BALB/c tumour cell lines induced by Gross leukaemia virus showed varying levels of H-$2K^d$ expression and H-$2D^d$ expression. In all cases, normal BALB/c spleen cells expressed substantially more H-2 antigen than did any of the tumour cell lines. Analysis of the H-2 restriction patterns exhibited by CTLs directed against these tumours suggested that the pattern of H-2 restriction might be directed by the quantitative modulation of H-2K or H-2D antigen expression by the tumour cells. That is, the H-2 restriction patterns of syngeneic CTLs directed against Gross MuLV-induced tumours were correlated with the quantity of H-$2K^d$ and H-$2D^d$ antigens expressed (Plata *et al.*, 1981).

### 7.2.4 Defective endogenous leukaemia viral sequences

Defective or incomplete viral or viral-like sequences have been discovered in the murine and human genomes and indirect studies suggest that they may play a role in MHC gene regulation. Two endogenous proviral-like sequences have been identified in the TL region of normal C57BL/10 mice (Meruelo, 1980; Meruelo & Kramer, 1981; Meruelo *et al.*, 1984). One sequence appeared to be a defective viral genome, while the other contained a 9 kb insertion that comprised sequences related to MuLV, bounded by VL30 LTRs which were retrovirus-like (Brown *et al.*, 1988). The Qa/Tla loci contain class I genes that are less polymorphic than H-2D, K and L, but there is a high degree of homology between the TL antigens and the classical MHC transplantation antigens (see Chapter 2). The function of the TL region of the mouse MHC is not known, but the TL antigens may function as restriction elements for TCR-γ/δ-expressing T cells. VL30-like elements have been isolated from the H-2 locus (Qa, K, TL and D regions) of several inbred mouse strains and viral sequences are also found adjacent to non-MHC histocompatibility loci and genes encoding several H/Ly antigens (Meruelo, 1980; Meruelo & Kramer, 1981; Blatt *et al.*, 1983; Wejman *et al.*, 1984; Rossomando & Meruelo, 1986). They may function in the generation of polymorphisms of H-2 genes between various strains. The possibility that the presence of these elements affects H-2 gene expression remains unproved (Choi & Meruelo, 1991).

Two endogenous retroviral LTRs were found in the human MHC complex locus HLA-DQ (Kanbhu, Falldorf & Lee, 1990). These elements exhibit greater than 90% homology to the LTRs of the human endogenous retrovirus HERV-K10. Endogenous retroviral genomes have been found in multiple locations within the region containing the murine non-classical class I genes (Pampeno & Meruelo, 1986). The putative enhancer regulatory sequences within the LTRs were conserved. These elements were not found in the DQ alleles of all haplotypes. The presence of these endogenous retroviral LTRs upstream of DQB1 may dissociate DQ expression in some cases from DR and DP expression.

It has been proposed that expression of endogenous viral gene products, particularly *env* products, on the cell surface could serve as histocompatibility antigens. Although the location of these viral sequences in the MHC is distinct from the major class I genes, there is some evidence that integrated retroviruses might be the equivalent of minor H genes (Brown *et al.*, 1988). A role for retroviruses in the generation of minor H gene polymorphisms and evolution has been suggested. If retroviruses played a role in the generation of MHC gene polymorphism, one would predict that whenever the retroviral family of genes is highly polymorphic in one species, the MHC genes would be expected to display significant polymorphism, with the converse also being true. This prediction was confirmed in the study of the genome of the Syrian hamster (Atherton *et al.*, 1984).

The role of the endogenous MuLVs in control of lymphocyte activation was studied by culture of murine spleen cells with specific anti-sense oligonucleotides complementary to retroviral sequences. Anti-sense oligonucleotides to endogenous MCF env gene translational initiation regions caused increases in RNA synthesis and increases in lymphocyte cell surface H-2 Ia and Ie expression relative to cells that received control oligonucleotides. The anti-sense oligonucleotides to xenotropic or ecotropic envelope sequences or to endogenous MCF non-envelope sequences had no effect. Therefore, endogenous MCF sequences may exert an inhibitory influence on the murine immune system (Krieg *et al.*, 1989).

### *7.2.5 Role of leukaemia virus induction of MHC antigens in leukaemogenesis*

It is significant that the cell surface proteins up-regulated by MuLV have been implicated in one or more T cell activation pathways (Koka *et al.*, 1991a,b). The TCR, CD3 and CD4 comprise the classical activation complex, while CD2 comprises an 'alternative' activation pathway. Stimulation of either MHC class I or class II molecules by mAbs has been reported also to induce markers of T lymphocyte activation (Poltronieri *et al.*, 1988; Cambier & Lehmann, 1989; Geppert *et al.*, 1989). The mechanisms by which the MuLVs induce leukaemogenesis are largely unknown. MuLV contains no classical oncogenes. MuLV promotes the induction of leukaemia in infected mice after a long latent period through a complex multi-step process involving the generation of novel recombinant retroviruses (Teich *et al.*, 1982). After infection by these thymotropic retroviruses and during the pre-leukaemic phase, the major cell population in the thymus bears high levels of H-2 antigens and viral gene products (Chazan & Haran-Ghera, 1976; Lee & Ihle, 1981). We and others have proposed that the increased levels of these lymphocyte cell surface proteins may facilitate the chronic immunostimulation that appears to be necessary for the leukaemogenic process during this phase. This immunostimulation may serve to provide the recombinant leukaemogenic viruses with a proliferating target cell population (Wilson *et al.*, 1987) or, perhaps, to enhance auto-stimulation of MuLV-infected, MuLV-specific T lymphocytes, producing a pre-neoplastic lymphoid hyperplasia (McGrath & Weissman, 1979; Plata *et al.*, 1981; McGrath, Tamura & Weissman, 1987; O'Neill *et al.*, 1987; Wilson *et al.*, 1987). Thus, the ability of MoMuLV to enhance expression of activation proteins in its target cells may be intrinsic to its ability to eventually generate lymphoid neoplasia. We have proposed that the fact that MuLVs enhance transcription and expression of a group of T cell surface proteins, all of which have been reported to be capable of transducing an activating signal to the lymphocyte, may, therefore, be relevant to the pathophysiological mechanisms whereby these viruses induce leukaemias and lymphomas. The ability of murine retroviruses also to *trans*-activate cellular genes encoding proto-oncogenes and transcription factors (such as c-*jun*), chemoattractant lymphokines (MCP-1) and enzymes that mediate tumour invasion, e.g. collagenase IV (Hendrix *et al.*, 1992), is also likely to be important in the tumorigenic process (Choi & Faller, 1994; 1995). This hypothesis will eventually be tested by the development of mutant MuLV incapable of such *tran*-activation and evaluation of their ability to cause pre-neoplastic lymphoid hyperplasia and eventual lymphoma or leukaemia.

### 7.3 Murine sarcoma viruses and viral oncogenes

Studies on a variety of human non-lymphoreticular tumours (Fleming *et al.*, 1981; Bhan & Des Marais, 1983; Whithwell *et al.*, 1984; Umpleby *et al.*, 1985; Momburg *et al.*, 1986), as well as in cell lines established in cultures from tumours (Trowsdale *et al.*, 1980; Lampson, Fisher & Whelan, 1983; Natali *et al.*, 1983; Doyle *et al.*, 1985), have revealed attenuation or virtual absence of HLA class I antigens. Although the role of oncogenes in the generation of all of these solid tumours is not clear, there appears to be a critical involvement of activated *ras* in specific human malignancies, especially bowel and lung cancer, in which concurrent loss of MHC expression is common (Doyle *et al.*, 1985; Umpleby *et al.*, 1985; Eliott *et al.*, 1986; Smith, 1991).

Cell lines from naturally occurring tumours may not reflect the initial stages of malignant transformation, since they may have been subjected to immunoselection during tumour progression *in vivo*. In addition, the normal cells that are usually used for comparison to the tumours are not the true progenitor cells from which the tumour arose. Finally, there is certainly variation in the expression of HLA genes in cells from various tissues. Consequently, control of MHC antigen expression by oncogenes has been more easily studied in model systems in culture, often employing the murine sarcoma viruses to transduce the oncogene. The level of CTL-mediated lysis of tumour cells both by MHC-restricted, tumour virus-specific CTLs and by CTLs directed against allogeneic MHC determinants is directly influenced by the level of MHC class I antigen expression on the surface of the tumour cells (Kuppers *et al.*, 1981; Plata *et al.*, 1981; Flyer *et al.*, 1983; 1985a,b). In many oncogene-transformed cells, the level of MHC protein expression is so dramatically down-regulated that the tumour viral antigens cannot be recognized by CTL (which require both syngeneic MHC antigens and bound viral peptides on the cell surface for recognition) or by CTL lines.

As discussed above, the expression of all three types of MHC class I antigens is increased up to 10-fold after MoMuLV infection and such cells are readily recognized by virus-specific CTL. However, co-infection of these cells with the v-*ras* or v-*mos* oncogene-containing sarcoma viruses caused a rapid and profound fall in the levels of MHC proteins on the surface of the transformed cell. Despite the high density of viral antigens expressed by these tumour cells, they were no longer recognized by allo-specific, virus-specific or tumour-specific CTLs (Flyer *et al.*, 1985a,b; 1986). Thus, the acutely transforming, oncogene-containing murine sarcoma viruses profoundly suppress MHC expression, while the non-transforming leukaemogenic murine viruses strongly induce MHC expression. The inhibition of MHC expression by v-*ras* or v-*mos* can be reversed by IFN-γ (Flyer *et al.*, 1985a,b). Treatment of MSV-infected fibroblasts with IFN-γ increased their susceptibility to lysis by both allogeneic and syngeneic CTLs.

### *7.3.1 MHC regulation by* ras *and* mos *oncogenes*

The effects of v-*mos* or v-Ki-*ras* oncogenes on MHC gene expression have been studied in cultured cells. Control of MHC class I antigen expression and response to IFN by viral oncogenes *in vitro* is complex and, in part, dependent on cellular density and culture conditions (Offermann & Faller, 1990). Either the v-*mos* or the v-Ki-*ras* oncogene can indirectly alter the enhancement of MHC class I antigen expression in response to IFN-γ. In v-Ki-*ras*- and v-*mos*-transformed cells, subconfluent cells have a greater increase in MHC class I antigen expression in response to IFN-γ than cells that have exceeded monolayer confluence. A serum factor also acts in the control of MHC expression in tumour and oncogene-transformed cells, and this has led to the identification of a novel mechanism of autocrine regulation of MHC class I gene expression by induction of IFN-β gene expression in transformed BALB/c-3T3 cells (Offermann & Faller, 1989). Under conditions of low serum levels (0.5 %), a 4- to 9-fold increase in cell surface levels of MHC class I antigens is induced in both v-Ki-*ras*- and v-*mos*-transformed cells but not in untransformed cells. These increases in MHC class I antigen levels result from induction of both the class I genes and the $\beta_2$-m genes. IFN-β mRNA is also induced by low-serum conditions in the transformed BALB/c-3T3 cells. Therefore, low-serum conditions lead to induction of IFN-β expression in the oncogene-transformed cells, which then directly mediates the autocrine enhancement of MHC class I gene expression (see also Real, Fliegel & Houghton, 1988). The serum factor responsible for MHC suppression is heat and acid stable and is sensitive to proteases.

Conversely, suppression of MHC gene expression in transformed cells in the presence of the serum factor is the result of inhibition of the endogenous IFN-β gene and IFN protein expression (Offermann & Faller, 1989). The suppression of IFN-β by this serum factor is relatively rapid (60–90 minutes). IFN-β mRNA levels correlate with the levels of IFN-β protein and MHC antigen, suggesting that the control of IFN and MHC protein by this serum factor is at the level of IFN-β mRNA transcription, a conclusion that was verified by nuclear 'run-on' transcription studies. This abnormal control of the IFN-β gene in oncogene-transformed cells differs from all previously reported situations in that the increased transcription of the IFN-β gene persisted in these cells as long as they remained in low-serum conditions. This is in marked contrast to the very transient up-regulation of transcription described for all the known inducers of IFN expression. Using transient expression of plasmids carrying nested deletions within the IFN-β promoter attached to the reporter gene CAT after transfection into v-Ki-*ras*- or v-*mos*-containing cells, the serum factor-responsive region of the IFN promoter has been mapped to within 77 bp of the start site. This 77 bp region thus defines a minimal upstream control region, which contains the previously defined positive regulatory domain I (PRDI), which extends from base pair –77 to

–64; PRDII, which extends from base pair –66 to –55; and the negative regulatory domain NRDI, which is located from base pair –63 to –36 (Du & Maniatis, 1992).

There are, therefore, strong parallels between control of MHC gene expression by the MoMuLV/MSV complex and the RadLV. Non-transforming infection by MoMuLV or RadLV increases MHC expression by a *trans*-activation mechanism. Conversely, transformation (acutely) by the oncogene-containing MSV or (over a period of time) by RadLV results in significant decreases in MHC class I expression. The mechanism of transformation by RadLV after a long latent period *in vivo* is not known. It remains possible that such transformation is the result of activation of endogenous oncogenes.

The modulation of MHC class I expression by oncogenes may depend in large part on the cell type analysed. MHC class I gene expression was examined in two epithelial differentiated cell lines and in one fibroblast cell line derived from rat thyroid before and after infection with various murine retroviruses containing transforming oncogenes (Fontana *et al.*, 1987; Racioppi *et al.*, 1988). The transforming genes studied included v-*ras*, v-*mos*, v-*src*, and the polyoma virus middle T oncogene. Infection of the thyroid fibroblast cells with the transforming oncogenes resulted in diminution of MHC class I expression. The two epithelial differentiated rat thyroid cell lines, however, behaved differently upon transformation by the oncogenes. One cell line showed a positive modulation of MHC class I antigen expression with v-Ha-*ras*, v-Ki-*ras*, v-*mos*, v-*src* and middle T. The other thyroid epithelial cell line showed a decrease in MHC class I after transformation with v-*mos* while v-Ha-*ras* and polyoma middle T caused almost complete loss of MHC class I expression. Examination of steady-state mRNA levels showed that in many cases the decreases in MHC class I expression were not accompanied by similar changes in mRNA levels or $\beta_2$-m transcript levels. Post-transcriptional mechanisms were, therefore, suggested to account for the changes in MHC class I expression observed in these cell lines.

MHC class I expression in lymphocytes containing *ras* or *mos* is altered in specific ways. Introduction of MuLV and MSV into primary, hapten-specific B lymphocytes or antigen-specific T lymphocytes resulted in established lines of murine T or B lymphocytes containing specific oncogenes (Lichtman *et al.*, 1986; 1987; 1995). For example, stimulation of the normal cells through the TCR or cell surface immunoglobulin resulted in transient (24 hour) up-regulation of class I. In the cell lines expressing the v-*ras* or v-*mos* oncogenes, however, this induction did not occur.

Activated *ras* oncogenes have also been reported to influence MHC class

II antigen expression, although again the magnitude and direction of the regulation varied depending on the type of cell studied. An effect of the *ras*-containing Kirsten murine sarcoma virus on class II antigens in mouse fibroblast cell lines has been reported (Maudsley & Morris, 1988). None of the fibroblasts examined constitutively expressed any detectable levels of class II antigen, but one line, C3H10T1/2, could be induced to express both H-2A and H-2E by IFN-γ. Infection of these cells with a v-Ki-*ras*-containing virus, with or without the helper virus, completely abolished the ability of IFN-γ to induce class II antigen and substantially reduced the induction of MHC class I antigens by IFN. The degree of reduction of class II expression varied among infected cell lines (Maudsley & Morris, 1989). In contrast, transformation of human melanocytes by Harvey murine sarcoma virus (containing v-Ha-*ras*) resulted in an increase in MHC class II antigens (Albino *et al.*, 1986). The effect of v-*ras* was specific in these melanocytes, and other oncogenes, including *myc*, *src*, and *fms*, did not induce class II antigen expression.

### *7.3.2 MHC regulation by the* src *oncogene*

Rous sarcoma virus (v-*src*)-transformed human fibroblasts cells demonstrated a profound (4- to 9-fold) reduction in cell surface MHC class I expression as well as cell surface $\beta_2$-m expression (Gogusev *et al.*, 1988). Immunoprecipitable MHC class I antigens and $\beta_2$-m from cell membranes or from total cell lysates was substantially decreased in the *src*-transformed human fibroblasts. Steady-state levels of MHC class I mRNAs were reduced in the transformed cells in a comparable pattern. The v-*src* gene product of the Rous sarcoma virus is a tyrosine kinase. Phosphorylation of the serine residues within the intracellular domain of the HLA class I heavy chains has been demonstrated for HLA-A and HLA-B antigens both *in vivo* and *in vitro* in normal as well as in EBV-transformed lymphocytes. H-2 proteins are also known to be phosphorylated *in vitro* at serines in the intracellular domain (Pober, Guild & Strominger, 1978; Rothbard *et al.*, 1980). The intracellular domains of the HLA gene products contain single tyrosine residues at amino acid 320. *In vitro* phosphorylation studies using purified pp60-v-*src* and purified, detergent-soluble HLA-A2 and HLA-B7 antigens demonstrated labelling at a single tyrosine residue of the intracellular domain (Guild, Erikson & Strominger, 1983), suggesting that the v-*src* oncogene product may directly or indirectly alter MHC antigen function by biochemical interaction with the class I antigen itself. The *src*-transformed human fibroblasts in the study of Gogusev *et al.* (1988), however, did not demonstrate phosphorylation of the tyrosine residues of the MHC class I heavy chains. Furthermore, there was no evidence for insertional mutagenesis or genomic rearrangement of the MHC genes in these cell lines. Therefore, the mechanism by which v-*src* reduces expression of MHC antigens remains unknown.

### *7.3.3 MHC regulation by the abl oncogene*

Abelson leukaemia virus arose from a recombination within the MoMuLV and the cellular proto-oncogene *abl*. Abelson leukaemia virus transforms haematopoietic cells of the B lymphocyte lineage (Goff *et al.*, 1980), particularly B cell precursors. Abelson leukaemia virus is a defective virus that requires a helper virus like MoMuLV. Cell lines transformed with Abelson leukaemia virus with MoMuLV as helper have shown decreased levels of H-2$D^d$ when passaged in culture (McMahon Pratt *et al.*, 1977). In contrast, passage *in vivo* resulted in cells expressing increased levels of H-2$D^d$. It, therefore, appears that selective pressures during the *in vivo* passage result in this modulation. Because the analysis of v-*abl*-induced changes in MHC class I gene expression relies on analysis of *abl*-induced tumours and because the Abelson murine leukaemia virus immortalizes immature pre-B cells, the possibility exists that the effects of v-*abl* on MHC class I expression are secondary and more directly related to the maturational state of the cell immortalized by the oncogenic virus. MHC class I gene expression is known to be repressed in a number of undifferentiated cell types, including embryonal cells, by a *trans*-dominant mechanism (see Chapter 4; Wan *et al.*, 1987).

H-2$K^b$ molecular-loss variants derived from the R8 cell line, which is a pre-B lymphosarcoma induced by Abelson murine leukaemia virus in an H-2B × H-2D $F_1$ mouse, have been analysed to establish the mechanism(s) of MHC repression (Zeff *et al.*, 1987). Variants were generated by mutagenesis with ethyl methane sulphonate (EMS) or ethyl nitrosourea (ENU) and selected with anti-H-2$K^b$ mAbs (Geier *et al.*, 1986) or by anti-H-2$K^b$-restricted CTLs lymphocytes (Sheil, Bevan & Sherman, 1986). Three separate mechanisms were found to give rise to the molecular-loss variants. One mechanism was deletion of the H-2$K^b$ gene (Zeff *et al.*, 1986a). Similar deletions have been found in parallel experiments in the H-2$D^d$ gene (Potter *et al.*, 1987). The second mechanism giving rise to the loss phenotype was somatic mutation (Zeff *et al.*, 1986a; 1990). Point mutations within the carboxy-terminal amino acids of the $\alpha_2$ domain resulted in diminished expression on the cell surface of the H-2$K^b$ antigen. This region of the gene encodes a portion of the polypeptide that directs MHC class I molecules through the exocytic pathway (see Chapter 1). The third H-2$K^b$ loss phenotype resulted from decreased levels of H-2$K^b$ mRNA, apparently because of decreased transcription or decreased mRNA stability (Zeff *et al.*, 1986b). There was also evidence in this particular loss variant for a *trans*-acting repressor acting on the mutated H-2K$^b$ gene.

### 7.3.4 MHC regulation by the fos oncogene

A number of studies have implicated the v-*fos* oncogene and its cellular homologue, c-*fos*, in the regulation of MHC gene expression. The introduction of the v-*fos* oncogene into 3LL mouse Lewis lung cancer cells, which had a low basal level of expression, resulted in increases in class I antigen expression and the loss of the highly metastatic potential of these clones (Eisenbach *et al.*, 1986). Analysis of highly metastatic and low metastatic clones derived from the 3LL line demonstrated that the weakly metastatic clone (A9) expressed high levels of both H-2K$^b$ and H-2D$^b$, while the highly metastatic clones (D122) expressed lower levels of H-2D$^b$ and extremely low levels of H-2K$^b$. The expression of H-2K$^b$ was also correlated with the presence of the c-*fos* mRNA transcript. IFN-$\gamma$ induced elevation of H-2K$^b$ antigens in the highly metastatic clones and there was a coincident decrease in the metastatic potential of the IFN-$\gamma$-treated cells. IFN-$\gamma$ was also found to induce the generation of c-*fos* transcripts within 30 minutes and the induction of H-2K$^b$ transcription in 1 to 2 hours. IFN-$\alpha$ and IFN-$\beta$, which were less effective in activating H-2K expression, were also less potent in inducing c-*fos* mRNA. Transfection of v-*fos* or human or murine c-*fos* genes into the D122 cells resulted in transcriptional activation of H-2K mRNA (Kushtai *et al.*, 1988). Elevated levels of cytoplasmic H-2 proteins and cell surface expression of H-2K$^b$ antigens were also noted in the transfected cells. The induction of H-2K$^b$ by *fos* correlated with a loss in the metastatic potential of the transfected cells (Kushtai, Feldman & Eisenbach, 1990). Differentiation inducers which turn on c-*fos* expression in lymphoid cell lines result in up-regulated expression of MHC class I (Barzilay *et al.*, 1987). Gel retardation assays using oligonucleotides corresponding to the promoter region of the H-2K$^b$ gene suggested a different pattern of DNA-binding proteins in the region between base pairs −161 and −163 from the cap site when comparing nuclear extracts from high-expressing versus low-expressing clones. Promoter mapping studies using chimaeric constructs in transient expression assays suggested that this same region can serve as a negative regulator of H-2K$^b$ gene expression. This region is flanked by an AP-1-binding site. The AP-1 molecule is made up of the *fos* protein and a member of the *jun* family of proteins or a pair of *jun* proteins (Feldman & Eisenbach, 1991).

In other studies, signals that induce differentiation in association with increased expression of MHC class I genes were shown to first induce the expression of the c-*fos* gene. The human cell line HL60, which originated from a promyelocytic leukaemia and which differentiates into a monocyte or a macrophage when treated with phorbol esters, expressed significantly higher levels of cell surface HLA antigens during the differentiation process. Treatment with dimethylsulphoxide (DMSO), however, induced differentiation along granulocytic lines, in association with a significant decrease in the expression of HLA molecules. Phorbol ester treatment of HL60 cells induced the expression of c-*fos* transcripts prior to the induction of HLA transcripts, whereas DMSO treatment did not induce expression of c-*fos*. Phorbol ester treatment of the human promonocytic leukaemia cell line U937, which differentiates along macrophage lines, induces transcription of c-*fos* initially, followed by increases in HLA gene and antigen expression. There is some correlation between differentiation agents associated with an increase in MHC products and a coincident increase in the expression of the c-*fos* gene in a phaeochromocytoma cell line. Conversely, when MHC positive cells of a murine erythroleukaemia cell line, which constitutively expressed c-*fos*, were induced to differentiate into proerythrocytes, which are H-2 negative, the disappearance of the c-*fos* gene products preceded the loss of MHC class I expression (Barzilay *et al.*, 1987).

## 7.4 Human leukaemia and immunodeficiency viruses

### *7.4.1 Human T cell leukaemia virus*

The human T cell leukaemia virus type I (HTLV-I) is associated with adult T cell leukaemia (ATLL), tropical spastic paraparesis or HTLV-I-associated myelopathy (TSP/HAM), and an increasing number of autoimmune-like diseases, including inflammatory arthropathies, alveolitis, Sjögren's syndrome and polymyositis (Hallsberg & Hafler, 1993). HTLV-I infection of $CD4^+$ or $CD8^+$ human CTL with known antigen specificity results in the establishment of functional CTL lymphocyte lines which propagate indefinitely in culture (Faller, Crimmins & Mentzer, 1988). In contrast to the uninfected cells, the infected T cell lines were independent of the need for antigen (target cell) stimulation as a requirement for proliferation and growth. The cell lines remained functionally identical to their uninfected parental CTL lymphocyte clones in their ability to specifically recognize appropriate target cells. Levels of MHC class II antigens were increased by 3.5- to 4-fold on the infected CTL lymphocyte clones. Expression of MHC class I antigens was increased by 2.5- to 4-fold on HTLV-I infected clones and on HTLV-I-infected cord blood lymphocytes. These increases were of the same magnitude as those found after IFN-γ stimulation of the uninfected clones or cord blood lymphocytes. Treatment of the infected CTL clones or cord blood lymphocytes with IFN-γ did not cause a further increase in MHC class I antigens. Analysis of the infected lymphocytes with HLA-A1- and HLA-B8-specific typing antisera demonstrated that the expression of both of these endogenous antigens was increased by the same magnitude as the increase demonstrated by the pan-reactive HLA antibody. Therefore, the enhancement of HLA expression was the result of enhancement of the expression of normally expressed HLA antigens rather than the induction of new class I-like antigens. Other lymphocyte cell surface antigens, including LFA-1, CD4, and CD8, were not upregulated in the infected CTLs. The HTLV-I-infected CTL clones exhibited higher levels of MHC class I-specific mRNA transcripts than did their uninfected counterparts. High levels or alterations in HLA expression have been noted on HTLV-I-induced tumours or in populations of T lymphocytes infected with the virus (Mann *et al.*, 1983). Interpretation of such observations is difficult, however, because the phenotype of the original cell, prior to the infection of the virus, could not be ascertained in such studies. In our studies, the use of well-characterized, clonal, untransformed T cells allowed demonstration of unequivocal and consistent increases in MHC class I expression after HTLV-I infection. Because the specific HLA type of these CTLs was known, it was clear that the increases in HLA expression were caused by specific increases in the levels of the endogenous, normal HLA antigens expressed by the CTLs, rather than to the expression of new or altered MHC class I-like antigens.

Induction of MHC class II genes has been reported in HTLV-I-producing tumours, and HTLV-I infection of cord blood lymphocytes results in the eventual generation of transformed cell lines that constitutively express high levels of MHC class II gene expression (Miyoshi *et al.*, 1981; Lando *et al.*, 1983; Depper *et al.*, 1984; Popovic *et al.*, 1984). HTLV-I-infected cells display proliferative responses in mixed lymphocyte cultures that are independent of their class II allele specificity (Suciu-Foca *et al.*, 1984). The aberrant regulation of class II antigens in HTLV-I-infected cells could contribute to the abnormal proliferation of those cells. The HTLV-I viral genome contains the *tax* gene, which encodes a *trans*-activator of the viral LTR, and this *trans*-activator gene is associated with the tumorigenic properties of the virus. The *tax* gene is capable of *trans*-activating certain cellular genes, including the IL-2 receptor, in addition to the viral LTR. It has been suggested that the *tax* gene of HTLV may be important in the activation of the MHC class II genes of T cells (Green, 1986; Mach *et al.*, 1986), but attempts to demonstrate this *in vitro* have been unsuccessful. The introduction of the *tax* gene of HTLV-II into HUT78 cells (a lymphoblastoid tumour cell line that expresses all three types of class II antigen) resulted in the down-regulation of these antigens at the level of mRNA. The magnitude of this repression was proportional to the amount of *tax* activity in the transfected cells (P. Koka, & D. V. Faller, unpublished results).

The association of HTLV-I infection with autoimmune-like disorders may be related to the ability of this virus to induce T cell activation antigens, production of cytokines and their receptors, MHC class I and II antigens and proteases. In addition to the *tax* element, we have also shown that the HTLV-I LTR encodes a similar transcript to that of the Mo-MuLV LTR. This HTLV-I transcript can *trans*-activate endogenous or transfected MHC class I, collagenase IV or MCP-1/JE genes (Choi & Faller, 1994; 1995).

In addition to adult T cell leukaemia, HTLV-I is also associated with neurological disorders. Normal cells of the CNS do not usually express MHC antigens and have been considered immunologically privileged. Transfection of an inducible *tax* gene (placed under the control of the metallothionein promotor and inducible with lipopolysaccharide or heavy metals) into rat glial cells resulted in the induction of MHC class I antigens in those cells after induction of the *tax* gene (Sawada *et al.*, 1990). There was no evidence that a soluble mediator-like interferon was involved in the up-regulation of the MHC antigens by the transfected gene. Co-transfection of a *tax*-expressing gene along with a chimaeric gene consisting of the promoter of the H-2L$^d$ class I gene driving the CAT gene suggested that transcriptional activation of MHC genes was mediated by expression of the *tax* gene. Whether this up-regulation of MHC expression occurs in human neural cells *in vitro* or *in vivo* after infection with HTLV-I remains to be determined, as does any possible correlation between the expression of class I antigens and those neurological disease processes associated with HTLV-I.

### *7.4.2 Human immunodeficiency virus (HIV)*

The effects of HIV on MHC class I and II expression have been studied *in vivo* and *in vitro*. *In vivo* analyses may be confounded by the unavoidable selection of cells for study that have not been lytically or cytopathically infected by the virus and by the presence of circulating inflammatory cytokines and autoreactive antibodies.

#### *7.4.2.1* In vivo *studies*

Levels of circulating HLA class I antigen in HIV-positive subjects are significantly higher than those found in control subjects (Puppo *et al.*,

1990). Soluble HLA class I antigen levels increase with disease progression: an increase of 2.5-fold occurring as disease severity progresses from CDC (Centers for Disease Control) classification Group 2 to Group 4. Soluble antigen levels correlated with HIV p24 antigen, IL-2 receptor and CD8 soluble antigen levels. The mechanisms responsible for release of class I antigens is unclear but may be secondary to shedding of HLA antigens during virus budding or in association with shed IL-2R and CD8, all of which are associated on the cell surface. It has been suggested that circulating HLA class I antigens might interfere with immune function and, thus, contribute to the pathogenesis of the immune deficiency. Increases in MHC class II expression have been observed in monocytes of patients with acquired immune deficiency syndrome (AIDS)-related complex compared with the levels found in normal individuals (Haegy *et al.*, 1984). However, by the time the entire AIDS disease develops, MHC class II expression on circulating monocytes is substantially less than that of healthy controls.

Homologous peptide sequences have been identified in the amino-terminal domain of human class II β chains and the carboxy-terminus of the HIV envelope (*env*) protein. These homologous regions are highly conserved among different DR and DQ alleles and also among different isolates of HIV. Monoclonal antibodies raised against the conserved HIV-derived peptides react strongly with the homologous class II-derived peptides. These antibodies also react with native MHC class II antigens expressed on human B cells and on mouse fibroblast L cell lines transfected with the genes encoding the α and β chains of human class II antigens. Sera from 36% of sero-positive AIDS patients contained antibodies that reacted against the class II-derived peptide (Golding *et al.*, 1988). Antibodies were present both at the early asymptomatic stage and also in later stages of the disease (Blackburn *et al.*, 1991). In another study, 28–48% of HIV-positive sera was found to contain antibodies that cross-reacted with the amino acid sequences conserved between HIV-1 and the class II polypeptide (Zaitseva *et al.*, 1992). These antibodies have been demonstrated to be functional. Antibody-containing sera inhibited the proliferative responses of normal CD4-bearing cells to tetanus toxoid *in vitro*. These antibodies could also generate an antibody-dependent cytotoxic cellular (ADCC) response against class II-bearing cells (Blackburn *et al.*, 1991). In a clinical study of patients with HIV infection, there was a positive correlation between lack of responsiveness to influenza and to tetanus in the presence of anti-class II antibodies (Zaitseva *et al.*, 1992). Therefore, there may be a role for this crossreactivity in autoimmune mechanisms as part of the AIDS complex.

*7.4.2.2* In vitro *studies*

The effects of HIV cytopathic infection on lymphoid cell lines or peripheral blood lymphocytes (PBL) in culture have been studied. HIV-1 infection of $CD4^+$ HeLa cells, H9 cells and peripheral T lymphocytes down-regulated MHC class I cell surface antigen expression. This effect was time dependent, requiring at least seven days for the effect to be detected, roughly paralleling the course of the viral infection in the cultures, as monitored by the expression of the p24 viral protein (protein of 24 kDa); the size of this effect varied from donor to donor. Levels of class I-specific mRNA were similar in infected and in uninfected cells, suggesting regulation at the post-transcriptional level or loss of MHC antigens by budding viruses. This down-regulation of HLA class I reduced the susceptibility of the infected cells to lysis by allogeneic CTLs (Kerkau *et al.*, 1989). Incubation of HIV-infected cells at 26 °C or treatment at 37 °C with peptides capable of binding to MHC class I proteins led to a reinduction of MHC class I levels on the surface of the infected cells comparable to levels found on uninfected cells. Thus, the mechanisms responsible for eventual stable association between MHC-binding peptides and MHC class I proteins may be affected by acute HIV infection, leaving the MHC class I antigens 'empty' (Kerkau *et al.*, 1992). In the absence of bound peptide, such MHC polypeptides either would not associate with $\beta_2$-m and, therefore, not be transported to the cell surface, or would be unstable at body temperature. In contrast, analysis of the human T lymphoid cell line CR-10 after non-cytopathic infection with HIV showed that levels of HLA-A2 and HLA-B7 antigens and mRNAs were unchanged. HIV-1 infection did result in gradual loss of cell-surface CD8, CD3 and CD2 expression. The same patterns of antigen loss were found in three T lymphoid cell lines that were cytopathically infected (Stevenson, Zhang & Volsky, 1987).

The structural proteins encoded by HIV may also modulate expression of MHC genes. The major component of the virus core, p24, inhibited the ability of IFN-$\gamma$ to induce HLA DR transcripts. This effect was abrogated by antibodies directed against p24. Levels of p24 comparable to those found in the serum of HIV-infected individuals was sufficient to affect the ability of IFN-$\gamma$ to induce MHC class II antigens (Nong *et al.*, 1991).

The simian immunodeficiency virus (SIV) is closely related to HIV and induces an AIDS-like syndrome in macaque monkeys. Cell populations infected *in vitro* with SIV exhibited increases in MHC class II antigen expression (Kannagi *et al.*, 1987). This expression was recognized by three different MHC class II-specific mAbs and was three to five times greater on the chronically SIV-infected cells than on uninfected human lymphocyte cell lines. This increase did not appear to be mediated by interferon. Levels of MHC class I and other T cell surface antigens were unaffected by infection. The increases in MHC class II antigen expression did not occur until days 22 to 30 after infection with SIV, long after the appearance of SIV antigens. MHC class II antigens appeared to be acquired by some virus particles while budding from cell membranes. Whether this acquisition was passive or the result of preferential interactions between viral structural proteins and MHC class II antigens is not known. It was postulated that, because MHC class II preferentially recognizes CD4, the expression of MHC class II on the surface of the virus might increase the tropism of the virus for $CD4^+$ cells.

In conclusion, the LTRs of the human retroviruses HTLV-I and HIV share with the murine MoMuLV LTR the ability to activate specific cellular genes (Choi & Faller, 1994; 1995). The LTRs of these viruses regulate expression of both endogenous and transfected genes, including the collagenase IV, MCP-1/JE and MHC genes. The relationship of this property to the life cycles and pathogenesis of these viruses remains to be determined.

## Acknowledgements

The author gratefully acknowledges the support of the National Cancer Institute and the American Cancer Society in this work.

## References

Albino, A.P., Houghton, A.N., Eisinger, M., Lee, J.S., Kantor, R.R., Oliff, A.T. & Old, L.J. (1986). Class II histocompatibility antigen expression in human melanocytes transformed by Harvey murine sarcoma virus (Ha-MSV) and Kirsten MSV retroviruses. *Journal of Experimental Medicine*, 164, 1710–1722.

Atherton, S.S., Streilein, R.D. & Streilein, J.W. (1984). Lack of polymorphism for C-type retrovirus sequences in the Syrian hamster. In *Advances in Gene Technology: Human Genetic Disorders*. ed. F. Ahmod, S. Black, J. Schultz, W.A. Scott & W.J. Whelon, pp. 128–147. Cambridge: ICSU Press.

Baldacci, P., Transy, C., Cochet, M., Penit, C., Israel, A. & Kourilsky, P. (1986). A *trans*-acting mechanism represses the expression of the major transplantation antigens in mouse hybrid thymoma cell lines. *Journal of Experimental Medicine*, 164, 677–694.

Barzilay, J., Kushtai, G., Plaksin, D., Feldmann, M. & Eisenbach, L. (1987). Expression of major histocompatibility class I genes in differentiating leukemic cells is temporally related to activation of c-*fos* proto-oncogene. *Leukaemia*, 1, 198–204.

Bhan, A.K. & Des Marais, C.L. (1983). Immunohistologic characterization of major histocompatibility antigens and inflammatory cellular infiltrate in human breast cancer. *Journal of the National Cancer Institute*, 71, 507–516.

Blackburn, R., Clerici, M., Mann, D., Lucey, D.R., Goedert, J., Golding, B., Shearer, G.M. & Golding, H. (1991). Common sequence in HIV 1 gp41 and HLA class II beta chains can generate crossreactive autoantibodies with immunosuppressive potential early in the course of HIV 1 infection. *Advances in Experimental Medicine and Biology*, 303, 63–69.

Blatt, C., Milham, K., Haas, M., Nesbitt, M.N., Harper, M.E. & Simon, M.I. (1983). Chromosomal mapping of the mink cell focus-inducing and xenotropic *env* gene family in the mouse. *Proceedings of the National Academy of Sciences of the USA*, 80, 6298–6302.

Brown, G.D., Choi, Y., Pameno, C. & Meruelo, D. (1988). Regulation of H-2 class I gene expression in virally transformed and infected cells. *CRC Critical Reviews in Immunology*, 8, 175–215.

Cambier, J.C. & Lehmann, K.R. (1989). Ia-mediated signal transduction leads to proliferation of primed B lymphocytes. *Journal of Experimental Medicine*, 170, 877–886.

Chazan, R. & Haran-Ghera, N. (1976). The role of thymus subpopulations in 'T' leukemia development. *Cellular Immunology*, 23, 356–375.

Choi, S.-Y. & Faller, D.V. (1994). The long terminal repeats of a murine retrovirus encode a *trans*-activator for cellular genes. *Journal of Biological Chemistry*, 269, 19691–19694.

Choi, S.-Y. & Faller, D.V. (1995). A transcript from the long terminal repeats of a murine retrovirus associated with *trans*-activation of cellular genes. *Journal of Virology*, in press.

Choi, S.Y., Koka, P., van de Mark, K. & Faller, D.V. (1991a). Murine leukemia viruses induce T cell activation antigens. *Journal of Cell Biology*, 115, 453a.

Choi, S.Y., Koka, P., van de Mark, K. & Faller, D.V. (1991b). Mechanism of transactivation of T-lymphocyte cellular genes by murine retroviral sequences. *Blood*, 78, 3.

Choi, Y.C. & Meruelo, D. (1991). Isolation of virus-like (VL30) elements from the Q10 and D regions of the major histocompatibility complex. *Biochemical Genetics*, 29, 91–101.

Cikes, M., Friberg, S. Jr & Klein, G. (1973). Progressive loss of H-2 antigens with concomitant increase of cell-surface antigen(s) determined by Moloney leukemia virus in cultured murine lymphoma. *Journal of the National Cancer Institute*, 50, 347–362.

Collins, J.K., Britt, W.J. & Cheseboro, B. (1980). Cytotoxic T lymphocyte recognition of gp70 on Friend virus-induced erythroleukemia cell clones. *Journal of Immunology*, 125, 1318–1324.

Decleve, A., Lieberman, M., Ihle, J.N. & Kaplan, H.S. (1977). Biological and serological characterisation of C-type RNA viruses isolated from C57BL/Ka strain of mice. II. Characterisation of isolates and their interactions *in vitro* and *in vivo*. In: *Radiation-induced Leukaemogenesis and Related Viruses: INSERM Symposium No. 4*, ed. J.F. Duplan, pp. 247–261. Amsterdam: Elsevier/North Holland Biomedical Press.

Depper, J.M., Leonard, W.J., Kronke, M., Waldmann, T.A. & Greene, W.C. (1984). Augmented T cell growth factor receptor expression in HTLV-I infected human leukemic T cells. *Journal of Immunology*, 133, 1691–1695.

Doyle, A., Martin, W.J., Funa, K., Gazdar, A., Carney, A., Martin, S.E., Linnoila, I., Cuttita, F., Mulshine, J., Bunn, P. & Minna, J. (1985). Decreased expression of class I histocompatibility antigens, protein and mRNA in human small-cell lung cancer. *Journal of Experimental Medicine*, 161, 1135–1151.

Du, W. & Maniatis, T. (1992). An ATF/CREB binding site protein is required for virus induction of the human interferon-β gene. *Proceedings of the National Academy of Sciences of the USA*, 89, 2150–2154.

Eisenbach, L., Kusthai, G., Plaksin, D. & Feldman, M. (1986). MHC genes and oncogenes controlling the metastatic phenotype of tumour cells. *Cancer Reviews*, 5, 1–18.

Eliott, B.E., Carlow, D.A., Rodricks, A.M. & Wade, A. (1986). Perspectives on the role of MHC antigens in normal and malignant cell development. *Advances in Cancer Research*, 53, 181–185.

Enjuanes, L., Lee, J.C. & Ihle, J.N. (1979). Antigenic specificities of the cellular immune response of C57BL/6 mice to the Moloney leukemia/sarcoma virus complex. *Journal of Immunology*, 122, 665–674.

Faller, D.V., Crimmins, M.A.V. & Mentzer, S.J. (1988). Human T-cell leukemia virus type I infection of $CD4^+$ or $CD8^+$ cytotoxic T-cell clones results in immortalization with retention of antigen specificity. *Journal of Virology*, 62, 2942–2950.

Faller, D.V., Flyer, D.C. & Wilson, L.D. (1987). Induction of class I MHC expression by murine leukaemia viruses. *Journal of Cellular Biochemistry*, 36, 297–310.

Feldman, M. & Eisenbach, L. (1991). MHC class I genes controlling the metastatic phenotype of tumour cells. *Seminars in Cancer Biology*, 2, 337–346.

Festenstein, H. (1989). Molecular features of the H-2 class I and Qa antigens expressed on Gross virus induced AKR leukaemias. *Journal of Immunogenetics*, 16, 329–333.

Festenstein, H., Allonzo, A., Ferluga, J. & Ma, B.L. (1983). Further molecular and functional evidence for the expression of H-2 allodeterminants on the AKR leukaemia K30. *Transplantation Proceedings*, 15, 2107.

Fleming, K.A., McMichael, A., Morton, J.A., Woods, J. & McGee, J.O'D. (1981). Distribution of HLA class I antigens in normal human tissue and in mammary cancer. *Journal of Clinical Pathology*, 34, 779–784.

Flyer, D.C., Burakoff, S.J. & Faller, D.V. (1983). Cytotoxic T lymphocyte recognition of transfected cells expressing a cloned retroviral antigen. *Nature (London)*, 305, 315–318.

Flyer, D.C., Burakoff, S.J. & Faller, D.V. (1985a). Expression and CTL recognition of cloned subgenomic fragments of Moloney murine leukemia virus in murine cells. *Survey of Immunological Research*, 4, 168–172.

Flyer, D.C., Burakoff, S.J. & Faller, D.V. (1985b). Retrovirus-induced changes in major histocompatibility complex antigen expression influence susceptibility to lysis by cytotoxic T lymphocytes. *Journal of Immunology*, 135, 2287–2292.

Flyer, D.C., Burakoff, S.J. & Faller, D.V. (1986). The immune response to Moloney murine leukemia virus induced tumors: induction of cytolytic T lymphocytes specific for both viral and tumour-associated antigens. *Journal of Immunology*, 137, 3968–3971.

Fontana, S., Del Vecchio, L., Racioppi, L., Carbone, E., Pinto, A., Colletta, G. & Zappacosta, S. (1987). Expression of major histocompatibility complex class I antigens in normal and transformed rat thyroid epithelial cell lines. *Cancer Research*, 47, 4178–4183.

Geier, S.S., Zeff, R.A., McGovern, D.M., Rajan, T.V. & Nathenson, S.G. (1986). An approach to the study of structure–function relationships of MHC class I molecules: isolation and serologic characterization of $H\text{-}2K^b$ somatic cell variants. *Journal of Immunology*, 137, 1239–1243.

Geppert, T.D., Wacholtz, M.C., Davis, L.S., Patel, S.S., Lightfoot, E. & Lipsky, P.E. (1989). Activation of human T cell clones and Jurkat cells by cross-linking class I MHC molecules. *Journal of Immunology*, 143, 3763–3772.

Goff, S., Gilboa, E., Witte, O.N. & Baltimore, D. (1980). Structure of the Abelson murine leukemia virus genome and the homologous cellular gene: studies with cloned viral DNA. *Cell*, 22, 777–785.

Gogusev, J., Teutsch, B., Moria, M.T., Mongiat, F., Hagenau, F., Suskind, G. & Rabotti, G.F. (1988). Inhibition of HLA class I antigen and mRNA expression induced by RSV in transformed human fibroblasts. *Proceedings of the National Academy of Sciences of the USA*, 85, 203–207.

Golding, H., Robey, F.A., Gates III, F.T., Linder, W., Beining, P.R., Hoffmann, T. & Golding, B. (1988). Identification of homologous regions in human immunodeficiency virus 1 gp41 and human MHC class II beta 1 domain. I. Monoclonal antibodies against the gp41-derived peptide and patients' sera react with native HLA class II antigens, suggesting a role for autoimmunity in the pathogenesis of acquired immune deficiency syndrome. *Journal of Experimental Medicine*, 167, 914–923.

Gomard, E., Duprez, V., Henin, Y. & Levy, J.P. (1976). H-2 region product as determinant in immune cytolysis of syngeneic tumour cells by anti-MSV T lymphocytes. *Nature (London)*, 260, 707–709.

Green, W. R. (1986). Expression of CTL-defined, AKR/Gross retrovirus-associated tumour antigens by normal spleen cells: control by Fv-1, H-2, and proviral genes and effect on anti-viral CTL generation. *Journal of Immunology*, 136, 308–312.

Green, W.R. & Phillips, J.D. (1986). Differential induction of H-2K versus H-2D class I major histocompatibility complex antigen expression by murine recombinant gamma interferon. *Journal of Immunology*, 137, 814–818.

Guild, B.C., Erikson, R.L. & Strominger, J.L. (1983). HLA-A2 and HLA-B2 antigens are phosphorylated *in vitro* by Rous sarcoma virus kinase ($pp60^{v\text{-}Src}$) at a tyrosine residue encoded in a highly conserved exon of the intracellular domain. *Proceedings of the National Academy of Sciences of the USA*, 80, 2894–2898.

Haegy, W., Kelley, V.E., Strom, T.B., Mayer, K., Shapiro, H.M., Mandel, R. & Finberg, R. (1984). Decreased expression of human class II antigens on monocytes from patients with acquired immune deficiency syndrome. *Journal of Clinical Investigation*, 74, 2089–2096.

Hanecak, R., Pattengale, P.K. & Fan, H. (1991). Deletion of a GC-rich region flanking the enhancer element within the long terminal repeat sequences alters the disease specificity of Moloney murine leukemia virus. *Journal of Virology*, 65, 5357–5363.

Hartley, J.W., Wolford, N.K., Old, L.J. & Rowe, W.P. (1977). A new class of murine leukemia viruses associated with development of spontaneous lymphomas. *Proceedings of the National Academy of Sciences of the USA*, 74, 789–792.

Hassan, Y., Priel, E., Segal, S., Hullihel, M. & Abond, M. (1990). Chemical-retroviral cooperative carcinogenesis and its molecular basis in NIH/3T3 cells. *Carcinogenesis*, 11, 2097/2102.

Hendrix, M.J., Seftor, E.A., Grogan, T.M., Seftor, R.E., Hersh, E.M., Boyse, E.A., Liotta, L.A., Stetler-Stevenson, W. & Ray, C.G. (1992). Expression of type-IV collagenase correlates with the invasion of human lymphoblastoid cell lines and pathogenesis in SCID mice. *Molecular and Cellular Probes*, 6, 59–65.

Henley, S.L., Wise, K.S. & Acton, R.T. (1984). Productive murine leukemia virus (MuLV) infection of EL-4 lymphoblastoid cells: selective elevation of H-2 surface expression and possible association of Thy-1 antigen with viruses. *Advances in Medicine and Experimental Biology*, 172, 365–381.

Hollsberg, P. & Hafler, D.A. (1993). Seminars of the Beth Israel Hospital, Boston – pathogenesis of diseases by human lymphotropic virus type I infection. *New England Journal of Medicine*, 328, 1173–1182.

Hood, L., Kronenberg, M. & Hunkapillar, T. (1985). T cell antigen receptors and the immunoglobulin supergene family. *Cell*, 40, 225–229.

Hui, K., Grosveld, F. & Festenstein, H. (1984). Rejection of transplantable AKR leukaemia cells following MHC DNA-mediated cell transformation. *Nature (London)*, 311, 750–752.

Kannagi, M., Kiyotaki, M., King, N.W., Lord, C.I. & Letvin, N.L. (1987). Simian immunodeficiency virus induces expression of class II major histocompatibility complex structures on infected target cells *in vitro*. *Journal of Virology*, 6, 1421–1426.

Kanbhu, S., Falldorf, P. & Lee, J.S. (1990). Endogenous retroviral long terminal repeats within the HLA-DQ locus. *Proceedings of the National Academy of Sciences of the USA*, 87, 4927–4231.

Katz, E., Peled, A. & Haran-Ghera, N. (1985). Changes in H-2 antigen expression on thymocytes during leukemia development by radiation leukemia virus. *Leukemia Research*, 9, 1219–1225.

Kawashima, K., Ikeda, H., Stokert, E., Takahashi, T. & Old, L.J. (1976). Age-related changes in cell surface antigens of preleukemic AKR thymocytes. *Journal of Experimental Medicine*, 144, 193–208.

Kerkau, T., Gernert, S., Kneitz, C. & Schimpl, A. (1992). Mechanism of MHC class I down-regulation in HIV-infected cells. *Immunobiology*, 184, 402–409.

Kerkau, T., Schmitt-Landgraf, R., Schimpl, A. & Wecker, E. (1989). Down-regulation of HLA class I antigens in HIV-1 infected cells. *AIDS Research and Human Retroviruses*, 5, 613–620.

King, W., Patel, M.D., Lobel, L.I., Goff, S.P. & Nguyen-Hun, M.C. (1985). Insertion mutagenesis of embryonal carcinoma cells by retroviruses. *Science*, 228, 554–558.

Klyczek, K.K. & Blank, K.J. (1989). Novel class I-like molecule expressed on a murine leukaemia virus-transformed cell line. *Cellular Immunology*, 118, 222–228.

Klyczek, K.K., Murasko, D.M. & Blank, K.J. (1987). Interferon-gamma, interferon-alpha/beta, and tumour necrosis factor differentially affect major histocompatibility complex class I expression in murine leukemia virus-induced tumor cell lines. *Journal of Cellular Immunology*, 139, 2641–2648.

Koka, P., Choi, S.Y., van de Mark, K. & Faller, D.V. (1991a). *Trans*-activation of genes encoding human T-lymphocyte cell surface activation proteins by retroviral sequences. *Clinical Research*, 39, 341A.

Koka, P., van de Mark, K. & Faller, D.V. (1991b). *Trans*-activation of genes encoding activation-associated human T-lymphocyte surface proteins by murine leukemia viruses. *Journal of Immunology*, 146, 2417–2425.

Krieg, A.M., Gause, W.C., Gourley, M.F. & Steinberg, A.D. (1989). A role for endogenous retroviral sequences in the regulation of lymphocyte activation. *Journal of Immunology*, 143, 2448–2451.

Kuppers, R.C., Ballas, Z.K., Green, W.R. & Henney, C.S. (1981). Quantitative appraisal of H-2 products in T cell-mediated lysis by allogeneic and syngeneic effector cells. *Journal of Immunology*, 127, 500–504.

Kushtai, G., Barzilay, J., Feldman, M. & Eisenbach, L. (1988). The c-fos proto-oncogene in murine 3LL carcinoma clones controls the expression of MHC genes. *Oncogene*, 2, 119–127.

Kushtai, G., Feldman, N. & Eisenbach, L. (1990). *c-fos* transfection of 3LL tumour cells turns on MHC gene expression and consequently reduces their metastatic competence. *International Journal of Cancer*, 45, 1131–1136.

Labeta, M.O., Fernandez, N., Reyes, A., Ferrara, P., Marelli, O., LeRoy, E., Houlihan, J. & Festenstein, H. (1989). Biochemical analysis of a novel H-2 class I-like glycoprotein expressed in five AKR (Gross Virus)-derived spontaneous T cell leukemias. *Journal of Immunology*, 143, 1245–1253.

Lampson, L.A., Fisher, C.A. & Whelan, J.P. (1983). Striking paucity of HLA-A, B, C and $\beta_2$-microglobulin on human neuroblastoma cell lines. *Journal of Immunology*, 130, 2471–2478.

Lando, Z.P., Sarin, P., Megson, M., Greene, W.C., Waldmann, T.A., Gallo, R.C. & Broder, S. (1983). Association of human T-cell leukaemia/lymphoma virus with the Tac antigen marker for the human T-cell growth factor receptor. *Nature (London)*, 305, 733–736.

Lee, J.C. & Ihle, J.N. (1981). Chronic immune stimulation is required for Moloney leukaemia virus-induced lymphomas. *Nature (London)*, 289, 407–409.

Lichtman, A.H., Lee, L. & Faller, D.V. (1995). Activated *ras* genes in murine T cells result in the stepwise progression of growth factor autonomy. *Cancer Molecular Biology*, in press.

Lichtman, A.H., Reynolds, D.H., Faller, D.V. & Abbas, A.K. (1986). Mature B lymphocytes immortalized by Kirsten sarcoma virus. *Nature (London)*, 324, 489–491.

Lichtman, A.H., Williams, M.E., Ohara, J., Paul, W.E., Faller, D.V. & Abbas, A.K. (1987). Retrovirus infection alters growth factor responses in T lymphocytes. *Journal of Immunology*, 138, 3276–3283.

Lonai, P. & Haran-Ghera, N. (1980). Genetic resistance to murine leukaemia induced by different radiation leukaemia-virus variants; a comparative study on the role of the H-2 complex. *Immunogenetics*, 1, 21–29.

Mach, B., Gorski, J., Rollini, P., Berte, C., Amaldi, I., Berdoz, J. & Ucla, C. (1986). Polymorphism and regulation of HLA class II genes of the major histocompatibility complex. *Cold Spring Harbor Symposium on Quantitative Biology*, 51, 67–74.

Mann, D.L., Popovic, M., Sarin, P., Murray, C., Reitz, M.S., Strong, D.M., Haynes, B.F., Gallo, R.C. & Blattner, W.A. (1983). Cells lines producing human T-cell lymphoma virus show altered HLA expression. *Nature (London)*, 305, 58–60.

Maudsley, D.J. & Morris, A.G. (1988). Kirsten murine sarcoma virus abolishes interferon $\gamma$-induced class II but not class I major histocompatibility antigen expression in a murine fibroblast line. *Journal of Experimental Medicine*, 167, 706–711.

Maudsley, D.J. & Morris, A.G. (1989). Regulation of IFN-gamma-induced host cell MHC antigen expression by Kirsten MSV and MLV. II. Effects on class II antigen expression. *Immunology*, 67, 26–31.

McGrath, M.S., Tamura, G. & Weissman, I.L. (1987). Receptor mediated leukemogenesis: murine leukemia virus interacts with BCL1 lymphoma cell surface IgM. *Molecular and Cellular Immunology*, 3, 227–242.

McGrath, M.S. & Weissman, I.L. (1979). AKR leukemogenesis: identification and biological significance of thymic lymphoma receptors for AKR retroviruses. *Cell*, 17, 65–75.

McMahon Pratt, D., Strominger, J., Parkman, R., Kaplan, D., Schwaber, J.,

Rosenberg, N. & Scher, C.D. (1977). Abelson virus-transformed lymphocytes: null cells that modulate H-2. *Cell*, 12, 683–690.

Mellor, A.L., Weiss, E.H., Kress, M., Jay, G. & Flavell, R.A. (1984). A polymorphic class I gene in the MHC. *Cell*, 36, 139–144.

Meruelo, D. (1979). A role for elevated H-2 antigen expression in resistance to RadLV-induced leukemogenesis: enhancement of effective tumour surveillance by killer lymphocytes. *Journal of Experimental Medicine*, 149, 898–909.

Meruelo, D. (1980). H-2D control of leukaemia susceptibility: mechanism and implications. *Journal of Immunogenetics*, 7, 81–90.

Meruelo, D., Kornreich, R., Rossomando, A., Pampeno, C., Boral, A., Silver, J.L., Buxbaum, J., Weiss, E.H., Devlin, J.J., Mellor, A.L., Flavell, R.A. & Pellicer, A. (1986). Lack of class I H-2 antigens in cells transformed by radiation leukemia virus is associated with methylation and rearrangement of H-2 DNA. *Proceedings of the National Academy of Sciences of the USA*, 83, 4504–4508.

Meruelo, D., Kornreich, R., Rossomando, A., Pampeno, C., Mellor, A.L., Weiss, E.H., Flavell, R.A. & Pellicer, A. (1984). Murine leukemia virus sequences are encoded in the murine major histocompatibility complex. *Proceedings of the National Academy of Sciences of the USA*, 81, 1804–1808.

Meruelo, D. & Kramer, J. (1981). H-2D control of radiation leukaemia virus induced neoplasia: evidence for interaction of viral and H-2 genomic information. *Transplantation Proceedings*, 13, 1858.

Meruelo, D., Lieberman, M., Ginzton, M., Deak, B. & McDevitt, H.O. (1977). Genetic control of radiation leukemia virus-induced tumorigenesis. I. Role of the murine major histocompatibility complex H-2. *Journal of Experimental Medicine*, 146, 1079–1087.

Meruelo, D., Nimelstein, J., Jones, P.P., Lieberman, M. & McDevitt, H.O. (1978). Increased synthesis and expression of H-2 antigens on thymocytes as a result of radiation leukemia virus infection: a possible mechanism for H-2 linked control of virus-induced neoplasia. *Journal of Experimental Medicine*, 147, 470–487.

Miyoshi, I., Kubonishi, I., Yoshimoto, S., Akagi, T., Ohtsuki, Y., Shiraishki, Y., Nagata, K. & Hinuma, Y. (1981). Type C particles in a cord T-cell line derived by co-cultivating normal human cord lymphocytes and human leukaemic cells. *Nature (London)*, 294, 770–771.

Momburg, F., Degener, T., Bacchus, E., Moldenhauer, G., Hammerling, G. & Moller, P. (1986). Loss of HLA-A,B,C and *de novo* expression of HLA-D in colorectal cancer. *International Journal of Cancer*, 37, 179–184.

Natali, P.G., Giacomini, P., Bigotti, A., Imai, K., Nicotra, M.R., Ng, A.K. & Ferrone, S. (1983). Heterogeneity in the expression of HLA and tumour-associated antigens by surgically removed and cultured breast carcinoma cells. *Cancer Research*, 43, 660–668.

Nobunaga, T., Brown, G.D., Morris, D.R. & Meruelo, D. (1992). A novel DNA binding activity is elevated in thymocytes expressing high levels of H-2D$^{d}$ after radiation leukemia virus infection. *Journal of Immunology*, 149, 871–879.

Nong, Y., Kandil, O., Tobin, E.H., Rose, R.M. & Remold, H.G. (1991). The HIV core protein p24 inhibits interferon-gamma-induced increase of HLA-DR and cytochrome b heavy chain mRNA levels in the human monocyte-like cell line THP1. *Cellular Immunology*, 132, 10–16.

Offermann, M.K. & Faller, D.V. (1989). Autocrine induction of major histocompatibility complex class I antigen expression results from induction of beta interferon in oncogene-transformed BALB/c-3T3 cells. *Molecular and Cellular Biology*, 9, 1969–1977.

Offermann, M.K. & Faller, D.V. (1990). Effect of cellular density and viral oncogenes on the major histocompatibility complex class I antigen response to γ-interferon in BALB-c/3T3 cells. *Cancer Research*, 50, 601–605.
O'Neill, H.C. (1989). Binding of radiation leukemia viruses to a thymic lymphoma involves some class I molecules on the T cell as well as the T cell receptor complex. *Journal of Molecular and Cellular Immunology*, 4, 213–23.
O'Neill, H.C., McGrath, M.S., Allison, J.P. & Weissman, I.L. (1987). A subset of T cell receptors associated with L3T4 molecules mediates C6VL leukemia cell binding of its cognate retrovirus. *Cell*, 49, 143–151.
Pampeno, C.L. & Meruelo, D. (1986). Isolation of a retrovirus-like sequence from the TL locus of the C57BL/10 murine major histocompatibility complex. *Journal of Virology*, 58, 296–306.
Peace, D.J., Chen, W., Nelson, H. & Cheever, M.A. (1991). T cell recognition of transforming proteins encoded by mutated *ras* proto-oncogenes. *Journal of Immunology*, 146, 2059–2065.
Plata, F., Tilkin, A.F., Levy, J.-P. & Lilly, F. (1981). Quantitative variations in the expression of H-2 antigens on murine leukemia virus-induced tumor cells can affect the H-2-restriction patterns of tumor-specific cytolytic T lymphocytes. *Journal of Experimental Medicine*, 154, 1795–1810.
Pober, J.S., Guild, B.C. & Strominger, J.L. (1978). Phosphorylation *in vivo* and *in vitro* of human histocompatibility antigens (HLA-A and HLA-B) in the carboxy-terminal intracellular domain. *Proceedings of the National Academy of Sciences of the USA*, 75, 6002–6006.
Poltronieri, L., Melloni, E., Rubini, M., Selvatici, R., Mazzilli, C., Baricordi, R. & Gandini, E. (1988). Anti-HLA class I monoclonal antibody effect on PKC kinetics in PHA activated human peripheral blood mononuclear and $E^+$ cells. *Biochemical and Biophysical Research Communications*, 156, 46–54.
Popovic, M., Flomenberg, N., Volkman, D.J., Mann, D., Fauci, A.S., Dupont, B. & Gallo, R.C. (1984). Alteration of T-cell functions by infection with HTLV-1 or HTLV-II. *Science*, 226, 459–462.
Potter, T.A., Zeff, R.A., Frankel, W. & Rajan, T.V. (1987). Mitotic recombination between homologous chromosomes generated H-2 somatic cell variants *in vitro*. *Proceedings of the National Academy of Sciences of the USA*, 84, 1634–1637.
Puppo, F., Orlandini, A., Ruzzenenti, R., Comuzio, S., Salamito, A., Farinelli, A., Stagnovo, R. & Indiveri, F. (1990). HLA class I soluble antigen serum levels in HIV-positive subjects – correlation with cellular and serological parameters. *Cancer Detection and Prevention*, 14, 321–323.
Racioppi, L., Carbone, E., Grieco, M., del Vecchio, L., Berlingieri, M.T., Fusco, A., Boncinelli, E., Zappacosta, S. & Fontana, S. (1988). The relationship of modulation of major histocompatibility complex class I antigens to retrovirus transformation in rat cell lines. *Cancer Research*, 48, 3816–3821.
Real, F., Fliegel, B. & Houghton, A.N. (1988). Surface antigens of human melanoma cells cultured in serum-free medium: induction of expression of major histocompatibility complex class II antigens. *Cancer Research*, 48, 686–693.
Rich, R.F., Gaffney, K.J., White, H.D. & Green, W.R. (1990). Differential up-regulation of H-2D versus H-2K class I major histocompatibility expression by interferon-gamma: evidence against a *trans*-acting allele-specific factor. *Journal of Interferon Research*, 10, 505–514.
Rommelaere, J., Faller, D.V. & Hopkins, N. (1978). Characterization and mapping of RNase Tl-resistant oligonucleotides derived from the genomes of Akv and

MCF murine leukemia viruses. *Proceedings of the National Academy of Sciences of the USA*, 75, 495–499.

Rossomando, A. & Meruelo, D. (1986). Viral sequences are associated with many histocompatibility genes. *Immunogenetics*, 23, 233–245.

Rothbard, J.B., Hopp, T.P., Edelman, G.M. & Cunningham, B.A. (1980). Structure of the heavy chain of the H-2K$^k$ histocompatibility antigen. *Proceedings of the National Academy of Sciences of the USA*, 77, 4239–4243.

Sawada, M., Suzumura, A., Yoshida, M. & Marunouchi, T. (1990). Human T-cell leukemia virus type I *trans*-activator induces class I major histocompatibility complex antigen expression in glial cells. *Journal of Virology*, 64, 4002–4006.

Sayre, P.H., Chang, H.C., Hussey, R.E., Brown, N.R., Richardson, N.E., Spagnoli, G., Clayton, N.K. & Reinherz, E.L. (1987). Molecular cloning and expression of T11 cDNAs reveal a receptor-like structure on human T lymphocytes. *Proceedings of the National Academy of Sciences of the USA*, 84, 2941–2945.

Schmidt, W. & Festenstein, H. (1982). Resistance to cell-mediated cytotoxicity is correlated with reduction of H-2K gene products in AKR leukaemia. *Immunogenetics*, 16, 257–264.

Shearer, G.M., Rehn, T.G. & Schmitt-Verhulst, A.M. (1976). Role of the murine major histocompatibility complex in the specificity of *in vitro* T-cell mediated lympholysis against chemically modified autologous lymphocytes. *Transplantation Reviews*, 29, 222–248.

Sheil, J.M., Bevan, M.J. & Sherman, L.A. (1986). Immunoselection of structural H-2K$^b$ variants: use of cloned cytolytic T cells to select for loss of a CTL-defined allodeterminant. *Immunogenetics*, 23, 52–59.

Smith, M.E.F. (1991). MHC class I expression in colorectal tumours. *Seminars in Cancer Biology*, 2, 17–23.

Sonnenfeld, G., Meruelo, D., McDevitt, H.O. & Merigan, T.C. (1981). Effect of type I and type II interferons on mouse thymocyte surface antigen expression: induction or selection? *Cellular Immunology*, 57, 427–439.

Stevenson, M., Zhang, X. & Volsky, D.J. (1987). Down-regulation of cell surface molecules during non-cytopathic infection of T cells with human immunodeficiency virus. *Journal of Virology*, 61, 3741–3748.

Suciu-Foca, N., Rubenstein, P., Popovic, M., Gallo, R.C. & King, D.W. (1984). Reactivity of HTLV-transformed human T-cell lines to MHC class II antigens. *Nature (London)*, 312, 275–277.

Teich, N., Wyke, J., Mak, T., Bernstein, A. & Hanry, W. (1982). Pathogenesis of retrovirus-induced disease. In *Molecular Biology of Tumor Viruses. RNA Tumor Viruses*, 2nd edn, ed. R.A. Weiss, N. Teich, H. Varmus & J. Coffin, pp. 785–998. New York: Cold Spring Harbor Press.

Trowsdale, J., Travers, P., Bodmer, W.F. & Patillo, R.A. (1980). Expression of HLA-A, -B and -C and $\beta_2$-microglobulin antigens in human choriocarcinoma cell lines. *Journal of Experimental Medicine*, 152, 11S–17S.

Umpleby, H., Heinemann, D., Symes, M. & Williamson, R. (1985). Expression of histocompatibility antigens and characterization of cell infiltrates in normal and neoplastic colorectal tissues of humans. *Journal of the National Cancer Institute*, 74, 1161–1168.

Vasmel, W.L., Sijts, E.J., Leupers, C.J., Matthews, E.A. & Melief, C.J. (1989). Primary virus-induced lymphomas evade T cell immunity by failure to express viral antigens. *Journal of Experimental Medicine*, 169, 1233–1254.

Vasmel, W.L.E., Zijlstra, M., Radaszkiewicz, T., Leupers, C.J., de Goede, R.E. & Melief, C.J. (1988). Major histocompatibility complex class II-regulated immunity to murine leukemia virus protects against early T- but not late B-cell lymphomas. *Journal of Virology*, 62, 3156–3166.

Wan, Y.-J., Orrison, B.M., Lieberman, R., Lazarovici, P. & Ozato, K. (1987). Induction of major histocompatibility class I antigens by interferon in undifferentiated F9 cells. *Journal of Cellular Physiology*, 130, 276–283.

Wejman, J.C., Taylor, B.A., Jenkins, N.A. & Copeland, N.G. (1984). Endogenous xenotropic murine leukemia virus-related sequences map to chromosomal regions encoding mouse lymphocyte antigens. *Journal of Virology*, 50, 237–247.

Weng, H., Choi, S.-Y. & Faller, D.V. (1995). The Moloney leukemia retroviral long terminal repeat *trans*-activates AP-1-inducible genes and AP-1 transcription factor binding. *Journal of Biological Chemistry*, in the press.

Whithwell, H.L., Hughes, H.P.A., Moore, M. & Ahmed, A. (1984). Expression of major histocompatibility antigens and leukocyte infiltration in benign and malignant human breast disease. *British Journal of Cancer*, 49, 161–172.

Wilson, L.D., Flyer, D.C. & Faller, D.V. (1987). Murine retroviruses control class I MHC expression via a *trans* effect at the transcriptional level. *Molecular and Cellular Biology*, 7, 2406–2415.

Zaitseva, M.B., Moshnikov, S.A., Kozhich, A.T., Frolova, H.A., Makarova, O.D., Pavlikov, S.P., Sidorovich, I.G. & Brondz, B.B. (1992). Antibodies to MHC class II peptides are present in HIV-1-positive sera. *Scandinavian Journal of Immunology*, 35, 267–273.

Zeff, R.A., Geier, S.S. & Nathenson, S.G. (1986a). Molecular loss variants of the murine major histocompatibility complex: non-expression of H-2K antigens associated with marked reduction in H-2K mRNA as determined by oligonucleotide hybridization analysis. *Journal of Immunology*, 137, 1366–1370.

Zeff, R.A., Gopas, J., Steinhauer, E., Rajan, T.V. & Nathenson, S.G. (1986b). Analysis of somatic cell H-2 variants to define the structural requirements for class I antigen expression. *Journal of Immunology*, 137, 897–903.

Zeff, R.A., Kumar, P.A., Mashimo, H., Nakagawa, M., McCue, B., Boriello, F., Kesari, K., Geliebter, J., Hemmi, S. & Pfaffenbach, G. (1987). Somatic cell variants of the murine major histocompatibility complex. *Immunologic Research*, 6, 133–144.

Zeff, R.A., Nakagawa, M., Mashimo, H., Gopas, J. & Nathenson, S.G. (1990). Failure of cell surface expression of a class I major histocompatibility antigen caused by somatic point mutation. *Transplantation*, 49, 803–808.

# 8

# Modulation of MHC class I antigen expression in adenovirus infection and transformation

G. ERIC BLAIR,[1] JOANNE L. PROFFITT[1]
and MARIA E. BLAIR ZAJDEL[2]
[1]University of Leeds and [2]Sheffield Hallam University

## 8.1 Introduction

Human adenoviruses (Ads) possess two well-studied mechanisms for modulation of MHC class I expression. The mechanism that is predominant in infected cells involves down-regulation of surface class I antigens caused by a block in the transport of class I heavy chains to the cell surface. This block is effected by a protein product of a viral early gene (termed E3) and seems to operate in cells infected with most virus serotypes. In adenovirus-transformed cells, the level of surface class I antigens can be either elevated or decreased depending on the serotype of the transforming adenovirus and is controlled mainly at the step of transcription of class I genes. Modulation of the rate of initiation of class I gene transcription in these cells is mediated by the product(s) of a different viral early gene (E1A). Interestingly, it is the expression of the E1A gene of highly oncogenic adenoviruses (such as Ad12) that results in down-regulation of class I transcription, providing a possible mechanism whereby oncogenic Ad12-transformed cells can escape host immune surveillance and form tumours. Adenovirus-infected and adenovirus-transformed cells are, therefore, interesting experimental systems for the study of class I modulation.

Adenovirus transformation and oncogenicity have been extensively reviewed (Knippers & Levine, 1989; Boulanger & Blair, 1991; Chinnadurai, 1992; Moran, 1993) and comprehensive reviews on virus structure, biology and pathogenicity are also available (Ginsberg, 1984; Doerfler, 1986; Horwitz, 1990a,b). The wider aspects of the immunobiology of adenoviruses have also been described (Wold & Gooding, 1991; Braithwaite *et al.*, 1993). In this chapter, the biology of the human adenoviruses relevant to immunomodulation will be briefly outlined and the cellular mechanisms that adenovirus proteins adapt to modulate expression of class I antigens will be described in detail.

## 8.2 Molecular pathology of adenoviruses

### 8.2.1 Adenoviruses and human disease

Adenoviruses were initially discovered in latently infected adenoids and have subsequently been shown to be widespread in several mammalian and avian species (Mautner, 1989). The adenoviruses cause about 30% of all viral respiratory diseases in humans (Horwitz, 1990b). They productively infect a variety of tissues and were one of the first human viruses shown to induce tumours in rodents and to transform rodent cells in culture. The 47 human Ad serotypes have been identified and classified into six subgenera (A–F) based on the oncogenicity of viruses in newborn rodents, their antigenic properties and DNA homology (Table 8.1). Viruses and DNA of all serotypes can transform rodent cells in culture; however, only members of subgenera A and B are oncogenic in newborn rodents, with a marked difference in the efficiency of tumour formation between the two subgenera (Table 8.1).

Adenoviruses from different subgenera appear to cause particular clinical symptoms in humans (reviewed by Straus, 1984). Thus, subgenus A viruses (such as Ad12) are shed in faeces and have been proposed to be involved in a gastrointestinal disorder, coeliac disease (Kagnoff, 1989), although their role in this disease has been questioned (Howdle *et al.*, 1989; Mahon *et al.*, 1991). Subgenus B viruses are commonly associated with mild respiratory diseases in children and with conjunctivitis. Subgenus C viruses (e.g. adenoviruses 1, 2, 5 and 6), which are frequently detected in humans, infect the respiratory tract at different sites, often causing cold-like symptoms. They also have the ability to establish persistent infections, especially in lymphoid tissues. Subgenus D viruses have been linked to keratoconjunctivitis, while subgenus E, comprising only one member (Ad4), has been associated with ocular infections, causing conjunctivitis. The subgenus F viruses, the so-called enteric adenoviruses Ad40 and Ad41, appear to be a significant cause of infantile diarrhoea.

### 8.2.2 Structure and assembly of adenoviruses

Human adenoviruses are non-enveloped viruses that display icosahedral symmetry. The virus particle consists of a nucleoprotein core that is formed from a linear double-stranded DNA molecule of approximately 36 kb complexed with two basic virus-coded proteins (polypeptides V and VII) in a chromatin-like structure. This is surrounded by an outer capsid that is composed of 252 protein morphological units (capsomers). The major capsid protein is the

Table 8.1. *Properties of human adenovirus serotypes of subgenera A–F*

| Subgenus | Serotype | DNA homology intra-subgenus (%) | Oncogenicity in newborn hamsters | Tropism/symptoms |
|---|---|---|---|---|
| A | 12, 18, 31 | 48–69 | High (tumours in most animals in four months) | Cryptic enteric infection |
| B | 3, 7, 11, 14, 16, 21, 34, 35 | 64–89 | Weak (tumours in a few animals in 14–18 months) | Respiratory disease; persistent infection of the kidney |
| C | 1, 2, 5, 6 | 99–100 | Nil | Respiratory disease persists in lymphoid tissue |
| D | 8, 9, 10, 13, 15, 17, 19, 20, 22–30, 32, 33, 36–39 | 94–99 | Nil | Kerato-conjunctivitis |
| E | 4 | | Nil | Conjunctivitis; respiratory disease |
| F | 40, 41 | 62–69 | Nil | Infantile diarrhoea |

hexon (polypeptide II), which is a homotrimer and accounts for the 240 capsomers located on the facets of the icosahedron. The capsomers located on the 12 vertices of the virus particle consist of the pentons, each of which comprises a pentameric structure (the penton base, containing five molecules of polypeptide III) into which the fibre homotrimer (polypeptide IV) is embedded. Other structural polypeptides ($III_a$, VI, VIII and IX) participate in protein–protein interactions, which are important in virion structure. A more detailed description of the structure of the adenovirus particle can be found in Stewart *et al.* (1991). As well as structural proteins, a number of non-structural proteins are encoded in the adenovirus genome, some of which (e.g. 100 kDa) are required for virus assembly (Cepko & Sharp, 1983). Certain viral structural proteins are synthesized as precursors (pVI, pVII and pVIII) that must be cleaved by a virus-coded endoproteinase (Webster, Russell & Kemp, 1989) for virus assembly to take place (Kemp, Webster & Russell, 1992).

### *8.2.3 Expression of the adenovirus genome*

The genome of human adenoviruses has been conventionally divided into 100 map units (m.u.) from the left end (Fig. 8.1) and encodes at least 20–30 polypeptides (Akusjarvi & Wadell, 1990). Expression of adenovirus genes in permissive (human) cells is temporally regulated. Genes that are expressed at early stages of infection, before the onset of viral DNA replication, are termed early (E). Proteins encoded by these genes are mainly involved in regulation of viral DNA replication and transcription. Genes that are active mainly after replication of viral DNA are termed late (L). The structural proteins incorporated into the virus particle are synthesized late in infection.

Adenovirus genes are expressed from transcription units, each of which is controlled by a single promoter. Eight transcription units are located on both strands of the viral DNA and their transcription is mediated by host cell RNA polymerase II (Fig. 8.1). Diversity in the number and sequence of proteins synthesized from each transcription unit is caused by multiple polyadenylation sites, differential splicing and multiple reading frames. Early genes are transcribed from unique promoters in each of the early transcription units: E1A, E1B, E2, E3 and E4.

The first viral gene to be expressed in the early phase of infection is E1A. Two major mRNAs, sedimenting at 13 S and 12 S, in Ad5-infected cells are transcribed from the E1A gene (Fig. 8.2*a*). These mRNAs are generated by differential splicing of a common RNA precursor and are translated into a family of phosphoproteins based on two polypeptides of 289 and 243 amino

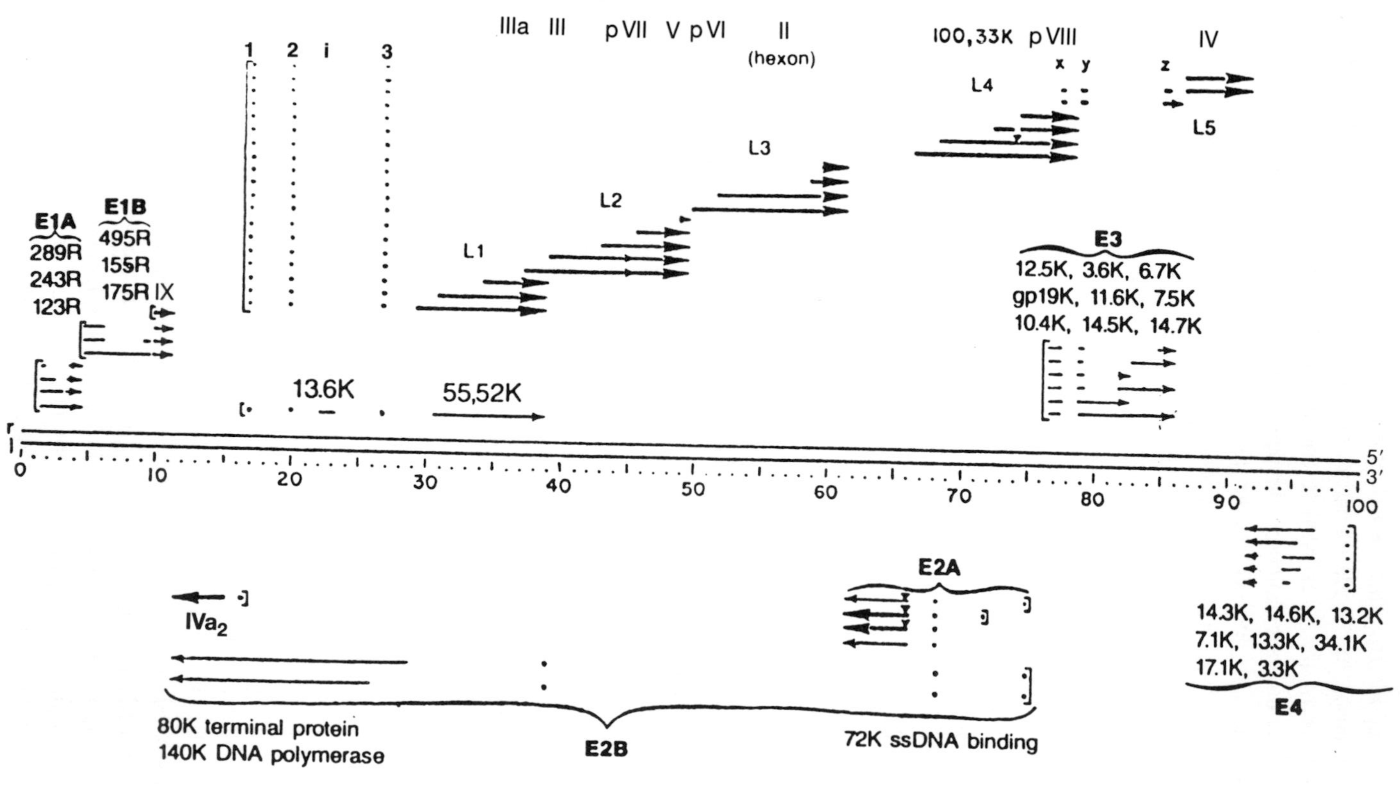


Fig. 8.1. Organization of the adenovirus type 2 genome. The 36 kb double-stranded DNA genome is denoted by two parallel lines, referring to the l strand from which leftward RNA transcripts are transcribed and r, the rightward reading strand. The 3′ ends of transcripts are denoted by the arrowhead. Proteins are denoted by the suffix R (residues), K (kDa) or are numbered or named (e.g. pV1 or hexon). The five early (E) regions are shown along with the major late (L) transcription unit, divided into late families of mRNAs, L1 to L5 each of which have a spliced leader consisting of the '1', '2' and '3' leader sequences. Modified from Akusjarvi & Wadell (1990).

acid residues (R), respectively. Comparison of E1A sequences from several viral serotypes led to the identification of three conserved regions (CR1, CR2 and CR3) in E1A proteins (Moran & Matthews, 1987; Braithwaite, Nelson & Bellett, 1991a). Mutational analysis has defined functions associated with each region (Lillie & Green, 1989) (Fig. 8.2*b*). CR3, which is unique to the 289R product, is required for transcriptional *trans*-activation of other viral early promoters (E1B, E2, E3 and E4), the major adenovirus late promoter (located at 16.3 m.u.) and certain cellular promoters, e.g. those of β-tubulin and heat shock protein genes (Boulanger & Blair, 1991). A zinc finger domain is contained within CR3 (Culp *et al.*, 1988), but E1A proteins do not possess sequence-specific DNA-binding properties. E1A-mediated *trans*-activation appears to proceed by heterodimer formation between E1A and members of the ATF transcription factor family (Liu & Green, 1994) as well as interaction between CR3 and the basal transcription factor TFIID (Boyer & Berk, 1993). The CR1 and CR2 domains, which are located in the amino-terminal region of E1A and are present in both the 289R and 243R products, are involved in transcriptional repression and transformation (Boulanger & Blair, 1991). Several promoters have been shown to be repressed by Ad5 E1A, including the SV40 and polyoma virus early promoters (Borelli, Hen & Chambon, 1984; Rochette-Egly, Fromental & Chambon, 1990) and the cellular fibronectin (Nakajima *et al.*, 1992), *neu* proto-oncogene (Yu *et al.*, 1990) and ferritin H subunit (Tsuji *et al.*, 1993) promoters. Ad5 E1A also appears to act as a repressor of differentiation in well-studied processes such as myogenesis (Webster, Muscat & Kedes, 1988). The CR1 and CR2 domains bind the tumour suppressor gene product p105-*RB*, which is a cellular gene product of the retinoblastoma susceptibility locus. Binding of E1A to p105-*RB* removes the growth-suppressing function of p105-*RB* and thus accelerates the growth of the adenovirus-transformed cell (reviewed in Boulanger & Blair, 1991). Furthermore, in normal, untransformed cells, p105-*RB* associates with a cellular transcription factor E2F and removal of p105-*RB* by binding to E1A activates E2F for transcriptional induction of a set of cellular genes that have a common function in DNA replication and metabolism (Nevins, 1992). Other proteins that bind to the amino-terminal region of E1A have been described, including a p107 protein, which binds to CR2, and a p300 protein, which binds at the amino-terminus to CR1. The p300 protein has been shown to have sequence-specific DNA-binding properties similar to those of the NF-κB/c-*rel* family (Rikitake & Moran, 1992; Abraham *et al.*, 1993; see also Chapter 4), and molecular cloning studies have suggested that p300 may also have properties of a transcriptional co-activator that interacts with basal transcription factors (Eckner *et al.*, 1994). The E1A proteins,

(a)

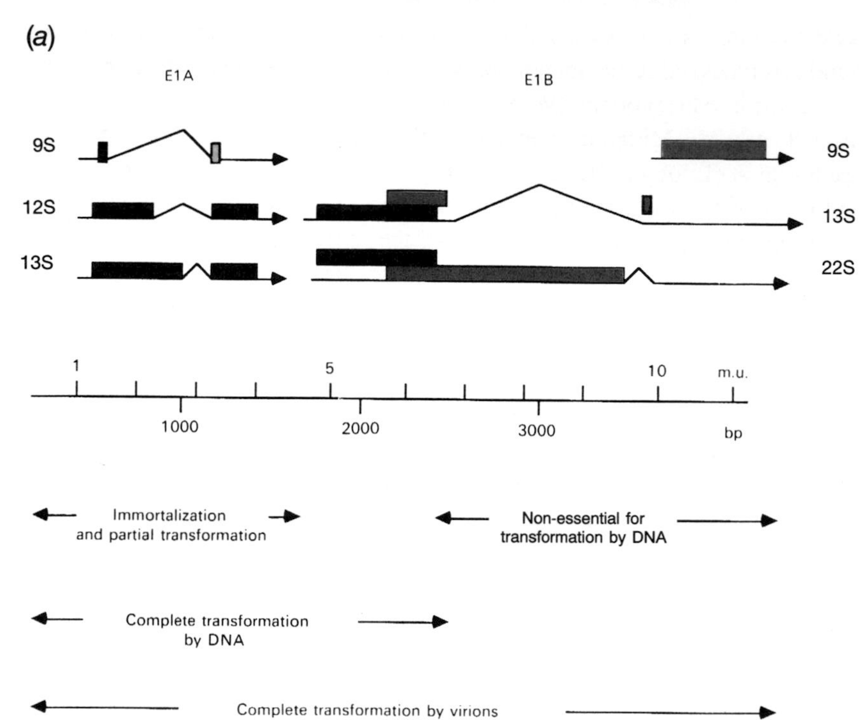
E1A
E1B
9S
12S
13S
9S
13S
22S
1
5
10
m.u.
1000
2000
3000
bp
Immortalization and partial transformation
Non-essential for transformation by DNA
Complete transformation by DNA
Complete transformation by virions

(b)

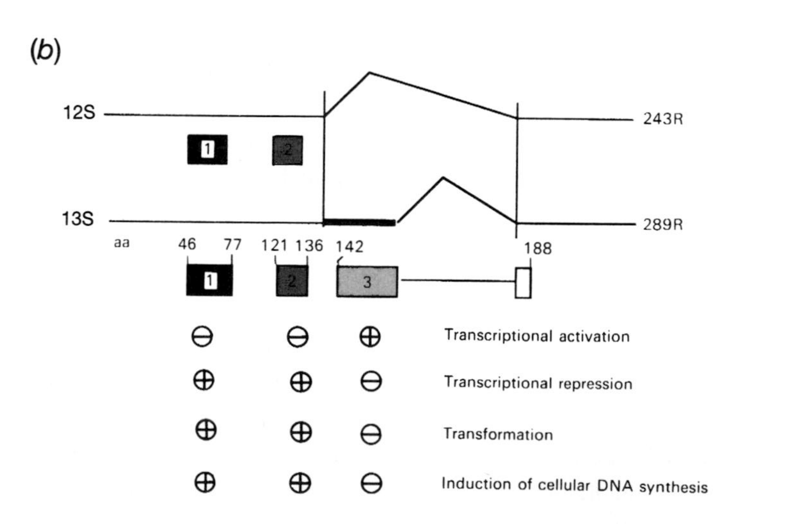
12S
243R
13S
289R
aa
46
77
121
136
142
188
1
2
3
Transcriptional activation
Transcriptional repression
Transformation
Induction of cellular DNA synthesis

therefore, are multifunctional and possess biological activities that can modify cellular mechanisms of transcription, cell cycle regulation, etc.

The E1B gene encodes two major mRNAs, sedimenting at 22 S and 13 S, that encode two unrelated polypeptides that function in viral infection and transformation (Fig. 8.2*a*). There are also minor E1B mRNAs and proteins (Boulanger & Blair, 1991). The larger E1B 58 kDa protein is synthesized only from the 22S mRNA and is involved in the transport of viral mRNA through an intranuclear compartment in infected cells (Leppard & Shenk, 1989). In infected cells, E1B 58 kDa protein forms a complex with the 34 kDa product of ORF6 of the viral E4 gene. Complex formation is necessary for mRNA transport and, ultimately, shut-off of host protein synthesis (Halbert, Cutt & Shenk, 1985). In Ad5-transformed and Ad5-infected cells, E1B 58 kDa protein forms a complex with the p53 tumour-suppressor gene product (Sarnow *et al.*, 1982; Braithwaite *et al.*, 1991b). At least one of the functions of the p53 protein is that of a transcription factor, possibly regulating expression of cell-cycle related genes (Farmer *et al.*, 1992). By analogy with the interaction of E1A with p105-*RB* described above, the function of the E1B 58 kDa protein may be to remove the growth-suppressing effects of p53 by complex formation. One problem with this proposal is that the corresponding E1B 55 kDa protein of Ad12 does not form a stable complex with p53 (Zantema *et al.*, 1985) and so the function of complex formation needs to be more carefully defined. A recent study has provided evidence for a transcriptional repressory function of the Ad5 E1B 58 kDa protein (Yew, Liu & Berk, 1994), which can be targeted to p53-responsive promoters via the interaction with p53. The smaller E1B 19 kDa protein is synthesized from both the 22S and 13S mRNAs. The best documented function of this protein is to protect viral and cellular DNA from degradation induced by infection (Boulanger & Blair, 1991). The 19 kDa protein also protects cells from apoptosis induced by E1A expression or other agents (White *et al.*,

Fig. 8.2. Organization of the transforming region of adenovirus type 2 and structure/functions of the E1A proteins.
(*a*) The transforming region, consisting of early regions E1A and E1B, spanning the leftmost 11 m.u., or approximately 3500 bp of DNA, of the Ad2 genome is shown. The 3′ ends of mRNAs are denoted by arrowheads and the ORFs for translation are depicted as shaded boxes. Functions of particular parts of the transforming region are indicated.
(*b*) The functions and domain structure of the Ad2 E1A proteins. Conserved regions (CR) 1, 2 and 3 of E1A oncoproteins are shown in different boxes and located on the 289R product of the 13S mRNA and the 243R product of the 12S mRNA, along with functions ascribed to these regions.
Reproduced, with permission, from Boulanger & Blair (1991).

1992; Debbas & White, 1993) and prevents TNF-α mediated cytolysis of human cells (Gooding *et al.*, 1991; see Section 8.5 below). The E1A and E1B regions thus have important roles in regulation of cell proliferation and in the response to cytokines. These regions also play important roles in immune recognition.

Early region 3, which is non-essential for virus growth in cell culture (Jones & Shenk, 1978), encodes at least six proteins, some of which have immunoregulatory functions and affect immune recognition of infected cells by the host (see Section 8.4 below).

The other viral early and late regions do not appear to have major roles in cellular immune recognition or regulation. The E2A region encodes a single-stranded DNA-binding protein that is required for replication of viral DNA. Two proteins are encoded by the E2B gene: the precursor to the terminal protein, which is covalently linked to the 5′ ends of adenovirus DNA, and a DNA polymerase (reviewed by Hay & Russell, 1989). Much less is known about the biochemistry of the proteins encoded by the E4 region, probably because this is the most complex of the early regions in terms of the number of ORFs. The function of the E4 34 kDa protein has been described above. Another E4 ORF encodes a 19 kDa protein that activates the E2A promoter by interacting with the E2F transcription factor (Nevins, 1991). Late genes are transcribed from a common major late promoter, located at 16.3 m.u. on the Ad5 genome (Fig. 8.1). Late mRNAs are post-transcriptionally processed in families (L1–L5) and encode structural proteins of the virus particle as well as late non-structural proteins. Two small RNAs, the VA RNAs I and II (viral-associated RNAs), which are transcribed by RNA polymerase III, do not code for proteins, but the major species (VA RNA I) is essential for virus infection and acts by inhibiting an interferon-induced protein kinase whose normal function is to shut down protein synthesis. Thus, the continued protein synthesis necessary for virus infection is ensured (Schneider & Shenk, 1987).

### 8.2.4 Adenovirus transformation and oncogenicity

When cells from rodents (rat, mouse or hamster) are infected by human adenovirus a different pattern of viral gene expression is observed to that which takes place in lytic infection of permissive human cells. Rodent cells can be either non- or semi-permissive for adenovirus replication. The infection can be either abortive, in which case the virus is eventually lost from the cells or, in a small proportion of cells, transformation occurs where a segment of viral DNA spanning at least the E1A and E1B genes is stably incorporated into cellular DNA. Adenovirus-transformed cells, therefore, constitutively express the E1A and E1B gene products. Proteins encoded by E1A and E1B have different functions in transformation and their co-operation is required to produce the fully transformed phenotype, which includes such properties as morphological alteration, unlimited growth in culture and, in the case of Ad12-transformed cells, oncogenicity in newborn rodents. Neither the E1A nor the E1B genes alone are sufficient for transformation. The E1A genes of either Ad12 or Ad5 can immortalize primary cells, albeit at low frequency, but without producing many of the features of fully adenovirus-transformed cells (Houweling, van den Elsen & van der Eb, 1980). Certain lines of Ad12 E1A-immortalized cells can be oncogenic in newborn animals (Gallimore *et al.*, 1984; Sawada *et al.*, 1985). Both major E1A gene products of Ad5 are required (along with E1B or activated *ras* genes) for transformation of primary rodent cells *in vitro* (Montell *et al.*, 1984), although in Ad12, only the larger E1A product appears to be necessary (Lamberti & Williams, 1990). Differences in requirements for expression of the major E1B products have also been noted, often depending on whether mutant viruses or plasmid DNA transfection was used to generate transformed cells (Fig. 8.2*a*). Ad5 E1B 58 kDa protein expression was found to be dispensable for DNA-mediated transformation, although mutant Ad5 viruses, such as hr6, that contain point mutations in the E1B 58 kDa protein, were transformation defective (Rowe & Graham, 1983; Rowe *et al.*, 1984). In Ad12 transformation, an opposite phenomenon has been described, where expression of the E1B 19 kDa protein was unnecessary for viral transformation (Edbauer *et al.*, 1988). Therefore, although Ad5 and Ad12 have overall similarities in the organization of their transforming regions, the mechanisms that they adopt to transform cells may be somewhat different. For a more detailed review of the biology of adenovirus transformation, see Boulanger & Blair (1991).

Primary cultures of human cells can be transformed by human adenoviruses at very low frequency, using DNA transfection techniques (Graham *et al.*, 1977; Gallimore *et al.*, 1984). Human adenoviruses, however, do not appear to be associated with common human cancers (Mackey, Rigden & Green, 1976; Green *et al.*, 1979), as judged by DNA hybridization analysis. These studies should be repeated with more sensitive techniques, such as PCR.

While all adenoviruses can transform primary rodent cells in culture, only certain adenoviruses are able to generate oncogenic transformed cells (Table 8.1). The subgenus A viruses (e.g. Ad12) induce tumours in newborn rodents with high frequency and short latency when purified virus particles are injected subcutaneously or intracranially. This contrasts with the subgenus C viruses (e.g. Ad2 or Ad5), which are non-oncogenic *in vivo*. The subgenus B viruses (e.g. Ad3 or Ad7) represent an intermediate case, being weakly oncogenic. The subgenus D virus, Ad9, induces benign mammary fibroadenomas in females of a particular strain of rats but is otherwise non-oncogenic (Ankerst *et al.*, 1974). The subgenus E and F viruses are also believed to be non-oncogenic. The oncogenicity of cloned adenovirus-transformed cell lines (tested by injection into syngeneic hosts) tends to reflect the oncogenicity of the virus serotype used for transformation. Thus subgenus A (i.e. Ad12) virus-transformed cells are highly oncogenic whereas subgenus C (i.e. Ad2 or Ad5) transformed cells are usually non-oncogenic. Certain Ad12-transformed rat and mouse cell lines form tumours in immunocompetent (i.e. adults) as well as newborn animals and, in certain cases, can form tumours in allogeneic as well as syngeneic hosts (Tanaka *et al.*, 1985). Immunosuppression of rats by administering anti-thymocyte serum allowed subgenus C virus-transformed cells (which are non-oncogenic in untreated animals) to form tumours (Harwood & Gallimore, 1975), pointing to a role for CTLs in immune surveillance. Many lines of subgenus C virus-transformed cells, which are non-oncogenic in syngeneic hosts, can induce tumours in congenitally athymic 'nude' mice or rats (Bernards & van der Eb, 1984), further implying a role for thymus-derived immunity in rejecting cells transformed by non-oncogenic adenovirus serotypes.

Although the genetic information required for cell transformation resides in the E1A and E1B regions of the viral genome (see above), the serotypic origin of the E1A gene determines the oncogenic phenotype of adenovirus-transformed cells. Using Ad5 viruses in which either the E1A or E1B gene was replaced with its Ad12 counterpart, it was shown that only cells transformed by viruses containing the Ad12 E1A gene could elicit tumours in syngeneic rats (Sawada *et al.*, 1985). Certain cell lines immortalized by the Ad12 E1A gene alone have been reported to be oncogenic (see above). The region of Ad12 E1A responsible for oncogenesis has been studied by both analysis of viral E1A mutants and comparative analysis of the DNA sequence of oncogenic adenoviruses. Kimelman *et al.* (1985) identified two amino acid sequences which were conserved between Ad12 E1A and the larger E1A protein of simian adenovirus 7 (SA7), which, like Ad12, generates oncogenic-transformed cells. Both Ad12 and SA7 are enteric viruses. These amino acid sequences are not present in E1A genes from non-oncogenic adenoviruses. The longer, alanine-rich sequence, which is located in exon 1 between CR2 and CR3 from base pair 897 to base pair 921 on the Ad12 sequence, is present in both the large and small E1A products and could form a determinant of oncogenicity, although this needs to be tested by mutational analysis.

Intriguingly, alanine-rich sequences are present in the transcriptional repressor domains of many *Drosophila* homeotic transcription factors (Licht *et al.*, 1990). Analysis of recombinant adenoviruses and plasmids containing chimaeric Ad5/Ad12 E1A genes suggested that the alanine-rich region does form at least one oncogenic determinant of Ad12 E1A (Telling & Williams, 1994; Jelinek, Pereira & Graham, 1994). However, transfer of the alanine-rich sequence into the CR2–CR3 spacer region of Ad5 conferred a moderate but not highly oncogenic phenotype on transformed cells (Jelinek *et al.*, 1994) indicating that there may be further oncogenic determinant(s) yet to be identified in Ad12 E1A. Murphy *et al.* (1987), studying an Ad12 mutant CS-1, which contained a 69 bp deletion in the E1A gene (from base pair 834–902), found it to be transformation defective and incapable of eliciting tumours in baby hamsters although it was replication competent. This sequence is located in the CR2 region and overlaps the alanine-rich sequence common to SA7, implying that exon 1 sequences including CR2 and the spacer region between CR2 and CR3 may determine oncogenicity of Ad12. Alternatively, the inability of Ad12 CS-1 to induce tumours could be simply a consequence of its defectiveness in transformation. A crucial question now to be resolved is if the alanine-rich sequence in Ad12 E1A functions as a transcriptional repressor of MHC class I gene expression and if this can be correlated with the oncogenic properties of Ad12 E1A.

The oncogenicity of subgenus A and D viruses may not be determined solely by functions encoded in the transforming (E1A and E1B) region (Bernards *et al.*, 1984; Sawada, Raska & Shenk, 1988), in particular a contribution of E4 has been ascribed to Ad9 tumorigenesis (Javier, Raska & Shenk, 1992). In conclusion, while domains of Ad12 E1A appear to be of major importance in generating virally induced tumours and oncogenic cell lines *in vitro*, the mechanism of tumour induction by virions *in vivo* appears less clear with possible requirements for additional viral genes to be expressed.

## 8.3 MHC class I antigen expression in adenovirus-transformed cells

### *8.3.1 Which immune effector cell(s) are responsible for rejecting cells transformed by non-oncogenic adenoviruses?*

The observation that MHC class I antigen expression was greatly reduced in highly oncogenic Ad12-transformed rat cells whereas non-oncogenic Ad5-transformed cells displayed normal, or elevated, class I levels (Schrier *et al.*, 1983) was an important step forward in deciding which immune cells are responsible for rejecting the latter group. The steady-state level of class I heavy chain mRNA was also found to be down-regulated in Ad12-transformed cells compared with Ad5-transformed cells. Repression of class I expression by Ad12 has since been described in human (Vasavada *et al.*, 1986; Grand *et al.*, 1987), mouse (Eager *et al.*, 1985; Haddada *et al.*, 1986) and hamster (Haddada *et al.*, 1986; Ackrill & Blair, 1988a) cells as well as in rat cells (Schrier *et al.*, 1983; Ackrill & Blair, 1988b; Katoh *et al.*, 1990). CTLs were shown to effectively kill rat cell lines expressing the Ad5 E1A gene but did not kill cells expressing the Ad12 E1A gene (Bernards *et al.*, 1983). This approach suggested that Ad12-transformed cells are oncogenic *in vivo* by a mechanism of evasion of host immune surveillance since they express very low levels of cell surface MHC class I antigens. In contrast Ad5-transformed cells are immunogenic since they display a processed viral antigen, probably a peptide fragment of E1A (Bellgrau, Walker & Cook, 1988; Kast *et al.*, 1989; Urbanelli *et al.*, 1989) bound to cell surface class I antigens and are recognized by CTLs and killed. In support of this hypothesis, influenza virus-infected Ad12-transformed mouse cells were much more resistant to lysis by influenza-specific CTLs than were infected Ad5-transformed cells (Yewdell *et al.*, 1988). Experiments performed by Tanaka *et al.* (1985) showed that oncogenicity of Ad12-transformed cells could be abrogated by expression of a transfected foreign class I gene. They introduced a functional, allogeneic class I gene into a highly oncogenic Ad12-transformed mouse cell line and found that the resulting transfected cells had greatly lowered tumorigenicity, increased levels of the appropriate class I antigens and increased sensitivity to T cell lysis. Up-regulation of surface class I antigens by IFN-γ treatment of Ad12-transformed cells reduced their tumorigenicity (Eager *et al.*, 1985; Hayashi *et al.*, 1985; Tanaka *et al.*, 1986). Together, these different experimental approaches provide persuasive evidence for a direct link between lowered levels of class I expression and the tumorigenicity of Ad12-transformed cells.

There are, however, some problems to be resolved. First, a broad survey of adenovirus-transformed hamster cells identified Ad2-transformed cell lines that displayed high or low levels of surface class I antigens yet were non-oncogenic (Haddada *et al.*, 1988). Tumorigenic Ad2- or Ad5-transformed hamster cell lines still displayed high levels of class I mRNA and proteins in tumour tissue and explanted cells (Ackrill & Blair, 1988a). It might, therefore, appear that, at least in the hamster, the level of class I expression in adenovirus-transformed cells is not a major factor in determining their oncogenicity. However, it has been noted that the hamster MHC does not display the polymorphism of human and mouse MHC genes (Darden & Streilein, 1983) and, therefore, cellular recognition of tumour cells may be different.

Furthermore, a series of experiments (Soddu & Lewis, 1992) that used an essentially identical approach to that of Tanaka *et al.* (1985) revealed that expression of allogeneic class I genes in Ad12-transformed BALB/c mouse cells increased (rather than reduced) their oncogenicity, a result that apparently conflicts with the previous study of Tanaka *et al.* (1985). Some salient differences in protocols between the two studies can be noted: firstly, different strains of mouse were used to generate transformed cells (and then transfectants) and, secondly, low-passage Ad12-transformed cells were used for transfection in the study by Soddu & Lewis (1992) whereas high-passage Ad12-transformed C57AT1 cells were used by Tanaka *et al.* (1985). Finally, Ad40- and Ad41-transformed rat cells have low levels of surface class I antigens, comparable to those on the surface of Ad12-transformed cells (Cousin *et al.*, 1991). Although the oncogenicity of the Ad40- and Ad41-transformed cells was not tested in these experiments, since it is assumed that the subgenus F viruses are non-oncogenic, this observation may cast doubt on a causal link between lowered levels of class I antigens and oncogenicity of adenovirus-transformed cells. Clearly, it will be important to resolve these issues in the future.

It might be argued that other immune effector cells, such as NK cells and activated macrophages may act on Ad5- and perhaps also in certain cases Ad12-transformed cells *in vivo*, as in the work of Soddu & Lewis (1992). Indeed the E1A gene of Ad12, but not Ad5, confers resistance to NK cytolysis (Braithwaite *et al.*, 1993; Sawada *et al.*, 1985; Kenyon & Raska, 1986). There is a paradox here, since increasing the levels of class I antigens on the surface of tumour cells reduces their ability to be killed by NK cells (Storkus *et al.*, 1989; see also Chapter 14). Therefore, it might be expected that Ad12-transformed cells would be more sensitive to NK cells than Ad5-transformed cells, although the reverse appears to be the case. Other cell-surface accessory molecules might modify interactions between NK cells and adenovirus-transformed cells.

### *8.3.2 What is the mechanism of down-regulation of MHC class I expression mediated by Ad12 E1A?*

Transfection of Ad12 E1A into a class I-expressing cell line generated transformants with lowered class I expression (Vaessen *et al.*, 1986; Katoh *et al.*, 1990), showing that down-regulation is an active, E1A-dependent process, rather than a secondary effect of transformation. In Ad12-transformed mouse cells, all class I loci are repressed (Vaessen *et al.*, 1986). Ad5 E1A abolishes the repressory effect of Ad12 E1A on class I expression; Ad12-transformed cells that also stably expressed the entire Ad5 E1 region showed normal levels of class I expression (Bernards *et al.*, 1983; Vaessen *et al.*, 1986). The creation of hybrid Ad12/Ad5 E1A genes has defined the first exon of Ad12 E1A as the region responsible for down-regulation of class I expression (Jochemsen, Bos & van der Eb, 1984). Therefore, a function encoded in the first exon of Ad12 E1A, which contains the CR1, CR2 and part of the CR3 domain of Ad12 (Braithwaite *et al.*, 1991a), is responsible for down-regulation of class I expression and the same region of Ad5 E1A can abolish the repressing effect of Ad12 on class I expression. Interestingly, exon 1 also contains sequences implicated in transformation (CR1 and CR2) and oncogenicity (the alanine-rich spacer between CR2 and CR3 in Ad12 E1A) as well as containing differences between Ad12 and Ad5 E1A at the amino-termini (Wang *et al.*, 1993). Although the product of the 13S but not the 12S mRNA was reported to drive down-regulation of class I in Ad12-transformed cells (Bernards *et al.*, 1983; Meijer *et al.*, 1991), work by R. Merrick and P. H. Gallimore (personal communication), where baby rat kidney cells were transformed by recombinant retroviruses expressing either the Ad12 13S or 12S mRNA, has shown that cells transformed by 12S alone have lowered levels of class I mRNA and protein compared with normal cells, although 13S-transformed lines had even lower levels of class I gene products. The precise role of the Ad12 13S and 12S products in down-regulation of class I expression, therefore, needs to be clarified.

The reduction in the level of class I gene expression found in Ad12-transformed cells is exerted mainly at the level of initiation of transcription (Ackrill & Blair, 1988b; Friedman & Ricciardi, 1988; Lassam & Jay, 1989; Meijer *et al.*, 1989; Katoh *et al.*, 1990). Conserved regulatory elements involved in transcriptional control of class I expression have been described in Chapter 3. The major regulatory element, the CRE, consists of three domains: region I, also termed enhancer A (an enhancer that contains the binding site for the NF-κB/H2TF1/c-*rel* family of transcription factors), located at base pairs −156 to −175 upstream from the start-site for transcription, at +1; region II (which binds factors of the nuclear hormone receptor superfamily), located at base pairs −175 to −203; and region III (whose function is less well defined), located at base pairs −161 to −189. One possibility is that Ad12 E1A protein(s) modify the expression or interaction of cellular transcription factors with region(s) of the CRE. In this respect, it is interesting to note that a nuclear factor (termed CRE2) present in Ad12-transformed cells, but at greatly reduced levels in Ad5-transformed cells, binds to the base pair −175 to −203 (region II) sequence of the H-$2K^b$ CRE (Ackrill & Blair, 1989; Ge *et al.*, 1992; Meijer *et al.*, 1992). Using multiple copies of the CRE2-binding site linked to a copy of the H-2K$^b$ region I sequence, Kralli *et al.* (1992) showed that this protein–DNA interaction could have functional significance in Ad12-transformed cells, since this combination of binding sites could down-regulate a heterologous SV40 basal promoter. As described in Chapter 4, region II binds transcription factors of the hormone receptor superfamily and a member of this superfamily, COUP-TF, has been found to be present at elevated levels in cells transformed with Ad12 compared with those tranformed with Ad5 and was serologically related to CRE2 (Liu *et al.*, 1994). Ackrill & Blair (1989) also showed that factor binding to the region I (H2TF1) element was greatly increased in Ad5- compared with Ad12-transformed cells, an observation since confirmed by others (Offringa *et al.*, 1990; Meijer *et al.*, 1992). In a study of Ad12-transformed cells transfected with the Ad5 E1A and E1B regions (generating stable cell lines that express normal levels of class I antigens), the level of H2TF1 element-binding activity increased while CRE2-binding activity remained unchanged (Meijer *et al.*, 1992). Furthermore, this was reflected by a decrease in enhancer activity of the H2TF1 site in cells expressing Ad12 13S but not Ad12 12S mRNA. This led to a proposal that lowered levels of H2TF1-binding activity in Ad12-transformed cells could explain down-regulation of class I transcription. Nielsch, Zimmer & Babiss (1991), studying a series of Ad5- and Ad12-transformed rat fibroblast (CREF) cell lines, discovered that Ad5- but not Ad12-transformed cells secreted low levels of IFN-β, leading to constitutive expression of nuclear interferon-stimulated gene factor 3 (ISGF3: see Chapter 4 and Section 10.5 below) and NF-κB. Overall, this could explain the higher level of class I transcription in Ad5-transformed cells. Other groups, however, have reported that Ad2 E1A represses the function of the H2TF1 site of the H-2K$^b$ class I gene (Rochette-Egly *et al.*, 1990) and the NF-κB site of the IL-6 gene (Janaswami *et al.*, 1992), which conflicts with the studies of Meijer *et al.* (1992) and Nielsch *et al.* (1991).

An alternative approach is that of systematically deleting portions of the 5′-flanking region of class I genes fused to a reporter gene, usually CAT, and searching for DNA sequences associated with down-regulation of reporter gene expression (Kimura *et al.*, 1986; Katoh *et al.*, 1990; Proffitt, Sharma & Blair, 1994). Using this approach, sequences in the H-2K$^b$ or H-2K$^{bm1}$ genes extending upstream to approximately base pair −2000 effectively down-regulated CAT activity when transfected into Ad12- but not Ad5-transformed cells. Katoh *et al.* (1990) have proposed that a sequence from base pair −1520 to −1840 in the H-2K$^{bml}$ gene is the target for Ad12 E1A repression. Regulatory sequences that have been subsequently located in this region are CAA repeats at base pair −1705 to −1725 and −1568 to

−1591 and a TATA-like sequence at −1767 to −1773 (Ozawa *et al.*, 1993; Tang *et al.*, 1995). Either CAA repeat with the TATA-like element was found to be insufficient for Ad12 E1A-mediated transcriptional down-regulation (Tang *et al.*, 1995). Proffitt *et al.* (1994) found that a sequence from base pair −1180 to −1440 of the mouse H-2K$^b$ gene could down-regulate a heterologous herpes simplex virus (HSV) thymidine kinase (tk) promoter. Deletion of this sequence from a CAT plasmid containing the entire 2 kb of 5′-flanking region of the H-2K$^b$ gene (including the basal promoter elements up to +4) resulted in a loss of down-regulation of CAT activity in Ad12-transformed cells. Similarly, in co-transfection assays, where either the deletion mutant or normal CAT plasmid is introduced into normal 3Y1 fibroblasts along with a plasmid expressing either Ad12 or Ad2 E1A, down-regulation is lost when the deletion mutant is co-transfected with the Ad12 expression plasmid (Fig. 8.3). Moreover, in the absence of the sequence from base pair −1180 to −1440, CAT activity is enhanced in the presence of Ad12 E1A. A similar observation was made by Katoh *et al.* (1990) and could be explained by a model in which a negative regulatory element, NRE (at either base pairs −1520 to −1840 or −1180 to −1440), responsive to Ad12 E1A over-rides the normal, positive regulatory function of the CRE. Proffitt *et al.* (1994) also detected specific protein binding to DNA fragments in the region base pair −1180 to −1440 using nuclear extracts of Ad12-, but not Ad5-transformed cells. In broad agreement with these results is the detection of an NRE responsive to Ad12 E1A between base pairs −237 and −1400 (Rotem-Yehudar, Schechter & Ehrlich, 1994a).

The mechanism of Ad12 E1A-mediated down-regulation of class I transcription is, therefore, still not clear. Four possible sites of action of E1A in the class I 5′-flanking region have been proposed (Fig. 8.4). Two of these sites are located within the CRE, namely modulation of H2TF1/NF-κB activity (Nielsch *et al.*, 1991; Meijer *et al.*, 1992) and CRE2 activity (Ge *et al.*, 1992; Kralli *et al.*, 1992). These proposals have the merit of adapting existing, well-documented regulatory elements by E1A. However, these conclusions have often been made by analysis of CAT expression driven by either single (Meijer *et al.*, 1992) or multiple copies of binding sites (Kralli *et al.*, 1992) linked to heterologous basal promoters, or by gel retardation analysis (Nielsch *et al.*, 1991), which could be viewed as unphysiological. In contrast, the studies by Katoh *et al.* (1990) and Proffitt *et al.* (1994) have used a careful mutational approach but have identified different E1A-responsive NREs for which there is, at present, no function in normal or non-virally transformed cells. More systematic, site-directed mutagenesis of putative regulatory elements in the class I promoter involved in Ad12 E1A-

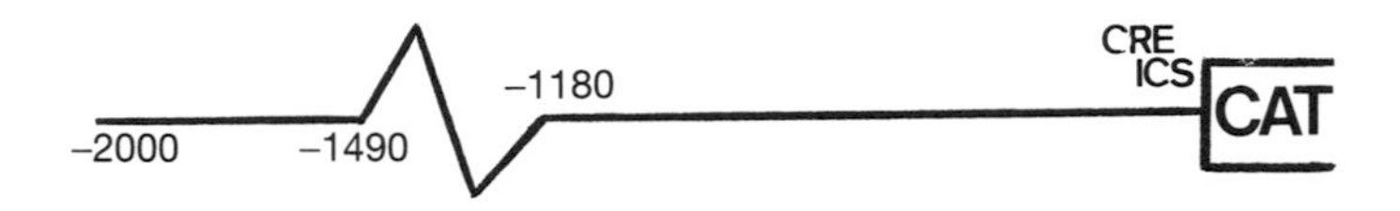


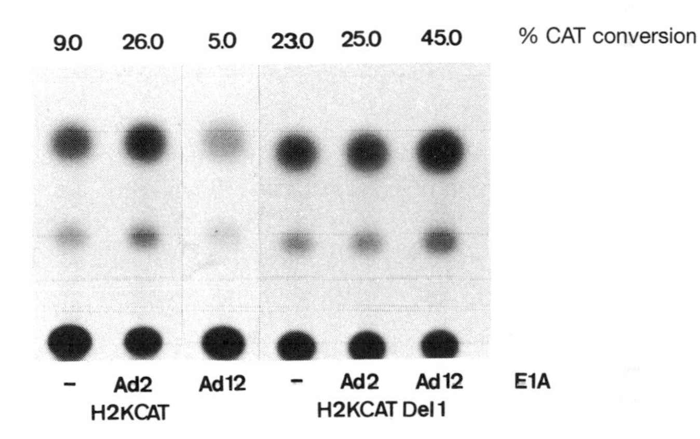


Fig. 8.3. The role of the upstream sequence from base pair –1100 to base pair –1400 of the mouse H-2K$^b$ class I promoter in transcriptional down-regulation mediated by Ad12 E1A. The CAT reporter gene constructs driven by either the entire 2 kb of 5′-flanking (promoter) region of the mouse H-2K$^b$ class I gene (H2KCAT) or a 5′-flanking region in which the approximate 300 bp region between base pair –1180 and –1490 was deleted by *Bcl*I digestion (H2KCAT Del I) were used in co-transfection experiments in rat 3YI fibroblast cells. Each CAT plasmid was co-transfected with either an Ad2 E1A or an Ad12 E1A expression plasmid or a control plasmid pAT153 (-). Following expression in transfected cells for 48 hours, cell lysates were prepared and CAT activity was assayed according to standard procedures (Proffitt *et al.*, 1994). The acetylated products of [$^{14}$C]chloramphenicol were separated by ascending thin layer chromatography (TLC) and CAT conversion, an indicator of promoter activity, was expressed as a percentage conversion of chloramphenicol following quantification of the TLC plate by phosphorimager analysis. Expression of Ad12, but not Ad2 E1A down-regulated CAT activity but this down-regulation was lost when the 300 bp region was deleted from the H-2K$^b$ promoter region.

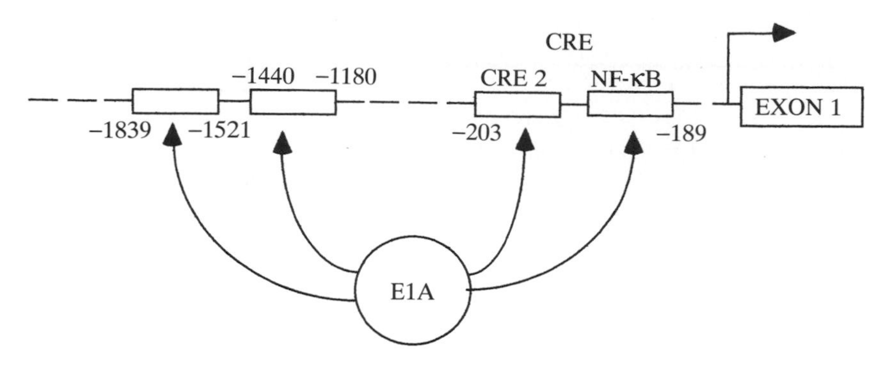


Fig. 8.4. Proposed sites of action of Ad12 E1A in the class I promoter. This diagram summarizes the sites of action of E1A in modulation of class I promoter activity in adenovirus-transformed cells, described in the text. Negative regulatory effects have been described for Ad12 E1A on upstream elements between base pairs −1180 to −1839 as well as on the CRE2 domain of the class I regulatory element (CRE). A positive function of Ad2 or Ad5 E1A has been noted on the level of NF-κB DNA-binding activity. Whether all, or some, of these mechanisms operate in adenovirus-transformed cells remains to be established.

mediated down-regulation might be expected to be fruitful as also might attempts at reproduction of E1A-mediated down-regulation in cell-free transcription systems, such as that described by Driggers *et al.* (1992). It has been reported that, at least in certain Ad12-transformed mouse cell lines, down-regulation of peptide transporter (TAP 1 and 2) gene expression may contribute to the overall reduction in the level of surface class I antigens (Rotem-Yehudar *et al.*, 1994b).

## 8.4 MHC class I antigen expression in adenovirus-infected cells

### *8.4.1 Transcriptional regulation of class I expression in infected cells*

The mechanism of Ad12 E1A-mediated down-regulation of class I expression can also be studied in semi-permissive rodent cells infected by Ad12. Such cultures can be expected to produce transformed foci one to two weeks following infection. Rosenthal *et al.* (1985) infected primary mouse embryo cell cultures with Ad5, Ad12 or a set of viral mutants with mutations in the transforming genes of either serotype and analysed class I mRNA sequences present in cytoplasmic RNA isolated at 50 hours after infection. Interestingly, rather than finding a selective effect of Ad12 in lowering the level of class I transcripts (as might be expected from studies on Ad12-transformed cells), either Ad5 or Ad12 infection led to elevated levels of cytoplasmic class I mRNA. *In vitro* transcription analysis in isolated nuclei showed that this elevation could be mainly accounted for by an increased rate of initiation of class I transcription. Viral mutants affecting the expression of the E1A or E1B genes of either serotype did not induce elevated levels of class I mRNA. Productive infection of primary human embryonic kidney cells by Ad12 did not result in alteration of class I protein levels (Grand *et al.*, 1987). The relationship of these studies on class I expression in infected cells to those in transformed cells is not yet clear, although they suggest that different mechanisms of regulation of class I expression may operate in cells infected and transformed with adenovirus.

### *8.4.2 Post-translational regulation of class I expression in infected cells*

The best-described control of class I expression in adenovirus-infected cells is mediated by a viral glycoprotein encoded in early region 3 (E3). E3 is located between 75.9–86.0 m.u. on the Ad2 genome (Fig. 8.1). It has been sequenced in Ad12 from subgenus A (Sprengel *et al.*, 1994); Ad3 (Signäs, Akusjarvi & Pettersson, 1986), Ad7 (Hong, Mullis & Engler, 1988; Hermiston *et al.*, 1993), Ad11 (Mei & Wadell, 1992) and Ad35 (Flomenberg, Chen & Horwitz, 1988) from subgenus B; Ad2 (Hérissé, Courtois & Galibert, 1980; Hérissé & Galibert, 1981) and Ad5 (Cladaras & Wold, 1985) from subgenus C; Ad4 from subgenus E (discussed in Horton *et al.*, 1990); and Ad40 from subgenus F (Davison *et al.*, 1993). In subgenus C adenoviruses, two major primary transcripts initiated from the E3 promoter (termed E3A and E3B) are differentially spliced into nine overlapping mRNAs. Seven proteins encoded by this region have been identified by immunoprecipitation of Ad2- or Ad5-infected cell extracts with specific antibodies.

The largest and most abundant product of the E3A transcription unit in Ad2 is a glycoprotein termed E3 19 kDa (because of the apparent molecular weight of its polypeptide) which exists in molecular complexes with MHC class I antigens of mouse, rat, monkey and human origin (Pääbo *et al.*, 1989). In most cases, these complexes can be detected by co-precipitation with antibodies directed against either the E3 19kDa protein, class I heavy chain or $\beta_2$-m (Kämpe *et al.*, 1983; Pääbo *et al.*, 1983).

The E3 19 kDa protein of Ad2 displays tenuous sequence homology to constant domains of members of the immunoglobulin superfamily (Chatterjee & Maizel, 1984). It is a transmembrane protein that is synthesized as a polypeptide chain consisting of 159 amino acid residues. A stretch of 17 amino acid residues at the amino-terminus functions as a signal for targeting to the membrane of the rough endoplasmic reticulum (RER). Immediately after translocation through the membrane of the RER, the signal sequence is cleaved off, leaving a mature protein of 142 amino acid residues that contains a 104 residue long intralumenal domain, followed by a 23 residue transmembrane region and a cytoplasmic carboxy-terminal tail of 15 amino acid residues (Hérissé *et al.*, 1980; Persson, Jörnvall & Zabielski, 1980). Carbohydrate groups are attached to Asn at positions 12 and 61 in the intralumenal domain. They always occur in a high-mannose form characteristic for pre-Golgi molecules (Kornfeld & Wold, 1981). Glycosylation is accompanied by an increase in the apparent molecular mass, the extent of which depends on the adenovirus subgenus (Pääbo, Nilsson & Peterson, 1986b). Thus the glycosylated form of the E3 19 kDa encoded by Ad2 (subgenus C) has an apparent molecular mass of 25 kDa, while that of Ad35 (subgenus B), which has only an additional seven residues, is 29 kDa (Flomenberg, Chen & Horwitz, 1987). The E3 19 kDa protein of Ad35 is more extensively glycosylated than Ad2 E3 19 kDa (four versus two high-mannose groups).

Analysis of the amino acid sequence of the ER lumenal domain, together with data from class I antigen binding and anti-19 kDa mAb reactivity, allowed it to be divided into three subdomains (Hermiston *et al.*, 1993). They have been termed the ER lumenal variable domain (ERL-VD; residues 1 to approximately 77 to 83), the ER lumenal conserved domain (ERL-CD; residues approximately 84–98) and the ER lumenal spacer domain (ERL-SD; residues approximately 99–107). Comparison, for example, of the amino acid sequence of the E3 19 kDa protein of Ad2 (subgenus C) and Ad7 (subgenus B) shows that there is 75% homology in the lumenal conserved domain and only 22% in the variable region.

The signal for retention of E3 19 kDa protein in the RER is located in the cytoplasmic, polar domain of the protein. Mutant forms of E3 19 kDa from which the transmembrane and cytoplasmic domains have been removed by directed mutagenesis were secreted from cells (Pääbo *et al.*, 1986a). Shortening of the cytoplasmic tail to seven amino acid residues resulted in the transport of the E3 19 kDa protein to the cell surface, although a five residue extension to the 15 amino acid residue long cytoplasmic domain did not alter the intracellular distribution of the protein (Pääbo *et al.*, 1987; Cox, Bennink & Yewdell, 1991). Attachment of amino acid residues corresponding to the last six carboxy-terminal amino acids of Ad2 E3 19 kDa (DEKKMP) to other type I transmembrane proteins conferred retention in the ER (Nilsson, Jackson & Peterson, 1989). It has been proposed, therefore, that a linear sequence of six amino acid residues at the extreme carboxy-terminus, which contains lysine residues in positions 3 and 4 from the carboxy-terminus, serves as a signal for retention of the E3 19 kDa protein in the ER (Pääbo *et al.*, 1987; Cox *et al.*, 1991). According to this model, the location of these two lysine residues is critical for the subcellular distribution of the E3 19 kDa protein (Jackson, Nilsson & Peterson, 1990). However, another group has suggested that the retention signal is a non-contiguous sequence in a complex spatial arrangement that involves the Ser–Phe–Ile located in the middle of the cytoplasmic domain and Met–Pro at the extreme carboxy-terminus. Lysines at positions 3 and 4 from the carboxy-terminus were considered to be of little or no importance (Gabathuler & Kvist, 1990). It has also been proposed that binding of the cytoplasmic domain to microtubule components, e.g. β-tubulin, plays a crucial role in the retention process (Dahllöf, Wallin & Kvist, 1991). The net effect of E3 19 kDa protein being sequestered in the ER and binding to class I antigens is a block in the transport of class I antigens to the cell surface. This is reflected in the carbohydrate structure of class I heavy chains, which retain an immature high mannose form sensitive to treatment with endo H, an enzyme that specifically cleaves asparagine-linked (N-linked) high-mannose oligosaccharides. In contrast, N-linked complex oligosaccharides, in which distal mannose residues are replaced by other sugars such as galactose and sialic acid, are resistant to endo H. The high-mannose carbohydrate structure is characteristic for glycoproteins located in the ER, whose carbohydrates are not further modified. For further modification to occur, transport to the Golgi apparatus is required.

The E3 19 kDa proteins or their equivalents from serotypes in adenovirus subgenera B, C, D and E have been shown to bind to class I antigens (Pääbo *et al.*, 1986b). Ad3, Ad11 and Ad34 of subgenus B, Ad2 and Ad5 of subgenus C, Ad9 and Ad19 of subgenus D and Ad4, the only member of subgenus E, all affect the post-translational glycosylation of class I antigens and their transport from the ER to the Golgi. However, it could not be demonstrated that the E3 19 kDa protein of all these serotypes co-immunoprecipitated with class I antigens. It has, therefore, been suggested that the Ad3 and Ad9 E3 19 kDa proteins that do not co-precipitate with class I antigens still bind to them but this interaction is abolished by detergent lysis and/or the immunoprecipitation procedure (Pääbo *et al.*, 1986b).

In cells infected with either Ad12 or Ad31, which belong to the highly oncogenic subgenus A, no viral proteins co-precipitated with the class I antigens and glycosylation of class I antigens was not affected (Pääbo *et al.*, 1986b). The published sequence of Ad12 DNA provides an explanation for this observation, since there is no ORF within the E3A region of this virus that would correspond to the 19 kDa protein (Sprengel *et al.*, 1994).

It has also been shown that the fastidious Ad40, which along with Ad41 constitutes subgenus F, also does not contain an ORF in E3A that would permit the synthesis of a protein analogous to the E3 19 kDa protein (Davison *et al.*, 1993). These findings raise interesting questions about the role of this protein in adenoviral pathogenicity.

Comparison of the DNA sequence of adenoviruses from different subgenera points to E3A as one of the two most variable regions of human adenovirus genomes, with significant divergence even within a subgenus (A. Bailey & V. Mautner, personal communication). Homology in the E3 19kDa amino acid sequences of serotypes from different subgenera is mainly confined to the conserved, relatively short (approximately 20 residues) region of the ER lumenal domain that has been termed ERL-CD, as described above.

The complex of E3 19 kDa protein with class I antigens can be formed both in lytically infected cells and in those adenovirus-transformed cells that contain and express early region 3 (Kämpe *et al.*, 1983; Severinsson & Peterson, 1985). The complex can also be detected in cells transfected with E3 19 kDa expression vectors, indicating that other adenovirus proteins are not required for its formation. The association between E3 19 kDa protein and class I heavy chains occurs in the ER immediately upon synthesis of both proteins. The binding is non-covalent and takes place when $\beta_2$-m is complexed with heavy chains, although $\beta_2$-m is not absolutely required for the interaction (Kämpe *et al.*, 1983; Burgert & Kvist, 1987). Both glycosylated and non-glycosylated species of the E3 19 kDa protein and heavy chains can form a complex, as shown by experiments using tunicamycin (Burgert & Kvist, 1987).

The ER lumenal domain of the Ad2 E3 19kDa protein is sufficient for binding to class I heavy chains (Pääbo *et al.*, 1986a). Studies with deletion mutants indicated that virtually the entire lumenal region (not only the domain termed ERL-CD, which is conserved between subgenera) is required for maximal binding (Hermiston *et al.*, 1993). From 13 in-frame virus mutants containing 4–12 amino acid deletions that spanned the lumenal region of Ad2 E3 19kDa, only one mutant, which had a deletion of residues 102 to 107 (covering approximately the spacer domain, ERL-SD), gave rise to a protein that retained the ability to bind class I antigens. Another group, however, reported that removing the corresponding region of the Ad35 E3 19 kDa protein completely abrogated class I antigen binding (Flomenberg *et al.*, 1992). This may reflect differences between Ad2 and Ad35 or differences in the methods used. However, in agreement with the data obtained for the Ad2 E3 19 kDa protein, the importance for binding of the conserved domain was confirmed; four non-conservative point mutations in the intersubgenus region of homology (ERL-CD) of Ad35 E3 19 kDa protein reduced or abolished binding to class I antigens.

Studies with hybrid H-2 class I heavy chains constructed by exchanging domains between $K^d$ and $K^k$ genes showed that the $\alpha_1$ and $\alpha_2$ domains of the heavy chain were essential for complex formation (Burgert & Kvist, 1987). The $\alpha_3$ domain of the class I heavy chain did not appear to be involved. This is interesting because processed peptides bind in the cleft formed by the $\alpha_1$ and $\alpha_2$ domains (see Chapter 1) and it has been reported that the E3 19 kDa protein of Ad2 does not block the association of peptides with class I antigens in the ER; therefore E3 19 kDa probably does not bind near the cleft (Cox *et al.*, 1990; 1991). A potentially interesting model has been proposed whereby the ERL-VD of E3 19 kDa protein is bound to the variable $\alpha_1$ and $\alpha_2$ domains of the class I heavy chain and ERL-CD is available for interaction with a highly conserved protein(s). In this model, the ERL-SD domain plays a structural role, allowing the ERL-VD and ERL-CD of E3 19 kDa protein to extend out from the membrane to facilitate association with the $\alpha_1$ and $\alpha_2$ domains of the class I heavy chain (Hermiston *et al.*, 1993). Further studies with hybrid heavy chains encoded by the $K^d$ and $K^k$ alleles established that the $\alpha_1$ and only the first half of the $\alpha_2$ domain of the class I heavy chain were required for association with Ad2 E3 19 kDa protein (Jefferies & Burgert, 1990). The same protein segment has been implicated in control of transport of the $K^d$ molecule out of the ER. Therefore, it was proposed that the E3 19 kDa protein blocks the passage of class I antigens out of the ER by binding to the region of the class I heavy chain involved in the transport to the Golgi apparatus. Insertion of the E3 19 kDa protein into the ER membrane is not necessary for complex formation; however, in some experiments with engineered mutants, increased avidity of binding was observed when the transmembrane and even the cytoplasmic domains were present in the 19 kDa protein (Pääbo *et al.*, 1986a; Gabathuler, Lévy & Kvist, 1990).

Allelic differences in the affinity of class I heavy chains for Ad2 or Ad5 E3 19 kDa protein have been detected in human and mouse systems. It has been calculated, for example, that human HLA-A2 antigens and the E3 19 kDa protein interact with a binding constant that is more than twice as high as that for HLA-B7, although co-precipitation with HLA-A2 could not be observed (Severinsson, Martens & Peterson, 1986). A marked difference has been reported in the E3 19 kDa-related decrease in the level of cell surface expression between HLA-A3 and HLA-B35, with HLA-A3 being more reduced than HLA-B35. This, presumably, reflects the higher affinity and, therefore, more efficient binding of HLA-A3 compared with that of HLA-B35 to E3 19 kDa protein (Körner & Burgert, 1994). These results lead to a hypothesis that class I antigens of the human HLA-B locus may generally bind less well than those of the A locus. However, caution is necessary as the affinity of the E3 19 kDa protein for only a small number of human class I antigens has been compared to date and the similarity in binding of the two A versus the two B alleles in both studies could be coincidental. The mouse H-2K$^d$ antigen associates with E3 19 kDa protein with a similar efficiency to human HLA-A2 or HLA-A3 molecules, whereas H-2K$^k$ does not bind the E3 19 kDa protein (Burgert & Kvist, 1987). By using cytotoxicity assays and flow cytometry, it has also been shown that H-2D$^b$, but not H-2K$^b$, binds to E3 19 kDa protein (Tanaka & Tevethia, 1988). Another group (who, in addition to the cytotoxicity assay, used co-immunoprecipitation and susceptibility of class I proteins to digestion with endo H) confirmed the observation that the H-2K$^d$ antigen binds efficiently to E3 19 kDa protein, unlike H-2K$^k$. They also established that the L$^d$ product binds to the viral protein but somewhat less efficiently than H-2K$^d$ and that the affinity of the D$^d$ product is even lower. However, transport of the D$^d$ product to the cell surface was delayed rather than being completely blocked. The D$^k$ antigen did not bind to the E3 19 kDa protein (Cox *et al.*, 1990).

CTLs, which are involved in elimination of virus-infected or tumour cells, identify target cells by cell contact-dependent recognition of processed viral or tumour antigens in the context of self-determinants present on cell surface class I antigens (see Chapter 1). Consequently, events that alter either the association of viral- or tumour-specific peptides with class I antigens or the presence of these complexes on the cell surface may modulate cellular immunity. Formation of a complex between the E3 19 kDa protein and class I antigens, which results in retention of class I molecules in the ER, is an example of a mechanism that could lead to a delay or escape of adenovirus-infected cells from the cellular immune system of the host. Several groups have reported reduced levels of surface class I antigens in immortalized fibroblasts, lymphocytes and epithelial cells transfected with E3 19 kDa expression vectors, accompanied by decreased susceptibility to lysis by CTLs (Burgert, Maryanski & Kvist, 1987; Körner & Burgert, 1994).

Relative resistance to lysis by CTLs that paralleled a decrease in cell surface class I antigens has also been observed in cells infected with Ad2 (Hermiston *et al.*, 1993). Deletions in E3 19 kDa protein that abolished its binding to class I antigens also abrogated the ability of this protein to prevent cytolysis by Ad-specific CTLs. In addition, expression of a mutant E3 19 kDa protein (lacking the cytoplasmic carboxy-terminal tail) at the cell surface did not alter the extent of cell lysis by antigen-specific CTLs (Cox *et al.*, 1991). However, it should be noted that another group has reported contradictory results in Ad2- and Ad5-infected cells (Routes & Cook, 1990). In this study, with the exception of the Ad5 early region 1 (E1)-transformed human cell line 293 (Graham *et al.*, 1977), Ad2 or Ad5 infection of fibroblastic, epithelial and lymphoid cells did not cause major decreases in surface class I antigens until the terminal stages of infection when cell death was imminent. Furthermore, newly synthesized class I molecules continued to be expressed on the surface of many cell types at times when infected cells contained large amounts of the E3 19 kDa protein. The protective effect of E3 19 kDa protein against lysis by CTLs was not observed in these infected cells. One possible explanation for these discrepancies could be the difference in HLA class I alleles expressed in the cells under study and consequent differences in complex formation with the E3 19 kDa protein. It has also been reported that this protein can inhibit the presentation of peptides generated in the cytosol without reducing the levels of cell surface class I antigens (Cox *et al.*, 1990). The process responsible was identified as the rate of exocytosis of class I molecules, not the level of cell surface expression.

Reduced levels of phosphorylation of class I heavy chains have been detected in cells infected by Ad2, but not in HSV- or CMV-infected cells (Lippé *et al.*, 1991). The functional consequences of this are not yet clear.

A fundamental role of the E3 region in viral pathogenicity has been shown by studies on infection of cotton rats by wild-type Ad5 and Ad5 E3 mutants (Ginsberg *et al.*, 1989). The cotton rat is a useful model for the study of adenovirus infection since it exhibits many of the characteristics of infection of humans by adenoviruses, including replication of the virus in epithelial cells of the lung. The studies by Ginsberg *et al.* (1989) showed, interestingly, that E3 mutants exhibited enhanced pathogenicity in cotton rats and also in a mouse model of adenovirus-induced pneumonia (Ginsberg *et al.*, 1991). This is perhaps the reverse of what might be expected, based on the hypothesis that the E3 protein 19 kDa mediates latent or persistent viral infection. However, Ginsberg *et al.* (1991) also showed increased inflammation, infiltration of monocytes and macrophages and local production of cytokines, such as IL-1, IL6, and TNF-$\alpha$, in mice infected with E3 mutants. This last aspect may be very important in control of viral infection and, as will be described below, E3 also encodes functions that regulate susceptibility of infected cells to cytokine-mediated lysis.

## 8.5 The effects of cytokines on MHC class I expression and immunomodulation in adenovirus infection and transformation

Human adenovirus infection is relatively insensitive to the anti-viral effects of IFNs. The presence of the Ad5 E1A gene abolished the IFN-$\alpha_2$-mediated inhibition of vesicular stomatitis virus (VSV) replication when human cells were co-infected with VSV and Ad5 (Anderson & Fennie, 1987). The VA-RNA1 also inhibits the activity of the IFN-induced double-stranded RNA protein kinase (see Section 8.2.3).

Infection of human cells with Ad5 containing a deletion in the E1A gene (dl312) induced the transcription of IFN-stimulated cellular genes (ISGs) (Reich *et al.*, 1988). However, the presence of a functional E1A gene in Ad5 suppressed the transcription of ISGs. Thus, Ad5 can interfere with an IFN-mediated anti-viral response directed against itself or a co-infecting virus. This work has been extended by analysis of Ad5 E1A-expressing human cell lines (Ackrill *et al.*, 1991; Gutch & Reich, 1991; Kalvakolanu *et al.*, 1991). In these cell lines, induction of expression of ISGs by IFN-α or IFN-γ was blocked by Ad5 E1A expression. Mutational analysis revealed that the CR1 domain was essential for Ad5 E1A-mediated repression of ISG transcription (Ackrill *et al.*, 1991). Further analysis showed that the key step in repression was a failure to activate a transcription factor ISGF3 required for transcription of ISGs (see Chapter 4 for a detailed description of the mechanism of activation of ISG transcription by IFNs). Similar experiments performed in Ad5-transformed rodent cells also showed that induction of MHC class I and 2–5 A synthetase transcription was abrogated by Ad2 or 5 E1A in a CR1-dependent manner, although activation of ISGF3 appeared unaffected (R. Yusof & G. E. Blair, unpublished results). This suggests that there may be multiple points in the IFN-signalling pathway at which E1A acts. In Ad12-transformed rodent cells, IFN-γ treatment restored levels of class I mRNA and proteins, at least in part (Hayashi *et al.*, 1985; Eager, Pfizenmaier & Ricciardi, 1989; Ackrill & Blair, 1990). This increase in class I gene expression was the result of an increase in the level of initiation of transcription and could be effected by IFN-α/β as well as IFN-γ (R. Yusof & G. E. Blair, unpublished results). Therefore, expression of Ad12 E1A does not interfere with the induction of ISGs by IFNs. The IFN-mediated increase in class I transcription in Ad12-transformed cells requires a functional IFN-response sequence (which overlaps the major class I regulatory element: see Chapter 4). The mechanism whereby class I transcription is augmented by IFNs in Ad12-transformed cells may be similar to that which operates in non-virally transformed cells (Blanar *et al.*, 1989). IFNs do not act by anti-viral effects on E1A or E1B proteins (or their mRNAs) present in adenovirus-transformed cells (Eager *et al.*, 1989; R. Yusof & G. E. Blair, unpublished results; A. M. Ackrill & G. E. Blair, unpublished results).

TNF-α also stimulates class I transcription in adenovirus-transformed cells (R. Yusof & G. E. Blair, unpublished results) although the mechanism whereby this is achieved is not yet understood. TNF-α has also been reported to act synergistically with IFN-γ to increase levels of class I antigens in adenovirus-transformed mouse cells (Eager *et al.*, 1989).

TNF-α also interferes with adenovirus infection. Rodent cells expressing the E1A gene of Ad5 or Ad12 acquired the property of sensitivity to the cytolytic action of TNF-α (Chen *et al.*, 1987; Duerksen-Hughes, Wold & Gooding, 1989). Further mutational analysis of the E1A gene has revealed that the CR1 domain (see Section 8.2.2), from amino acid residues 36 to 60, is required to induce susceptibility to TNF-α cytolysis in infected mouse C3HA fibroblasts (Duerksen-Hughes *et al.*, 1991).

Not all rodent cells expressing E1A are sensitive to TNF-α cytolysis (Kenyon, Dougherty & Raska, 1991; R. Yusof & G. E. Blair, unpublished results) and there is also no selectivity in TNF-α killing for Ad2 or Ad5 E1A-expressing cells compared with those expressing Ad12 E1A (Chen *et al.*, 1987; Kenyon *et al.*, 1991). There are also variations in sensitivity to TNF-α killing exhibited by different clones of E1A-expressing cells derived from the same parental clonal rat fibroblast cell line (Vanhaesebroeck *et al.*, 1990). TNF-α-mediated cytolysis might, therefore, have differential effects depending on the cell type and constitute a relatively non-specific defence against adenovirus infection. Release of TNF-α by host inflammatory cells at sites of infection might serve to limit the spread of the virus.

Adenoviruses also possess a response to the anti-viral effect of TNF-α. The studies described above were performed with cells that expressed only E1A (or sometimes along with E1B). If rodent cells are infected by wild-type adenoviruses, they are resistant to TNF-α killing, but if viral mutants such as Ad5 d1309 (which has a deletion of the E3 region) are used for infection, then the infected cells are susceptible to TNF-α lysis (Gooding *et al.*, 1988). The ORF in early region 3 (see Section 8.2.2), which confers resistance to TNF-α encodes a 14.7 kDa protein. Rodent cell lines constitutively expressing the 14.7 kDa protein resisted TNF-α cytolysis following adenovirus infection or other anti-cellular treatments (such as incubation with cycloheximide) that would normally sensitize the untransfected parental cell line to TNF-α lysis (Horton *et al.*, 1991). Expression of the 14.7 kDa protein was not capable of inducing resistance to TNF-α in all cell lines. NCTC-929 mouse cells were sensitive to TNF-α and remained so even when relatively high levels of the 14.7 kDa protein were expressed (Gooding *et al.*, 1991a). The 14.7 kDa protein (or a close structural analogue) is present in subgenera A–E where it protects infected cells from TNF-α-mediated killing (Horton *et al.*, 1990). Deletion of the E3 region of Ad4 also conferred TNF-α sensitivity upon mutant-infected cells (Horton *et al.*, 1990). The 14.7 kDa polypeptide has been detected in nuclear and cytoplasmic fractions of infected cells (Gooding *et al.*, 1991a), although its mechanism of action is at present unknown. Other proteins encoded in E3 can substitute for the E3 14.7 kDa protein in protecting infected mouse cell lines from TNF-α lysis (Gooding *et al.*, 1991b). The E3 10.4 kDa and E3 14.5 kDa proteins are both required for protection of several mouse cell lines against TNF-α. A molecular complex between the 10.4 kDa and 14.5 kDa proteins has been detected by immunoprecipitation analysis (Gooding *et al.*, 1991a). These studies were performed in rodent cells. In human cells, E3 has no protective effect against TNF-α cytolysis, whereas the E1B 19 kDa protein performs this function in several human cell lines (Gooding *et al.*, 1991b). E1B 19 kDa protein has multiple locations within adenovirus-infected human cells being present in the nucleus, ER (White, Blose & Stillman, 1984; Blair-Zajdel *et al.*, 1985) and the plasma membrane (Persson, Katze & Philipson, 1982; Smith, Gallimore & Grand, 1989). There are, therefore, two early regions, E1B and E3, specifying four proteins (E1B 19 kDa, E3 14.7 kDa, E3 14.5 kDa and E3 10.4 kDa) that are involved in the adenovirus defence against TNF-α. Although some, if not all, of these viral proteins may have multiple functions in infection, the coding capacity of the viral genome devoted to this protective function underlines the probable importance of TNF-α in combating early stages of adenovirus infection.

## 8.6 Conclusions

Studies on adenovirus-transformed and adenovirus-infected cells have proved useful not only in understanding the biology of adenoviruses but also in providing an insight into control mechanisms regulating expression of class I antigens. Further analysis of the mechanism of Ad12 E1A-mediated down-regulation of class I transcription in transformed cells may reveal pathways of class I regulation that are altered in naturally occurring tumour cells which have lost expression of class I antigens. Equally, mapping of the site(s) on Ad12 E1A oncoproteins responsible for down-regulation of class I expression may provide the means to test whether this property of E1A is also directly linked to viral oncogenicity. A more detailed understanding of the mode of Ad12 E1A-mediated down-regulation of class I transcription may also have practical benefits, for example in transplantation, where levels of class I expression need to be manipulated.

Another mechanism utilized by most adenoviruses to reduce surface levels of class I antigens, involving a product of a different gene, the E3 19 kDa protein, is better understood. However, the molecular basis of the interaction between the E3 19 kDa protein and class I heavy chain remains to be clarified. In particular the identification of the contact points between these proteins requires more detailed studies, perhaps using site-directed mutagenesis of amino acids that differ between class I alleles. Developments that have highlighted the importance of calnexin and TAPs in the function of class I antigens have also raised the question of how the interaction of these proteins with class I antigens is affected by binding to the E3 19 kDa protein (Ou *et al.*, 1993; Ortmann, Androlewicz & Cresswell, 1994). Clinical research on the levels of class I antigens in tissues of patients suffering acute adenovirus infection may be expected to establish whether down-regulation of surface class I antigens is important in viral disease *in vivo*. Finally, as in the case of the Ad12 E1A proteins, a more detailed understanding of the mechanism of down-regulation of surface class I antigens by the E3 19 kDa protein may also form the basis for the design of therapies to reduce surface class I antigens, e.g. in organ transplantation or treatment of autoimmune diseases.

## Acknowledgements

We thank Diane Baldwin and Mani Tummala for help in preparing this Chapter and the Medical Research Council and the Yorkshire Cancer Research Campaign for research support.

## References

Abraham, S.E., Lobo, S., Yaciuk, P., Wang, H.-G.H. & Moran, E. (1993). p300 and p300-associated proteins are components of TATA-binding protein (TBP) complexes. *Oncogene*, 8, 1639–1647.

Ackrill, A.M. & Blair, G.E. (1988a). Expression of hamster MHC class I antigens in transformed cells and tumours induced by human adenoviruses. *European Journal of Cancer Clinical Oncology*, 24, 1745–1750.

Ackrill, A.M. & Blair, G.E. (1988b). Regulation of major histocompatibility class I gene expression at the level of transcription in highly oncogenic adenovirus transformed rat cells. *Oncogene*, 3, 483–487.

Ackrill, A.M. & Blair, G.E. (1989). Nuclear proteins binding to an enhancer element of the major histocompatibility class I promoter: differences between highly oncogenic and nononcogenic adenovirus-transformed rat cells. *Virology*, 172, 643–646.

Ackrill, A.M. & Blair, G.E. (1990). Interferon-γ regulation of major histocompatibility class I gene expression in rat cells containing the adenovirus 12 E1A oncogene. *Virology*, 174, 325–328.

Ackrill, A.M., Foster, G.R., Laxton, C.D., Flavell, D.M., Stark, G.R. & Kerr, I.M. (1991). Inhibition of the cellular response to interferons by products of the adenovirus type 5 E1A oncogene. *Nucleic Acids Research*, 19, 4387–4393.

Akusjarvi, G. & Wadell, G. (1990). Adenovirus. In *Genetic Maps. Locus Maps of*

*Complex Genomes*, 5th edn, ed. S.J. O'Brian, pp. 1.98–1.01. Cold Spring Harbor, New York: Cold Spring Harbor Laboratory Press.

Anderson, K.P. & Fennie, E. (1987). Adenovirus early region 1A modulation of interferon antiviral activity. *Journal of Virology*, 61, 787–795.

Ankerst, J. Jonsson, N., Kjellen, L., Norrby, E. & Sjogren, H.O. (1974) Induction of mammary fibroadenomas in rats by adenovirus type 9. *International Journal of Cancer*, 13, 286–290.

Bellgrau, D., Walker, T.A. & Cook, J.L. (1988). Recognition of adenovirus E1A gene products on immortalised cell surfaces by cytotoxic T lymphocytes. *Journal of Virology*, 62, 1513–1519.

Bernards, R., de Leeuw, M.G.W., Vaessen, M.J., Houweling, A. & van der Eb, A.J. (1984). Oncogenicity by adenovirus is not determined by the transforming region only. *Journal of Virology*, 50, 847–853.

Bernards, R., Schrier, P.I., Houweling, A., Bos, J.L., van der Eb, A.J., Zijlstra, M. & Melief, C.J.M. (1983). Tumorigenicity of cells transformed by adenovirus 12 by evasion of T cell immunity. *Nature (London)*, 305, 776–779.

Bernards, R. & van der Eb, A.J. (1984). Adenovirus: transformation and oncogenicity. *Biochimica et Biophysica Acta*, 783, 187–204.

Blair-Zajdel, M.E., Barker, M.D., Dixon, S.C. & Blair, G.E. (1985). The use of monoclonal antibodies to study the proteins specified by the transforming region of human adenoviruses. *Biochemical Journal*, 225, 649–655.

Blanar, M.A., Baldwin, A.S. Jr, Flavell, R.A. & Sharp, P.A. (1989). A gamma-interferon-induced factor that binds the interferon response sequence of the MHC class I gene, H-$2K^b$. *EMBO Journal*, 8, 1139–1144.

Borelli, E., Hen, R. & Chambon, P. (1984). Adenovirus 2 E1A products repress enhancer-induced stimulation of transcription. *Nature (London)*, 312, 608–612.

Boulanger, P.A. & Blair, G.E. (1991). Expression and interactions of human adenovirus oncoproteins. *Biochemical Journal*, 275, 281–299.

Boyer, T.G. & Berk, A.J. (1993). Functional interaction of adenovirus E1A with holo-TFIID. *Genes and Development*, 7, 1810–1823.

Braithwaite, A.W., Bellett, A.J.D., Müllbacher, A., Blair, G.E. & Zhang, X. (1993). Human adenoviruses: genetics and immunobiology. In *Viruses and the Cellular Immune Response*, ed. Thomas, D.B., pp. 389–428. New York: Marcel Dekker.

Braithwaite, A.W., Blair, G.E., Nelson, C.C., McGovern, J. & Bellett, A.J.D. (1991b). Adenovirus E1B 58 kD binds to p53 during infection of rodent cells: evidence for an N-terminal binding site on p53. *Oncogene*, 6, 781–787.

Braithwaite, A.W., Nelson, C.C. & Bellett, A.J.D. (1991a). E1A revisited: the case for multiple cooperative *trans*-activation domains. *New Biologist*, 3, 18–26.

Burgert, H.G. & Kvist, S. (1987). The E3/19K protein of adenovirus type 2 binds to the domains of histocompatibility antigens required for CTL recognition. *EMBO Journal*, 6, 2019–2026.

Burgert, H.G., Maryanski, Y.L. & Kvist, S. (1987). 'E3/19K' protein of adenovirus type 2 inhibits lysis of cytolytic T lymphocytes by blocking cell-surface expression of histocompatibility class I antigens. *Proceedings of the National Academy of Sciences of the USA*, 84, 1356–1360.

Cepko, C. & Sharp, P.A. (1983). Analysis of Ad5 hexon and 100K *ts* mutants using conformation-specific monoclonal antibodies. *Virology*, 129, 137–154.

Chatterjee, D. & Maizel, J.V. Jr (1984). Homology of adenoviral E3 glycoprotein with HLA-DR heavy chain. *Proceedings of the National Academy of Sciences of the USA*, 81, 6039–6043.

Chen, M.J., Holskin, B., Strickler, J., Gorniak, J., Clark, M.A., Johnson, P.J.,

Mitcho, M. & Shalloway, D. (1987). Induction by E1A oncogene expression of cellular susceptibility to lysis by TNF. *Nature (London)*, 330, 581–593.

Chinnadurai, G. (1992). Adenovirus E1A as a tumor suppressor gene. *Oncogene*, 7, 1255–1258.

Cladaras, C. & Wold, W.S.M. (1985). DNA sequence of the early region E3 transcription unit of adenovirus 5. *Virology*, 140, 28–43.

Cousin, C., Winter, N., Gomes, S.A. & d'Halluin J.-C. (1991). Cellular transformation by E1 genes of enteric adenovirus. *Virology*, 181, 277–287.

Cox, J.H., Bennink, J.R. & Yewdell, J.W. (1991). Retention of adenovirus E19 glycoprotein in the endoplasmic reticulum is essential to its ability to block antigen presentation. *Journal of Experimental Medicine*, 174, 1629–1637.

Cox, J.H., Yewdell, J.W., Eisenlohr, L.C., Johnson, P.R. & Bennink, J.R. (1990). Antigen presentation requires transport of MHC class I molecules from the endoplasmic reticulum. *Science*, 247, 715–718.

Culp, J.S., Webster, L.C., Friedman, D.J., Smith, C.L., Huang, W.-J., Wu, F.Y.-H., Rosenberg, M. & Ricciardi, R.P. (1988). The 289 amino acid E1A protein of adenovirus binds zinc in a region that is important for *trans*-activation. *Proceedings of the National Academy of Sciences of the USA*, 85, 6450–6454.

Dahllöf, B., Wallin, M. & Kvist, S. (1991). The endoplasmic reticulum retention signal of the E3/19K protein of adenovirus-2 is microtubule binding. *Journal of Biological Chemistry*, 266, 1804–1808.

Darden, A.G. & Streilein, J.W. (1983). Syrian hamsters possess class I-like molecules detectable with xenoantisera but not alloantisera. *Transplantation Proceedings*, XV, 145–148.

Davison, A.J., Telford, E.A.R., Watson, M.M., McBride, K. & Mautner, V. (1993). The DNA sequence of adenovirus type 40. *Journal of Molecular Biology*, 234, 1308–1316.

Debbas, M. & White, E. (1993). Wild-type p53 mediates apoptosis by E1A, which is inhibited by E1B. *Genes and Development*, 7, 546–554.

Doerfler, W. (ed.) (1986). *Adenovirus DNA: The Viral Genome and its Expression.* The Hague: Martinus Nijhoff.

Driggers, P.H., Elenbaas, B.A., An, J.B., Lee, I.J. & Ozato, K. (1992). Two upstream elements activate transcription of a major histocompatibility complex class I gene *in vitro*. *Nucleic Acids Research*, 20, 2533–2540.

Duerksen-Hughes, P.J., Hermiston, T.W., Wold, W.S. & Gooding, L.R. (1991). The amino-terminal portion of CD1 of the adenovirus E1A proteins is required to induce susceptibility to tumor necrosis factor cytolysis in adenovirus-infected mouse cells. *Journal of Virology*, 65, 1236–1244.

Duerksen-Hughes, P.J., Wold, W.S.M. & Gooding, L.R. (1989). Adenovirus E1A renders infected cells sensitive to tumor necrosis factor. *Journal of Immunology*, 143, 4193–4200.

Eager, K.B., Pfizenmaier, K. & Ricciardi, R.P. (1989). Modulation of major histocompatibility complex (MHC) class I genes in adenovirus 12 transformed cells: interferon-γ-increases class I expression by a mechanism that circumvents E1A induced expression and tumor necrosis factor enhances the effect of interferon-γ. *Oncogene*, 4, 39–44.

Eager, K.B., Williams, J., Breiding, D., Pan, S., Knowles, B., Appella, E. & Ricciardi, R.P. (1985). Expression of histocompatibility antigens H-2K, -D and -L is reduced in adenovirus 12-transformed mouse cells and is restored by interferon γ. *Proceedings of the National Academy of Sciences of the USA*, 82, 5525–5529.

Eckner, R., Ewen, M.E., Newsome, D., Gerdes, M., De Caprio, J., Lawrence,

J.B. & Livingston, D.M. (1994). Molecular cloning and functional analysis of the adenovirus E1A-associated 300 kD protein (p300) reveals a protein with the properties of a transcriptional adaptor. *Genes and Development*, 8, 869–884.

Edbauer, D., Lamberti, C., Tong, J. & Williams, J. (1988). Adenovirus type 12 E1B 19-kilodalton protein is not required for oncogenic transformation in rats. *Journal of Virology*, 62, 3265–3273.

Farmer, G., Bargonetti, J., Zhu, H., Friedman, P., Prywes, R. & Prives, C. (1992). Wild-type p53 activates transcription *in vitro*. *Nature (London)*, 358, 83–86.

Flomenberg, P.R., Chen, M. & Horwitz, M.S. (1987). Characterisation of a major histocompatibility complex class I antigen-binding glycoprotein from adenovirus type 35, a type associated with immunocompromised hosts. *Journal of Virology*, 61, 3665–3671.

Flomenberg, P.R., Chen, M. & Horwitz, M.S. (1988). Sequence and genetic organization of adenovirus type 35 early region 3. *Journal of Virology*, 62, 4431–4437.

Flomenberg, P., Szmulewicz, J., Gutiernez, E. & Lupatkin, H. (1992). Role of the adenovirus E3 19 K conserved region in binding major histocompatibility complex class I molecules. *Journal of Virology*, 66, 4478–4483.

Friedman, D.J. & Ricciardi, R.P. (1988). Adenovirus type 12 E1A represses accumulation of MHC class I mRNAs at the level of transcription. *Virology*, 165, 303–305.

Gabathuler, R. & Kvist, S. (1990). The endoplasmic reticulum retention signal of the E3/19K protein of adenovirus type 2 consists of three separate amino acid segments at the carboxy terminus. *Journal of Cell Biology*, 111, 1803–1810.

Gabathuler, R., Lévy, F. & Kvist, S. (1990). Requirements for the association of adenovirus type 2 E3 19K wild-type and mutant proteins with HLA antigens. *Journal of Virology*, 64, 3679–3685.

Gallimore, P.H., Byrd, P.J., Grand, R., Whitaker, J., Breiding, D. & Williams, J. (1984). An examination of the transforming and tumor-inducing capacity of a number of adenovirus type 12 early region/host range mutants and cells transformed by fragments of the Ad12 E1 region. *Cancer Cells*, 2, 339–348.

Ge, R., Kralli, A., Weinmann, R. & Ricciardi, R.P. (1992). Down-regulation of the major histocompatibility class I enhancer in adenovirus 12-transformed cells is accompanied by an increase in factor binding. *Journal of Virology*, 66, 6969–6978.

Ginsberg, H.S. (ed.) (1984). *The Adenoviruses*. New York: Plenum Press.

Ginsberg, H.S., Lundholm-Beauchamp, U., Horswood, R.L., Pernis, B., Wold, W.S.M., Chanock, R.M. & Prince, G.A. (1989). Role of early region 3 (E3) in pathogenesis of adenovirus disease. *Proceedings of the National Academy of Sciences of the USA*, 86, 3823–3827.

Ginsberg, H.S., Moldawer, L., Sehgal, P., Redington, M., Kilian, P., Chanock, R.M. & Prince, G.A. (1991). A mouse model for investigating the molecular pathogenesis of adenovirus pneumonia. *Proceedings of the National Academy of Sciences of the USA*, 88, 1651–1655.

Gooding, L.R., Aquino, L., Duerksen-Hughes, P.J., Day, D., Horton, T.M., Yei, S. & Wold, W.S.M. (1991b). The E1B-19K protein of group C adenoviruses prevents cytolysis by tumor necrosis factor of human cells but not mouse cells. *Journal of Virology*, 65, 3083–3094.

Gooding, L.R., Elmore, L.W., Tollefson, A.E., Brady, H.A. & Wold, W.S.M. (1988). A 14,700 MW protein from the E3 region of adenovirus inhibits cytolysis by tumor necrosis factor. *Cell*, 53, 341–346.

Gooding, L.R., Ranheim, T.S. Tollefson, A.E., Aquino, L. Duerksen-Hughes, P., Horton, T.M. & Wold, W.S.M. (1991a). The 10.4K and 14.5K proteins encoded by region E3 of adenovirus function together to protect many but not all mouse cell lines against lysis by tumor necrosis factor. *Journal of Virology*, 65, 4114–4123.

Graham, F.L., Smiley, J., Russell, W.C. & Nairn, R. (1977). Characteristics of a human cell line transformed by DNA from human adenovirus type 5. *Journal of General Virology*, 36, 59–74.

Grand, R.J.A., Rowe, M., Byrd, P.J. & Gallimore, P.H. (1987). The level of expression of class I MHC antigens in adenovirus-transformed human cell lines. *International Journal of Cancer*, 40, 213–219.

Green, M., Wold, W.S.M., Mackey, J.K. & Rigden, P. (1979). Analysis of human tonsil and cancer DNAs and RNAs for DNA sequences of group C (serotypes 1, 2, 5 and 6) human adenoviruses. *Proceedings of the National Academy of Sciences of the USA*, 76, 6606–6610.

Gutch, M.J. & Reich, N.C. (1991). Repression of the interferon signal transduction pathway by the adenovirus E1A oncogene. *Proceedings of the National Academy of Sciences of the USA*, 88, 7913–7917.

Haddada, H., Lewis, A.M. Jr, Sogn, J.A., Coligan, J.E., Cook, J.L., Walker, T.A. & Levine, A.S. (1986). Tumorigenicity of hamster and mouse cells transformed by adenovirus types 2 and 5 is not influenced by the level of class I major histocompatibility antigens expressed on the cells. *Proceedings of the National Academy of Sciences of the USA*, 83, 9684–9688.

Haddada, H., Sogn, J.A., Coligan, J.E., Carbone, M., Dixon, K., Levine, A.S. & Lewis, A.M. Jr (1988). Viral gene inhibition of class I major histocompatibility antigen expression: not a general mechanism governing the tumorigenicity of adenovirus type 2-, adenovirus type 12- and simian virus 40-transformed Syrian hamster cells. *Journal of Virology*, 62, 2755–2761.

Halbert, D.N., Cutt, J.R. & Shenk, T. (1985). Adenovirus early region 4 encodes functions required for efficient DNA replication late gene expression and host cell shutoff. *Journal of Virology*, 42, 30–41.

Harwood, L.M.J. & Gallimore, P.H. (1975). A study of the oncogenicity of adenovirus type 2-transformed rat embryo cells. *International Journal of Cancer*, 16, 498–508.

Hay, R.T. & Russell, W.C. (1989). Recognition mechanisms in the synthesis of animal viral DNA. *Biochemical Journal*, 258, 3–16.

Hayashi, H., Tanaka, K., Jay, F., Khoury, G. & Jay, G. (1985). Modulation of the tumorigenicity of human adenovirus 12-transformed cells by interferon. *Cell*, 43, 263–267.

Hérissé, J. & Galibert, F. (1981). Nucleotide sequence of EcoRIE fragment of adenovirus 2 genome. *Nucleic Acids Research*, 9, 1229–1240.

Hérissé, J., Courtois, G. & Galibert, F. (1980). Nucleotide sequence of the EcoRID fragment of adenovirus 2 genome. *Nucleic Acids Research*, 8, 2173–2192.

Hermiston, T.W., Tripp, R.A., Sparrer, J., Gooding, L.R. & Wold, W.S.M. (1993). Deletion mutation analysis of the adenovirus type 2 E3-gp19K protein: identification of sequences within the endoplasmic reticulum lumenal domain that are required for class I antigen binding and protection from adenovirus-specific cytotoxic T lymphocytes. *Journal of Virology*, 67, 5289–5298.

Hong, J.S., Mullis, K.G. & Engler, J.A. (1988). Characterisation of the early region 3 and fiber genes of Ad7. *Virology*, 167, 545–553.

Horton, T.M., Ranheim, T.S., Aquino, L., Kusher, D.I., Saha, S.K., Ware, C.L.,

Wold, W.S.M. & Gooding, L.R. (1991). Adenovirus E3 14.7 K protein functions in the absence of other adenovirus proteins to protect transfected cells from tumor necrosis factor cytolysis. *Journal of Virology*, 65, 2629–2639.

Horton, T.M., Tollefson, A.E., Wold, W.S.M. & Gooding, L. (1990). A protein serologically and functionally related to the group C E3 14, 700-dalton protein is found in multiple adenovirus serotypes. *Journal of Virology*, 64, 1250–1255.

Horwitz, M.S. (1990a). Adenoviridae and their replication. In *Fields' Virology*, 2nd edn, ed. B.N. Fields & D.M. Knipe, pp. 1679–1722. New York: Raven Press.

Horwitz, M.S. (1990b). Adenoviruses. In *Fields' Virology*, 2nd edn, ed. B.N. Fields & D.M. Knipe, pp. 1723–1747. New York: Raven Press.

Houweling, A., van den Elsen, P.J. & van der Eb, A.J. (1980). Partial transformation of primary rat cells by the leftmost 4.5% fragment of adenovirus DNA. *Virology*, 105, 537–550.

Howdle, P.D., Blair Zajdel, M.E., Smart, C., Trejdosiewicz, L.K., Losowsky, M.S. & Blair, G.E. (1989). Lack of serologic response to an E1B protein of adenovirus 12 in coeliac disease. *Scandinavian Journal of Gastroenterology*, 24, 282–286.

Jackson, M.R., Nilsson, T. & Peterson, P.A. (1990). Identification of a consensus motif for retention of transmembrane proteins in the endoplasmic reticulum. *EMBO Journal*, 9, 3153–3162.

Janaswami, P.M., Kalvakolanu, D.V.R., Zhang, Y. & Sen, G.C. (1992). Transcriptional repression of interleukin-6 gene by adenoviral E1A proteins. *Journal of Biological Chemistry*, 267, 24886–24891.

Javier, R., Raska, K. Jr & Shenk, T. (1992). Requirement for the adenovirus type 9 E4 region in production of mammary tumors. *Science*, 257, 1267–1271.

Jefferies, W.A. & Burgert, H.-A. (1990). E3/19K from Ad2 is an immunosubversive protein that binds to a structural motif regulating the intracellular transport of major histocompatibility complex class I proteins. *Journal of Experimental Medicine*, 172, 1653–1664.

Jelinek, T., Pereira, D.S. & Graham, F.L. (1994). Tumorigenicity of adenovirus-transformed rodent cells is influenced by at least two regions of adenovirus type 12 early region 1A. *Journal of Virology*, 68, 888–896.

Jochemsen, A.G., Bos, J.C. & van der Eb, A.J. (1984). The first exon of region E1A genes of adenoviruses 5 and 12 encodes a separate functional domain. *EMBO Journal*, 3, 2923–2927.

Jones, N. & Shenk, T. (1978). Isolation of deletion and substitution mutants of adenovirus type 5. *Cell*, 13, 181–188.

Kagnoff, M.F. (1989). Celiac disease: adenovirus and alpha gliadin. *Current Topics in Microbiology and Immunology*, 145, 67–78.

Kalvakolanu, D.V.R., Bandyopadhyay, S.K., Harter, M.L. & Sen, G.C. (1991). Inhibition of interferon-inducible gene expression by adenovirus E1A proteins: block in transcriptional complex formation. *Proceedings of the National Academy of Sciences of the USA*, 88, 7459–7463.

Kämpe, O., Bellgrau, D., Hammerling, U., Lind, P., Pääbo, S., Severinsson, L. & Peterson, P.A. (1983). Complex formation of class I transplantation antigens and a viral glycoprotein. *Journal of Biological Chemistry*, 258, 10594–10598.

Kast, W.M., Offringa, R., Peters, P.J., Voordouw, A.C., Meloen, R.H., van der Eb, A.J. & Melief, C.J.M. (1989). Eradication of adenovirus E1-induced tumors by E1A-specific cytotoxic T lymphocytes. *Cell*, 59, 603–614.

Katoh, S., Ozawa, K., Kondoh, S., Soeda, E., Israel, A., Shiroki, K., Fujinaga, K.,

Itakura, K., Gachelin, G. & Yokoyama, K. (1990). Identification of sequences responsible for positive and negative regulation by E1A in the promoter of H-2K$^{bm1}$ class I MHC gene. *EMBO Journal*, 9, 127–135.

Kemp, G.D., Webster, A. & Russell, W.C. (1992). Proteolysis is a key process in virus replication. *Essays in Biochemistry*, 27, 1–16.

Kenyon, D.J., Dougherty, J. & Raska, K. Jr (1991). Tumorigenicity of adenovirus-transformed cells and their sensitivity to tumor necrosis factor α and NK/LAK cytolysis. *Virology*, 180, 818–821.

Kenyon, D.J. & Raska, K. Jr (1986). Region E1A of highly oncogenic adenovirus 12 in transformed cells protects against NK but not LAK cytolysis. *Virology*, 155, 644–654.

Kimelman, D., Miller, J.S., Porter, D. & Roberts, B.E. (1985). E1A regions of the human adenoviruses and of the highly oncogenic simian adenovirus 7 are closely related. *Journal of Virology*, 53, 399–409.

Kimura, A., Israel, A., Le Bail, O. & Kourilsky, P. (1986). Detailed analysis of the mouse H-2K$^b$ promoter: enhancer-like sequences and their role in the regulation of class I gene expression. *Cell*, 44, 261–272.

Knippers, R. & Levine, A.J. (eds.) (1989). Transforming proteins of DNA tumor viruses. *Current Topics in Microbiology and Immunology*, 144, 1–284.

Körner, H. & Burgert, H.-G. (1994). Down-regulation of HLA antigens by the adenovirus type 2 E3/19K protein in a T-lymphoma cell line. *Journal of Virology*, 68, 1442–1448.

Kornfeld, R. & Wold, W.S.M. (1981). Structures of the oligosaccharides of the glycoprotein coded by early region E3 of adenovirus 2. *Journal of Virology*, 40, 440–449.

Kralli, A., Ge, R., Graeven, U., Ricciardi, R.P. & Weinmann, R. (1992). Negative regulation of the major histocompatibility class I enhancer in adenovirus type 12-transformed cells via a retinoic acid response element. *Journal of Virology*, 66, 6979–6988.

Lamberti, C. & Williams, J. (1990). Differential requirement for adenovirus type 12 E1A gene products in oncogenic transformation. *Journal of Virology*, 64, 4997–5007.

Lassam, N. & Jay, G. (1989). Suppression of MHC class I RNA in highly oncogenic cells occurs at the level of transcription initiation. *Journal of Immunology*, 143, 3792–3797.

Leppard, K.N. & Shenk, T. (1989). The adenovirus E1B 55 kD protein influences mRNA transport via an intranuclear effect on RNA metabolism. *EMBO Journal*, 8, 2329–2336.

Licht, J.D., Grossel, M.J., Figge, J. & Hansen, U.M. (1990). *Drosophila* kruppel protein is a transcriptional repressor. *Nature (London)*, 346, 76–79.

Lillie, J.W. & Green, M.R. (1989). Transcriptional activation by the adenovirus E1A protein. *Nature (London)*, 338, 39–44.

Lippé, R., Luke, E., Kuah, Y.T., Lomas, C. & Jefferies, W.A. (1991). Adenovirus infection inhibits the phophorylation of major histocompatibility complex class I proteins. *Journal of Experimental Medicine*, 174, 1159–1166.

Liu, X., Ge, R., Westmoreland, S., Cooney, A.J., Tsai, S.Y., Tsai, M.-J. & Ricciardi, R.P. (1994). Negative regulation by the R2 element of the MHC class I enhancer in adenovirus 12 transformed cells correlates with high levels of COUP-TF binding. *Oncogene*, 9, 2183–2190.

Liu, F. & Green, M.R. (1994). Promoter targeting by adenovirus E1A through interaction with different cellular DNA-binding domains. *Nature (London)*, 368, 520–525.

Mackey, J.K., Rigden, P.M. & Green, M. (1976). Do highly oncogenic group A human adenoviruses cause human cancer? Analysis of human tumors for adenovirus 12 transforming DNA sequences. *Proceedings of the National Academy of Sciences of the USA*, 73, 4657–4661.

Mahon, J., Blair, G.E., Wood, G.M., Scott, B.B., Losowsky, M.S. & Howdle, P.D. (1991). Is persistent adenovirus 12 infection involved in coeliac disease? A search for viral DNA using the polymerase chain reaction. *Gut*, 32, 1114–1116.

Mautner, V. (1989). Adenoviridae: In *Andrewes' Viruses of Vertebrates*, 5th edn, ed. J.S. Porterfield, pp. 249–282. London: Ballière Tindall.

Mei, Y.-F. & Wadell, G. (1992). The nucleotide sequence of adenovirus type 11 early 3 region: comparison of genome type Ad11p and Ad11a. *Virology*, 191, 125–133.

Meijer, I., Boot, A.J.M., Mahibin, G., Zantema, A. & van der Eb, A.J. (1992). Reduced binding activity of the transcription factor NF-κB accounts for the MHC class I repression in adenovirus 12 E1-transformed cells. *Cellular Immunology*, 145, 56–65.

Meijer, I., Jochemsen, A.G., de Wit, C.M., Bos, J.L., Morello, D. & van der Eb, A.J. (1989). Adenovirus type 12 E1A down regulates expression of a transgene under control of a major histocompatibility complex class I promoter: evidence for transcriptional control. *Journal of Virology*, 63, 4039–4042.

Meijer, I., van Dam, H., Boot, A.J.M., Bos, J.L., Zantema, A. & van der Eb, A.J. (1991). Co-regulated expression of *jun*B and MHC class I genes in adenovirus-transformed cells. *Oncogene*, 6, 911–916.

Montell, C., Courtois, G., Eng, C. & Berk, A.J. (1984). Complete transformation by adenovirus 2 requires both E1A proteins. *Cell*, 36, 951–961.

Moran, E. (1993). DNA tumour virus transforming proteins and the cell cycle. *Current Opinions in Genetics and Development*, 3, 63–70.

Moran, E. & Matthews, R.B. (1987). Multiple functional domains in the adenovirus E1A gene. *Cell*, 48, 177–178.

Murphy, M., Opalka, B., Sajaczkowski, R. & Schulte-Holthausen, H. (1987). Definition of a region required for transformation of E1A of adenovirus 12. *Virology*, 159, 49–56.

Nakajima, T., Nakamura, T., Tsunoda, S., Nakada, S. & Oda, K. (1992). E1A-responsive elements for repression of rat fibronectin gene transcription. *Molecular and Cellular Biology*, 12, 2837–2846.

Nevins, J.R. (1991). Transcriptional activation by viral regulatory proteins. *Trends in Biochemical Sciences*, 16, 435–439.

Nevins, J.R. (1992). E2F: a link between the Rb tumor suppressor and viral oncoproteins. *Science*, 258, 424–429.

Nielsch, U., Zimmer, S. & Babiss, L. (1991). Changes in NF-κB and ISGF3 DNA binding activities are responsible for differences in MHC and β-interferon gene expression in Ad5- versus Ad12-transformed cells. *EMBO Journal*, 10, 4169–4175.

Nilsson, T., Jackson, M. & Peterson, P.A. (1989). Short cytoplasmic sequences serve as retention signals for transmembrane proteins in the endoplasmic reticulum. *Cell*, 58, 707–718.

Offringa, R., Gebel, S., van Dam, H., Timmers, M., Smits, A., Zwart, R., Stein, B., Boot, J.L., van der Eb, A.J. & Herrlich, P. (1990). A novel function of the transforming domain of E1A: repression of AP-1 activity. *Cell*, 62, 527–538.

Ortmann, B., Androlewicz, M.J. & Cresswell, P. (1994). MHC class I/

$\beta_2$-microglobulin complexes associate with TAP transporters before peptide binding. *Nature (London)*, 368, 864–867.

Ou, W.-J., Cameron, P.H., Thomas, D.Y. & Bergeron, J.J.M. (1993). Association of folding intermediates of glycoproteins with calnexin during protein maturation. *Nature (London)*, 364, 771–776.

Ozawa, K., Hagiwara, H., Tang, X., Saka, F., Kitabayashi, I., Shiroki, K., Fujinaga, K., Israel, A., Gachelin, G. & Yokoyama, K. (1993). Negative regulation of the gene for H-2K$^b$ class I antigen by adenovirus 12-E1A is mediated by a CAA repeated element. *Journal of Biological Chemistry*, 268, 27258–27268.

Pääbo, S., Bhat, B.M., Wold, W.S.M. & Peterson, P.A. (1987). A short sequence in the COOH-terminus makes an adenovirus membrane glycoprotein a resident of the endoplasmic reticulum. *Cell*, 50, 311–317.

Pääbo, S., Nilsson, T. & Peterson, P.A. (1986b). Adenoviruses of subgenera B, C, D and E modulate cell-surface expression of major histocompatibility complex class I antigens. *Proceedings of the National Academy of Sciences of the USA*, 83, 9665–9669.

Pääbo, S., Severinsson, L., Anderson, M., Martens, I., Nilsson, T. & Peterson, P.A. (1989). Adenovirus proteins and MHC expression. *Advances in Cancer Research*, 52, 151–163.

Pääbo, S., Weber, F., Kamye, O., Schaffner, W. & Peterson, P.A. (1983). Association between transplantation antigens and a viral membrane protein synthesized from a mammalian expression vector. *Cell*, 35, 445–453.

Pääbo, S., Weber, F., Nilsson, T., Schaffner, W. & Peterson, P.A. (1986a). Structural and functional dissection of an MHC class I antigen-binding adenovirus glycoprotein. *EMBO Journal*, 5, 1921–1927.

Persson, H., Jörnvall, H. & Zabielski, J. (1980). Multiple mRNA species for the precursor to an adenovirus-encoded glycoprotein: identification and structure of the signal sequence. *Proceedings of the National Academy of Sciences of the USA*, 77, 6349–6353.

Persson, H., Katze, M.G. & Philipson, L. (1982). Purification of a native membrane-associated adenovirus tumor antigen. *Journal of Virology*, 42, 905–917.

Proffitt, J.L., Sharma, E. & Blair, G.E. (1994). Adenovirus 12-mediated down-regulation of the major histocompatibility complex (MHC) class I promoter: identification of a negative regulatory element responsive to Ad12 E1A. *Nucleic Acids Research*, 22, 4779–4788.

Reich, N., Pine, R., Levy, D. & Darnell, J.E. Jr (1988). Transcription of interferon-stimulated genes is induced by adenovirus particles but is suppressed by E1A products. *Journal of Virology*, 62, 114–119.

Rikitake, Y. & Moran, E. (1992). DNA-binding properties of the E1A-associated 300-kilodalton protein. *Molecular and Cellular Biology*, 12, 2826–2836.

Rochette-Egly, C., Fromental, C. & Chambon, P. (1990). General repression of enhansor activity by the adenovirus 2 E1A proteins. *Genes and Development*, 4, 137–150.

Rosenthal, A., Wright, S., Quade, K., Gallimore, P., Cedar, H. & Grosveld, F. (1985). Increased MHC H-2K gene transcription in cultured mouse embryo cells after adenovirus infection. *Nature (London)*, 315, 579–581.

Rotem-Yehudar, R., Schechter, H. & Ehrlich, R. (1994a). Transcriptional regulation of class I-major histocompatibility complex genes transformed in murine cells is mediated by positive and negative regulatory elements. *Gene*, 144, 265–270.

Rotem-Yehudar, R., Winograd, S., Sela, S., Coligan, J.E. & Ehrlich, R. (1994b).

Downregulation of peptide transporter genes in cell lines transformed with the highly oncogenic adenovirus 12. *Journal of Experimental Medicine*, 180, 477–488.

Routes, J.M. & Cook, J.L. (1990). Resistance of human cells to the adenovirus E3 effect on class I MHC antigen expression. *Journal of Immunology*, 144, 2763–2770.

Rowe, D.T., Branton, P.E., Yee, S.P., Bacchetti, S.E. & Graham, F.L. (1984). Establishment and characterisation of hamster cell lines transformed by restriction endonuclease fragments of adenovirus 5. *Journal of Virology*, 49, 674–685.

Rowe, D.T. & Graham, F.L. (1983). Transformation of rodent cells by DNA extracted from transformation-defective adenovirus mutants. *Journal of Virology*, 46, 1039–1044.

Sarnow, P., Ho, Y., Williams, J. & Levine, A.J. (1982). Adenovirus E1B 58 kD tumor antigen and SV40 large tumor antigen are physically associated with the same 54 kD cellular protein in transformed cells. *Cell*, 28, 387–394.

Sawada, Y., Fohring, B., Shenk, T.E. & Raska, K. Jr (1985). Tumorigenicity of adenovirus-transformed cells: region E1A of adenovirus 12 confers resistance to natural killer cells. *Virology*, 147, 413–421.

Sawada, Y., Raska, K. Jr & Shenk, T. (1988). Adenovirus type 5 and adenovirus type 12 recombinant viruses containing heterologous E1 genes are viable, transform rat cells, but are not tumorigenic in rats. *Virology*, 166, 281–284.

Schneider, R.J. & Shenk, T. (1987). Impact of virus infection on host cell protein synthesis. *Annual Review of Biochemistry*, 56, 317–332.

Schrier, P.I., Bernards, R., Vaessen, R.T.M.J., Houweling, A. & van der Eb, A.J. (1983). Expression of class I major histocompatibility antigens is switched off by highly oncogenic adenovirus 12 in transformed rat cells. *Nature (London)*, 305, 771–775.

Severinsson, L., Martens, I. & Peterson, P.A. (1986). Differential association between two human MHC class I antigens and adenoviral glycoprotein. *Journal of Immunology*, 137, 1003–1009.

Severinsson, L. & Peterson, P.A. (1985). Abrogation of cell surface expression of human class I transplantation antigens by an adenovirus protein in *Xenopus laevis* oocytes. *Journal of Cell Biology*, 101, 540–547.

Signäs, C., Akusjarvi, G. & Pettersson, U. (1986). Region E3 of human adenoviruses: differences between the oncogenic adenovirus 3 and the non-oncogenic adenovirus 2. *Gene*, 50, 173–184.

Smith, K.A., Gallimore, P.H. & Grand, R.J.A. (1989). The expression of Ad12 E1B 19K protein on the surface of adenovirus-transformed and infected human cells. *Oncogene*, 4, 489–497.

Soddu, S. & Lewis, A.M. Jr (1992). Driving adenovirus type 12-transformed BALB/C mouse cells to express high levels of class I major histocompatibility complex proteins enhances, rather than abrogates, their tumorigenicity. *Journal of Virology*, 66, 2875–2884.

Sprengel, J., Schmitz, B., Heus-Neitzel, D., Zock, C. & Doerfler, W. (1994). Nucleotide sequence of human adenovirus type 12 DNA: comparative functional analysis. *Journal of Virology*, 68, 379–389.

Stewart, P.L., Burnett, R.M., Cyrklaff, M. & Fuller, S.D. (1991). Image reconstruction reveals the complex molecular organization of adenovirus. *Cell*, 67, 145–154.

Storkus, W.J., Alexander, J., Payne, J.A., Dawson, J.R. & Cresswell, P. (1989). Reversal of natural killing susceptibility in target cells expressing transfected

class I HLA genes. *Proceedings of the National Academy of Sciences of the USA*, 86, 2361–2364.
Straus, S.E. (1984). Adenovirus infections in humans. In *The Adenoviruses*, ed. H.S. Ginsberg, pp. 451–496. New York: Plenum Press.
Tanaka, K., Isselbacher, K.J., Khoury, G. & Jay, G. (1985). Reversal of oncogenesis by the expression of a major histocompatibility complex class I gene. *Science*, 228, 26–30.
Tanaka, K., Hayashi, H., Hamada, C., Khoury, G. & Jay, G. (1986). Expression of major histocompatibility class I antigens as a strategy for potentiation of immune recognition of tumor cells. *Proceedings of the National Academy of Sciences of the USA*, 83, 8723–8727.
Tanaka, K. & Tevethia, S.S. (1988). Differential effect of adenovirus 2 E3/19K glycoprotein on the expression of H-2K$^b$ and H-2D$^b$ class I antigens and H-2K$^b$ and H-2D$^b$ restricted SV40-specific CTL-mediated lysis. *Virology*, 165, 357–366.
Tang, X., Li, H.-O., Sakatsume, O., Ohta, T., Tsutsui, H., Smit, A.F.A., Horikoshi, M., Kourilsky, P., Israel, A., Gachelin, G. & Yokoyama, K. (1995). Cooperativity between an upstream TATA-like element mediates E1A-dependent negative repression of the H-2K$^b$ class I gene. *Journal of Biological Chemistry*, 270, 2327–2336.
Telling, G.C. & Williams, J. (1994). Constructing chimeric type 12/type 5 adenovirus E1A genes and using them to identify an oncogenic determinant of adenovirus type 12. *Journal of Virology*, 68, 877–887.
Tsuji, Y., Kwak, E., Saika, T., Torti, S.V. & Torti, F.M. (1993). Preferential repression of the H-subunit of ferritin by adenovirus E1A in NIH 3T3 mouse fibroblasts. *Journal of Biological Chemistry*, 268, 7270–7275.
Urbanelli, D., Sawada, Y., Raskova, J., Jones, N.C., Shenk, T. & Raska, K. Jr (1989). C-terminal domain of the adenovirus E1A oncogene product is required for the induction of cytotoxic T lymphocytes and tumor-specific transplantation immunity. *Virology*, 173, 607–614.
Vaessen, R.T.M.J., Houweling, A., Israel, A., Kourilsky, P. & van der Eb, A.J. (1986). Adenovirus E1A-mediated regulation of class I MHC expression. *EMBO Journal*, 5, 335–341.
Vanhaesebroeck, B., Timmers, H.T.M., Pronk, G.J., van Roy, F., van der Eb, A.J. & Fiers, W. (1990). Modulation of cellular susceptibility to the cytotoxic/cytostatic action of tumor necrosis factor by adenovirus E1 gene expression is cell type-dependent. *Virology*, 176, 362–368.
Vasavada, R., Eager, K.B., Barbanti-Brodano, G., Caputo, A. & Ricciardi, R.P. (1986). Adenovirus type 12 early region 1A proteins repress class I HLA expression in transformed human cells. *Proceedings of the National Academy of Sciences of the USA*, 83, 5257–5261.
Wang, H.-G. H., Yaciuk, P., Ricciardi, R.P., Green, M., Yokoyama, Y. & Moran, E. (1993). The E1A products of oncogenic adenovirus serotype 12 include amino-terminally modified forms able to bind the retinoblastoma protein but not p300. *Journal of Virology*, 67, 4804–4813.
Webster, A., Russell, W.C. & Kemp, G.D. (1989). Characterisation of the adenovirus proteinase: substrate specificity. *Journal of General Virology*, 70, 3225–3234.
Webster, K.A., Muscat, G.E.O. & Kedes, L. (1988). Adenovirus E1A products suppress myogenic differentiation and inhibit transcription from muscle-specific promoters. *Nature (London)*, 332, 553–557.
White, E., Blose, S.H. & Stillman, B.W. (1984). Nuclear envelope localization of

an adenovirus tumor antigen maintains the integrity of cellular DNA. *Molecular and Cellular Biology*, 4, 2865–2875.

White, E., Sabbatini, P., Debbas, M., Wold, W.S.M., Kusher, D.I. & Gooding, L.R. (1992). The 19-kilodalton adenovirus E1B transforming protein inhibits programmed cell death and prevents cytolysis by tumor necrosis factor alpha. *Molecular and Cellular Biology*, 12, 2570–2580.

Wold, W.S.M. & Gooding, L.R. (1991). Region 3 of adenovirus: a cassette of genes involved in host immunosurveillance and virus–cell interactions. *Virology*, 184, 1–8.

Yew, P.R., Liu, X, & Berk, A.J. (1994). Adenovirus E1B oncoprotein tethers a transcriptional repression domain to p53. *Genes and Development*, 8, 190–202.

Yewdell, J.W., Bennink, J.R., Eager, K.B. & Ricciardi, R.P. (1988). CTL recognition of adenovirus-transformed cells infected with influenza virus: lysis by anti-influenza CTLs parallels adenovirus 12-induced suppression of class I MHC molecules. *Virology*, 162, 236–238.

Yu, O., Suen, T.-C., Yan, D.-H., Chang, L.S. & Hung, M.-C. (1990). Transcriptional repression of the *neu* proto-oncogene by the adenovirus 5 E1A gene products. *Proceedings of the National Academy of Sciences of the USA*, 87, 4499–4503.

Zantema, A., Schrier, P.I., Davis-Olivier, A., van Laar, T., Vaessen, R.T.M.J. & van der Eb, A.J. (1985). Adenovirus serotype determines association and localisation of the large E1B tumor antigen with cellular tumor antigen p53 in transformed cells. *Molecular and Cellular Biology*, 5, 3084–3091.

# 9

# MHC Expression in HPV-associated cervical cancer

JENNIFER BARTHOLOMEW[1], JOHN M. TINSLEY[2] and PETER L. STERN[1]
[1]Christie Hospital NHS Trust, Manchester and [2]John Radcliffe Hospital, Oxford

## 9.1 Introduction

Cervical cancer and pre-cancer form a disease continuum ranging from cervical intraepithelial neoplasia (CIN) through microinvasion to invasive carcinoma; about 70% of the tumours are squamous and 30% are adeno- and adenosquamous carcinomas (Buckley & Fox, 1989). Most tumours are thought to develop from an area of intraepithelial neoplasia within the transformation zone (Coppelston & Reid, 1967). Cervical cancer is estimated to be the second most common female cancer with approximately 500 000 new cases per annum worldwide (Parkin, Laara & Muir, 1980). Sexually transmitted infections are recognized as one of the major risk factors and the active agents are thought to be specific types of human papillomavirus (HPV) (Munoz *et al.*, 1992).

Papillomaviruses are small DNA viruses associated with benign and malignant proliferative lesions of cutaneous epithelium. Over 60 different types of papillomavirus have been described and they can be segregated into groups distinguished by DNA sequence homology and the specific lesions with which they are associated (de Villiers, 1989). HPV 6 and 11 are found most commonly in cervical condyloma, benign lesions that tend to regress spontaneously, and low-grade CIN. HPV 16 and 18 are the types most commonly associated with high-grade CIN lesions and invasive carcinoma of the cervix. The viruses will replicate only in specific differentiation stages of epithelia, which limits the use of *in vitro* culture methods for producing HPV. To circumvent this, molecular biological techniques have been utilized extensively to characterize HPV. Cloning of the viral DNA into recombinant DNA vectors has allowed the determination of the complete nucleotide sequences of several types of HPV. The genome can be divided into three functional regions: early, E, late, L, and upstream regulatory region, URR (Sousa, Dostatni & Yaniv, 1990). The early region has several open reading frames

(E1–E7 ORF) encoding information for viral replication and cellular transformation, and the late genes L1 and L2 encode the capsid proteins that are essential for vegetative viral replication. The URR is a non-coding region containing the origin of DNA replication and regulatory elements like promoters and enhancers. The genes frequently retained in cervical carcinomas are HPV16/18 E6 and E7, which have been established by *in vitro* studies as being important in transformation (Matlashewski, 1989; zur Hausen, 1989; DiMaio, 1991). How the viruses promote the development of cervical cancer is not known, although it seems likely that infection with high-risk HPV types is a prerequisite for the development of carcinoma of the cervix (Munoz *et al.*, 1992). In many cases, the continued growth of the tumour is probably related to the expression of the E6 and/or E7 oncogene products, which function by interfering with the tumour suppressor functions of p53 (Levine, Momand & Finlay, 1991) and *Rb* (Dyson *et al.*, 1989), respectively.

The putative viral aetiology of cervical neoplasia suggests that immunological intervention might be a useful approach to the control and prevention of this condition. The expression of specific viral proteins through the various stages of the natural history of cervical cancer may evoke immunological consequences of relevance to the progression, or not, of the disease. Appropriate immunization could potentially elicit specific antibodies to viral antigens that could prevent infection, while cell-mediated effectors would specifically eliminate virally infected cells. There is some evidence to support an active role of the immune system in preventing the development of HPV-associated malignancies. Immunocompromised individuals, such as allograft recipients, cancer and AIDS patients, exhibit an increased incidence of HPV-associated lesions (Kinlen *et al.*, 1979; Schneider, Kay & Lee, 1983; Halpert *et al.*, 1985; Koss, 1987). It is likely that the T cell-mediated arm of the immune response would be of primary importance in any surveillance of tumour cells.

The recognition of antigen by T cells is restricted by the polymorphic products of the MHC. Thus, any viral tumour antigens would be recognized as processed peptides that bind either to particular MHC class I or to class II products, which restrict $CD8^+$ (mainly cytotoxic) T cells or $CD4^+$ (mainly helper) T cells, respectively (see Chapters 1 and 3). The expression of MHC polymorphic products and the normal functioning of peptide processing and transport pathways are critical factors in any potential immune recognition of virus-infected cells (Monaco, 1992; Neefjes & Ploegh, 1992). This emphasizes the need to investigate the tumour expression of MHC products as well as putative tumour antigens such as oncofoetal or virus-related molecules.

Several studies have examined the expression of MHC class I and II antigens in cervical lesions and any possible relationship between the presence of HPV and changes in MHC expression. A significant proportion of the tumours show down-regulation of MHC class I antigens, which would prevent effective cytotoxic T cell activity against potential viral antigens (Connor & Stern, 1990). For example, in this study, 16% of the biopsies showed complete or heterogeneous loss of HLA expression as judged by reactivity with antibodies recognizing monomorphic determinants of class I heavy chain bound to $\beta_2$-m. In addition, other biopsies showed a loss in expression of particular allelic products, giving a minimum estimate of carcinomas of the cervix exhibiting altered HLA class I expression of approximately 45%. However, there was no direct correlation between the presence of HPV DNA and altered class I expression. The biological consequences of such changes are indicated by the demonstration that the earlier-stage cancers that showed HLA class I down-regulation had a significantly poorer clinical outcome (Connor *et al.*, 1993).

Altered MHC class II antigen expression has been documented in several types of carcinoma where the tumour is derived from HLA class II-negative tissue and this up-regulation can be associated with both better and poorer prognosis for the patients (Ruiter, Brocker & Ferrone, 1986; Esteban *et al.*, 1990). It has also been established that the majority of squamous carcinomas expresses MHC class II antigens, although normal cervical squamous epithelium is HLA class II negative (Glew *et al.*, 1992). Again, there was no apparent correlation between the class II phenotype and the presence of HPV 16 DNA in the cervical cancer specimens. It would, therefore, seem that HPV does not directly influence MHC expression. However, the pathogenesis of cervical cancer is a multifactorial and multistage process and a relationship between MHC class I and/or class II expression and HPV infection may be evident in the evolution of pre-malignant disease. In order to evaluate further the possible influence of HPV on MHC expression, human keratinocytes transfected with the HPV 16 genome have been investigated for expression of various cell surface molecules that regulate immune interactions.

## 9.2 HPV expression in keratinocytes

A foetal human keratinocyte cell line SVD2 immortalized with an origin minus SV40 (simian virus 40) mutant plus a number of integrated copies of HPV 1 (Burnett & Gallimore, 1983) was transfected with either a pSV2*neo* plasmid (generating clones Ai and Aii) or a *KIV*-HPV16e construct plus pSV2*neo* (generating clones *IV*B and *IV*D). The *KIV*-HPV16e construct con-

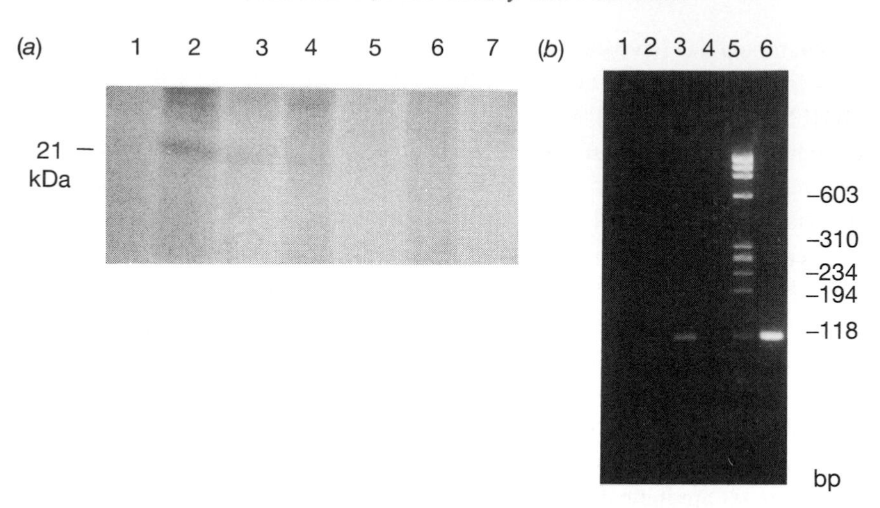


Fig. 9.1. (*a*) Immunoprecipitation of HPV 16 E7 protein from [$^{35}$S]cysteine-labelled keratinocytes. Cells were labelled for 4 hours with 500 μCi/ml of [$^{35}$S]cysteine, lysed in 0.5% NP40/0.1% SDS/Tris-buffered saline (TBS) (pH 8.1) containing 1 mM phenylmethylsulphonyl fluoride, centrifuged at 20 000 x *g* and phenylmethylsulphonyl fluoride immunoprecipitated for 1 hour at 4 °C with 5 μl of each antiserum. Washed immunoprecipitates were analysed on 15% polyacrylamide/SDS gels under reducing conditions, washed in Amplify (Amersham) for 20 minutes and dried. Gels were exposed to Kodak XAR film for 14 days. Lanes 1–5 are HPV 16-transfected SVD2 *IVD* cells and lanes 6–7 are neomycin-resistant SVD2 Ai cells. Lanes 1 and 7, normal mouse serum (NMS); lanes 2 and 6, pooled mAbs to E7; lanes 3 and 4, anti-E7 positive human sera; lane 5, seronegative human serum. The E7 band resolves with an apparent molecular mass of 21 kDa.
(*b*) PCR detection of HPV 16 E6 transcripts. Complementary DNAs were prepared by reverse transcription of total cellular RNA using a first strand synthesis kit and oligo(dT) primers (Stratagene). Each cDNA (1 μl) was then amplified using HPV 16 E6-specific primers complementary to bases 421–440 and 521–540. Amplification was performed over 40 cycles using a Biometra dry heat block using the following conditions: 94 °C, 30 s; 50 °C, 90 s; 72 °C, 2 minutes. Products were analysed by 3% agarose gel electrophoresis and visualized by ethidium bromide staining under UV light. Markers were ϕX174 RF DNA digested with *Hae*III (Promega). Lane 1, blank; lane 2, cDNA of Ai cells; lane 3, cDNA of *IVD* cells; lane 4, water; lane 5, ϕX174 markers; lane 6, 1 μg Caski genomic DNA.

tains the bovine keratin *IV* promoter plus 2.2 kb upstream regulatory sequences and the HPV 16 sequences are 8 bp 5′ to the second TATA box. Northern blot analysis of RNA using a full-length HPV 16 probe confirmed the presence of HPV 16 transcripts. The transfected cell lines were analysed for the expression of HPV 16 oncogene products.

The expression of the E7 protein was investigated by immunoprecipitation from HPV 16-transfected cells labelled with [$^{35}$S]cysteine for 4 hours in cysteine-free medium using several different antibodies against E7 (Fig. 9.1*a*).

The HPV 16-transfected clones expressed E7 polypeptides, which were immunoprecipitated with a pool of mAbs to E7 (Stacey *et al.*, 1993) and two human sera with anti-E7 activity, as judged from reactivity in Western blotting to an MS2–E7 fusion protein (Ghosh *et al.*, 1993) but not to a seronegative serum.

It has not been possible to demonstrate the presence of the E6 oncogene product in the transfected cells by immunoprecipitation, probably because of lack of an appropriate serological reagent that recognizes the native conformation of the protein. Therefore, immunoprecipitates from [$^{35}$S]cysteine-labelled cell lysates with a rabbit polyclonal anti-E6 serum raised against an E6–MS2 fusion protein (Stacey *et al.*, 1992) failed to detect a specific 18 kDa E6 product on SDS–PAGE analysis. This antiserum has subsequently been shown to react with a linear epitope within the E6 protein since it does not recognize a more authentic, native form of E6 (than the MS2 fusion protein) expressed by a recombinant baculovirus system (Stacey *et al.*, 1992). The possible expression of HPV 16 E6 was further studied at the transcriptional level. Complementary DNA prepared by reverse transcription of total RNA from each cell line was amplified using PCR and specific primers complementary to bases 421–440 and 521–540 within the E6 ORF (Young *et al.*, 1989). The predicted 120 bp PCR product derived from E6 transcripts was detected in the HPV 16-transfected keratinocytes and Caski cells but not the *neo*-transfected control keratinocytes. It is clear that the HPV 16 transfected keratinocyte cell lines are expressing specific viral oncogene products that are associated with cervical neoplasia. The expression of HLA class I and class II antigens of the paired cell lines was then analysed by radioimmunobinding assay using a variety of specific mAbs.

## 9.3 MHC antigen expression in HPV 16-transfected keratinocytes

### *9.3.1 MHC class I expression*

The W6/32 mAb recognizes a monomorphic determinant of class I heavy chains in association with $\beta_2$-m and binding was comparable in both HPV-transfected and control lines, with slightly higher levels detected in both the transfected clones (Fig. 9.2). The level of class I expression detected using W6/32 following a 48 hour treatment with IFN-$\gamma$ was increased in all lines but was not significantly different in the keratinocytes with or without HPV transfection. Binding of the mAb HC10, which recognizes unassociated or dissociating class I heavy chains (Neefjes & Ploegh, 1988; Gillet *et al.*, 1990), is proportionately higher than W6/32 in the HPV 16-transfected cells compared with the control cell lines (Fig. 9.2 shows results for clones *IV*B

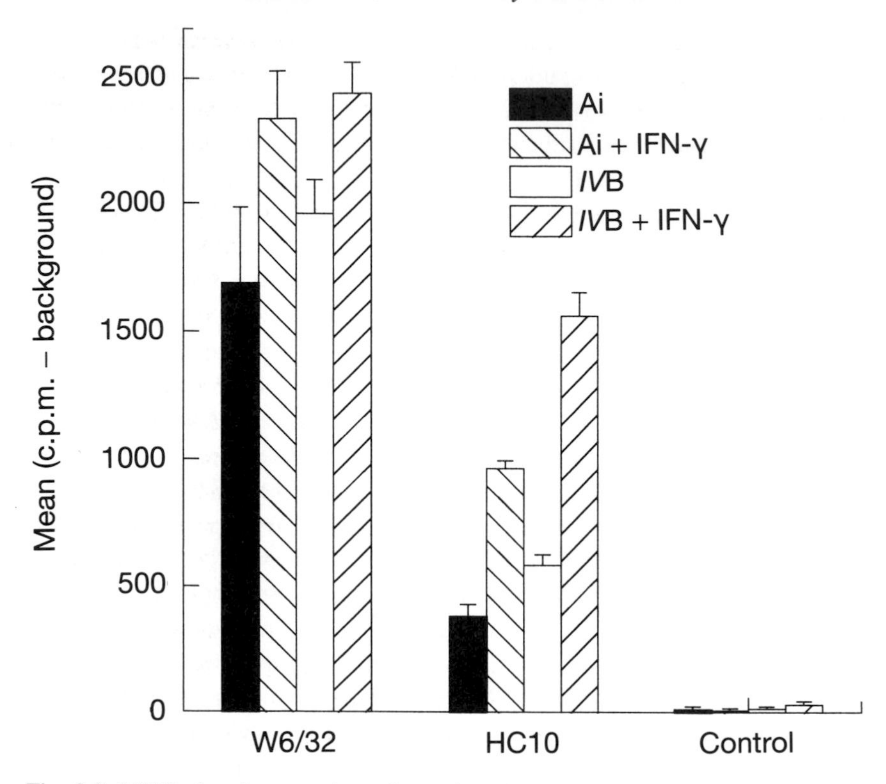


Fig. 9.2. MHC class I expression of transfected keratinocytes assessed by a radiobinding assay of clones Ai and *IV*B with and without IFN-γ treatment. Cells were seeded into 96 well plates in DMEM (Dulbecco's modified Eagle's medium) supplemented with 10% fetal calf serum and 500 μg/ml G418, 48 hours prior to assay and were 85% confluent at assay. Cell numbers were assessed following trypsinization of individual wells. After washing, cells were incubated with primary mAb for 1 hour at 4 °C in 100 μl HBSS containing 0.5% bovine serum albumin (BSA), 0.1% $NaN_3$, washed twice in this medium followed by addition of [$^{125}$I]-labelled $(Fab)_2$ sheep anti-mouse antiserum (50 000 c.p.m.) for 1 hour. Following three washes, the binding of each mAb was analysed by harvesting the cells in 0.5 M NaOH and gamma counting. Results are expressed as mean c.p.m. of triplicate wells minus background (second antibody only). Anti-HLA antibodies were W6/32 and HC10; mAb 11.4.1, against H-$2^k$, was an irrelevant mouse mAb. IFN-γ treatment was for 48 hours at 1000 U/ml where indicated.

versus Ai). This proportional difference is maintained following IFN treatment, which results in an increase in HC10 binding in all cell lines.

Any alteration of MHC expression that interferes with heavy chain–peptide interactions and subsequent stabilization by $\beta_2$-m would have direct consequences for potential immune recognition of tumour target antigens, which are presented to specific T cells as peptides associated with class I antigens.

The increased free heavy chain expression detected by HC10 mAb in the HPV-transfected cells could be consistent with an altered ability of some heavy chain molecules to associate with $\beta_2$-m (Neefjes & Ploegh, 1988). A precedent for viral modulation of functional MHC expression by inhibition of the transport of peptide-loaded MHC class I molecules into the medial-Golgi compartment has been described (Del Val *et al.*, 1992). During the early phase of murine cytomegalovirus (CMV) gene expression, the MHC class I heavy chain glycosylation remains in an endoglycosidase H (endo H)-sensitive form, suggesting a block at the ER/cis-Golgi compartment. The effect of early gene expression by CMV is to inhibit further glycosylation of MHC class I heavy chains so preventing the transport of these proteins through the Golgi compartment and interfering with the natural pathway for the processing and presentation of CMV-derived antigenic peptides (see Chapter 12). Adenoviruses use a related strategy for effects on intracellular transport of MHC class I antigens (see Chapter 8). A retention signal is contained within the six carboxy-terminal amino acid residues of the E3 19 kDa protein of Ad2 and this mediates the ER/cis-Golgi retention of complexes between E3-19K and MHC class I antigens, which is also reflected in the endo H sensitivity of the class I heavy chains (Andersson *et al.*, 1985; Cox, Bennink & Yewdell, 1991).

To locate any possible effect on post-translational modification of HLA class I antigens in the HPV 16-transfected keratinocytes, the susceptibility of MHC proteins to endo H digestion was determined. MHC class I heavy chains co-translationally acquire a high-mannose core of N-linked oligosaccharides in the ER. Endo H preferentially cleaves immature N-linked oligosaccharides characteristic of glycoproteins that have not reached the medial-Golgi compartment. Further processing of the oligosaccharide chain by enzymes located in the medial-Golgi compartment leads to the fully mature glycoproteins and renders the glycan structure resistant to endo H digestion. Maturation of MHC class I antigens was studied by pulse–chase experiments. Cells were labelled for 15 minutes [$^{35}$S]methionine and chased for intervals up to 90 minutes post-labelling. Cells were lysed in buffer containing 0.5% NP40 and immunoprecipitated with W6/32 or HC10. Each immunoprecipitate was divided and treated with endo H, or mock treated, for 18 hours at 37 °C and the glycosylation state assessed following separation in 12% SDS–PAGE under reducing conditions. Comparison of endo H-treated and untreated immunoprecipitates from either HPV-transfected or control keratinocytes showed that class I heavy chains retain their sensitivity to endo H up to 60 minutes post-label but, by 90 minutes, fully glycosylated mature class I antigens are produced. Figure. 9.3 shows the time course of full

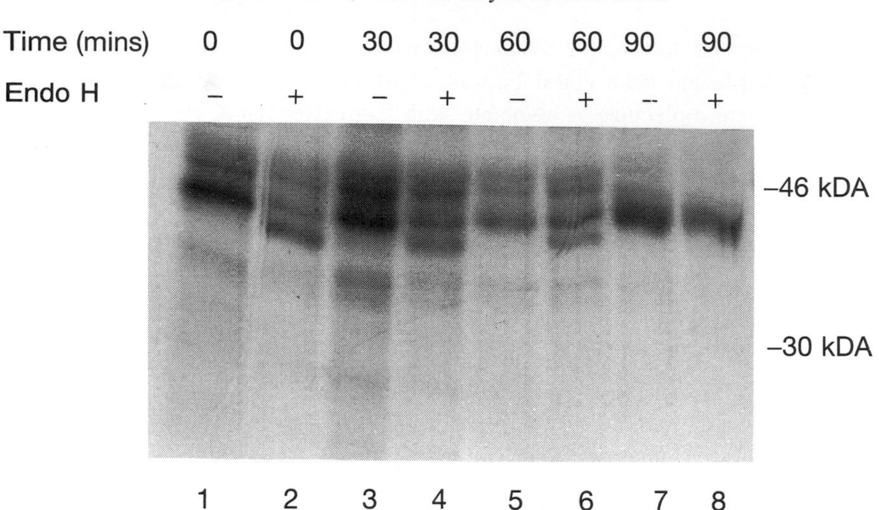


Fig. 9.3. Endoglycosidase H (endo H) sensitivity of metabolically labelled MHC class I heavy chains. Keratinocytes were radiolabelled with [$^{35}$S]methionine (500 μCi/ml) for 15 minutes; 5 mM unlabelled methionine was added and cells were harvested at 0, 30, 60 and 90 minutes post-labelling, lysed in 0.5% NP40/TBS (pH 8.1) and immunoprecipitated with W6/32 or HC10. Samples were divided, treated with endo H (2 U/sample, Boehringer-Mannheim), or mock treated, for 18 hours at 37 °C. The HC10 precipitates from *IVD* cells are shown resolved on a 12% polyacrylamide gel under reducing conditions. Lanes 1, 3, 5, 7 were untreated; lanes 2, 4, 6, 8 were treated with endo H.

glycosylation of heavy chains immunoprecipitated by HC10 from *IVD* cells; similar results were obtained with heavy chains precipitated by W6/32. It, therefore, appears that HPV 16 expression does not interfere in the production of mature glycosylated class I antigens in these keratinocyte cell lines.

### *9.3.2 MHC class II expression*

An analysis by radio-binding assays using several mAbs recognizing monomorphic determinants of HLA class II failed to detect any expression by either the HPV 16-transfected or control keratinocyte cell lines. This lack of HLA class II expression might be analogous to the normal squamous epithelium of the cervix, which is class II negative. However, following IFN-γ treatment for 48 hours, there was differential induction of expression of HLA class II antigens by the HPV-transfected keratinocytes but not by the control cells (Fig. 9.4). TNF-α enhanced the HLA class II induction by IFN-γ but had no effect alone (results not shown). The levels were low but the observations were reproducible and demonstrable with antibodies that recognize

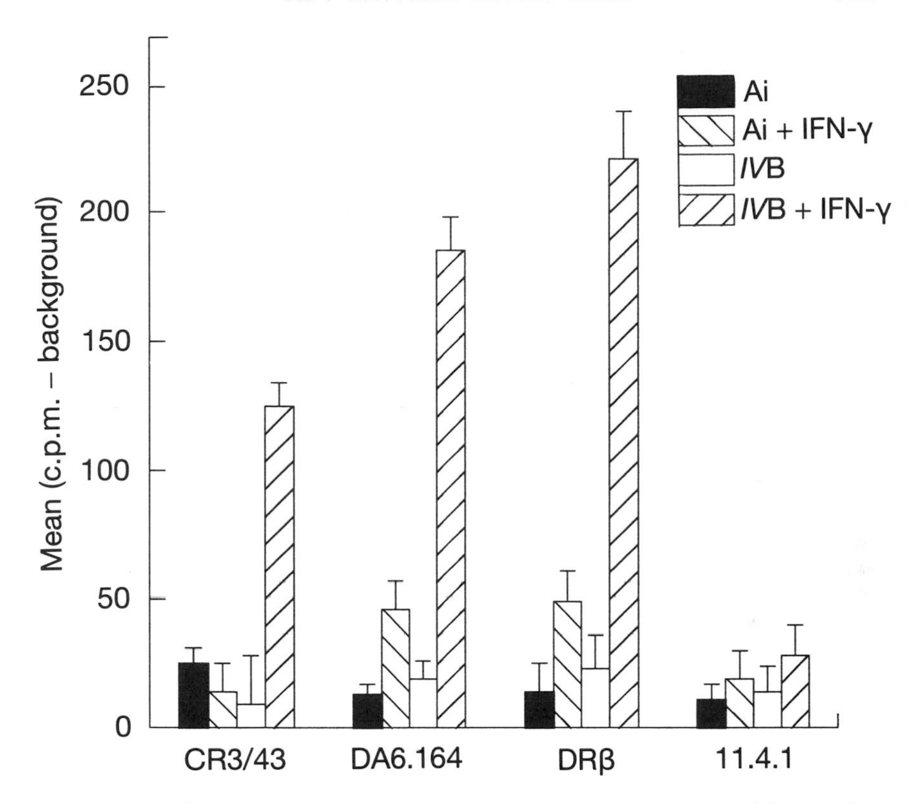


Fig. 9.4. MHC class II expression in transfected keratinocytes assessed by a radio-binding assay. Methods were as in Fig. 9.2. mAbs were CR3/43, which recognizes a monomorphic determinant of HLA-DR, HLA-DQ and HLA-DP (Dako), anti-HLA-DR mAbs DRβ (Becton Dickinson) and DA6.164 and mAb 11.4.1, an irrelevant mouse mAb control.

HLA-DR but not HLA-DQ or HLA-DP. These observations were confirmed by fluorescence-activated cell sorting (FACS) analysis using HLA-DR-specific antibodies; only a relatively small proportion of the IFN-γ treated HPV 16-transfected cells were labelled. It has not proved possible to identify these antigens by immunoprecipitation from either surface- or metabolically labelled cells, presumably because of the very low levels of expression. Northern blot analysis with an HLA-DRβ chain probe indicated that specific transcripts of MHC class II were present at low levels in RNA isolated from both IFN-induced HPV-transfected cells and the parent keratinocyte cells. No transcripts could be detected in the keratinocyte lines without IFN-γ induction. The specificity of HLA-DR serological reactivity of the HPV transfectants appears not to result from differential transcription between

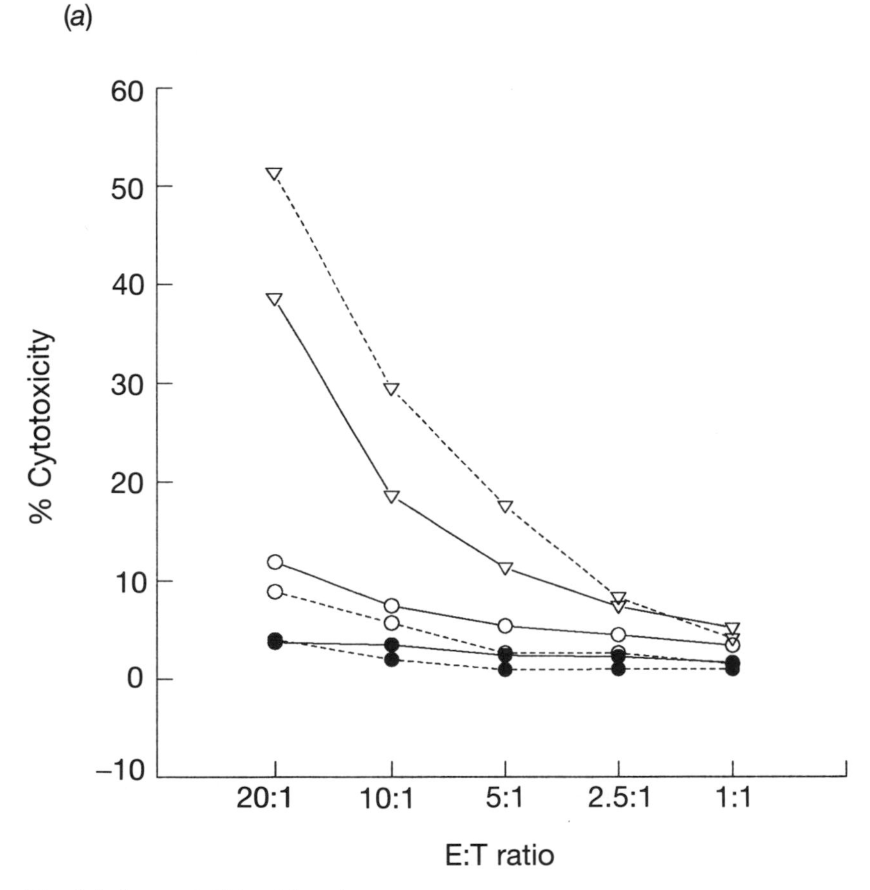


Fig. 9.5. Susceptibility of keratinocytes to NK, LAK and activated macrophages in $^{51}Cr$ release assay. (*a*) Macrophages were isolated from PBL by adherence to plastic for 1 hour. These cells and the non-adherent cells were treated with IFN-α (1000 U/ml) for 18 hours. Target cells were Ai cells (solid lines) and *IVD* cells (dotted lines); macrophage (○—○), IFN-α-activated macrophages (●—●) and IFN-α-activated NK effectors (▽—▽).

these and control transfected cell lines (although this needs to be confirmed by RNAse protection analysis). Overall, these data indicate that the presence of the HPV 16 genome, as well as transcription of both E6 and E7 genes and E7 translation products, in these keratinocytes does not have a gross affect on HLA class I or class II expression. However, there may be an influence of HPV 16 on the MHC class II expression in response to IFN-γ, although the mechanism remains to be elucidated.

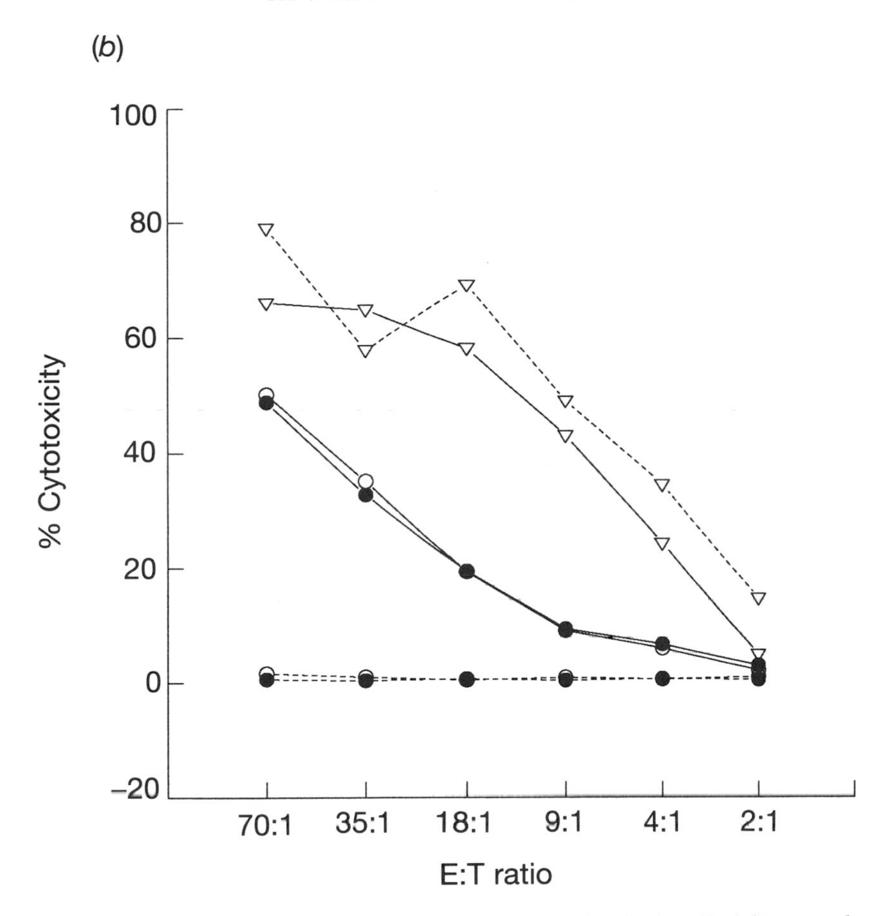


Fig. 9.5. (*b*) NK and LAK activity was assessed as previously described (Burt *et al.*, 1991). LAK (solid lines) and NK (dotted lines) effectors versus K562 (NK target) or Daudi (LAK target) (▽—▽); Ai (○—○); or *IVD* (●—●).

### *9.3.3 Cell-mediated lysis*

The sensitivity of the keratinocyte lines to different cytotoxic effectors has been assessed because alterations in MHC expression can modulate susceptibility to effectors like NK cells (Ljunggren & Karre, 1990). The sensitivity of HPV 16- and control-transfected keratinocyte lines to NK cytolysis, lymphokine-activated killer (LAK) cytolysis and interferon-activated macrophage killing has been assessed in a short-term $^{51}$Cr release assay. Both control and HPV 16-transfected keratinocytes were equally sensitive to activated NK and LAK killing and there were no differences in the low-level killing by activated macrophages from peripheral blood (Fig. 9.5). It appears that there is no parallel with HPV 16-transformed 3T3 lines, which have been shown to be susceptible to killing by activated macrophages (Denis, Chadee & Matlashewski, 1989), a sensitivity that is apparently associated with the expression of HPV 16 E7 (Banks *et al.*, 1991).

## 9.4 Conclusions

HPV 16-transfected SVD2 keratinocytes showed no major down-regulation of HLA class I expression. These cells also showed no change in glycosylation of their HLA class I heavy chains, which resulted in continued endo H sensitivity, a mechanism used by other viruses to prevent the presentation of viral peptides by class I antigens in infected cells (Andersson *et al.*, 1985; Cox *et al.*, 1991; Del Val *et al.*, 1992; see Chapters 8 and 12). Banks *et al.* (1991) have stated that expression of HPV 16 E7 in mouse 3T3 cells did not down-regulate MHC class I transcription, which is another strategy used by certain oncogenic adenoviruses to repress MHC class I expression in transformed cells (Schrier *et al.*, 1983; see Chapter 8). These observations are consistent with a lack of correlation of HPV 16 DNA in tumour biopsies and immunohistological detection of MHC class I down-regulation in cervical squamous carcinomas (Connor & Stern, 1990; Cromme *et al.*, 1993). The latter study showed, using RNA *in situ* hybridization, that neoplastic cells containing actively transcribing HPV 16 E7 exhibit both normal and aberrant MHC class I expression. It, therefore, seems unlikely that HPV 16 E7 plays a significant role in the modulation of MHC class I expression in cervical carcinoma.

It is possible that the acquisition of the HLA class I-negative phenotype by the tumours is a relatively late occurrence in the multistep evolution of cervical cancer. Studies of class I expression in pre-malignant CIN lesions have shown a very low frequency of complete down-regulation (Glew *et al.*, 1993). It should be noted that the biological behaviour of CIN lesions is very heterogeneous and cannot be predicted by a single histological observation. In general, it is assumed that only a minority of CIN lesions show progression to invasive cancer (Richart, 1987). Thus only a small proportion of CIN lesions examined will represent progressive CIN. It will be important to examine the relationship between HPV infection and MHC expression in patients exhibiting regressive, progressive or persistent CIN disease (Cromme *et al.*, 1993).

The HPV 16-transfected SVD2 keratinocytes did show modulation of the interferon inducibility of serologically detectable cell surface expression of

HLA-DR. This appeared to act post-transcriptionally but needs to be confirmed using additional clones of HPV 16-immortalized keratinocytes. The SVD2 keratinocyte line is perhaps unusual in that class II expression is not readily induced by IFN-γ treatment; this may be a reflection of its foetal origin. The IFN-mediated HLA class II inducibility of the target cells infected by HPV in the transformation zone of the cervix is also not known. In contrast to normal cervical squamous epithelium, over 80% of squamous carcinomas of the cervix express HLA class II antigens (Glew *et al.*, 1992). There is no apparent correlation between the HLA class II phenotype and the presence of HPV 16 DNA in the specimens. However, a relationship between MHC class II expression and HPV infection might be evident in the evolution of pre-malignant disease.

An immunohistological study of 104 colposcopic biopsies (Glew *et al.*, 1993) has established that HLA class II expression occurred in a significant proportion of squamous epithelia showing histological evidence of wart virus infection and CIN I to III. Similarly, vaginal, vulval and peri-anal intraepithelial neoplasias are frequently HLA class II positive. However, there was no correlation between the detection of high-risk HPV DNA (types 16, 18, 31 and 33) by PCR and the MHC class II phenotype of the lesion. This suggests that altered HLA class II expression is neither a consequence nor a prerequisite for HPV infection. Whether class II expression confers additional antigen-presenting capacity to HPV-infected, dyskaryotic and malignant cells remains to be determined. The up-regulation of HLA class II expression in cervical lesions may have immunosuppressive consequences. For example, human keratinocytes derived from the cervix have been shown to induce tolerance in HLA-DR restricted T cells (Bal *et al.*, 1990).

It has been shown that cytotoxic T cells can recognize HPV 16 E6 and E7 in mice (Chen *et al.*, 1991; 1992). The MHC restriction of T cell recognition of antigen functions at the molecular level by selection of short peptide sequences for binding to HLA class I and this is influenced by the polymorphic sites (Bjorkman *et al.*, 1987; Yu Rudensky *et al.*, 1991). In cervical lesions with HLA class I expression, specific HLA allelic products might present HPV-derived peptides to the immune system providing the peptide processing and transport pathway is functional in HPV-infected cells. The loss of MHC class I expression in a significant proportion of squamous carcinomas of the cervix may be the result of selection to evade specific immunosurveillance. It will be important to establish the nature and origin of peptides presented by MHC class I antigens in the keratinocytes that are infected by HPV.

Using biochemical techniques (Rötzschke *et al.*, 1990), the MHC-peptide profiles of HPV 16-transfected versus the control keratinocyte cell lines have been compared. In preliminary experiments, peptides eluted from purified MHC complexes have been fractionated by reverse-phase HPLC and several unique peptide peaks identified. When control clones Ai and Aii profiles were compared, they were identical and the differences between the HPV 16-transfected and control peptide profiles were reproducible. The sensitivity of the detection system indicates that the quantity of peptides recovered is consistent with that expected (Christinck *et al.*, 1991) and $10^9$ cells yields sufficient for sequencing analysis. The several different HPLC fractionated peaks specifically identified from HPV 16-transfected cells are being further resolved and sequenced and will ultimately be assigned to the known HLA alleles of the SVD2 line. It will be of central importance to establish whether keratinocytes can present HPV-derived peptides in MHC class I complexes. It is important to realize that even if this were shown to be the case in this model system, it will still be necessary to discover what peptides, if any, are actually presented by cervical carcinomas.

The elegant work of Boon and colleagues has established that CTLs can be directed against peptides derived from normal cell products that carry point mutations (Lurquin *et al.*, 1989), are differentially glycosylated (Szikora *et al.*, 1990) or are simply over-expressed (van den Eynde *et al.*, 1991; van der Bruggen *et al.*, 1991). Since high-risk type HPV infection can clearly influence normal cellular protein expression, for example p53 and *Rb*, then the target peptides for specific CTL in HPV-associated cervical lesions might be derived from normal keratinocyte proteins. Finally, the role of other effector cell types, like NK and activated macrophages, in the control of cervical malignancy must also be considered. The results presented here show no increased susceptibility to IFN-activated macrophages by the transfected keratinocytes, as is the case with rodent cells expressing HPV 16 E7 (Denis *et al.*, 1989; Banks *et al.*, 1991). However, the assay system, the source of lymphocytes (PBL versus bone marrow) and precise activation conditions were different. In this study, the keratinocyte cell lines were very sensitive to both activated NK and LAK cells. The demonstration that exogenous MHC class I specific peptides can abrogate the class I/self-peptide inhibition of NK-mediated killing (Ljunggren & Karre, 1990; Liao *et al.*, 1991; Chadwick *et al.*, 1992) suggests that the immune system has two complementary strategies for exploiting the intracellular binding and transport of class I antigens: one leads to display of specific 'foreign' determinants and the other to the concomitant loss of self determinants. It is clear that surface expression of either foreign or self peptides by HLA-A, HLA-B or HLA-C antigens by keratinocytes may influence the immune surveillance of HPV-associated cervical lesions.

## Acknowledgements

This work was supported by the Cancer Research Campaign. The assistance of D. Burt and I. Illingworth is gratefully acknowledged.

## References

Andersson, M., Paabo, S., Nilsson, T. & Peterson, P.A. (1985). Impaired intracellular transport of class I MHC antigens as a possible means for adenoviruses to evade immune surveillance. *Cell*, 43, 215–222.

Bal, V., McIndoe, A., Denton, G., Hudson, D., Lombardi, G., Lamb, J. & Lechler, R. (1990). Antigen presentation by keratinocytes induces tolerance in human T cells. *European Journal of Immunology*, 20, 1893–1897.

Banks, L., Moreau, F., Vousden, K., Pim, D. & Matlashewski, G. (1991). Expression of the human papillomavirus E7 oncogene during cell transformation is sufficient to induce susceptibility to lysis by activated macrophages. *Journal of Immunology*, 146, 2037–2042.

Bjorkman, P.J., Saper, M.A., Samraoui, B., Bennet, W.S., Strominger, J.L. & Wiley, D.C. (1987). The foreign antigen binding site and T cell recognition regions of class I histocompatibility antigens. *Nature (London)*, 329, 506–512.

Buckley, C.H. & Fox, H. (1989). Carcinoma of the cervix. In *Recent Advances in Histopathology* (No. 14), ed. P.P Anthony & R.M.N. Macsween, pp. 63–78. London: Churchill Livingstone.

Burnett, T.S. & Gallimore, P.H. (1983). Introduction of cloned human papillomavirus la DNA into rat fibroblasts: integration, *de novo* methylation and absence of cellular morphological transformation. *Journal of General Virology*, 64, 1509–1520.

Burt, D., Johnston, D., Rinke de Wit, T.F., van den Elsen, P. & Stern, P.L. (1991). Cellular immune recognition of HLA-G expressing choriocarcinoma cell line Jeg-3. *International Journal of Cancer*, Suppl. 6, 117–122.

Chadwick, B.S., Sambhara, S.R., Sasakura, Y. & Miller, R.G. (1992). Effects of class I MHC binding peptide on recognition by natural killer cells. *Journal of Immunology*, 149, 3150–3156.

Chen, L.P., Mizuno, M.T., Singhal, M.C., Hu, S.-L., Galloway, D.A., Hellstrom, I. & Hellstrom, K.E. (1992). Induction of cytotoxic lymphocytes specific for a syngeneic tumour expressing the E6 oncoprotein of human papillomavirus type 16. *Journal of Immunology*, 148, 2617–2621.

Chen, L.P., Thomas, E.K., Hu, S.L., Hellstrom, I. & Hellstrom, K.E. (1991). Human papillomavirus type 16 nucleoprotein E7 is a tumour rejection antigen. *Proceedings of the National Academy of Sciences of the USA* 88, 110–114.

Christinck, E.R., Luscher, M.A., Barber, B.H. & Williams, D.B. (1991). Peptide binding to class I MHC on living cells and quantitation of complexes required for CTL lysis. *Nature (London)*, 352, 67–70.

Connor, M.E., Davidson, S.E., Stern, P.L., Arrand, J.R. & West, C.M.L. (1993).

Evaluation of multiple biologic parameters in cervical carcinoma: high macrophage infiltration in HPV-associated tumours. *International Journal of Gynecological Cancer*, 3, 103–109.

Connor, M.E. & Stern, P.L. (1990). Loss of MHC class I expression in cervical carcinomas. *International Journal of Cancer*, 46, 1029–1034.

Coppelston, M. & Reid, B.L. (1967). *Preclinical Carcinoma of the Cervix Uteri*, pp. 321. London: Pergamon Press.

Cox, J.H., Bennink, J.R. & Yewdell, J.W. (1991). Retention of adenovirus E19 glycoprotein in the endoplasmic reticulum is essential to its ability to block antigen presentation. *Journal of Experimental Medicine*, 174, 1629–1637.

Cromme, F.V., Meijer, C.J.L.M., Snijders, P.J.F., Uijterline, A., Kenemans, P., Helmerhorst, T., Stern, P.L., van den Brule, A.J.C. & Walboomers, J.M.M. (1993). Analysis of MHC class I and II expression in relation to presence of HPV genotypes in premalignant and malignant cervical lesions. *British Journal of Cancer*, 67, 1372–1380.

Denis, M., Chadee, K. & Matlashewski, G.J. (1989). Macrophage killing of human papillomavirus type 16-transformed cells. *Virology*, 170, 342–345.

Del Val, M., Hengel, H., Hacker, H., Hartlaub, U., Ruppert, T., Lucin, P. & Koszinowski, U.H. (1992). Cytomegalovirus prevents antigen presentation by blocking the transport of peptide-loaded MHC class I molecules into the medial-Golgi compartment. *Journal of Experimental Medicine*, 176, 729–738.

de Villiers, E.M. (1989). Heterogeneity of the human papillomavirus group. *Journal of Virology*, 63, 4895–4903.

DiMaio, D. (1991). Transforming activity of bovine and human papillomaviruses in cultured cells. *Advances in Cancer Research*, 56, 133–159.

Dyson, N., Howley, P.M., Munger, K. & Harlow, E. (1989). The human papillomavirus 16 E7 oncoprotein is able to bind to the retinoblastoma gene product. *Science*, 243, 934–936.

Esteban, F., Ruiz-Cabello, F., Conchas, A., Perez-Ayala, M., Sanchez-Rozas, J.A. & Garrido, F. (1990). HLA-DR expression in association with excellent prognosis in squamous cell carcinoma of the larynx. *Clinical and Experimental Metastases*, 8, 4319–4328.

Gillet, A.C., Perarnan, B., Mercier, P. & Lemonnier, F.A. (1990). Serological analysis of the dissociation process of HLA-B and C class I molecules. *European Journal of Immunology*, 20, 759–764.

Ghosh, A.K., Smith, N.K., Stacey, S.N., Glew, S.S., Connor, M.E., Arrand, J.R. & Stern, P.L. (1993). Serological responses to HPV16 in cervical dysplasia and neoplasia: correlation of antibodies to E6 with cervical cancer. *International Journal of Cancer*, 53, 1–6.

Glew, S.S., Connor, M.E., Snijders, P.J.F., Stanbridge, C.M., Buckley, C.H., Walboomers, J.M.M., Meijer, C.J.L.M. & Stern, P.L. (1993). HLA expression in preinvasive cervical neoplasia in relationship to human papillomavirus infection. *European Journal of Cancer*, 29A, 1963–1970.

Glew, S.S., Duggan-Keen, M., Cabrera, T. & Stern, P.L. (1992). HLA class II antigen expression in HPV associated cervical cancer. *Cancer Research*, 52, 4009–4016.

Halpert, R., Fruchter, R.G., Seclus, A., Butt, K., Boyce, J.G. & Sillman, F.H. (1985). HPV and lower genital tract neoplasia in renal transplant patients. *Obstetrics and Gynecology*, 68, 251–258.

Kinlen, L.J., Sheil, A.G.R., Peto, J. & Doll, R. (1979). Collaborative UK-Australian study of cancer in patients treated with immunosuppressive drugs. *British Medical Journal*, 2, 1461–1466.

Koss, L.G. (1987). Cytologic and histologic manifestations of human papillomavirus infection of the female genital tract and their clinical significance. *Cancer*, 60, 1942–1950.

Levine, A.J., Momand, J. & Finlay, C.A. (1991). The p53 tumour suppressor gene. *Nature (London)*, 351, 453–456.

Liao, N.S., Bix, M., Zilstra, R., Jaenisch, R. & Raulet, D. (1991). MHC class I deficiency: susceptibility to natural killer (NK) cells and impaired NK activity. *Science*, 253, 199–202.

Ljunggren, H.G. & Karre, K. (1990). In search of the 'missing self': MHC molecules and NK recognition. *Immunology Today*, 11, 237–244.

Lurquin, C., van Pel, A., Mariame, B., de Plaen, E., Szikora, J.P., Janssens, C., Reddehase, M.J., Lejeune, J. & Boon, T. (1989). Structure of the gene of tum$^-$ transplantation antigen P91A: the mutated exon encodes a peptide recognized with $L^d$ by cytolytic T cells. *Cell*, 58, 293–303.

Matlashewski, G. (1989). The cell biology of human papillomaviruses. *AntiCancer Research*, 9, 1447–1556.

Monaco, J.J. (1992). A molecular model of MHC class I-restricting antigen processing. *Immunology Today*, 13, 173–179.

Munoz, N., Bosch, F.X., Shah, K.V. & Meheus, A. (1992). *The Epidemiology of Human Papillomavirus and Cervical Cancer*. Lyon: IARC Scientific Publications, No. 119.

Neefjes, J.J. & Ploegh, H.L. (1988). Allele and locus specific differences in cell surface expression and the association of HLA class I heavy chain with $\beta_2$-microglobulin: differential effects of inhibition of glycosylation on class I subunit association. *European Journal of Immunology*, 18, 801–810.

Neefjes, J.J. & Ploegh, H.L. (1992). Intracellular transport of MHC class II molecules. *Immunology Today*, 13, 179–183.

Parkin, D.M., Laara, E. & Muir, C.S. (1980). Estimates of the world frequency of sixteen major cancers in 1980. *International Journal of Cancer*, 41, 184–197.

Richart, R.M. (1987). Causes and management of cervical intraepithelial neoplasia. *Cancer*, 60, 1951–1959.

Rötzschke, O., Falk, K., Deres, K., Schild, H., Norda, M., Metzger, J., Jung, G. & Rammensee, H.-G. (1990). Isolation and analysis of naturally processed viral peptides as recognized by cytotoxic T cells. *Nature (London)*, 348, 252–254.

Ruiter, D.J., Brocker, E.B. & Ferrone, S. (1986). Expression and susceptibility to modulation by interferons of HLA class I and II antigens on melanoma cells. Immunohistochemical analysis and clinical relevance. *Journal of Immunogenetics*, 13, 229–234.

Schneider, V., Kay, S. & Lee, H.M. (1983). Immunosuppression as a high-risk factor in the development of condyloma acuminata and squamous neoplasia of the cervix. *Acta Cytologica*, 27, 220–224.

Schrier, P.I., Bernards, R., Vaessen, R.T.M.J., Houweling, A. & van der Eb, A.J. (1983). Expression of class I major histocompatibility antigens switched off by highly oncogenic adenovirus 12 in transformed rat cells. *Nature (London)*, 305, 771–775.

Sousa, R., Dostatni, N. & Yaniv, M. (1990). Control of papillomavirus gene expression. *Biochimica et Biophysica Acta*, 1032, 19–37.

Stacey, S.N., Bartholomew, J.S., Ghosh, A.K., Stern, P.L., Mackett, M. & Arrand, J.R. (1992). Expression of human papillomavirus type 16 E6 protein by recombinant baculovirus and use for detection of anti-E6 antibodies in human sera. *Journal of General Virology*, 73, 2337–2345.

Stacey, S.N., Ghosh, A.K., Bartholomew, J.S., Tindle, R.W., Stern, P.L., Mackett,

M. & Arrand, J.R. (1993). Expression of human papillomavirus type 16 E7 protein by recombinant baculovirus and use for the detection of E7 antibodies in sera from cervical carcinoma patients. *Journal of Medical Virology*, 40, 14–21.

Szikora, J.-P., van Pel, A., Brichard, V., Andre, M., van Baren, N., Henry, P., de Plaen, E. & Boon, T. (1990). Structure of the gene of tum- transplantation antigen P35B: presence of a point mutation in the antigenic allele. *EMBO Journal*, 9, 1041–1050.

van der Bruggen, P., Traversi, C., Chomez, P., Lurquin, C., de Plaen, E., van den Eynde, B., Knuth, A. & Boon, T. (1991). A gene encoding an antigen recognized by cytolytic T lymphocytes on a human melanoma. *Science*, 254, 1643–1647.

van den Eynde, B., Lethe, B., van Pel, A., de Plaen, E. & Boon, T. (1991). The gene coding for a major tumor rejection antigen of tumor P815 is identical to the normal gene of syngeneic DBA/2 mice. *Journal of Experimental Medicine*, 173, 1373–1384.

Young, L.S., Bevan, I.S., Johnson, M.A., Blomfield, P.I., Bromidge, T., Maitland, N.J. & Woodman, C.B.J. (1989). The polymerase chain reaction: a new epidemiological tool for investigating cervical human papillomavirus infection. *British Medical Journal*, 298, 14–18.

Yu Rudensky, A., Preston-Hurlburt, P., Hong, S.C., Barlow, A. & Janeway, C.A. (1991). Sequence analysis of peptides bound to MHC class II molecules. *Nature (London)*, 353, 622–627.

zur Hausen, H. (1989). Papillomaviruses in anogenital cancer as a model to understand the role of viruses in human cancer. *Cancer Research*, 49, 4677–4681.

# 10

# Inhibition of the cellular response to interferon by hepatitis B virus polymerase

GRAHAM R. FOSTER
*St Mary's Hospital, London*

## 10.1 Introduction

Infection with hepatitis B virus (HBV) can cause either an acute or a chronic hepatitis (Nielsen *et al.*, 1971). Chronic infection is a major health problem affecting over 250 million people worldwide (Ganem & Varmus, 1987). Without treatment, chronic infection with HBV progresses to cirrhosis and/or hepatocellular carcinoma in over 50% of infected patients. It is still not clear why some infected individuals develop a relatively mild, self-limiting hepatitis whilst others suffer from a prolonged, chronic infection. Recent research suggests that interferon-induced expression of MHC antigens on the surface of infected hepatocytes may play a key role in determining the outcome of infection with HBV.

## 10.2 Viral clearance in acute HBV infection

The majority of healthy adults infected with HBV develops a brief hepatitis, which resolves within a few months and is followed by elimination of the virus. A small percentage of infected adults (less than 5%) and the majority of infected neonates and children (greater than 90%) develop a chronic infection. Research in patients and chimpanzees acutely infected with HBV has provided an insight into the mechanisms underlying the normal eradication of HBV and these studies suggest that induction of MHC expression, by type I interferon (i.e. IFN-α/β), is a key factor in virus eradication.

Following infection with HBV, most adults show a marked increase in the serum concentration of circulating IFN (Ikeda, Lever & Thomas, 1986; Kato, Nakagawa & Kobayashi, 1986; Pignatelli *et al.*, 1986). This enhanced IFN production leads to an increase in the expression of MHC class I antigens on the surface of hepatocytes. Resting hepatocytes express very low levels of MHC antigens (Pignatelli *et al.*, 1986), but the IFN surge that accompanies an acute HBV infection leads to an increase in the expression of these proteins. The increase in hepatic expression of MHC antigens is associated with an inflammatory infiltrate in the liver (Mondelli *et al.*, 1982) that contains $CD8^{+}$ T cells. During this period, CTLs directed against HBV antigens can be detected in the peripheral blood (Penna *et al.*, 1971), suggesting that an immunological response against HBV proteins has developed. The increase in hepatic MHC expression and the development of CTLs against HBV proteins is associated with an increase in the severity of the hepatitis, i.e. there is an increase in the rate of destruction of hepatocytes. Presumably, infected liver cells that express MHC antigens are able to present viral antigens to circulating lymphocytes which are able to recognize and lyse these cells. This immunologically mediated destruction of infected hepatocytes leads to elimination of infected cells and ultimately to eradication of the virus. Hence, in an acute HBV infection, viral clearance depends upon IFN induction of MHC antigens and the development of an appropriate immunological response.

## 10.3 Viral persistence in chronic HBV infection

A small percentage of adults infected with HBV does not develop the IFN surge that accompanies an acute infection (Ikeda *et al.*, 1986; Kato *et al.*, 1986). The mechanism underlying this failure to produce IFN is not yet clear, but there is evidence to show that one of the viral proteins, namely the HBV core protein, can inhibit the production of IFN (Twu & Schloemer, 1989). In patients who do not produce IFN, there is no activation of IFN-inducible genes, including MHC class I antigens, and hence infected hepatocytes expressing low, unstimulated levels of MHC antigens are unable to present viral antigens to the immune system and, therefore, there is no effective anti-viral response.

## 10.4 Interferon treatment in chronic HBV infections

Attempts to overcome the relative IFN deficiency in patients with chronic HBV infection have been made by treating infected subjects with exogenous IFN. A large number of clinical trials have shown that IFN therapy eliminates HBV in about 50% of patients (Jacyna & Thomas, 1990). In those patients who do respond to IFN, serological and histological studies suggest that the mechanism of viral eradication is similar to that involved in viral clearance during an acute infection with HBV: in responder patients there is a marked increase in the serum concentration of IFN and viral clearance is preceded by an increase in MHC antigen expression on the surface of liver cells (Pignatelli *et al.*, 1986). This increase in hepatocyte MHC expression is associated with an increase in the severity of the hepatitis, as is observed in patients with acute infections, and presumably IFN stimulates an immune response that eliminates infected hepatocytes. However, CTLs directed against HBV antigens have not been detected in patients who respond to IFN, suggesting that either they are formed in relatively small numbers or that other mechanisms operate during viral clearance in IFN-treated patients.

## 10.5 Non-response to interferon: viral inhibition of the cellular response to interferon

Although many patients respond to therapy with exogenous IFN, a significant proportion fails to do so. A number of studies have shown that non-responder patients fail to show the expected increase in IFN-inducible protein expression after IFN administration (Pignatelli *et al.*, 1986), indicating that these patients do not respond to IFN (Thomas, Pignatelli & Lever, 1986). In human cell lines transfected with multimers of the HBV genome, there is a reduced response to IFN, as assessed by production of the IFN-inducible protein $\beta_2$-m (Onji *et al.*, 1990), suggesting that HBV itself may be able to inhibit the response to IFN.

To examine this phenomenon in more detail, we have studied the effects of the individual HBV genes on the cellular response to IFN (Foster *et al.*, 1991; 1993). HBV contains four ORFs encoding two structural proteins (core and surface proteins) and two non-structural proteins, the *trans*-activating protein (protein X) and the viral polymerase protein (POL). Each of these ORFs was cloned into a eukaryotic expression vector and transfected into HeLa cells along with an IFN-inducible reporter construct containing the human 6–16 gene promoter linked to bacterial CAT. Transfected cells were treated with IFN for 48 hours and CAT activity was then assayed. The results of these experiments are summarized in Fig. 10.1. Co-transfection with constructs expressing HBV core, surface and X proteins did not substantially affect the cellular response to IFN, but transfection with a construct that expressed HBV POL greatly reduced the response. This reduction was not seen with constructs containing the POL gene in the reverse orientation (rPOL), suggesting that production of POL is required for inhibition of the cellular response to IFN. To analyse the specificity of this inhibitory effect further, cells were co-transfected with the POL or rPOL constructs and a construct in which the CAT gene was regulated by the SV40 early promoter.

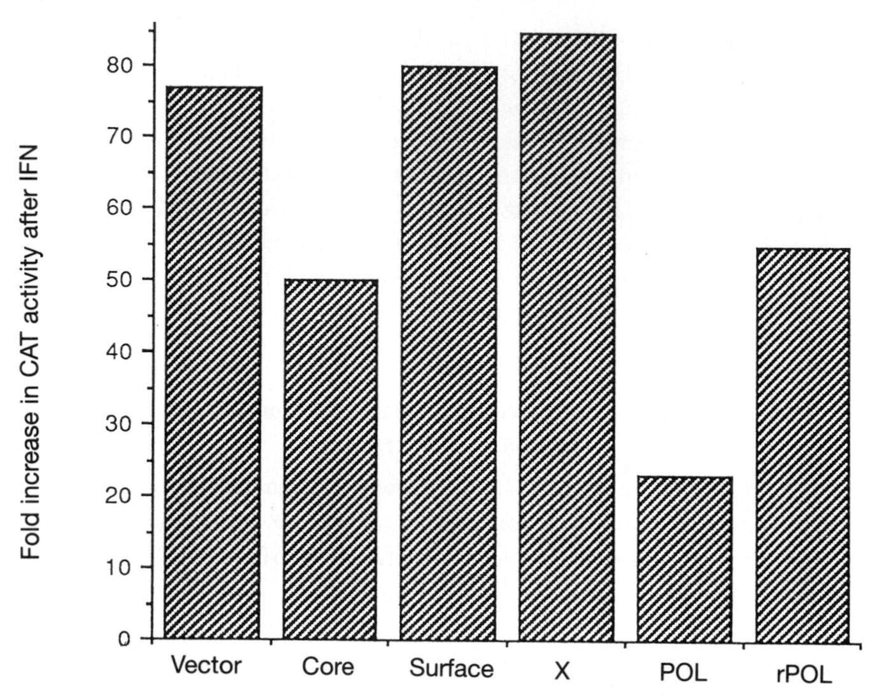


Fig. 10.1. Effects of individual HBV ORFs on expression of an interferon-inducible CAT construct. (rPOL is POL in the reverse orientation.) Cells were transfected with an HBV construct and a reporter construct in the molar ratio 15:1. Results are expressed as the fold increase in CAT activity after IFN treatment for 48 hours. Each bar represents the mean of four separate experiments.

Co-transfection with POL did not reduce the activity of this reporter construct, showing that the inhibitory effect of POL was specific for IFN-inducible promoters (results not shown).

To confirm the results of these experiments in stable cell lines, the cell line 2fTGH (Pellegrini *et al.*, 1989) was used. This cell line contains the bacterial *gpt* gene under the control of the IFN-inducible 6–16 promoter. If these cells are treated with IFN, then *gpt* is expressed and if cells expressing *gpt* are grown in the presence of 6-thioguanine, the 6-thioguanine is converted into a toxic metabolite and the cells die. Hence 2fTGH cells contain an IFN-inducible toxin and can be used to select out clones of cells that do not respond to IFN. Since the frequency of spontaneous IFN resistance in 2fTGH cells is very low ($< 1 \times 10^8$) these cells can be used to identify factors that inhibit the response to IFN. Any factor that inhibits the cellular response to IFN should increase the frequency of IFN resistance in 2fTGH cells and clones of cells that express the inhibitor should be obtained after selection in IFN and 6-thioguanine.

Table 10.1. *Frequency of survival in 6-thioguanine plus IFN of stably transfected 2fTGH cells*

| Test construct | Transfected clones | Transfectants surviving in 6-thioguanine plus IFN |
|---|---|---|
| Core | 330 | 1[a] |
| Surface | 180 | 0 |
| X | 300 | 0 |
| rPOL | 880 | 1[a] |
| POL | 1060 | 18 |

Cells were transfected with a construct expressing neomycin and an HBV protein-expressing construct. Transfectants were selected in G418 and then selected in IFN plus 6-thioguanine.
[a] Subsequent analysis showed that these clones responded normally to IFN, suggesting that their survival in the selective medium was the result of loss of the transfected *gpt* construct, as has been previously described (Pellegrini *et al.*, 1989).
Six of the eighteen POL transfectants that survived the selection were tested for their response to IFN and in three the response was reduced.

2fTGH cells were transfected with a variety of constructs containing various HBV genes and the frequency of IFN-resistant transfectants determined, by growth in 6-thioguanine plus IFN. The results of these experiments are summarized in Table 10.1. Cell lines that did not respond to IFN were only isolated from clones transfected with the HBV POL gene. These cells expressed HBV POL protein (as assessed by direct staining of the cells with an antibody directed against the terminal protein domain of HBV POL) and had a reduced response to IFN (Fig. 10.2). Cell lines that expressed HBV POL grew extremely slowly (doubling time for expressing cells was 5 days, compared with 24 hours for parental cells) and when the selection for IFN resistance (6-thioguanine and IFN) was withdrawn, expression of POL was rapidly lost. Cells that no longer expressed HBV POL had a normal response to IFN. Hence expression of HBV POL appears to be toxic to cells, but cells expressing this protein can be obtained by selection for failure to respond to IFN. If the selection is withdrawn, the cells readily lose POL expression. It is unlikely that the reduction in the cellular response to IFN in cells expressing POL is the result of general toxicity of POL, since the levels of mRNAs for constitutively expressed genes (such as γ-actin and glyceraldehyde-3-phosphate dehydrogenase) were unimpaired in cells expressing POL (Fig. 10.2).

To determine the clinical significance of these effects, we examined the effects of HBV POL in liver biopsies taken from patients with chronic HBV infections (Foster *et al.*, 1993). In liver biopsies taken from patients who

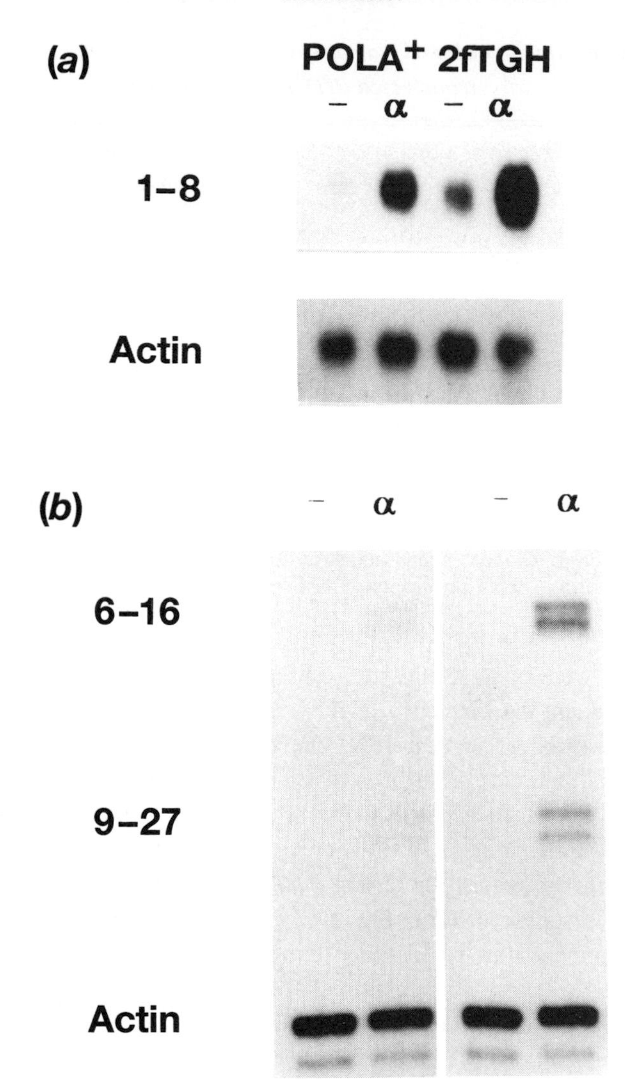


Fig. 10.2. Response to IFN of a cell line that expresses the HBV POL protein. (*a*) Northern blot analysis and (*b*) RNase protection assay using probes for the IFN-inducible genes 1–8, 6–16 and 9–27 in cells that express HBV POL ($POLA^+$) and parental cells (2fTGH). Cells were untreated (–) or treated with IFN-α (500 IU/ml) (α) for 6 hours. Similar results were observed with a number of other IFN-inducible genes.

were not receiving IFN therapy, the IFN-inducible protein $\beta_2$-m could not be detected on the surface of hepatocytes (using either monoclonal or polyclonal antibodies directed against this protein). However, in patients who were receiving IFN therapy, some 40% of the hepatocytes expressed detectable amounts of $\beta_2$-m. Hence, IFN induction of $\beta_2$-m can be used to determine whether a particular hepatocyte has responded to IFN. Biopsies from five patients with chronic HBV infections, who were receiving IFN treatment, were doubly stained with antibodies against $\beta_2$-m and with an antibody directed against the terminal protein domain of the HBV POL protein. A median of 44% of uninfected cells responded to IFN, i.e. expressed $\beta_2$-m, whereas only 8% of cells expressed HBV POL also expressed $\beta_2$-m. The statistically significant decrease in the number of cells co-expressing POL and $\beta_2$-m suggests that, in patients, expression of HBV POL is associated with a failure of hepatocytes to respond to IFN, suggesting that it may inhibit the cells' ability to respond to IFN during chronic HBV infection.

If inhibition of the cellular response to IFN by HBV POL is a major determinant of the outcome of IFN therapy, then one would predict that subjects who do not respond to IFN should express more POL than those who do respond. This hypothesis has recently been tested (Foster *et al.*, 1993). Pretreatment liver biopsies from 28 patients with chronic HBV infections were stained with antibodies against HBV POL and against the HBV core protein. In liver biopsies taken from the 15 patients who subsequently did respond to treatment, there was a significant decrease in the number of cells expressing HBV POL when compared with the 13 patients who did not respond to treatment. There was no significant difference in the expression of HBV core protein in the two groups of patients (Fig. 10.3). Hence overexpression of HBV POL is associated with a failure to respond to IFN therapy.

## 10.6 Conclusions

Eradication of HBV depends upon induction of MHC antigens by IFN and the development of an immune response directed against HBV proteins displayed on the surface of infected hepatocytes. In patients who are acutely infected with HBV, endogenous IFN production is sufficient to enhance hepatocyte MHC antigen expression and activate the immune system. In chronically infected patients, endogenous IFN is inadequate and induction of MHC proteins does not occur. In patients who are chronically infected with HBV, treatment with exogenous IFN can induce an appropriate immune response in some patients, but many subjects do not respond. In patients who do not respond to therapy, overexpression of the HBV POL protein may act as a

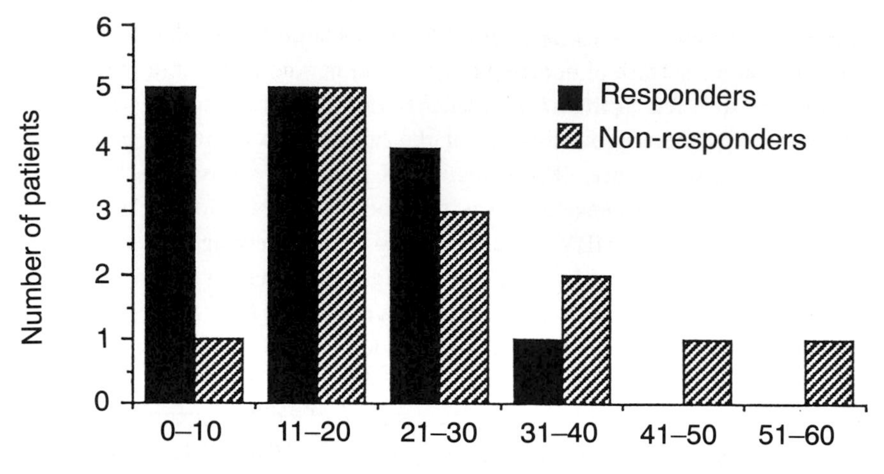


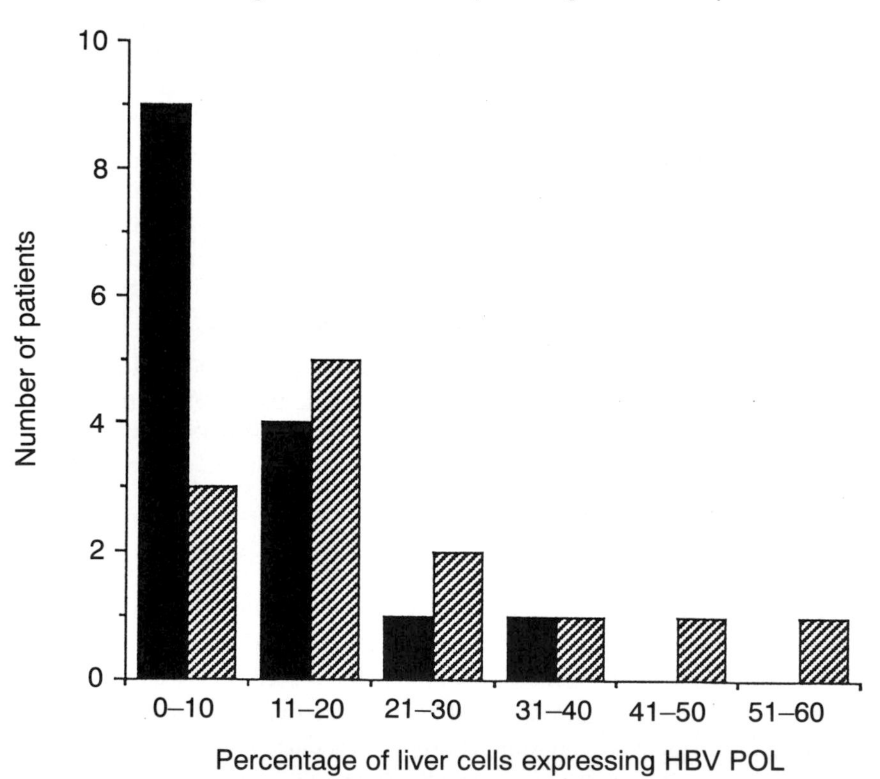


Fig. 10.3. Expression of HBV POL and core proteins in pre-treatment liver biopsies from patients who did or did not respond to IFN therapy. Modified from Foster *et al.* (1993).

viral inhibitor of the cellular response to IFN and thus prevent presentation of viral proteins to the immune system. Further studies will be required to determine how HBV POL inhibits the response to IFN and to determine why some patients overexpress this anti-interferon protein.

## References

Foster, G.R., Ackrill, A.M., Goldin, R.D., Kerr, I.M., Thomas, H.C. & Stark, G.R. (1991). Expression of the terminal protein region of hepatitis B virus inhibits cellular responses to interferons α and γ and double-stranded RNA. *Proceedings of the National Academy of Sciences of the USA*, 88, 2888–2892.

Foster, G.R., Goldin, R.D., Hay, A., McGarvey, M.J., Stark, G.R. & Thomas, H.C. (1993). Expression of the terminal protein of hepatitis B virus is associated with a failure to respond to interferon therapy. *Hepatology*, 17, 757–762.

Ganem, D. & Varmus, H.E. (1987). The molecular biology of the hepatitis B viruses. *Annual Review of Biochemistry*, 56, 651–693.

Ikeda, T., Lever, A.M.L. & Thomas, H.C. (1986). Evidence for a deficiency of interferon production in patients with chronic hepatitis B virus infection acquired in adult life. *Hepatology*, 6, 962–965.

Jacyna, M. & Thomas, H.C. (1990). Antiviral therapy: hepatitis B. *British Medical Bulletin*, 46, 368–382.

Kato, Y., Nakagawa, H. & Kobayashi, K. (1986). Interferon production by peripheral lymphocytes in HBsAg positive liver disease. *Hepatology*, 6, 645–647.

Mondelli, M.U., Vergani, G.M., Alberti, A., Vergani, D., Portmann, B., Eddleston, A.L. & Williams, R. (1982). Specificity of T lymphocyte cytotoxicity to autologous hepatocytes in chronic hepatitis B virus infection: evidence that T cells are directed against HBV core antigen expressed on hepatocytes. *Journal of Immunology*, 129, 2773–2780.

Nielsen, J.O., Dietrichson, O., Elling, P. & Christofferson, P. (1971). Incidence and meaning of persistence of Australia antigen in patients with acute viral hepatitis: development of chronic hepatitis. *New England Journal of Medicine*, 285, 1157–1160.

Onji, M., Lever, A.M.L., Saito, I. & Thomas, H.C. (1990). Defective response to interferons in cells transfected with the hepatitis B virus genome. *Hepatology*, 9, 92–96.

Pellegrini, S., John, J., Shearer, M., Kerr, I.M. & Stark, G.R. (1989). Use of a selectable marker regulated by alpha interferon to obtain mutations in the signalling pathway. *Molecular and Cellular Biology*, 9, 4605–4612.

Penna, A., Chisari, F.V., Bertoletti, A., Missale, G., Fowler, P., Giuberti, T., Fiaccadori, F. & Ferrari, C. (1991). Cytotoxic T lymphocytes recognize an HLA-A2 restricted epitope within the hepatitis B virus nucleocapsid antigen. *Journal of Experimental Medicine*, 174, 1565–1570.

Pignatelli, M., Waters, J., Brown, D., Lever, A.M.L., Iwarson, S., Schaff, Z., Gerety, R. & Thomas, H.C. (1986). HLA class I antigens on the hepatocyte membrane during recovery from acute hepatitis B virus infection and during interferon therapy in chronic hepatitis B virus infection. *Hepatology*, 6, 349–353.

Thomas, H.C., Pignatelli, M. & Lever, A.M.L. (1986). Homology between HBV

DNA and a sequence regulating the interferon-induced anti-viral response: possible mechanism of persistent infection. *Journal of Medical Virology*, 19, 63–70.

Twu, R.S. & Schloemer, R.H. (1989). The transcription of the human β-interferon gene is inhibited by hepatitis B virus. *Journal of Virology*, 63, 3065–307.

# 11

# Cellular adhesion molecules and MHC antigens in cells infected with Epstein–Barr virus: implications for immune recognition

MARTIN ROWE[1] and MARIA G. MASUCCI[2]
[1]University of Birmingham and [2]Karolinska Institute

## 11.1 Introduction: the CTL response to Epstein–Barr virus

Epstein–Barr virus (EBV) is a γ-herpesvirus that is carried by more than 90% of adults worldwide as an asymptomatic persistent infection. This virus is the causative agent of infectious mononucleosis, a benign lymphoproliferative disease and is also implicated in the pathogenesis of an increasing number of malignant diseases, including: Burkitt's lymphoma (BL), nasopharyngeal carcinoma and Hodgkin's disease (Miller, 1990). The importance of cellular immunity, and especially of CTLs, in controlling the potentially harmful consequences of EBV infection is underscored by the increased incidence of EBV-driven lymphoproliferations in organ transplant recipients receiving immunosuppressive therapy and in AIDS patients (Thomas, Allday & Crawford, 1991).

The pathogenic potential of EBV is well illustrated by the ease with which the virus can infect resting B lymphocytes *in vitro*, causing growth-transformation and the establishment of lymphoblastoid cell lines (LCLs) with infinite doubling capacity. This *in vitro* infection model provided the first clue as to how the virus is controlled *in vivo*. If circulating lymphocytes from EBV-seropositive donors are infected with EBV *in vitro* without first removing or inactivating the T lymphocytes, then the initial proliferation of EBV-infected B lymphocytes is followed by a regression at two to three weeks post-infection because of reactivation of EBV-specific memory T lymphocytes that eliminate the virus-infected cells (Rickinson *et al.*, 1981). Activated T cells isolated from such cultures contain CTL populations that are HLA class I-restricted and specific for EBV-encoded antigens.

Elucidation of the precise viral antigen specificity of this CTL response has been difficult because of the complexity of gene expression in LCLs. Whilst LCLs are generally not permissive for complete virus replication, the form of 'latent' infection displayed in these lines is characterized by expression of an unusually large number of virally encoded proteins, including: two latent membrane proteins, LMP1 and LMP2, and six nuclear antigens, EBNA1, EBNA2, EBNA3A, EBNA3B, EBNA3C and EBNA-LP (Kieff & Leibowitz, 1990). In principle, all of these latent gene products could provide target sequences for recognition by EBV-specific CTLs. Recent progress in demonstrating the specificity of these CTLs has resulted largely from the availability of a series of recombinant vaccinia viruses that has allowed the efficient expression of each of the latent proteins individually in autologous EBV-negative target cells, such as fibroblasts or mitogen-stimulated lymphoblasts. Using these recombinant vaccinia-infected autologous targets, the EBV-specificity has been determined for CTL populations generated from several healthy individuals representing a wide range of HLA class-I restrictions. To date, with the notable exception of EBNA1, all of the latent-infection proteins have been identified as the source of target peptides for EBV-specific CTL responses (Murray *et al.*, 1990; 1992; Gavioli *et al.*, 1992; Khanna *et al.*, 1992).

The number of EBV-specific memory CTLs present in the circulation of healthy individuals carrying the virus is persistently high, in the region of 1 in $10^3$ T lymphocytes and is much greater than the reported frequency of CTL precursors specific for other viruses in humans (Bourgault *et al.*, 1991). Therefore, infection with EBV is highly immunostimulatory and yet, paradoxically, the potent immune response never completely eliminates the virus in the healthy infected individual. It is widely accepted that the mechanism of virus persistence involves a reservoir of non-dividing, latently infected cells that is resistant to the potent virus-specific CTL response; the CTL response itself would be chronically stimulated by LCL-like EBV-infected B cells that are continuously generated from this latent reservoir.

## 11.2 Alternative EBV – host cell interactions

*In vivo*, the host cell range of EBV is not only restricted to B lymphocytes. In particular, certain epithelial cells are also important target cells for this virus (Wolf, zur Hausen & Becker, 1973; Sixbey *et al.*, 1984; Wolf, Haus & Wilmes, 1984). The current balance of evidence suggests that infection of epithelial cells usually results in virus replication and cell death, whereas the normal site of EBV persistence is in the lymphoid system where infection is largely latent and non-permissive for virus replication. It, therefore, follows that there must be a form of latent infection in B lymphocytes whereby the virus can persist and not be recognized by CTLs. The first indication of what form this particular virus–host interaction might take came from studies with the EBV-positive malignant cells of BL.

Burkitt's lymphoma is a childhood tumour that occurs at high incidence (5–10 cases per 100 000 children) in regions of equatorial Africa and Papua New Guinea where infection with *Plasmodium falciparum* malaria is holoendemic. Outside these 'endemic' regions, sporadic cases of BL occur worldwide at 50- to 100-fold lower incidence. The tumours of endemic BLs are almost always EBV positive, but only about 15% of the sporadic tumours carry the virus. Despite differences in their association with EBV, the two types of BL share many clinical features and all the tumours possess a characteristic chromosomal translocation involving chromosome 8 in the region of the c-*myc* oncogene and either chromosome 14 in the region of the Ig heavy chain gene or, less commonly, chromosomes 2 or 22 in the region of the Ig light chain genes. The clinical and genetic features of BL have been reviewed in detail elsewhere (Magrath, 1990).

All BL tumours are phenotypically distinct from normal EBV-positive LCLs with respect to cell surface markers (Favrot *et al.*, 1984; Rowe *et al.*, 1985; Rooney *et al.*, 1986b; Gregory *et al.*, 1988; Rincon, Prieto & Patarroyo, 1992). Therefore, BL cells characteristically express two markers, CD10 and CD77, that are not normally expressed on LCLs. Conversely, LCLs constitutively express several markers of B cell activation, including CD23, CD30, CD39, CD70 and the family of adhesion molecules (LFA1, CD11a/18; LFA3, CD58; and ICAM-1, CD54) that are expressed only weakly, if at all, on BL cells *in vivo*. In addition, whereas early-passage BL lines grow as a carpet of uniformly spherical cells showing little tendency to aggregate, LCLs are composed of larger cells with a more blastoid morphology and characteristically grow as multicellular aggregates. Upon serial passaging *in vitro*, many BL lines undergo a 'phenotypic drift' towards the cell surface marker expression and growth phenotype of LCLs. A widely adopted nomenclature has been used to distinguish those lines that retain a biopsy-like, or 'group I' cellular phenotype ($CD10^+$, $CD77^+$; $CD23^-$, $CD30^-$, $CD39^-$, $CD70^-$) from those that have acquired an LCL-like, or 'group III' cellular phenotype ($CD10^-$, $CD77^-$; $CD23^+$, $CD30^+$, $CD39^+$, $CD70^+$). It should be emphasized that the BL lines that undergo phenotypic drift do retain the characteristic chromosomal translocations and the monoclonal immunoglobulin expression displayed by the original tumour, thus ruling out the possibility that the above-mentioned phenomenon is simply the result of overgrowth of the cultures by normal EBV-transformed LCLs.

These results assumed greater significance when it was realized that BL biopsy cells and early-passage lines (i.e. those displaying a group I cellular phenotype) show a remarkably restricted pattern of EBV gene expression, where EBNA1 is expressed but the other five EBNAs and the LMPs are down-regulated (Masucci *et al.*, 1987; Rowe *et al.*, 1987). This form of latent infection displayed in BL is now referred to as 'latency I' (Lat I), whilst the more expressive form of infection displayed by LCLs is referred to as 'latency III' (Lat III). Where progression of BL lines from a group I to a group III cellular phenotype is observed, it is regularly associated with a concomitant broadening of viral gene expression from Lat I to Lat III. In certain circumstances, B lymphocytes may display yet another form of infection, Lat II, in which the LMPs are expressed along with EBNA1, but the other five EBNAs remain down-regulated. The Lat II form of latent infection has been induced experimentally in a subpopulation of cells in BL cultures (Rowe *et al.*, 1992), whilst *in vivo* this form of infection is observed in some AIDS lymphomas, in a proportion of nasopharyngeal carcinomas and in the Reed–Sternberg cells in all cases of EBV-positive Hodgkin's disease (Young & Rowe, 1992).

## 11.3 Modulation of cellular phenotype by EBV

The phenotypic drift that is associated with progression from a Lat I to a Lat III form of infection in EBV-positive BL lines *in vitro* shares many important features with the process of transformation of normal resting B lymphocytes following infection with EBV. Therefore, this BL model has been used as a basis for understanding the role of individual EBV latent genes in effecting cellular transformation. Several studies have used EBV-negative BL lines, established from rare sporadic cases of BL, as target cells for gene-transfection studies in which an effect of individual latent genes upon the cellular phenotype was observed (Wang *et al.*, 1987; 1988; 1990; Aman *et al.*, 1990). Naturally there are limitations to this experimental model, the most obvious being that the EBV-negative lines are malignant cells and, therefore, some important effects of viral genes on cell proliferation may be masked. Nevertheless, these gene-transfection studies have shown that: (a) EBNA2 can specifically induce expression of CD21 and CD23, (b) EBNA3C can specifically induce CD21, albeit less efficiently than EBNA2, and (c) LMP1 can induce a plethora of changes, including down-regulation of the BL-associated marker CD10 and up-regulation of CD21, CD23, CD39, CD40, CD44, ICAM-1, LFA-1 and LFA-3. In addition, LMP1 and EBNA2 can co-operate to induce higher levels of CD23 than either LMP1 or EBNA2 alone; this co-operative effect is apparently specific for CD23 and occurs because LMP1 and EBNA2 induce different mRNA species, each of which encode the same CD23 protein (Wang *et al.*, 1990).

The above assay system implicates LMP1 as having a pivotal role in inducing phenotypic changes in B lymphocytes that are potentially important with respect to immune recognition. Whilst the effects of LMP1 upon expression of CD21, CD23, CD39, CD40 and CD44 are somewhat variable, depending on the particular target cell line chosen for transfection, the down-regulation of CD10 and up-regulation of the adhesion molecules are regular features with all the B cell lines studied. Consistent with the up-regulation of the adhesion molecules, both homotypic adhesions amongst LMP1-transfected cells and heterotypic adhesions between LMP1-transfected cells and activated T lymphocytes are enhanced. However, in the BJAB lymphoma line, which already expresses high levels of ICAM-1 and LFA-1, transfection of LMP1 has little effect upon the levels of these proteins and yet homotypic adhesions mediated via ICAM-1–LFA-1 interactions are markedly enhanced (Cuomo *et al.*, 1990; Wang *et al.*, 1990). This result is consistent with other data suggesting that high levels of these adhesion molecules are not necessarily sufficient for formation of homotypic conjugates and that functional activation of LFA-1 may also be required (Dustin & Springer, 1989; Rousset *et al.*, 1989; Salcedo *et al.*, 1991; Valmu *et al.*, 1991).

Transfection of the LMP1 gene into B cell lines also results in up-regulation of HLA class I and TAP1 and TAP2, which are key components of the endogenous antigen-processing pathway (Rowe *et al.*, 1995). These properties of LMP1 suggest that it has a major role in increasing the immunogenicity of EBV-infected cells. LMP1 is an integral membrane phosphoprotein that shares some features with receptor molecules (Kieff & Leibowitz, 1991), but the biochemical mechanisms through which LMP1 mediates its pleotropic effects are incompletely understood. Activation of the nuclear transcription factor NF-κB is implicated as a key signalling event (Laherty *et al.*, 1992; Huen *et al.*, 1995), and LMP1 may also engage signalling proteins for the TNF receptor family (Mosialos *et al.*, 1995). There are many similarities between LMP1 and IFN-γ in their effects upon CD40, ICAM-1, HLA class I, HLA class II, TAP1 and TAP2, suggesting some convergence between the LMP1 and IFN-γ signalling pathways (Rowe *et al.*, 1995).

## 11.4 BL as a model of escape from EBV-specific immunosurveillance

Whilst EBV has the potential to growth-transform normal B lymphocytes (and possibly other cell types), in so doing the virus expresses viral gene products that are known targets for CTL responses and it concomitantly induces a cellular phenotype that is conducive to immune recognition. This form of latent infection is, therefore, unlikely to provide the basis for persistent infection in healthy individuals. However, the studies on BL cells have demonstrated that there are other forms of virus–host cell interactions that have the potential to evade recognition by EBV-specific CTLs. Indeed, functional assays have shown that CTLs from BL patients are able to control outgrowth of normal EBV-transformed B lymphocytes (Whittle *et al.*, 1984; Rooney *et al.*, 1985a) and yet *in vivo* they are clearly ineffective in preventing the outgrowth of tumours. Furthermore, when a panel of BL lines was compared for sensitivity to lysis by EBV-specific CTLs from HLA-matched donors, none of the early-passage BL lines, which displayed a biopsy-like group I cellular phenotype, was killed (Rooney *et al.*, 1985b; Rowe *et al.*, 1986; Torsteinsdottir *et al.*, 1986). In contrast, paired LCLs established by *in vitro* infection of normal B lymphocytes from the same patients were efficiently lysed by the CTLs, as were some of the BL lines that had drifted towards a group III cellular phenotype. From these data, we can conclude that BL tumours *in vivo* are resistant to EBV-specific CTLs.

The particular form of latency that EBV establishes in the tumour cells of BL *in vivo* is characterized by at least two features that could contribute to the observed immune escape: the essential target viral antigens are not expressed and neither are two accessory molecules, LFA3 and ICAM-1. These two adhesion molecules mediate conjugate formation with CTLs by associating with CD2 and LFA-1 respectively, on the effector cells. However, recent studies have questioned the importance of expression of LFA-3 and ICAM-1 in determining sensitivity of target cells to lysis by CTLs (Masucci *et al.*, 1992; Griffin *et al.*, 1992; Khanna *et al.*, 1993). Several BL lines that display a group I cellular phenotype and express low or undetectable levels of LFA-3 and ICAM-1 were assayed for the ability to form conjugates with activated T cells and also for their susceptibility to lysis by allospecific CTLs. Conjugate formation was not observed in these lines, although infection with an LFA-3-expressing recombinant vaccinia virus restored the ability to form conjugates via the CD2/LFA3 adhesion pathway (Griffin *et al.*, 1992). However, all of the LFA-3-negative BL lines were killed by allospecific CTLs, in some cases as efficiently as the corresponding normal LCL. In other cases where the lysis of the BL line was slightly less efficient than the corresponding LCL, infection with an LFA-3-expressing recombinant vaccinia virus did enhance the lysis to LCL levels, but there was never a dramatic sensitization to allospecific lysis commensurate with the obvious LFA-3-mediated improvement in conjugate formation (Griffin *et al.*, 1992). Other studies have extended these observations to target cell recognition by EBV-specific CTLs, showing that efficient lysis of group I BL lines was obtained following incubation with synthetic peptides corresponding to the CTL epitopes from EBV-encoded proteins; again, LFA-3 and ICAM-1 were apparently dispensible (Gavioli *et al.*, 1992; Khanna *et al.*, 1993). These unexpected findings may be explained by the fact that, whilst these BL lines show down-regulated expression of LFA-3 and ICAM-1, they express high levels of another member of the adhesion molecule family, ICAM-2, which can provide alternative adhesion pathways for recognition by CTLs (de Fougerolles *et al.*, 1991; Seth *et al.*, 1991 Khanna *et al.*, 1993).

Whatever the importance of adhesion molecules in the immune recognition of BL cells, it is obvious that if the viral targets for CTLs are not expressed then the tumour cells will escape recognition by EBV-specific CTLs. Current evidence suggests that the malignant cells of BL do not express any of the latent gene products that are demonstrable targets for EBV-specific CTLs. However, the crucial issue is whether or not EBNA1, the only viral protein expressed in BL, can elicit CTL responses. To date, no EBNA1-specific CTLs have been demonstrated, suggesting that it cannot provide target peptides for CTLs. This intriguing possibility is supported by the demonstration that the Gly–Ala-rich internal repeat region of EBNA1 inhibits processing and MHC class I presentation of a known immunodominant epitope from EBNA3B in chimaeric EBNA1 molecules (Levitskaya *et al.*, 1995). This observation highlights a previously unknown mechanism of viral escape from CTL surveillance and supports the view that the resistance of EBNA1-expressing cells to CTL-mediated rejection is a critical requirement for EBV persistence and pathogenesis.

## 11.5 HLA class I expression by BL tumour cells

Other factors that are likely to contribute to the immune escape of BL cells include the expression of HLA class I antigens and other components of the endogenous antigen-processing pathway. It is well established that BL cells can show a reduced cell surface expression of HLA class I antigens compared with that on normal EBV-transformed LCLs (Rooney *et al.*, 1985a,b; 1986a; Torsteinsdottir *et al.*, 1986; 1988). Figure 11.1 shows FACS profiles obtained using the W6/32 antibody to detect cell surface HLA expression on a group I BL line and a paired LCL from the same patient. The drift of BL cells from a group I to a group III cellular phenotype is also associated with increased HLA class I expression (see Fig. 11.2 and Table 11.2, below) and gene transfection studies have identified the LMP1 EBV gene as being

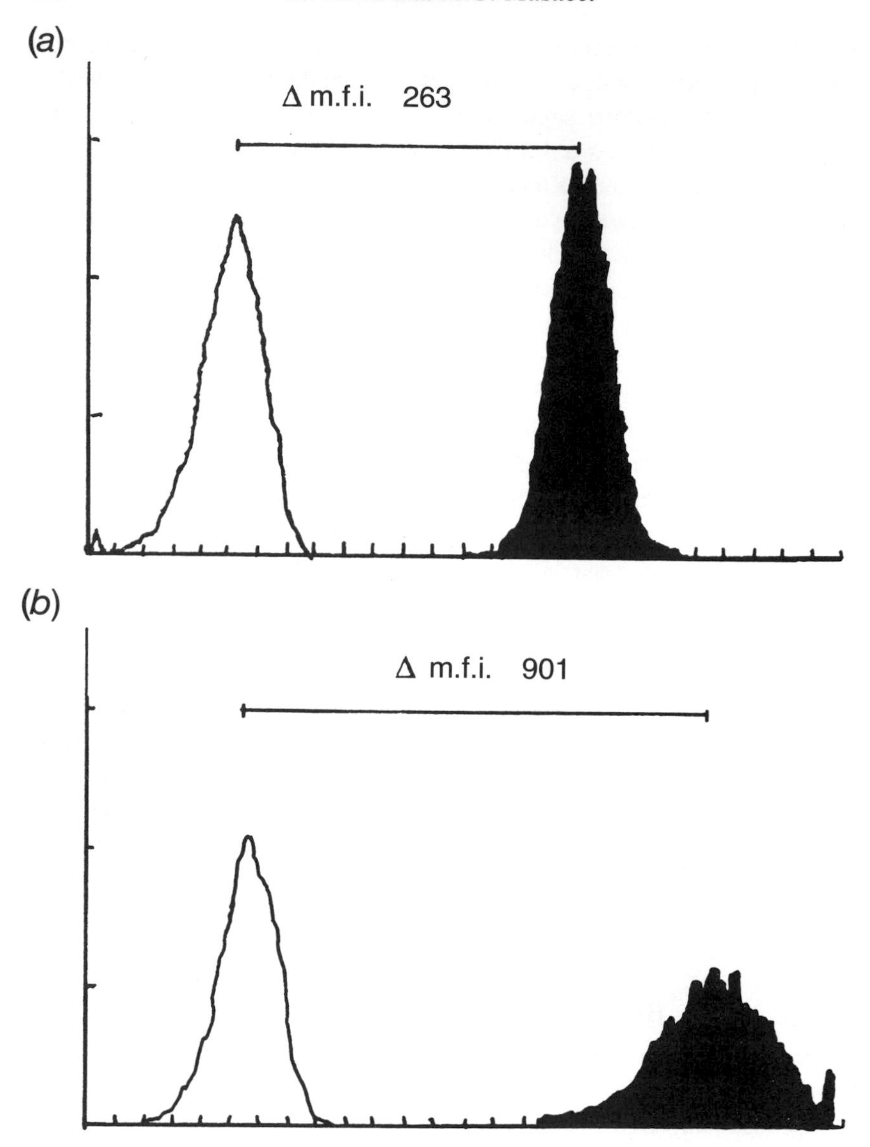


Fig. 11.1. Fluorescent antibody cell sorting profiles showing the relative expression of HLA class I antigens (*a*) on the Mutu-BL all line displaying a group I cellular phenotype and (*b*) on a normal EBV-transformed LCL established from the same patient. The solid profiles represent the fluorescence staining obtained with the W6/32 mAb specific for a common framework determinant of HLA class I, and the hollow profiles represent the background staining obtained with an irrelevant control antibody. Fluorescence intensity was measured on an arithmetic scale and the mean fluorescence intensity of W6/32 staining above background staining is indicated (Δ m.f.i.) for each line.

responsible for this effect (Rowe *et al.*, 1995). BL lines displaying a group I cellular phenotype typically express around threefold to fivefold less HLA class I antigens than do group III BL lines or normal EBV-transformed LCLs (Torsteinsdottir *et al.*, 1988; Masucci *et al.*, 1989; Rowe *et al.*, 1995). However, the significance of this observation is unclear, since the levels of HLA class I are still substantial and sufficient to render the tumour cells sensitive to lysis by allospecific CTLs (Rooney *et al.*, 1985b; Torsteinsdottir *et al.*, 1986) and also by EBV-specific CTL if the appropriate peptide epitopes are provided (Khanna *et al.*, 1993).

Of particular interest is the selective down-regulation of specific class I alleles that has been observed in many BL lines, first described for a series of BL lines derived from HLA-A11-positive patients (Torsteinsdottir *et al.*, 1986, 1988; Masucci *et al.*, 1987). In these initial studies, it was observed that the BL lines were resistant to lysis by A11-allospecific CTLs that killed paired normal LCLs from the same patients. The phenomenon was apparently restricted to HLA-A11 since the same BL lines were killed by CTLs specific for other class I alleles. Furthermore, immunofluorescent staining for cell surface HLA class I showed that the CTL resistance correlated with a substantially decreased reactivity with the anti-A11 mAb AUF5.13 relative to the reactivity with the W6/32 mAb specific for a common class I framework determinant. The selective down-regulation of HLA-A11 in BL appears to be a common feature since it has been observed in all BL lines derived from five HLA-A11-positive patients. The defect occurs at the level of transcription, since HLA-A11-specific mRNA cannot be detected, but in some of the BL lines expression can be partially restored by treatment with IFN-γ (Imreh *et al.*, 1995).

The phenomenon of allele-selective down-regulation of HLA class I has been studied using a biochemical method involving immunoprecipitation with the W6/32 monomorphic class I antibody and separation by one-dimensional isoelectric focussing (Masucci *et al.*, 1989; Andersson *et al.*, 1991). This technique has allowed detailed analysis of the expression of a wider range of HLA class I alleles in a large panel of paired BL/LCL lines and it is now clear that allele-selective down-regulation is not confined to HLA-A11. As illustrated in Fig. 11.2 and summarized in Table 11.1, selective down-regulation has been observed for several HLA class I alleles, although only HLA-A11 is consistently affected. Expression of some of the affected alleles could be induced with IFN-γ, but other alleles were resistant; this pattern of responsiveness to IFN-γ was not cell line related since different alleles within the same cell line responded differently (Andersson *et al.*, 1991). As with the more general down-regulation of HLA class I alleles described earlier,

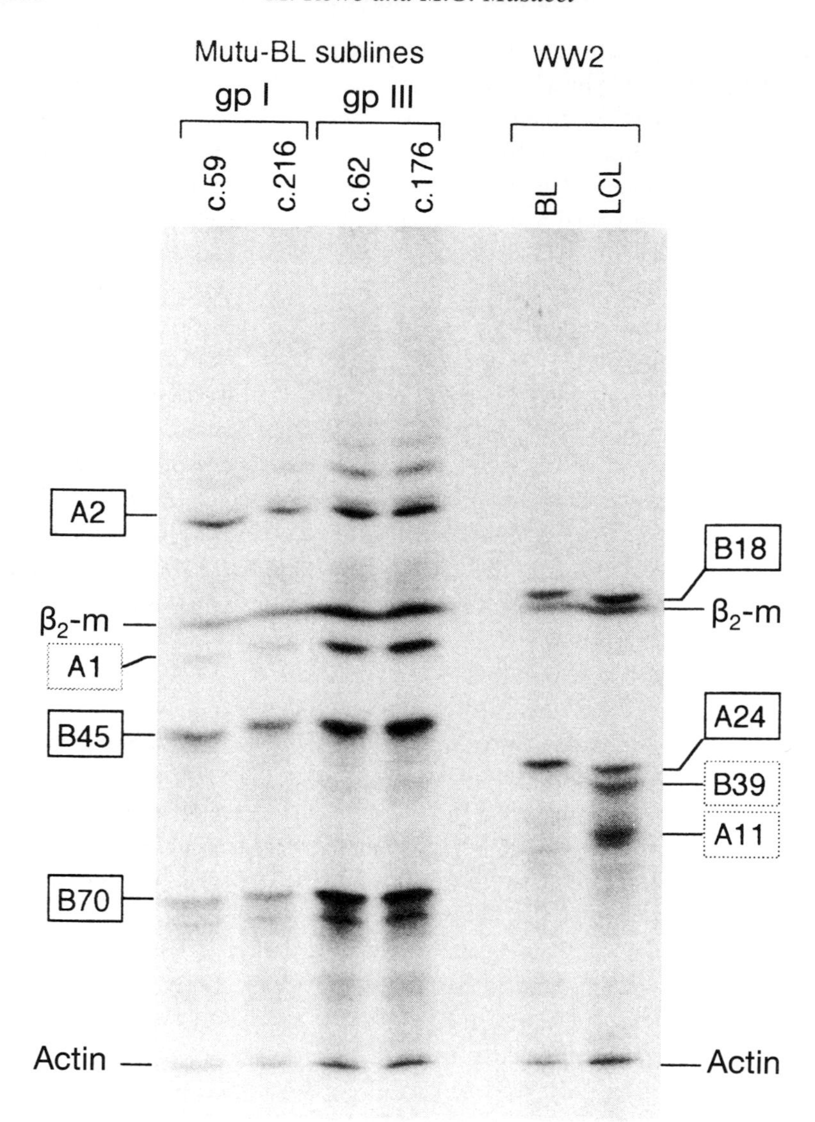


Fig. 11.2. One-dimensional isoelectric focussing analysis of HLA class I expression on BL lines and EBV-transformed LCLs. Cells were metabolically labelled with [$^{35}$S]methionine, solubilized and immunoprecipitated with the W6/32 monoclonal antibody. The precipitated HLA was digested with neuraminidase and then separated on an IEF polyacrylamide gel containing 9 M urea (Masucci *et al.*, 1989). In the experiment shown, two group I sublines of the Mutu-BL show a selective down-regulation of HLA-A1 compared with two group III sublines of the same tumour and the group I WW2-BL line shows a selective loss of HLA-A11 compared with a normal EBV-transformed LCL from the same patient.

Table 11.1. *Selective down-regulation of HLA alleles in BL lines revealed by isoelectric focussing analysis*

| HLA | Selective down-regulation[a] |
|---|---|
| A1 | 1/3 |
| A2 | 0/5 |
| A3 | 1/2 |
| A11 | 5/5 |
| A23 | 0/1 |
| A24 | 0/2 |
| A30 | 0/2 |
| A31 | 0/1 |
| A32 | 0/2 |
| A33 | 0/1 |
| Aw69 | 1/3 |
| | |
| Cw4 | 3/4 |
| Cw5 | 1/2 |
| Cw6 | 0/1 |
| Cw7 | 2/3 |
| Cw8 | 1/1 |
| Cw9 | 0/1 |
| | |
| B7 | 0/1 |
| B8 | 0/1 |
| B14 | 0/1 |
| B18 | 0/4 |
| Bw22 | 0/2 |
| B27 | 0/2 |
| B35 | 0/3 |
| B39 | 1/2 |
| B44 | 0/2 |
| B45 | 0/1 |
| B49 | 0/2 |
| B62 | 0/1 |
| Bw70 | 0/2 |

[a] The data indicate the number of BL patients in which down-regulation of a particular allele was observed relative to the other HLA class I alleles expressed on the same BL line. This selective down-regulation, which may represent a partial or a complete loss of expression, is distinct from the general reduction of all HLA class I antigens observed in group I BL lines relative to paired LCLs.

Table 11.2. *Relative expression of HLA class I heavy chains in phenotypically distinct subclones of the Mutu-BL line*

| | Total HLA class I expression[a] | Ratio of heavy chain/$\beta_2$-m[b] | | | |
|---|---|---|---|---|---|
| | | HLA-A1 | HLA-A2 | HLA-B45 | HLA-Bw70 |
| *Group III lines*[c] | | | | | |
| Parental | 100 | 1.04 | 1.41 | 1.17 | 0.90 |
| Clone 62 | 113 | 0.71 | 1.57 | 0.79 | 0.89 |
| Clone 99 | 155 | 0.78 | 1.46 | 0.64 | 0.91 |
| Clone 176 | 120 | 0.84 | 1.55 | 0.70 | 0.81 |
| Mean | | 0.84 | 1.49 | 0.83 | 0.88 |
| *Group I lines* | | | | | |
| Clone 59 | 9 | 0.00 | 0.80 | 0.92 | 0.70 |
| Clone 145 | 24 | 0.09 | 0.69 | 0.86 | 0.69 |
| Clone 148 | 14 | 0.4 | 1.30 | 1.50 | 0.94 |
| Clone 179 | 21 | 0.35 | 1.52 | 1.09 | 1.10 |
| Mean | | 0.21 | 1.08 | 1.09 | 0.85 |

[a] Calculated from the sum of the specific HLA class I heavy chain bands measured by densitometry of one representative isoelectric focussing autoradiogram. The sum of the areas measured in the parental Mutu-BL line was arbitrarily assigned as 100, and the results obtained with the other lines were expressed relative to this value.
[b] The expression of the individual HLA class I heavy chains was calculated as the ratio between amount of the heavy chain and amount of $\beta_2$-m determined by densitometry measurements of the respective bands on the representative isoelectric focussing autoradiogram.
[c] The Mutu-BL line was subcloned during early passage, and phenotypically distinct subclones were selected and characterized (Gregory *et al.*, 1990). The uncloned 'parental' line and many of the derived subclones drifted to a group III cellular phenotype, but several subclones were established that stably retained the group I cellular phenotype of the original biopsy cells.

the allele-selective down-regulation was predominantly observed in BL lines that had maintained a biopsy-like cellular phenotype and a shift towards an LCL-like phenotype was accompanied by up-regulation of HLA expression. This phenomenon is best illustrated in studies with sublines of the EBV-positive Mutu-BL line (Gregory, Rowe & Rickinson, 1990), where biopsy-like group I sublines express up to about fivefold less HLA class I antigens and show a selective loss of HLA-A1 when compared with sublines showing a group III cellular phenotype (Masucci, 1990; see Fig. 11.2 and Table 11.2).

Another aspect of MHC modulation implicated in the immune escape of BL has recently come to light from studies on the antigen-processing pathways of BL lines (Khanna *et al.*, 1994). Infection of the EBV-negative line

BL30 with an EBNA3A-expressing recombinant vaccinia virus failed to sensitize the BL cells to lysis by CTLs specific for a peptide epitope of EBNA3A, although the cells were efficiently lysed following pre-incubation with exogenous synthetic 9-mer peptide corresponding to the EBNA3A CTL epitope. Sensitization to lysis could be achieved, however, by transfecting BL30 cells with a recombinant vector containing the EBNA3A epitope sequence linked to a localization signal sequence for the endoplasmic reticulum. These results suggest that BL30 cells are unable to transport peptides from the site of protein processing in the cytoplasm to the site of HLA class I assembly in the endoplasmic reticulum (see Chapter 1). This interpretation is consistent with the finding that BL30 cells show a transcriptional deficiency of two transporter genes, TAP1 and TAP2, that are localized within the HLA class II region of chromosome 6 (see Chapter 2). Results from a wider panel of BL lines suggest that reduced expression of TAP relative to paired LCLs from the same patients is a regular feature of group I BL cells (Rowe *et al.*, 1995). Interestingly, expression of EBV-encoded LMP1 upregulates both TAP1 and TAP2; it is not surprising, considering the effects that LMP1 has upon TAP, HLA class I and adhesion molecule expression, that LMP1 restores the ability of BL tumour cells to process endogenous antigen and be recognized by CTLs.

## 11.6 Conclusions

At least three features of EBV-positive BL tumours may have a bearing on their resistance to EBV-specific CTLs: the restricted latent viral gene expression, the down-regulation of adhesion molecules and the down-regulation of MHC genes. Contrary to earlier expectations, the reduced expression of the LFA-3 and ICAM-1 adhesion molecules may not be of great significance in the context of sensitivity of BL cells to CTLs. The importance of HLA expression may be more subtle than was originally perceived; a general reduction in expression of all class I alleles may be of little consequence, but the more marked down-regulation of selected alleles is of potential importance. Likewise, perturbations of the antigen processing and transport pathways may selectively impair the presentation of certain antigens.

Finally, it is pertinent to consider that the mechanism of immune-escape by BL cells may be analogous to the mechanism of virus latency in the healthy infected host. This is supported by recent studies showing that, in the blood of healthy infected individuals, EBV DNA is found in a subpopulation of $CD19^{+}CD23^{-}$ cells (Miyashita *et al.*, 1995) and that EBV gene expression appears to be restricted to EBNA1, or to EBNA1 plus LMP2, in a non-cycling B cell subpopulation (Tierney *et al.*, 1994; Chen *et al.*, 1995). We would predict that these cells are not recognized by EBV-specific CTLs and that upon activation to an LCL-like cell the acquired proliferative potential is tempered by an LMP1-induced sensitivity to CTL lysis.

## References


Aman, P., Rowe, M., Kai, C. *et al.* (1990). Effect of the EBNA-2 gene on the surface antigen phenotype of transfected EBV-negative B-lymphoma lines. *International Journal of Cancer*, 45, 77–82.

Andersson, M.L., Stam, N.J., Klein, G., Ploegh, H.L. & Masucci, M.G. (1991). Aberrant expression of HLA class-I antigens in Burkitt's lymphoma cells. *International Journal of Cancer*, 47, 544–550.

Bourgault, I., Gomez, A., Gomard, E. & Levy, J.P. (1991). Limiting-dilution analysis of the HLA restriction of anti-Epstein–Barr virus-specific cytolytic T lymphocytes. *Clinical and Experimental Immunology*, 84, 501–507.

Chen, F., Zou, J.-Z., di Rienzo, L. *et al.* (1995). A subpopulation of latently EBV infected B-cells resemble Burkitt's lymphoma (BL) cells in expressing EBNA1 but not EBNA2 or LMP1. *Journal of Virology*, in press.

Cuomo, L., Trivedi, P., Wang, F., Winberg, G., Klein, G. & Masucci, M. (1990). Expression of the Epstein–Barr virus (EBV)-encoded membrane antigen (LMP) increases the stimulatory capacity of EBV-negative B lymphoma lines in allogeneic mixed lymphocyte cultures. *European Journal of Immunology*, 20, 2293–2299.

de Fougerolles, A.R., Stacker, S.A., Schwarting, R. & Springer, T.A. (1991). Characterization of ICAM-2 and evidence for a third counter-receptor for LFA-1. *Journal of Experimental Medicine*, 174, 253–267.

Dustin, M.L. & Springer, T.A. (1989). T-cell receptor cross-linking transiently stimulates adhesiveness through LFA-1. *Nature (London)*, 341, 619–624.

Favrot, M.C., Philip, I., Philip, T., Portoukalian, J., Doré, J.F. & Lenoir, G.M. (1984). Distinct reactivity of Burkitt's lymphoma cell lines with eight monoclonal antibodies correlated with ethnic origin. *Journal of the National Cancer Institute*, 73, 841–847.

Gavioli, R., De Campos, L.P., Kurilla, M.G., Kieff, E., Klein, G. & Masucci, M.G. (1992). Recognition of the Epstein–Barr virus-encoded nuclear antigens EBNA-4 and EBNA-6 by HLA-A11-restricted cytotoxic T lymphocytes. Implications for down-regulation of HLA-A11 in Burkitt lymphoma. *Proceedings of the National Academy of Sciences of the USA*, 89, 5862–5866.

Gregory, C.D., Murray, R.J., Edwards, C.F. & Rickinson, A.B. (1988). Down regulation of cell adhesion molecules LFA-3 and ICAM-1 in Epstein–Barr virus-positive Burkitt's lymphoma underlies tumour cell escape from virus-specific T cell surveillance. *Journal of Experimental Medicine*, 167, 1811–1824.

Gregory, C.D., Rowe, M. & Rickinson, A.B. (1990). Different Epstein–Barr virus (EBV)-B cell interactions in phenotypically distinct clones of a Burkitt lymphoma cell line. *Journal of General Virology*, 71, 1481–1495.

Griffin, H., Rowe, M., Murray, R., Brooks, J. & Rickinson, A. (1992). Restoration of the LFA-3 adhesion pathway in Burkitt's lymphoma cells using an LFA-3 recombinant vaccinia virus – consequences for T-cell recognition. *European Journal of Immunology*, 22, 1741–1748.

Imreh, M., Zhang, Q.-J., de Campos-Lima, P.O. *et al.* (1995). Mechanisms of allele-selective down-regulation of HLA class I in Burkitt's lymphoma. *International Journal of Cancer*, in press.

Khanna, R., Burrows, S.R., Argaet, V. & Moss, D.J. (1994). Endoplasmic reticulum signal sequence facilitated transport of peptide epitopes restores immunogenicity of an antigen processing defective tumour cell line. *International Immunology*, 6, 639–645.

Khanna, R., Burrows, S.R., Kurilla, M.G. *et al.* (1992). Localization of Epstein–Barr virus cytotoxic T-cell epitopes using recombinant vaccinia – implications for vaccine development. *Journal of Experimental Medicine*, 176, 169–176.

Khanna, R., Burrows, S.R., Suhrbier, A. *et al.* (1993). Epstein–Barr virus peptide epitope sensitization restores human cytotoxic T cell recognition of Burkitt's lymphoma cells. Evidence for a critical role for ICAM2. *Journal of Immunology*, 150, 5154–5162.

Kieff, E. & Liebowitz, D. (1990). Epstein–Barr virus and its replication. In *Fields' Virology*, 2nd edn, Vol. 2, ed. B.N. Fields & D.M. Knipe, pp. 1889–1920. New York: Raven Press.

Laherty, C.D., Hu, H.M., Opipari, A.W., Wang, F. & Dixit, V.M. (1992). Epstein–Barr virus LMP1 gene product induces A20 zinc finger protein expression by activating nuclear factor κB. *Journal of Biological Chemistry*, 267, 24157–24160.

Levitskaya, J., Coram, M., Levitsky, V. *et al.* (1995). The internal repeat region of the Epstein–Barr virus nuclear antigen-1 is a *cis*-acting inhibitor of antigen processing. *Nature (London)*, in press.

Magrath, I. (1990). The pathogenesis of Burkitt's lymphoma. *Advances in Cancer Research*, 55, 133–270.

Masucci, M.G. (1990). Cell phenotype dependent down-regulation of MHC class I antigens in Burkitt's lymphoma cells. *Current Topics in Microbiology and Immunology*, 166, 309–316.

Masucci, M.G., Stam, N., Torsteinsdottir, S., Neefjes, J.J., Klein, G. & Ploegh, H.L. (1989). Allele-specific down-regulation of MHC class I antigens in Burkitt's lymphoma cell lines. *Cellular Immunology*, 120, 396–400.

Masucci, M.G., Torsteinsdottir, S., Colombani, J., Brautbar, C., Klein, E. & Klein, G. (1987). Down-regulation of class-I HLA antigens and of the Epstein–Barr virus (EBV)-encoded latent membrane protein (LMP) in Burkitt's lymphoma lines. *Proceedings of the National Academy of Sciences of the USA*, 84, 4567–4572.

Masucci, M.G., Zhang, Q.-J., de Campos Lima, P.O. *et al.* (1992). Immune-escape of Epstein–Barr virus (EBV) carrying Burkitt's lymphoma: *in vitro* reconstitution of sensitivity to cytotoxic T cells. *International Immunology*, 4, 1283–1292.

Miller, G. (1990). Epstein–Barr virus. Biology, pathogenesis, and medical aspects. In *Fields' Virology*, 2nd edn, Vol. 2, ed. B.N. Fields & D.M. Knipe, pp. 1921–1958. New York: Raven Press.

Miyashita, E.M., Yang, B., Lam, K.M.C., Crawford, D.H. & Thorley-Lawson, D.A. (1995). A novel form of Epstein–Barr virus latency in normal B cells in vivo. *Cell*, 80, 593–601.

Murray, R.J., Kurilla, M.G., Brooks, J.M. *et al.* (1992). Identification of target antigens for the human cytotoxic T-cell response to Epstein–Barr virus

(EBV) – implications for the immune control of EBV-positive malignancies. *Journal of Experimental Medicine*, 176, 157–168.

Murray, R.J., Kurilla, M.G., Griffin, H.M. *et al.* (1990). Human cytotoxic T cell responses against Epstein–Barr virus nuclear antigens demonstrated using recombinant vaccinia viruses. *Proceedings of the National Academy of Sciences of the USA*, 87, 2906–2910.

Rickinson, A.B., Moss, D.J., Wallace, L.E. *et al.* (1981). Long-term T-cell-mediated immunity to Epstein–Barr virus. *Cancer Research*, 41, 4216–4221.

Rincon, J., Prieto, J. & Patarroyo, M. (1992). Expression of integrins and other adhesion molecules in Epstein–Barr virus-transformed B lymphoblastoid cells and Burkitt's lymphoma cells. *International Journal of Cancer*, 51, 452–458.

Rooney, C.M., Edwards, C.F., Lenoir, G.M., Rupani, H. & Rickinson, A.B. (1986a). Differential activation of cytotoxic responses by Burkitt's lymphoma (BL)-cell lines: relationship to the BL-cell surface phenotype. *Cellular Immunology*, 102, 99–112.

Rooney, C.M., Gregory, C.D., Rowe, M. *et al.* (1986b). Endemic Burkitt's lymphoma: phenotypic analysis of tumour biopsy cells and of the derived tumour cell lines. *Journal of the National Cancer Institute*, 77, 681–687.

Rooney, C.M., Rickinson, A.B., Moss, D.J., Lenoir, G.M. & Epstein, M.A. (1985a). Cell-mediated immunosurveillance mechanisms and the pathogenesis of Burkitt's lymphoma. In *Burkitt's Lymphoma: A Human Cancer Model*, vol. 60, ed. G.M. Lenoir, G.T. O'Conor & C.L.M. Olweny, pp. 249–264. Lyon: International Agency for Research on Cancer.

Rooney, C.M., Rowe, M., Wallace, L.E. & Rickinson, A.B. (1985b). Epstein–Barr virus-positive Burkitt's lymphoma cells not recognized by virus-specific T-cell surveillance. *Nature (London)*, 317, 629–631.

Rousset, F., Billaud, M., Blanchard, D. *et al.* (1989). IL-4 induces LFA-1 and LFA-3 expression on Burkitt's lymphoma cell lines. *Journal of Immunology*, 143, 1490–1498.

Rowe, M., Khanna, R., Jacob, C.A. *et al.* (1995). Restoration of endogenous antigen processing in Burkitt's lymphoma cells by Epstein–Barr virus latent membrane protein-1: coordinate upregulation of peptide transporters and HLA class I antigen expression. *European Journal of Immunology*, in press.

Rowe, M., Lear, A., Croom-Carter, D., Davies, A.H. & Rickinson, A.B. (1992). Three pathways of Epstein–Barr virus (EBV) gene activation from EBNA1-positive latency in B lymphocytes. *Journal of Virology*, 66, 122–131.

Rowe, M., Rooney, C.M., Rickinson, A.B. *et al.* (1985). Distinctions between endemic and sporadic forms of Epstein–Barr virus-positive Burkitt's lymphoma. *International Journal of Cancer*, 35, 435–441.

Rowe, D.T., Rowe, M., Evan, G.I., Wallace, L.E., Farrell, P.J. & Rickinson, A.B. (1986). Restricted expression of EBV latent genes and T-lymphocyte-detected membrane antigen in Burkitt's lymphoma cells. *EMBO Journal*, 5, 2599–2607.

Rowe, M., Rowe, D.T., Gregory, C.D. *et al.* (1987). Differences in B cell growth phenotype reflect novel patterns of Epstein–Barr virus latent gene expression in Burkitt's lymphoma. *EMBO Journal*, 6, 2743–2751.

Salcedo, R., Fuerstenberg, S.M., Patarroyo, M. & Winberg, G. (1991). The Epstein–Barr virus BNLF-1 membrane protein LMP1 induces homotypic adhesion mediated by CD11a/CD18 in a murine B-cell line, mimicking the action of phorbol ester. *Journal of Virology*, 65, 5558–5563.

Seth, R., Salcedo, R., Patarroyo, M. & Makgoba, M.W. (1991). ICAM-2 peptides mediate lymphocyte adhesion by binding to CD11a/CD18 and CD49d/CD29 integrins. *FEBS Letters*, 282, 193–196.

Sixbey, J.W., Nedrud, J.G., Raab-Traub, N., Hanes, R.A. & Pagano, J.S. (1984). Epstein–Barr virus replication in oropharyngeal epithelial cells. *New England Journal of Medicine*, 310, 1225–1230.

Thomas, J.A., Allday, M. & Crawford, D.H. (1991). Epstein–Barr virus-associated lymphoproliferative disorders in immunocompromised individuals. *Advances in Cancer Research*, 57, 329–380.

Tierney, R.J., Steven, N., Young, L.S. & Rickinson, A.B. (1994). Epstein–Barr virus latency in blood mononuclear cells: analysis of viral gene transcription during primary infection and in the carrier state. *Journal of Virology*, 68, 7374–7385.

Torsteinsdottir, S., Brautbar, C., Ben Bassat, H., Klein, E. & Klein, G. (1988). Differential expression of HLA antigens on human B-cell lines of normal and malignant origin: a consequence of immune surveillance or a phenotype vestige of the progenitor cells? *International Journal of Cancer*, 41, 913–919.

Torsteinsdottir, S., Masucci, M.G., Ehlin-Henriksson, B. *et al.* (1986). Differentiation dependent sensitivity of human B-cell derived lines to major histocompatibility complex-restricted T-cell cytotoxicity. *Proceedings of the National Academy of Sciences of the USA*, 83, 5620–5640.

Valmu, L., Autero, M., Siljander, P., Patarroyo, M. & Gahmberg, C.G. (1991). Phosphorylation of the beta-subunit of CD11/CD18 integrins by protein kinase C correlates with leukocyte adhesion. *European Journal of Immunology*, 21, 2857–2862.

Wang, F., Gregory, C.D., Rowe, M. *et al.* (1987). Epstein–Barr virus nuclear protein 2 specifically induces expression of the B cell activation antigen CD23. *Proceedings of the National Academy of Sciences of the USA*, 84, 3452–346.

Wang, F., Gregory, C.D., Sample, C. *et al.* (1990). Epstein–Barr virus latent membrane protein (LMP-1) and nuclear proteins 2 and 3C are effectors of phenotypic changes in B lymphocytes: EBNA2 and LMP-1 cooperatively induce CD23. *Journal of Virology*, 64, 2309–2318.

Wang, D., Liebowitz, D., Wang, F. *et al.* (1988). Epstein–Barr virus latent infection membrane protein alters the human B-lymphocyte phenotype: deletion of the amino terminus abolishes activity. *Journal of Virology*, 62, 4173–4184.

Whittle, H.C., Brown, J., Marsh, K. *et al.* (1984). T-cell control of Epstein–Barr virus-infected B cells is lost during *P. falciparum malaria. Nature (London)*, 312, 449–450.

Wolf, H., Haus, M. & Wilmes, E. (1984). Persistence of Epstein–Barr virus in the parotid gland. *Journal of Virology*, 51, 795–798.

Wolf, H., zur Hausen, H. & Becker, V. (1973). EB-viral genomes in epithelial nasopharyngeal carcinoma cells. *Nature (London)*, 244, 245–247.

Young, L.S. & Rowe, M. (1992). Epstein–Barr virus, lymphomas and Hodgkin's disease. *Seminars in Cancer Biology*, 3, 273–284.

# 12

# Effect of human cytomegalovirus infection on the expression of MHC class I antigens and adhesion molecules: potential role in immune evasion and immunopathology

JANE E. GRUNDY
*Royal Free Hospital School of Medicine*

## 12.1 Introduction

Human cytomegalovirus (CMV) is a ubiquitous agent that rarely causes disease in healthy individuals but is an important pathogen in the immunocompromised or the immunologically immature foetus. Individuals at high risk are those with deficiencies in cell-mediated immunity and it is, therefore, believed that in the immunocompetent host the cell-mediated immune response is the most important form of defence against CMV (Grundy, 1991). In the immunocompromised, CMV infection can be associated with a wide range of symptoms. The most serious complication in allogeneic transplant recipients is CMV interstitial pneumonitis, which is associated with a high mortality (Meyers, Flournoy & Thomas, 1982), and the virus remains the most important infectious cause of death following bone marrow transplantation. In AIDS patients, CMV can cause sight-threatening retinitis (Collaborative DHPG Treatment Study Group, 1986) and serious disease throughout the gastrointestinal tract. Congenital CMV infection can be associated with disseminated disease, including in the CNS, and the virus is an important cause of deafness and mental retardation (Alford *et al.*, 1990). Thus, CMV causes disease in many organ systems in various patient populations. However, even in the immunocompetent host, the virus is not eradicated from the body following primary infection and can persist in a latent form throughout life. Hence, whilst the normal host immune response is usually able to prevent disease associated with primary CMV infection, the virus ultimately evades the host response and becomes latent.

The sites of latency of CMV and the mechanisms by which the virus evades host responses are largely unknown. CMV has been found to suppress immune responses both *in vivo* and *in vitro* (reviewed in Griffiths & Grundy, 1987), and this may aid the establishment of latency. Low-passage clinical isolates of CMV are very cell associated and cell to cell spread of the virus *in vivo* might help it evade neutralizing antibody. In addition, the most immunogenic proteins of CMV are internal components of the virion and, accordingly, antibodies against these proteins are not protective. This may aid the virus by deflecting the humoral immune response away from the more important envelope glycoproteins. Furthermore, competition between non-neutralizing antibodies and neutralizing antibodies for antigenic sites on the envelope glycoproteins may reduce the efficacy of any neutralizing antibodies present. The induction of Fc receptors on the surface of CMV-infected cells (Furukawa *et al.*, 1975) may prevent antibody- and complement-mediated lysis and antibody-dependent cellular cytotoxicity of infected cells by the binding of non-CMV-specific antibody via its Fc portion. As with many viruses, antigenic variation in the neutralizing epitopes of CMV envelope glycoproteins has been described (Darlington *et al.*, 1991) and may contribute to evasion of antibody recognition following reinfection with new CMV strains.

CTL recognition of virally infected cells requires the recognition by the T cell receptor of processed viral peptides in the peptide binding groove of the HLA class I heavy chains (Townsend *et al.*, 1986; see Chapter 1). In addition, interaction between accessory molecules on the T cell and the target cell are required. The main interactions are between CD2 and LFA-1 on the T cell and LFA-3 and ICAM-1 on the target cell, respectively (Springer, 1990). In addition, the CD8 antigen on the T cell interacts directly with the $\alpha_3$ domain of the class I heavy chain on the target cell and if the target cell is also of leukocyte origin, there is a reciprocal interaction between LFA-1 on the target cell and ICAM-1 on the T cell. Virus-induced changes in the expression of HLA class I or adhesion molecules may affect the ability of CTLs to kill virally infected target cells and changes in adhesion molecule expression may also affect lysis by NK cells. We have, therefore, studied the effect of CMV infection of human embryo lung fibroblasts *in vitro* on the cell surface expression of HLA class I, LFA-3 and ICAM-1, and related the changes observed to leukocyte binding and target cell lysis.

## 12.2 Effect of CMV infection on the expression of HLA class I on the surface of the infected cell

We found that CMV induced a progressive decrease in the expression of HLA class I antigens on the surface of the infected cell from 24 hours post-infection (Hutchinson, Eren & Grundy, 1991; Barnes & Grundy, 1992). Flow cytometric analysis indicated that both the number of cells expressing HLA class I and the mean fluorescent intensity of HLA class I expression

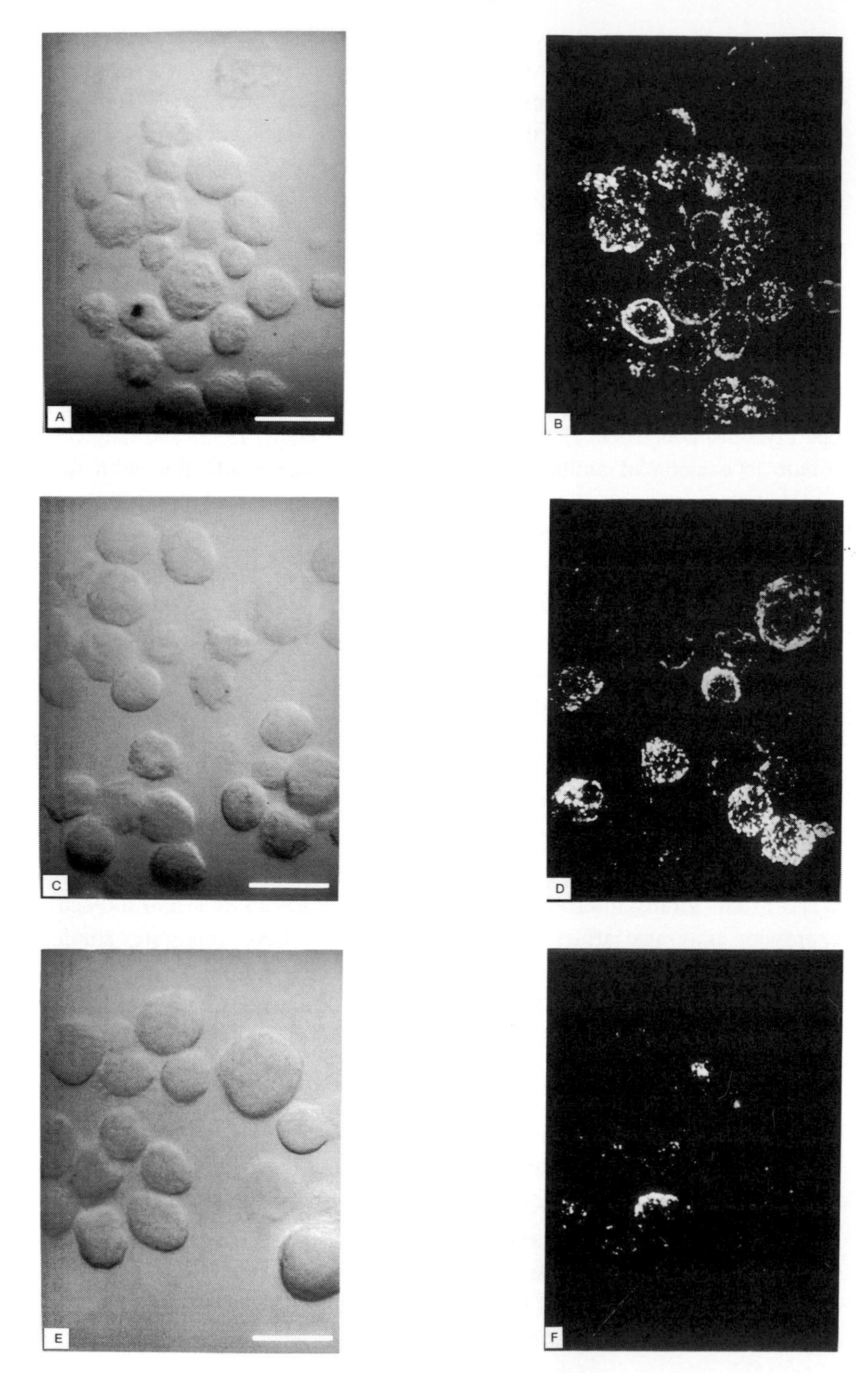
A
B
C
D
E
F

decreased, such that by late stages of infection (day 5) the majority of infected cells had no detectable cell surface class I (Barnes & Grundy, 1992; Hutchinson *et al.*, 1991). The same results were obtained whether antibodies against the mature class I heterodimer (W6/32 and PA 2.6) or antibodies specific for the HLA class I light chain ($\beta_2$-m) were used (Hutchinson *et al.*, 1991; Barnes & Grundy, 1992). This effect can be seen in Fig. 12.1, which shows that the loss of expression of mature HLA class I antigens can be discerned at day 1 post-infection and is virtually complete by day 5 post-infection. We conclude that, at late stages of infection, infected cells would be expected to be poor targets for CTLs. Decreased cell surface expression of class I antigens has now also been found following infection of mouse fibroblasts with murine CMV (Campbell *et al.*, 1992; Del Val *et al.*, 1992; Campbell & Slater, 1994).

## 12.3 Effect of CMV infection on the cell surface expression of HLA class I on bystander uninfected cells

The down-regulation of HLA class I is seen on the infected cell surface; however, at viral doses that initially result in less than 100% infection, a mixture of infected and uninfected cells is present and IFN-$\beta$, released from infected fibroblasts, can up-regulate class I on the bystander uninfected cells. This effect can be seen in Fig. 12.2, where the virus dose used resulted in infection of only about 40% of the cells by 24 hours, as determined by expression of a viral immediate early protein using flow cytometry. At this time, a similar percentage of cells had reduced cell surface class I expression, whilst the remainder (the uninfected cells) had normal levels of class I. By day 3 post-infection, two distinct peaks of cells could be observed, one expressing reduced HLA class I, corresponding to the percentage of cells expressing viral antigen, and another peak in which cells displayed a

---

Fig. 12.1. Expression of cell surface HLA class I antigens in CMV-infected human embryo lung fibroblasts. Human embryo lung fibroblasts were infected with CMV strain AD169 and analysed by confocal scanning laser microscopy for the cell surface expression of HLA class I. Immunofluorescent staining was performed on live cells using the mAb W6/32, and the cells were subsequently fixed with paraformaldehyde. Conventional phase-contrast images (left panels) and fluorescent images (right panels) were collected simultaneously. Uninfected cells are shown in the top row, cells at day 1 post-infection in the middle row and at day 5 post-infection in the bottom row. At day 1 post-infection, a substantial number of cells visible in the phase-contrast image have barely detectable expression of HLA class I and by day 5 it can be seen that the majority of infected cells has virtually no detectable expression of cell surface class I.

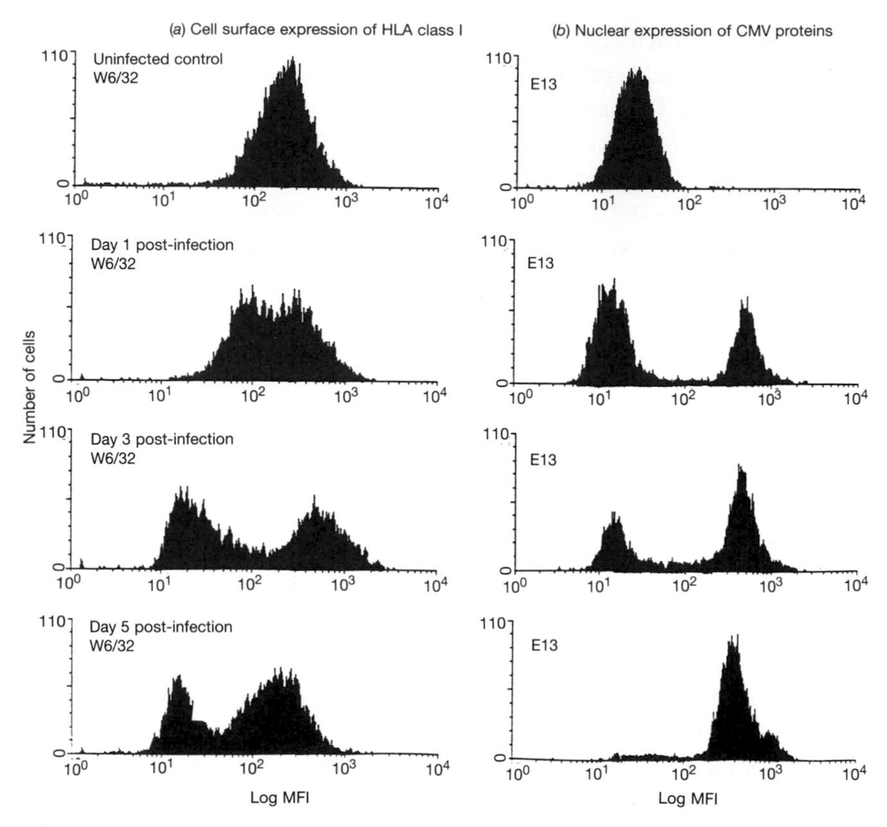


Fig. 12.2. Kinetics of HLA class I down-regulation in CMV-infected human embryo lung fibroblasts. Human embryo lung fibroblasts were infected with CMV strain AD169 and at daily intervals the cell surface expression of HLA class I was quantified by flow cytometry using mAb W6/32 (*a*). In parallel at each time point, cells were permeabilized and analysed by flow cytometry for the nuclear expression of the CMV immediate early proteins IE1 and IE2 (*b*), using mAb E13. The flow cytometric profiles show the mean fluorescence intensity (MFI) on the $x$ axis on a logarithmic scale and the number of cells on the $y$ axis on a linear scale. By day 1, only 43% of cells were expressing CMV proteins, and a similar proportion of cells (47%) had reduced cell surface expression of HLA class I (MFI 94.0 compared with MFI of 566.0 for the remaining 53% of cells in the infected culture on day 1, and an MFI of 443.6 for uninfected cells). On day 3 post-infection, two populations can be discerned, one with low class I (MFI 41.2) and the other with markedly increased class I (MFI 890.9). By day 5 the proportion of cells expressing CMV proteins had increased to 96%, and the overall level of class I expression was reduced compared with uninfected cells, although two peaks of class I expression could be discerned. The MFI of the left peak, representing 35% of the cells, was 19.3, whilst the remaining 65% of cells had an MFI of 225.7. Therefore, the cells initially infected by day 1 had virtually no cell surface class I by day 5, whilst the cells that became infected subsequently (by day 3) now had reduced cell surface expression of class I compared with the control uninfected cells.

markedly higher mean fluorescent intensity than the uninfected cells (Fig. 12.2), which presumably reflected the effect of IFN-β released from infected cells on the uninfected cells. At later stages, the infection had spread: by day 5 most cells were expressing viral antigen and the level of cell surface class I was reduced on all the cells (Fig. 12.2). By day 7, most cells had been infected with CMV for several days and almost all cells were virtually devoid of surface HLA class I (data not shown). In other studies using an even lower virus inoculum, which resulted in only 10% of cells being infected, the predominant overall effect was of increased class I (Grundy *et al.*, 1988). However, this increase could be abrogated by the addition of neutralizing antibody to IFN-β and, therefore, presumably represented IFN-induction of HLA class I on the predominantly uninfected cell population (Grundy *et al.*, 1988). A stimulatory effect on HLA class I expression has been reported following CMV infection of endothelial cells (van Dorp *et al.*, 1989), in conditions where only 10–20% of cells were infected, again probably reflecting interferon release by infected cells up-regulating HLA class I on the uninfected cells that comprised the majority of cells in the culture. Two populations, one with increased class I and the other with decreased class I, were observed following CMV infection of human aortic smooth muscle cells (Hosenpud, Chou & Wagner, 1991). Experiments performed with viral inocula that result in less than 100% initial infection are, therefore, difficult to interpret, unless double-labelling for viral antigen is used in order to allow the distinction between effects on the infected cell itself and bystander effects mediated by cytokine release on uninfected cells in the same culture. Whilst the former is of importance for CTL recognition of the infected cell, the latter effect might play a role in pathogenesis *in vivo*, for example by provoking host responses to alloantigens in transplant recipients.

## 12.4 Intracellular expression of HLA class I in CMV-infected cells

Despite the decreased class I expression on the infected cell surface, class I antigens could still be detected in the cytoplasm of these cells. When infected cells were permeabilized with acetone, expression of class I HLA and $\beta_2$-m could be detected from 48 hours post-infection (Barnes & Grundy, 1992). Optical sectioning of infected cells by confocal scanning laser microscopy demonstrated that the intracellular distribution of class I antigens was altered following CMV infection and suggested that class I molecules were present in a compartment adjacent to the nucleus, which could represent the Golgi apparatus (Barnes & Grundy, 1992). These studies were performed with antibodies against the mature heterodimer, such as PA2.6, as well as antibodies such as BBM.1 and L368 that react with both free and heavy-chain-bound $\beta_2$-m (Barnes & Grundy, 1992). We concluded that in CMV-infected cells, mature HLA class I molecules are synthesized but fail to be transported to the cell surface (Barnes & Grundy, 1992; Hutchinson *et al.*, 1991).

### 12.5 Synthesis of HLA class I antigens in CMV-infected cells

Browne *et al.* (1990) reported that HLA class I antigens were not synthesized in CMV-infected cells despite normal levels of mRNA, which would appear to contradict the hypothesis that the down-regulation of class I antigens on the surface of CMV-infected cells was the result of altered transport and not lack of synthesis. However, Burns *et al.* (1993) reported that the CMV major immediate early 1 gene *trans*-activated HLA class I genes, suggesting that synthesis of class I may even be increased by CMV infection. This apparent conflict has been partially resolved by the findings of two other groups. Yamashita *et al.* (1993) studied the synthesis of the HLA class I heavy chain and $\beta_2$-m by Western blotting and found that both the heavy chain and $\beta_2$-m were synthesized normally in infected cells at 24, 48 or 72 hours post-infection. They could immunoprecipitate both heavy chain and $\beta_2$-m with an anti-$\beta_2$-m antibody (ST1–2C), although they failed to do so with the antibody W6/32, and concluded that the defect occurred post-translationally, either at the level of correct complex formation or intracellular transport. Since they could detect the heterodimer with ST1–2C but not with W6/32, they postulated that the conformation of the molecule might be altered such that it was not recognized well by W6/32. As Browne *et al.* (1990) had used the antibody W6/32 in their experiments, this may explain their lack of detection of class I synthesis. Beersma, Biljlmakers & Ploegh (1993) subsequently reported that the stability of the heavy chain was affected by CMV infection, such that class I antigens detected by W6/32 in lysates of infected cells were rapidly degraded at 37 °C, whilst this did not happen in uninfected cell lysates. In agreement with the data of Yamashita *et al.* (1993), they found that synthesis of the heavy chain was unaffected by CMV infection (Beersma *et al.*, 1993). However, they found that following a 10 minute pulse-label, most of the heavy chain in uninfected cells was found in a W6/32 reactive complex with $\beta_2$-m, whilst in infected cells only 50% was complexed and heavy chain could be detected by a rabbit antiserum against free heavy chains. Pulse-chase experiments suggested that W6/32 reactive molecules were not stable in CMV-infected cells. Surprisingly, this effect was not seen in infected mouse L cells transfected with human heavy chain and human $\beta_2$-m, where complexes between murine heavy chain and human $\beta_2$-m were stable (Beersma *et al.*, 1993). However, since Yamashita *et al.* (1993) could detect the human HLA class I heterodimer with ST1–2C but not with W6/32, and the detection of the murine–human heterodimer used a mAb specific for H-2K$^k$ (Beersma *et al.*, 1993), it is difficult to distinguish between a conformational alteration in the heterodimer and its degradation in the case of the human heavy chain–human $\beta_2$-m combination. Clearly, if class I antigens are altered or unstable in infected cells, then caution will be needed in future studies in selecting the antibodies and the conditions (particularly the temperature) used for immunoprecipitations. Yamashita *et al.* (1993) used 4 °C and overnight incubation with the mAb, whilst Beersma *et al.* (1993) used 1 hour incubations and did not specify the temperature. Interestingly, the latter group found that incubation of infected cell lysates with uninfected cell lysates did not affect the detection of W6/32-reactive class I antigens in the uninfected cells, suggesting that induction of proteases was not responsible for degradation of class I in lysates of infected cells. Whether or not the apparent lack of reactivity of class I antigens in detergent lysates of CMV-infected cells is the result of their instability or altered conformation remains unclear. The state of class I antigens in CMV-infected cells *in vitro* under physiological conditions and the precise mechanism whereby class I antigens are synthesized but not transported to the cell surface remains to be ascertained.

Studies with murine CMV have shown that class I antigens are synthesized in the infected cell but are retained in the endoplasmic reticulum/cis-Golgi in an endo H-sensitive form (Del Val *et al.*, 1992). There did not appear to be degradation or altered conformation of the class I complexes in the murine system; however, the antibodies used may not be as conformationally sensitive as W6/32. It is not clear whether in the human system class I antigens are endo H sensitive or resistant. Beersma *et al.* (1993) did attempt to address this question in pulse-chase experiments, but at the appropriate chase time necessary to show conversion from the immature endo H-sensitive form to the mature endo H-resistant form in the uninfected cells, they could not detect class I antigens using W2/32 in infected cells; therefore, the question remains unanswered. The murine and human systems both demonstrate that class I antigens are synthesized in infected cells and in both systems altered transport of class I antigens appears to be the mechanism of the down-regulation at the cell surface. Differences may occur in the fate of the heterodimer within the infected cell in the two species, or the results may simply reflect the use of haplotype-specific antibodies in the murine system and monomorphic antibodies in the human system.

## 12.6 Role of the CMV HLA class I homologue in the down-regulation of cell surface class I following CMV infection

The CMV genome has been found to contain a gene (UL-18) with about 20% homology to the HLA class I heavy chain (Beck & Barrell, 1988). It was, therefore, possible that the binding of $\beta_2$-m by the CMV heavy chain homologue in the cytoplasm of CMV-infected cells prevented transport of class I antigens to the cell surface. Indeed, when the CMV homologue was expressed in BHK cells via a recombinant vaccinia virus and the cells co-infected with a second vaccinia recombinant carrying the human $\beta_2$-m gene, a 68 kDa protein product of the CMV UL-18 gene could be precipitated together with $\beta_2$-m by the antibody BBM.1 (Browne *et al.*, 1990), suggesting that the CMV class I homologue could form a complex with $\beta_2$-m. However, a protein product of this gene has not yet been identified in CMV-infected cells (Browne, Churcher & Minson, 1992) and we have been unable to precipitate any viral protein together with $\beta_2$-m or class I antigens using a panel of mAbs, including BBM.1. In addition, the down-regulation of class I antigens can be seen by 24 hours post-infection (Hutchinson *et al.*, 1991; Barnes & Grundy, 1992), whilst it might, perhaps, be expected that the CMV class I homologue (UL-18), a glycoprotein, would be produced at late times in infection. Furthermore, decreased cell surface expression of class I was seen in cells treated with an inhibitor of viral DNA synthesis (Yamashita *et al.*, 1993) that blocks synthesis of late glycoproteins, suggesting that a CMV immediate early or early gene product (and not a late glycoprotein) was involved.

Immunoprecipitation studies using a deletion mutant of CMV lacking the UL-18 gene have been performed (Browne *et al.*, 1992). However, as mentioned above, this group could not precipitate class I antigens with W6/32 in their system following infection with the wild-type virus (Browne *et al.*, 1990) and similiar findings were observed with the UL-18 deletion mutant (Browne *et al.*, 1992). Using the antibody BBM.1, $\beta_2$-m could be precipitated from cells infected with both the wild type and the deletion mutant. In the light of the work of Yamashita *et al.* (1993), referred to above, further studies with antibodies against the class I heterodimer other than W6/32 are needed to study class I synthesis in cells infected with the deletion mutant. In addition, the cell surface expression of class I was not analysed on cells infected with the deletion mutant. Therefore, it is difficult to draw conclusions about the role of the UL-18 gene from these studies.

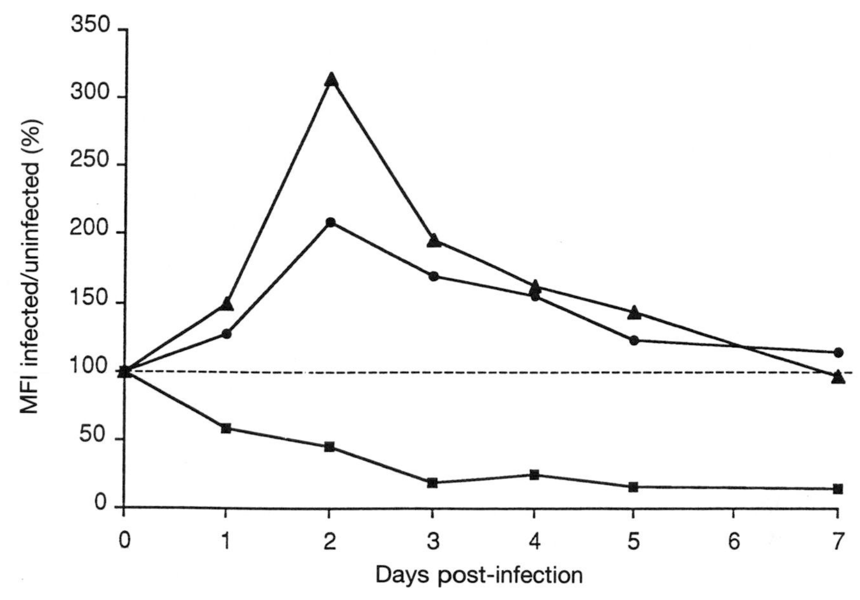


Fig. 12.3. Effects of CMV infection of fibroblasts on the cell surface expression of LFA-3(●), ICAM-1(▲) and HLA class 1(■). Antigen expression was quantified at various times post-infection with CMV by flow cytometry. The data shown represent the mean fluorescence intensity (MFI) of the level of expression observed on infected cells expressed as a percentage of that found on uninfected cells at each time point. The dotted line represents the baseline uninfected cell level. HLA class I expression was analysed using the antibody W6/32. Figure reprinted with permission from Grundy *et al.* (1993).

## 12.7 Effect of CMV infection on the cell surface expression of adhesion molecules

In contrast to the down-regulation of HLA class I on the infected cell surface, CMV infection induced enhanced cell surface expression of the adhesion molecules LFA-3 and ICAM-1 (Hutchinson *et al.*, 1991; Grundy & Downes, 1993) (Fig. 12.3). This effect was observed from two to four days post-infection, with a peak at the third day. At later stages of infection when infected cells were displaying a cytopathic effect, the expression of these adhesion molecules returned to the uninfected cell level, contrasting with the almost complete loss of cell surface HLA class I at this stage of infection (Hutchinson *et al.*, 1991; Barnes & Grundy, 1992, and Fig. 12.3). Therefore, it is unlikely that the altered transport of class I antigens in infected cells was a result of a general effect of CMV infection on the transport of proteins to the cell surface.

## 12.8 Effect of CMV infection on leukocyte adhesion

The increased expression of cell surface adhesion molecules on CMV-infected cells was accompanied by increased adherence of peripheral blood leukocytes to the infected cells (Grundy, Pahal & Akbar, 1993). This effect was mediated by the CD2-positive lymphocyte subset and mAbs against LFA-3 could block the binding, suggesting that the CD2-LFA-3 interaction was responsible for the increased adhesion. The time of increased lymphocyte adhesion correlated with the maximal enhancement of expression of LFA-3 on the infected cells. Effects on adhesion mediated by LFA-1–ICAM-1 were much lower and more variable (Grundy *et al.*, 1993); however, the peripheral blood leukocytes used in these studies were not activated and the affinity of LFA-1 for ICAM-1 increases following activation (Moingeon *et al.*, 1991). Hence, it is possible that the increased expression of ICAM-1 on infected cells might affect LFA-1-mediated binding of activated leukocytes. Increased leukocyte adhesion could be transferred to uninfected cells by supernatants from infected cells or co-cultures of leukocytes and infected cells (Grundy *et al.*, 1993), suggesting that, *in vivo*, soluble factors induced by CMV infection could increase leukocyte adherence to surrounding uninfected cells, thereby possibly contributing to increased inflammation.

## 12.9 Recognition of CMV-infected cells by MHC-restricted cytotoxic cells

CMV infection of cells can be divided into three phases, namely immediate early, early and late, representing phases of sequential viral gene expression. Several groups have studied the lysis of CMV-infected cells by MHC-restricted CTLs during the infectious cycle. Infected fibroblasts are good targets for MHC-restricted cytotoxicity at immediate early stages of infection, but become poor targets at late stages of infection. The major immediate early protein of human CMV, which is expressed by 4 hours post-infection, has been shown to be an important target for CTL recognition (Borysiewicz *et al.*, 1983; 1988a,b) and the major immediate early protein of murine CMV has also been shown to be a predominant CTL target (Koszinowski *et al.*, 1987; Reddehase *et al.*, 1987). Interestingly, recent work with human CMV has shown that the proteins of the incoming virus particle are also predominant CTL targets, thus the infected cell could be killed by CTLs in the absence of any viral protein synthesis (Riddell *et al.*, 1991). However, when late proteins of human CMV such as glycoprotein B are expressed via recombinant vaccinia viruses, they can act as good CTL targets (Borysiewicz *et al.*,

1988b). This suggests that the reason that late-stage-infected cells are poor CTL targets in the natural cycle of CMV infection is not because peptides of late proteins cannot be presented by HLA class I to CTLs, but that the altered transport of class I antigens may result in a deficient presentation of these peptides at these late stages of infection.

Defective presentation of peptides to CTLs has been described during infection with murine CMV by two laboratories (Campbell *et al.*, 1992; Del Val *et al.*, 1992). Both groups believed that a murine CMV early gene product was involved in the defective peptide presentation; however, whilst one group believed this to be because of the blocking of peptide-loaded class I antigens in the Golgi, the other group concluded that the defect was unrelated to reduced cell surface class I expression (Campbell *et al.*, 1992). In the former case, the peptide studied was the murine CMV major immediate early peptide whilst in the latter the peptide was from SV40 T antigen. Thus, CMV may have evolved more than one mechanism of evading CTL recognition of infected cells.

## 12.10 Recognition of CMV-infected cells by non-MHC restricted cytotoxic cells

In contrast to MHC-restricted lysis when CMV-infected fibroblasts are good targets at immediate early times post-infection, infected cells are good targets for non-MHC restricted killing at early and late stages of infection (Borysiewicz *et al.*, 1985). At these times, we have found that HLA class I expression is reduced, although adhesion molecule expression is high. A reciprocal relationship between HLA class I expression and NK cell lysis has been described in many systems, such that cells with low class I are often good NK cell targets. At times when CMV-infected cells were good targets for NK cell recognition, increased effector cell/target cell binding was observed (Borysiewicz *et al.*, 1985). This also correlated with the times at which we found increased expression of LFA-3 and ICAM-1 on the infected cells (Grundy & Downes, 1993) and increased adherence of CD2-positive lymphocytes (Grundy *et al.*, 1993), a subset of which includes NK cells. Therefore, the correlation in susceptibility to NK cell lysis of CMV-infected cells may be with enhanced adhesion molecule expression and increased effector cell adherence rather than decreased HLA class I expression.

It should be noted that caution must be exercised in viewing the data on NK cell lysis of cells infected with human CMV, since in the majority of cases 18–20 hours $^{51}Cr$ release assays were used (Starr *et al.*, 1984; Borysiewicz *et al.*, 1985; Bandyopadhyay *et al.*, 1987), rather than the conventional 4 hour assays. In our experience, non-MHC restricted killing of CMV-infected fibroblasts by *in vivo* activated cells from bone marrow transplant recipients did not occur in 4 hour $^{51}$Cr release assays, whilst the same effector cells were capable of killing lymphoblastoid target cells infected with Epstein–Barr virus in that time period (Duncombe *et al.*, 1992). When CMV-infected fibroblasts were co-cultivated with peripheral blood leukocytes from these patients, substantial augmentation of cytokine production ensued, which was not seen in co-cultures with uninfected fibroblasts (Duncombe *et al.*, 1990). Therefore it is likely that in the 18 hour $^{51}$Cr release assays, cytokine release contributes to activation of the non-MHC-restricted effector cells in cultures with the CMV-infected fibroblasts, but not with the uninfected cells. The resultant lysis of the infected cells and not the uninfected cells may therefore, reflect differences in the activation of effector cells rather than, or as well as, in recognition of the target cells. Indeed in our studies with leukocytes from the bone marrow transplant recipients referred to above, the addition of IL-2 to the cultures resulted in equivalent killing of both infected and uninfected fibroblasts in 4 hour $^{51}$Cr release assays (Duncombe *et al.*, 1991). A requirement for HLA DR-positive accessory cells for NK cell lysis of CMV-infected target cells has been demonstrated by others in 18 hour assays (Bandyopadhyay *et al.*, 1986) and this group has also found that the addition of IL-2 or IFN-$\alpha$ or IFN-$\beta$ could enhance NK cell lysis of both infected and uninfected target cells (Bandyopadhyay *et al.*, 1987). Others have also shown that the time period for target cell lysis could be reduced to 6 hour by pretreatment of the effector cells with IFN (Borysiewicz *et al.*, 1985), although these authors still found increased lysis of infected cells compared with uninfected target cells.

We do not have clear evidence that the loss of HLA class I during CMV infection, or the possible appearance of unknown NK cell recognition structures on the infected cell, actually results in increased susceptibility of the infected cell to NK cell lysis. Activation of effector cells, increased adhesion molecule expression and increased effector-target binding also appear to be involved. Despite the complexity of these interactions *in vitro*, NK cells have been shown to play an important role in defence against murine CMV infection *in vivo* (Shanley, 1990; Welsh *et al.*, 1991).

## 12.11 Conclusions

CMV infection clearly results in a dramatic loss of cell surface HLA class I antigens, concomitant with increased expression of the cell surface adhesion molecules LFA-3 and ICAM-1. The decreased expression of cell surface class

I correlates with decreased susceptibility to CTL lysis late in infection, whilst the adhesion molecule increase is associated with greater binding of CD2-positive lymphocytes to the infected cells, an effect that can be transferred to uninfected cells by soluble factors released from CMV-infected cells. These changes in expression of immunologically important cell surface molecules may both aid the virus to escape cell-mediated immunity and establish latency and contribute to increased inflammation and immunopathology. Understanding the mechanisms by which CMV induces such changes will help in designing strategies to control or prevent CMV infection and in establishing the role of the virus in pathogenesis of disease *in vivo*.

## Acknowledgements

The contribution of Luci MacCormac and Dr Annette Manning to the data presented in this review is gratefully acknowledged. Much of the research work described originating from this laboratory was supported by the Wellcome Trust.

## References

Alford, C.A., Stagno, S., Pass, R.F. & Britt, W.J. (1990). Congenital and perinatal cytomegalovirus infections. *Reviews of Infectious Diseases*, 12 (Suppl 7), S745–S753.

Bandyopadhyay, S., Miller, D.S., Matsumoto-Kobayashi, M., Clark, S.C. & Starr, S.E. (1987). Effects of interferons and interleukin 2 on natural killing of cytomegalovirus-infected fibroblasts. *Clinical and Experimental Immunology*, 67, 372–382.

Bandyopadhyay, S., Perussia, B., Trinchieri, G., Miller, D.S. & Starr, S.E. (1986). Requirement for HLA-$DR^+$ accessory cells in natural killing of cytomegalovirus-infected fibroblasts. *Journal of Experimental Medicine*, 164, 180–195.

Barnes, P.D. & Grundy, J.E. (1992). Down-regulation of the class I HLA heterodimer and $\beta_2$-microglobulin on the surface of cells infected with cytomegalovirus. *Journal of General Virology*, 73, 2395–2405.

Beck, S. & Barrell, B.G. (1988). Human cytomegalovirus encodes a glycoprotein homologous to MHC class I antigens. *Nature (London)*, 331, 269–272.

Beersma, M.F.C., Biljlmakers, M.J.E. & Ploegh, H.L. (1993). Human cytomegalovirus down-regulates HLA class I expression by reducing the stability of class I H chains. *Journal of Immunology*, 151, 4455–4464.

Borysiewicz, L.K., Graham, S., Hickling, J.K., Mason, P.D. & Sissons, J.G. (1988a). Human cytomegalovirus-specific cytotoxic T cells: their precursor frequency and stage specificity. *European Journal of Immunology*, 18, 269–275.

Borysiewicz, L.K., Hickling, J.K., Graham, S., Sinclair, J., Cranage, M.P., Smith, G.L. & Sissons, J.G. (1988b). Human cytomegalovirus-specific cytotoxic T cells. Relative frequency of stage-specific CTL recognizing the 72-kD immediate early protein and glycoprotein B expressed by recombinant vaccinia viruses. *Journal of Experimental Medicine*, 168, 919–931.

Borysiewicz, L.K., Morris, S., Page, J.D. & Sissons, J.G. (1983). Human cytomegalovirus-specific cytotoxic T lymphocytes: requirements for *in vitro* generation and specificity. *European Journal of Immunology*, 13, 804–809.

Borysiewicz, L.K., Rodgers, B., Morris, S., Graham, S. & Sissons, J.G. (1985). Lysis of human cytomegalovirus infected fibroblasts by natural killer cells: demonstration of an interferon-independent component requiring expression of early viral proteins and characterization of effector cells. *Journal of Immunology*, 134, 2695–2701.

Browne, H., Churcher, M. & Minson, T. (1992). Construction and characterization of a human cytomegalovirus mutant with the UL18 (class I homolog) gene deleted. *Journal of Virology*, 66, 6784–6787.

Browne, H., Smith, G., Beck, S. & Minson, T. (1990). A complex between the MHC class I homologue encoded by human cytomegalovirus and $\beta_2$-microglobulin. *Nature (London)*, 347, 770–772.

Burns, L.J., Waring, J.F., Reuter J.J., Stinski, M.F. & Ginder, G.D. (1993). Only the HLA class I gene minimal proximal elements are required for transactivation by human cytomegalovirus immediate early genes. *Blood*, 81, 1558–1566.

Campbell, A.E & Slater, J.S. (1994). Down-regulation of major histocompatibility complex class I synthesis by murine cytomegalovirus early gene expression. *Journal of Virology*, 68, 1805–1811.

Campbell, A.E., Slater, J.S., Cavanaugh, V.J. & Stenberg, R.M. (1992). An early event in murine cytomegalovirus replication inhibits presentation of cellular antigens to cytotoxic T lymphocytes. *Journal of Virology*, 66, 3011–3017.

Collaborative DHPG Treatment Study Group (1986). Treatment of serious cytomegalovirus infections with 9-(1,3-dihydroxy-2-propoxymethyl) guanine in patients with AIDS and other immunodeficiencies. *New England Journal of Medicine*, 314, 801–805.

Darlington, J., Super, M., Patel, K., Grundy, J.E., Griffiths, P.D. & Emery, V.C. (1991). Use of polymerase chain reaction to analyse sequence variation within a major neutralizing epitope of glycoprotein B (gp58) in clinical isolates of human cytomegalovirus. *Journal of General Virology*, 72, 1985–1989.

Del Val, M., Hengel, H., Hacker, H., Hartlaub, U., Ruppert, T., Lucin, P. & Koszinowski, U.H. (1992). Cytomegalovirus prevents antigen presentation by blocking the transport of peptide-loaded major histocompatibility complex class I molecules into the medial-Golgi compartment. *Journal of Experimental Medicine*, 176, 729–738.

Duncombe, A.S., Grundy, J.E., Oblakowski, P., Prentice, H.G., Gottlieb, D.J., Roy, D.M., Reittie, J.E., Bello-Fernandez, C., Hoffbrand, A.V. & Brenner, M.K. (1992). Bone marrow transplant recipients have defective MHC-unrestricted cytotoxic responses against cytomegalovirus in comparison with Epstein–Barr virus: the importance of target cell expression of lymphocyte function-associated antigen 1 (LFA-1). *Blood*, 79, 3059–3066.

Duncombe, A.S., Grundy, J.E., Prentice, H.G. & Brenner, M.K. (1991). Activated killer cells may contribute to cytomegalovirus induced hypoplasia after bone marrow transplantation. *Bone Marrow Transplantation*, 7, 81–87.

Duncombe, A.S., Meager, A., Prentice, H.G., Grundy, J.E., Heslop, H.E., Hoffbrand, A.V. & Brenner, M.K. (1990). Gamma-interferon and tumor necrosis factor production after bone marrow transplantation is augmented by exposure to marrow fibroblasts infected with cytomegalovirus. *Blood*, 76, 1046–1053.

Furukawa, T., Hornberger, E., Sakuma, S. & Plotkin, S.A. (1975). Demonstration of immunoglobulin G receptors induced by human cytomegalovirus. *Journal of Clinical Microbiology*, 2, 332–336.

Griffiths, P.D. & Grundy, J.E. (1987). Molecular biology and immunology of cytomegalovirus. *Biochemical Journal*, 241, 313–324.

Grundy, J.E. (1991). The immune response to cytomegalovirus. In: *Progress in Cytomegalovirus Research*, ed. M.P. Landini, pp. 143–155. Amsterdam: Elsevier.

Grundy, J.E., Ayles, H.M., McKeating, J.A., Butcher, R.G., Griffiths, P.D. & Poulter, L.W. (1988). Enhancement of class I HLA antigen expression by cytomegalovirus: role in amplification of virus infection. *Journal of Medical Virology*, 25, 483–495.

Grundy, J.E. & Downes, K.L. (1993). Up-regulation of LFA-3 and ICAM-1 on the surface of fibroblasts infected with cytomegalovirus. *Immunology*, 78, 405–412.

Grundy, J.E., Pahal, G.S. & Akbar, A.N. (1993). Increased adherence of CD2 positive peripheral blood lymphocytes to cytomegalovirus infected fibroblasts is blocked by anti-LFA-3 antibody. *Immunology*, 78, 413–418.

Hosenpud, J.D., Chou, S.W. & Wagner, C.R. (1991). Cytomegalovirus-induced regulation of major histocompatibility complex class I antigen expression in human aortic smooth muscle cells. *Transplantation*, 52, 896–903.

Hutchinson, K., Eren, E. & Grundy, J.E. (1991). Expression of ICAM-1 and LFA-3 following cytomegalovirus infection. In: *Progress in Cytomegalovirus Research*, ed. M.P. Landini, pp. 267–270, Amsterdam: Elsevier.

Koszinowski, U.H., Keil, G.M., Schwarz, H., Schickedanz, J. & Reddehase, M.J. (1987). A nonstructural polypeptide encoded by immediate-early transcription unit 1 of murine cytomegalovirus is recognized by cytolytic T lymphocytes. *Journal of Experimental Medicine*, 166, 289–294.

Meyers, J.D., Flournoy, N. & Thomas, E.D. (1982). Nonbacterial pneumonia after allogeneic marrow transplantation: a review of ten years' experience. *Reviews of Infectious Diseases*, 4, 1119–1132.

Moingeon, P.E., Lucich, J.L., Stebbins, C.C., Recny, M.A., Wallner, B.P., Koyasu, S. & Reinherz, E.L. (1991). Complementary roles for CD2 and LFA-1 adhesion pathways during T cell activation. *European Journal of Immunology*, 21, 605–610.

Reddehase, M.J., Mutter, W., Munch, K., Buhring, H.J. & Koszinowski, U.H. (1987). CD8-positive T lymphocytes specific for murine cytomegalovirus immediate-early antigens mediate protective immunity. *Journal of Virology*, 61, 3102–3108.

Riddell, S.R., Rabin, M., Geballe, A.P., Britt, W.J. & Greenberg, P.D. (1991). Class I MHC-restricted cytotoxic T lymphocyte recognition of cells infected with human cytomegalovirus does not require endogenous viral gene expression. *Journal of Immunology*, 146, 2795–2804.

Shanley, J.D. (1990). *In vivo* administration of monoclonal antibody to the NK 1.1 antigen of natural killer cells: effect on acute murine cytomegalovirus infection. *Journal of Medical Virology*, 30, 58–60.

Springer, T.A. (1990). Adhesion receptors of the immune system. *Nature (London)*, 346, 425–434.

Starr, S.E., Smiley, L., Wlodaver, C., Friedman, H.M., Plotkin, S.A. & Barker, C.

(1984). Natural killing of cytomegalovirus-infected targets in renal transplant recipients. *Transplantation*, 37, 161–164.

Townsend, A.R., Rothbard, J., Gotch, F.M., Badahur, G., Wraith, D. & McMichael, A.J. (1986). The epitopes of influenza nucleoprotein recognized by cytotoxic T lymphocytes can be defined with short synthetic peptides. *Cell*, 44, 959–968.

van Dorp, W.T., Jonges, E., Bruggeman, C.A., Daha, M.R., van Es, L.A. & van der Woude, F.J. (1989). Direct induction of MHC class I, but not class II, expression on endothelial cells by cytomegalovirus infection. *Transplantation*, 48, 469–472.

Welsh, R.M., Brubaker, J.O., Vargas-Cortes, M. & O'Donnell, C.L. (1991). Natural killer (NK) cell response to virus infections in mice with severe combined immunodeficiency. The stimulation of NK cells and the NK cell-dependent control of virus infections occur independently of T and B cell function. *Journal of Experimental Medicine*, 173, 1053–1063.

Yamashita, Y., Shimokata, K., Mizuno, S., Yamaguguchi, H. & Nishiyama, Y. (1993). Down-regulation of the surface expression of class I MHC antigens by human cytomegalovirus. *Virology*, 193, 727–736.

# 13

# Oncogenes and MHC class I expression

LUCY T. C. PELTENBURG and PETER I. SCHRIER
*University Hospital, Leiden*

## 13.1 Introduction: immune recognition of tumour cells

In the late 1980s it became clear that cancer cells develop by multiple genetic alterations (reviewed in Weinberg, 1989; Bishop, 1991). These alterations include activation of proto-oncogenes as well as inactivation of tumour suppressor genes. Proto-oncogenes, more commonly called oncogenes, exert important functions in cell proliferation and differentiation and their activity is usually tightly controlled to ensure a minimal risk of inappropriate activity. Activation of oncogenes in animal and human tumours may occur through several mechanisms including amplification, elevated expression and point mutations. These genetic alterations usually result in an altered activity of the oncogene-encoded protein and this contributes to uncontrolled proliferation.

In the light of the crucial role that HLA class I antigens play in the interaction of altered self antigens or viral antigens with CTLs (see Chapter 1), one would expect that antigens specifically present in tumours are presented by HLA class I molecules. Such antigens might be viral antigens in the case of virally induced tumours or altered self proteins in the case of tumours induced by xenobiotics like carcinogens or radiation. In numerous animal and human tumours, mutations in oncogenes have been shown to be responsible for the tumorigenic properties of the tumour cell. These mutations are potential targets for recognition by the immune system of the host. Candidate oncogene and tumour suppressor proteins are *ras* and p53 proteins, respectively; these can be activated by various mutations and are involved in many forms of human cancer (reviewed in Bos, 1989; Levine, Momand & Finlay, 1991).

For human melanomas, similar observations point to the existence of different tumour-specific antigens that can be discerned by HLA class I-restricted CTLs (Knuth *et al.*, 1989; van den Eynde *et al.*, 1989; Degiovanni *et al.*, 1990). In two independent studies (Wölfel *et al.*, 1989; Crowley *et al.*, 1991), HLA-A2 could be assigned as the allele presenting a tentative tumour-specific peptide. Recently, a gene designated MAGE-1, encoding a self antigen that was specifically recognized by autologous CTLs was isolated from a human melanoma cell line (van der Bruggen *et al.*, 1991; Traversari *et al.*, 1992b). A MAGE-1-derived nonapeptide is recognized on the HLA-A1 molecule by CTLs (Traversari *et al.*, 1992a). The combined findings so far indicate that T cell responses against self antigens on tumour cells can occur and that MHC antigens play a crucial role in the presentation of these antigens to T cells. From this point of view, it is evident that modulation of MHC class I expression will have severe consequences for the interaction of immune effector cells with tumour cells. The observation that several activated oncogenes are capable of down-modulating MHC class I expression is intriguing, because potential recognition of tumour-specific peptides by CTLs is thereby reduced or even abolished. In this chapter the expression and regulation of MHC class I by the *myc* oncogene will be reviewed. Work performed on MHC class I regulation by the adenovirus E1A oncogene provided an important model for the study of class I regulation by nuclear oncogenes. Although the subject of class I regulation by E1A is extensively reviewed in Chapter 8, the major features of this system relevant to our understanding of *myc* regulation of class I expression will be briefly reviewed. The potential effects of this regulation for the elimination of malignant cells by the immune system will also be discussed.

## 13.2 Modulation of MHC class I expression

### *13.2.1 Regulation of MHC class I expression in adenovirus-transformed cells*

The first discovery that transforming genes may affect MHC class I expression was made several years ago when the reactivity of heteroantisera raised in mice against transformed rat cells was studied (Schrier *et al.*, 1983). The sera raised against non-oncogenic Ad5-transformed rat cells recognized an epitope absent on oncogenic Ad12-transformed cells (see Chapter 8). This epitope turned out to be an as yet unidentified MHC class I molecule non-covalently linked to $\beta_2$-m. In addition, rat alloantisera detected a dramatic difference between the two types of cell, indicating that MHC class I expression is switched off in the oncogenic Ad12-transformed cells, but not in the non-oncogenic Ad5-transformed cells. Transformation with the viral E1A oncogene was sufficient to establish the down-modulation of the class I heavy chain mRNA and protein (Schrier *et al.*, 1983; Vaessen *et al.*, 1986). The effect was also seen in murine and human cells transformed by

Ad12 (Eager *et al.*, 1985; Vaessen *et al.*, 1986; Vasavada *et al.*, 1986; Grand *et al.*, 1987). The differential effect of transformation by Ad12 E1A and by Ad5 E1A on down-modulation of MHC class I expression (Schrier *et al.*, 1983) might explain why Ad5-transformed cells are not oncogenic, in contrast to Ad12-transformed cells, which are highly oncogenic: because of low MHC class I expression, the latter may evade T cell immunity (Bernards *et al.*, 1983). This effect can be reversed: transfection of an exogenous H-2 gene into the oncogenic cells in order to raise MHC class I expression results in reduction of their oncogenic potential (Tanaka *et al.*, 1985; 1986). An Ad5-specific MHC class I-restricted CTL clone recognizing an E1A-encoded peptide has been isolated and shown to be capable of efficiently eliminating Ad5 E1A-induced tumours in nude mice (Kast *et al.*, 1989). Comparable Ad12-specific CTLs have not yet been found, even after MHC class I expression on the stimulator cells was increased (W. M. Kast, personal communication). The poor binding of Ad12 E1A peptide(s) to MHC class I antigens might play a role in this phenomenon. This is favoured by the notion that the amino acid sequence of the major Ad5 MHC class I-binding peptide (Kast *et al.*, 1989; Kast & Melief, 1991) is unique and not present in the Ad12 E1A protein (W. M. Kast, personal communication). Since the oncogenicity of Ad12-transformed cells can be reversed by transfection of exogenous MHC class I genes, it can be assumed that such a peptide presentation defect can be (partially) overcome by high MHC class I expression.

### *13.2.2 Regulation of MHC class I expression in tumour cells*

Down-modulation of MHC class I expression has been shown in many animal and human tumours (reviewed in Schrier & Peltenburg, 1993; see also Chapters 14 and 15). Murine tumours often are devoid of MHC class I antigens and losses seem not to be restricted to particular tumour types. Well-known examples are the B16 melanoma cell line (Nanni *et al.*, 1983; Taniguchi, Kärre & Klein, 1985), the methylcholanthrene-induced T10 fibrosarcoma (Katzav *et al.*, 1983) and the spontaneous Lewis lung carcinoma (Isakov *et al.*, 1983). In addition, tumours of haemopoietic origin, e.g. the AKR lymphoma (Hui, Grosveld & Festenstein, 1984), can be H-2 negative. Other experimental tumours lacking MHC class I expression have also been described (Garrido *et al.*, 1986; Bahler *et al.*, 1987; Nishimura *et al.*, 1988). In all these cases, the precise genetic alterations leading to transformation into tumour cells are not known and, therefore, no link between a particular genetic defect and the modulation of MHC class I expression can be established.

## 13.3 Regulation of MHC class I expression by *myc*

### *13.3.1 The* myc *oncogenes*

The *myc* proto-oncogenes family consists of at least six genes, c-, N-, L-, P-, R- and B-*myc* (Schwab *et al.*, 1983; Nau *et al.*, 1985; Alt *et al.*, 1986; Ingvarsson *et al.*, 1988). Here, attention will be focussed on c-*myc* and N-*myc*, because these two genes have been shown to be capable of altering HLA class I expression. These genes consist of three exons in humans, encoded on chromosome 8 (c-*myc*) and chromosome 2 (N-*myc*). Exon 1 is non-coding, exons 2 and 3 code for a nuclear 65–67 kDa phosphoprotein that complexes with another cellular protein, termed Max, to form a DNA-binding complex that interacts with the consensus nucleotide sequence CACGTG (Blackwell *et al.*, 1990; Blackwood & Eisenman, 1991). The precise function of the *myc* proteins is not known, but they play a role in the regulation of the cell cycle, in particular at the onset of proliferation and in differentiation of certain cell lineages (Cole, 1986; 1991; Lüscher & Eisenman, 1990). The mechanism by which these processes are effected usually involves elevated expression of the *myc* gene, probably leading to enhanced rates of transcription of a variety of other cellular genes involved in proliferation and cell growth. Expression of c-*myc* can be activated in many cell types, while the expression of N-*myc* seems restricted to only certain lineages derived from the neuronal crest. The two genes are expressed in distinct embryonal compartments during development (Stanton *et al.*, 1992). The expression of N-*myc* is often correlated with a particular state of differentiation (Thiele, Reynolds & Israel, 1985; Stanton, Schwab & Bishop, 1986; Zimmerman *et al.*, 1986). Mice homozygous for a targeted disruption of the N-*myc* gene die pre-natally, which demonstrates its essential role during development (Stanton *et al.*, 1992; Charron *et al.*, 1992). Strikingly, in these mice, the c-*myc* gene is expressed in neuroepithelial tissue, a site where it is normally not present, which indicates the existence of cross-regulation of the expression of the two *myc* genes (Stanton *et al.*, 1992).

Activation of the c-*myc* oncogene has been found in numerous forms of human cancer. Prominent examples are colon carcinoma (Viel *et al.*, 1990) and lung carcinoma (Bergh, 1990). Usually, the activation consists of augmented expression, which is often a result of gene amplification. Activation of c-*myc* by chromosomal translocations involving the enhancer sequences of immunoglobulin genes occurs in Burkitt's lymphoma (Klein, 1983). In addition, in other types of B cell malignancy, *myc* rearrangements are involved, often in concert with activation of the *bcl-2* gene (Klein, 1991).

The N-*myc* gene is primarily involved in neuroblastoma (Schwab *et al.*, 1983; Seeger *et al.*, 1985) and small cell lung cancer (Nau *et al.*, 1986; Bergh, 1990). The gene is often amplified, and in neuroblastoma the degree of amplification is inversely correlated with progression-free survival (Seeger *et al.*, 1985). In certain cases of neuroblastoma, including neuroepithelioma, c-*myc* was found to be activated, coinciding with low or absent N-*myc* expression (Rosolen *et al.*, 1990; Versteeg *et al.*, 1990), again suggesting that their expression is mutually exclusive.

### *13.3.2 MHC class I down-modulation by* myc

Remarkably, in a number of tumour types where *myc* has often been found to be activated, low expression of HLA class I antigens was demonstrated. These tumour types include breast carcinomas, Burkitt's lymphomas, cervical carcinomas, colorectal carcinomas, melanomas and, most prominently, neuroblastomas and small cell lung carcinomas (reviewed in Schrier & Peltenburg, 1993). For most tumour types, low HLA class I expression and *myc* activation were found in independent studies, but in neuroblastoma and melanoma, the two phenomena were found in the same individual tumours (Bernards, Dessain & Weinberg, 1986; Versteeg *et al.*, 1988). This suggests the possibility that elevated expression of c-*myc* or N-*myc* can regulate HLA class I expression. This idea is further supported by the finding that the c-*myc* and N-*myc* genes have demonstrated functional homology with the adenovirus E1A gene. Firstly, either of these genes can co-operate with an activated *ras* oncogene in transforming primary rodent cells (Land, Parada & Weinberg, 1983; Ruley, 1983; Schwab *et al.*, 1985; Land *et al.*, 1986). Secondly, both E1A and *myc* proteins are capable of binding to the retinoblastoma gene product (Whyte *et al.*, 1988; Rustgi, Dyson & Bernards, 1991) and to the transcription factor TBP (Lee *et al.*, 1991; Hateboer *et al.*, 1993). This indicates that the products of the *myc* genes and the E1A gene interact with similar target structures in the cell and may, therefore, have similar functional effects, other than the transformation of primary cells *per se*.

One of the peculiar properties of E1A proteins is their capability to down-modulate MHC class I expression along with transformation. The question, therefore, arises whether c-*myc* or N-*myc* can also down-modulate MHC class I expression. A number of independent studies in melanoma and neuroblastoma tumours and cell lines reveals that this is indeed the case. In a panel of melanoma cell lines, an inverse correlation between c-*myc* expression and HLA class I expression was found (Versteeg *et al.*, 1988; 1989a,b; Schrier *et al.*, 1991). In a series of neuroblastoma cell lines, the same phenomenon was found for the N-*myc* gene (Bernards *et al.*, 1986; Bernards & Lenardo, 1990; Gross, Beck & Favre, 1990; Versteeg *et al.*, 1990; Sugio, Nakagawara & Sasazuki, 1991). A proof for a direct role of c-*myc* and N-*myc* in switching off MHC class I expression came from transfection experiments, in which functional *myc* genes were transfected into tumour lines with low expression of *myc*, i.e. melanoma cell lines with c-*myc* (Versteeg *et al.*, 1988; 1989a,b) and a rat neuroblastoma cell line with N-*myc* (Bernards *et al.*, 1986). Transfected clones with high expression of the *myc* genes showed low MHC class I expression at the mRNA as well as at the protein level. The expression of $\beta_2$-m was not affected in the melanoma cell lines but was low in one of the two transfected rat neuroblastoma cell lines. These experiments showed a direct correlation between elevated *myc* expression and low MHC class I expression.

### *13.3.3 Locus-specific down-modulation by* c-myc

When the assay for HLA class I expression was performed with locus-specific probes, surprisingly, only down-modulation of HLA-B genes was noted in melanoma cell lines (Versteeg *et al.*, 1989a). Down-modulation of HLA-A occurred to a much lesser extent and down-regulation of HLA-C mRNA expression was found only in one cell line. Therefore, the down-regulation of HLA-C might be allele specific rather than locus specific. The low HLA-B expression is not an effect of establishment of the cell lines or long-term culture, because in two patients whose tumour material was available, the selective abrogation of HLA class I expression was also found in sections of the original tumour using locus-specific antibodies in immunohistochemistry (Versteeg *et al.*, 1989a). Moreover, the selective loss of HLA-B was also apparent after transfection of c-*myc* into two different cell lines with low endogenous c-*myc* expression: in these cases HLA-B8 and HLA-Bw62 proteins were down-regulated, while expression of HLA-A1 and HLA-A2 remained largely unaltered (Versteeg *et al.*, 1989a). In various transfected clones, the level of HLA class I down-modulation precisely correlated with the amount of c-*myc* protein analysed on Western blots (Peltenburg, Dee & Schrier, 1993). The expression of HLA class I is regulated at the transcriptional level (see Chapters 4 and 8) and, therefore, it seems likely that the action of c-*myc* is limited to specific sequences in the promoter of the HLA-B genes that differ in the HLA-A promoter.

Locus-specific regulation of HLA-B alleles has been found in several other human tumours, in particular in bladder carcinoma (Nouri *et al.*, 1990; 1991) and renal cell carcinoma (P. I. Schrier, unpublished results). In other tumour types, such as Burkitt's lymphoma (Andersson *et al.*, 1991) and colorectal carcinoma (Soong & Hui, 1991; Momburg *et al.*, 1989), several HLA-A alleles are selectively down-modulated. In none of these cases has a direct correlation with activation of c-*myc* been reported. In small cell lung carcinomas (Marley *et al.*, 1989) and in renal tumours of the Wilms' type (Shaw *et al.*, 1988; Maitland *et al.*, 1989) a correlation between c-*myc* expression and HLA class I down-modulation was found. However, in these cases, no locus-specific analysis was performed. A locus-specific study on v-*myc* transfected human monoblastic U937 showed allele-specific down-modulation of HLA class I: only HLA-A3 was unaffected by the v-*myc* over-expression (Larsson *et al.*, 1992).

No inverse correlation between high c-*myc* expression and HLA class I expression was found in other forms of human cancer (Minafra *et al.*, 1989; Dahllöf, 1990; Redondo *et al.*, 1991; Soong, Oei & Hui, 1991). An obvious explanation for the discrepancy might be that the analyses were usually not performed with locus-specific probes and, therefore, low expression of specific alleles might have been overlooked. An alternative explanation might be that the effects of *myc* expression on MHC class I expression are lineage specific. It can be hypothesized that the regulatory suppressive factors mediating the effect of *myc* are merely present in certain tissues or certain differentiation lineages. In this context, it may be relevant that tumours in which an effect of c-*myc* or N-*myc* was often found, i.e. neuroblastoma, small cell lung carcinoma and melanoma, are all of neuroendocrine origin, and it may be speculated that expression of the suppressive factors is only permitted in these types of cell. Even more specifically, expression may be limited to a particular stage of differentiation or development at which the cells were arrested during tumorigenesis. This may explain why in two neuroblastoma cell lines with low N-*myc* expression, HLA class I expression could not be down-regulated by transfection of the N-*myc* gene (Feltner *et al.*, 1989). Another possibility to reconcile the inconsistencies might be that the level of *myc* protein in the tumour cell lines or in the transfected cell lines is not sufficiently high to bring about the HLA class I down-modulating effect. Our own experience is that, in the case of c-*myc*, the extent of down-modulation clearly depends on the level of c-*myc* expression. The function of c- and N-*myc* genes is in many respects similar. With respect to MHC class I regulation, however, these genes act differently: N-*myc* down-regulates all HLA loci, while c-*myc* acts only on HLA-B loci. Indeed, differences in transcriptional regulation have been found: N-*myc* acts via modulation of binding of the transcription factor KBF1 to a domain of the class I regulatory element, termed enhancer A (see Chapter 4), while c-*myc* exerts its effect through another regulatory sequence in the HLA class I promoter (see below).

### *13.3.4 Mechanism of down-regulation by* myc

In most cases where the mechanism of down-regulation of MHC class I in tumour cells has been studied in detail, expression turned out to be regulated at the level of transcription. A number of studies favour the notion that suppression occurs through a major *cis*-regulatory element of MHC class I genes, termed either enhancer A or region I of the class I regulatory element, CRE (see Chapter 4). Firstly, in AKR leukaemias, the absence of H-2K$^k$ antigens is correlated with a decrease of binding of transcription factor H2TF1 (also termed KBF1) to enhancer A (Henseling *et al.*, 1990). Secondly, the locus-specific down-regulation of HLA class I mRNA in colorectal carcinoma seems to involve changes in protein binding to the enhancer A region (Soong *et al.*, 1991), although in additional studies on these cell lines other promoter regions were implicated (Soong & Hui, 1992). Thirdly, in a panel of 23 human tumour cell lines, low levels of KBF1 and NF-κB binding to enhancer A, correlating with low HLA class I expression, were found (Blanchet *et al.*, 1992). However, in all these cases the (presumptive oncogene-encoded) factor responsible for down-regulation has not been identified. A similar mechanism of transcriptional regulation is involved in the down-modulation of MHC class I expression in neuroblastoma cells where N-*myc* is activated. As shown by gel retardation assays, using the core enhancer A-binding site as a probe, elevated N-*myc* expression appeared to correlate with reduced binding of KBF1 to enhancer A in both human neuroblastoma cell lines and rat neuroblastoma transfectants (Lenardo *et al.*, 1989). As described in Chapter 4, the KBF1 protein is a homodimer of the p50 protein, which also forms part of the p50–p65 heterodimeric transcription factor NF-κB. The reduction in enhancer-binding activity by N-*myc* was shown to be caused by suppressing the level of p50 mRNA, because re-expression of p50 mRNA at high levels in neuroblastoma cells that over-express N-*myc* led to restoration of activity of the MHC class I enhancer and to re-expression of MHC class I antigens (van 't Veer, Beijersbergen & Bernards, 1993).

In contrast to N-*myc*, which has been shown to be capable of suppressing the expression of all MHC class I loci in neuroblastoma, activation of c-*myc* in human melanoma results in down-modulation of predominantly HLA-B locus products (Versteeg *et al.*, 1989a). As N-*myc* and c-*myc* are both members of the *myc* family of proteins, our initial studies on the mechanism of regulation of HLA class I by c-*myc* focussed on the enhancer A region that is influenced by N-*myc*. The binding of regulatory proteins to this conserved palindromic sequence, present in whole-cell extracts of c-*myc* transfectants, the original melanoma cell line IGR39D and in a panel of melanoma cell lines, was determined *in vitro* (Peltenburg *et al.*, 1993). No (inverse) correlation between binding activity of KBF1 or NF-κB and c-*myc* expression could be established in gel retardation assays. This suggested that the enhancer A sequence was not involved in down-modulation of HLA class I genes by c-*myc*. These findings were confirmed by studying the functional activity of the enhancer A core sequence in transient transfection assays with the bacterial CAT gene as a reporter (Peltenburg *et al.*, 1993). Transfection of CAT constructs controlled by a minimal chicken conalbumin promoter with and without the KBF1-binding site in our panel of c-*myc* transfectants confirmed the observation that the activity of this enhancer does not correlate with c-*myc* expression. These experiments, thus, definitively excluded a role for the enhancer A core region in down-regulation of HLA-B by c-*myc*. This conclusion is in agreement with the available DNA sequences of HLA-A and HLA-B genes, showing no specific differences among the respective enhancer A regions that could explain the HLA-B-specific down-modulation by c-*myc*. The mechanism of transcriptional repression of HLA-B by c-*myc* must involve a domain in the 0.6 kb promoter region of the regulated gene, as was demonstrated by the decreased activity of this region in luciferase reporter assays in c-*myc* transfectants (Peltenburg *et al.*, 1993).

Other genes have been demonstrated to be down-regulated by c-*myc*: the LFA-1 adhesion molecule (Inghirami *et al.*, 1990) and, at the level of transcription, the mouse pro-alpha 2 (I) collagen (Yang *et al.*, 1991) as well as the human c-*neu* oncogene (Suen & Hung, 1991). The last two appear to be regulated by c-*myc* through a region just upstream of the transcriptional initiation site (+1) in these genes. A parallel can be drawn between regulation of MHC class I expression and down-modulation of these two genes: the down-regulation of HLA-B by the oncogene c-*myc* seems to involve a dominant regulatory element located in the region surrounding the general transcription elements (Peltenburg & Schrier, 1994). Whereas the effect on class I expression induced by N-*myc* is regulated through enhancer A, the effect exerted by c-*myc* on class I expression appears to be achieved in an entirely different manner, as is implied by the experiments analysing enhancer A activity in human melanoma cells (see Fig. 13.1). Although on the basis of the structural homology of c-*myc* and N-*myc*, similar mechanisms for down-modulation of MHC class I might be expected, this is apparently not the case. This conclusion is not necessarily in conflict with other data on c-*myc* and N-*myc*: the apparent differences in tissue-specific expression of the *myc* proteins (Jakobovitz *et al.*, 1985; Stanton *et al.*, 1986) argue against an identical function of c-*myc* and N-*myc*, which is consistent with our data on MHC class I regulation.

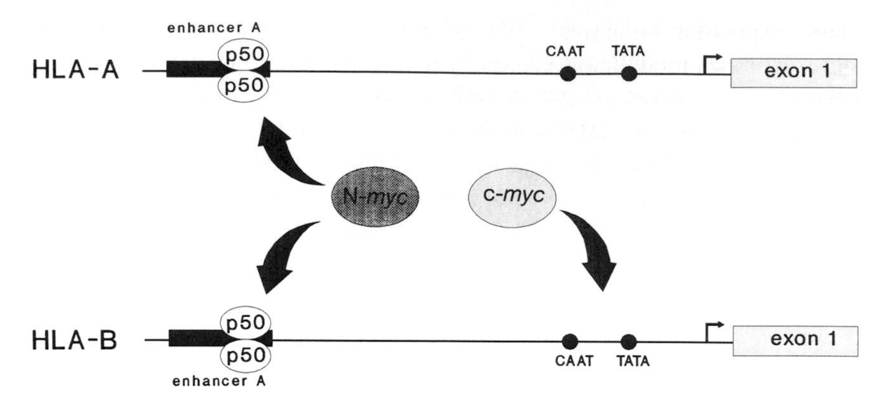


Fig. 13.1. Proposed mechanism of down-regulation of HLA class I by c-*myc* and N-*myc*. The N-*myc* oncogene regulates MHC class I expression through inactivation of the core region of enhancer A. This is accomplished by down-modulation of the mRNA encoding the p50 protein (van 't Veer *et al.*, 1993). The c-*myc* protein is unable to influence the activity of this enhancer (Peltenburg *et al.*, 1993). A role for the HLA-B proximal promoter elements in down-regulation by c-*myc* has been indicated (Peltenburg & Schrier, 1994).

## 13.4 Biological consequences of *myc* activation

So far, it has become clear that MHC class I expression is affected in many tumours and that oncogenes, in particular N-*myc* and c-*myc*, can be responsible for this effect. In animal models, the tumorigenic properties of cells may depend on the expression of particular H-2 alleles. No consistent effect of H-2 modulation on tumorigenicity, however, can be assigned: sometimes the tumorigenicity and metastatic capability is abrogated by elevated H-2 expression (Hui *et al.*, 1984; Wallich *et al.*, 1985; Plaksin *et al.*, 1988; Porgador, Feldman & Eisenbach, 1989). In other cases, the metastatic capacity of the cells increased after H-2 transfection (Gopas *et al.*, 1989). The data on an *in vivo* effect of HLA class I modulation are sparse and no clear conclusion can be drawn; in some tumour types, e.g. colon carcinoma, an apparent lack of correlation between HLA class I expression and malignant potential has been found, whereas in other types, e.g. B cell lymphomas and melanomas, such correlation could be made. The last two are considered particularly immune reactive and this might be the reason why, in these cases, low HLA class I expression is correlated with poor prognosis: low HLA class I expression will largely diminish the efficacy of eradication of neoplastic cells by CTLs.

The selective down-modulation of particular loci or alleles by the *myc* oncogenes may provide a hint as to the functional implications of low MHC class I expression on tumour cells. At first sight, one could conclude that the selective down-modulation is the result of a selection process by which T cells eliminate tumour cells expressing a particular combination of HLA class I alleles and peptide. Cells lacking expression of these HLA class I alleles, because of c-*myc* overexpression, will survive. According to this view, immune escape from CTLs is caused by c-*myc*, implying that c-*myc* activation is associated with progression of the disease, rather than being involved in the origin of the tumour. *myc* oncogenes, however, are capable of inducing neoplastic transformation, albeit in combination with other oncogenes. Therefore, c-*myc* activation in melanoma may very well be a tumour-initiating step. In that case, HLA class I alleles are concomitantly down-modulated in the resulting newly formed tumour cells, and one might wonder what the rationale is for such a coupling. One explanation is suggested by the notion that low MHC class I expression may lead to lower tumorigenic or metastatic properties of tumour cells because of increased sensitivity to NK cells. It has been shown in a number of experimental tumour models that the sensitivity of tumour cells to NK cells is determined by the level of expression of MHC class I antigens on the tumour cells. This phenomenon has led to the formulation of the 'missing self' hypothesis by Kärre and co-workers (reviewed by Ljunggren & Kärre, 1990). Briefly, this states that permanently surveying NK cells scan somatic cells for MHC class I expression. If MHC class I is not present, the NK cell will kill the target. Two mechanisms can be envisaged (Ljunggren & Kärre, 1990). Firstly, the 'target interference' model, where MHC class I protects a tentative NK target from interaction with the NK cell. Secondly, the 'effector inhibition' model according to which after interaction of NK cell and target cell and recognition of MHC class I, the lytic hit is blocked. If MHC class I is absent the target cell is killed. The advantage of such a surveillance system is that the immune system need not be primed after appearance of a foreign or altered epitope present in the target cell, as in the case of a T cell response, but will immediately attack and eliminate an aberrant MHC class I-negative cell. In this view, low MHC class I expression on a tumour cell is not the result of an escape from immune surveillance but rather an intrinsic property of the tumour cell, as in the case of melanomas where low MHC class I expression is effected by c-*myc* activation.

Several lines of evidence suggest that in human tumours modulation of MHC class I expression alters the NK sensitivity of the tumour cells. Firstly, the level of expression of HLA class I is inversely correlated with NK sensitivity in several lymphoblastoid cell lines (Storkus *et al.*, 1987; Ohlen *et al.*, 1989; Harel-Bellan *et al.*, 1991). Secondly, lysis of a number of human tumour cell lines by NK cells can be blocked by antibodies specific for HLA class I (Lobo & Spencer, 1989; Maio *et al.*, 1991). These cell lines include B and T lymphoma, small cell lung carcinoma and melanoma. Thirdly, perhaps the most convincing evidence is that transfection of HLA class I heavy chain or $\beta_2$-m genes into cell lines completely devoid of HLA class I expression resulted in NK resistance (Quillet *et al.*, 1988; Shimizu & DeMars, 1989; Storkus *et al.*, 1989b; Maio *et al.*, 1991). The effect was dose dependent, i.e. the effect on NK sensitivity correlated with the expression of the transfected gene. In addition, the identity of the gene was important: not all HLA class I alleles exerted the effect (Storkus *et al.*, 1989a) and murine genes had no effect in human systems (Storkus *et al.*, 1989b). This suggests a specific interaction of a receptor structure on NK cells with HLA class I antigens.

In the experiments discussed so far, the tumour cells were devoid of the products of all HLA class I loci. As discussed previously, the regulation of HLA class I expression by c-*myc* is locus specific: in melanoma cell lines with high endogenous c-*myc* expression, or in cell lines transfected with c-*myc*, only HLA-B alleles are down-modulated. This means that the unaffected HLA-A allele products are still abundantly present at the cell surface. Notwithstanding, the c-*myc*-transfected cells have a much higher sensitivity to NK cells than their parental counterparts (Versteeg *et al.*, 1989b). This is not the result of an effect of c-*myc* transfection but is directly related to HLA-B down-modulation, since super-transfection of these cells with an HLA-B gene reverses their NK sensitivity to the level of the original cell line (Peltenburg *et al.*, 1992). These experiments indicate that the selective down-modulation of HLA-B alleles is recognized by NK cells, again suggesting a mechanism involving some sort of specific receptor on NK cells discriminating between the various alleles.

This view is supported by the findings of Storkus *et al.* (1991), showing that amino acid alterations in the peptide-binding groove of an HLA class I antigen, which normally do not protect against NK lysis, result in protection against NK lysis when the HLA class I molecule is transfected in an NK-sensitive target. Further evidence for involvement of the antigen-binding site of the HLA class I molecule was provided by pre-treatment with synthetic peptides of cells expressing individual class I alleles (Storkus *et al.*, 1992). In each case in which peptide had been shown to bind to the protective class I molecule, peptide pulsing resulted in increased sensitivity to NK-mediated lysis. This strongly suggests that a tentative NK receptor interacts with the peptide-binding groove. Combining the data, protecting and non-protecting HLA allelic products can be discriminated, HLA-A3, Aw68, Aw69, B5, B8 and B27 being protecting alleles (Shimizu & DeMars, 1989; Versteeg *et al.*, 1989b; Storkus *et al.*, 1991; Peltenburg *et al.*, 1992). In two independent studies, HLA-A1 and HLA-A2 expression apparently does not protect (Versteeg *et al.*, 1989b; Storkus *et al.*, 1991), while in another study these alleles do have at least some effect on NK susceptibility (Shimizu & DeMars, 1989). These results can be reconciled by assuming that other factors, in addition to HLA class I, may contribute to the eventual susceptibility of a tumour target to lysis by NK cells.

## 13.5 Conclusions

The link between c-*myc* activation and regulation of MHC class I expression plays an important role in determining the sensitivity of tumour cells to the immune system of the host. However, it should be kept in mind that the major function of c-*myc* lies in the regulation of cell growth and differentiation in normal cells. Although the precise biological function of the coupling between c-*myc* expression and MHC class I expression in normal cells is not known, this phenomenon may, in principle, provide a mode of immune surveillance against tumour cells with activated *myc* genes.

## References

Alt, F.W., Depinho, R., Zimmerman, K., Legouy, E., Hatton, K., Ferrier, P., Tesfaye, A., Yancopoulos, G. & Nisen, P. (1986). The human *myc* gene family. *Cold Spring Harbor Symposium on Quantitative Biology*, 51 (Pt 2), 931–941.

Andersson, M.L., Stam, N.J., Klein, G., Ploegh, H.L. & Masucci, M.G. (1991). Aberrant expression of HLA class I antigens in Burkitt's lymphoma cells. *International Journal of Cancer*, 47, 544–550.

Bahler, D.W., Frelinger, J.G., Harwell, L.W. & Lord, E.M. (1987). Reduced tumorigenicity of a spontaneous mouse lung carcinoma following H-2 gene transfection. *Proceedings of the National Academy of Sciences of the USA*, 84, 4562–4566.

Bergh, J.C.S. (1990). Gene amplification in human lung cancer – the *myc* family genes and other proto-oncogenes and growth factor genes. *American Review of Respiratory Disease*, 142, S20–S26.

Bernards, R., Dessain, S.K. & Weinberg, R.A. (1986). N-*myc* amplification causes down-modulation of MHC class I antigen expression in neuroblastoma. *Cell*, 47, 667–674.

Bernards, R. & Lenardo, M. (1990). Molecular events during tumour progression in neuroblastoma. In: *Oncogenes in Cancer Diagnosis*, ed. C.R. Bartram, K. Munk, & M. Schwab, pp. 86–93. Basel: Karger.

Bernards, R., Schrier, P.I., Houweling, A., Bos, J.L., van der Eb, A.J., Zijlstra, M. & Melief, C.J.M. (1983). Tumorigenicity of cells transformed by

adenovirus type 12 by evasion of T cell immunity. *Nature (London)*, 305, 776–779.
Bishop, J.M. (1991). Molecular themes in oncogenesis. *Cell*, 64, 235–248.
Blackwell, T.K., Kretzner, L., Blackwood, E.M., Eisenman, R.N. & Weintraub, H. (1990). Sequence-specific DNA binding by the c-*myc* protein. *Science*, 250, 1149–1151.
Blackwood, E.M. & Eisenman, R.N. (1991). Max – a helix-loop-helix zipper protein that forms a sequence-specific DNA-binding complex with *myc*. *Science*, 251, 1211–1217.
Blanchet, O., Bourge, J.F., Zinszner, H., Israel, A., Kourilsky, P., Dausset, J., Degos, L. & Paul, P. (1992). Altered binding of regulatory factors to HLA class I enhancer sequence in human tumor cell lines lacking class I antigen expression. *Proceedings of the National Academy of Sciences of the USA*, 89, 3488–3492.
Bos, J.L. (1989). *Ras* oncogenes in human cancer: a review. *Cancer Research*, 49, 4682–4689.
Charron, J., Malynn, B.A., Fisher, P., Stewart, V., Jeannotte, L., Goff, S.P., Robertson, E.J. & Alt, F.W. (1992). Embryonic lethality in mice homozygous for a targeted disruption of the N-*myc* gene. *Genes and Development*, 6, 2248–2257.
Cole, M.D. (1986). The *myc* oncogene: its role in transformation and differentiation. *Annual Review of Genetics*, 20, 361–384.
Cole, M.D. (1991). Myc meets its Max. *Cell*, 65, 715–716.
Crowley, N.J., Darrow, T.L., Quinn-Allen, M.A. & Seigler, H.F. (1991). MHC-restricted recognition of autologous melanoma by tumor-specific cytotoxic T cells – evidence for restriction by a dominant HLA-A allele. *Journal of Immunology*, 146, 1692–1699.
Dahllöf, B. (1990). Down-regulation of MHC class I antigens is not a general mechanism for the increased tumorigenicity caused by c-*myc* amplification. *Oncogene*, 5, 433–435.
Degiovanni, G., Hainaut, P., Lahaye, T., Weynants, P. & Boon, T. (1990). Antigens recognized on a melanoma cell line by autologous cytolytic T lymphocytes are also expressed on freshly collected tumour cells. *European Journal of Immunology*, 20, 1865–1868.
Eager, K.B., Williams, J., Breiding, D., Pan, S., Knowles, B., Appella, E. & Ricciardi, R.P. (1985). Expression of histocompatibility antigens H-2K, D and L is reduced in adenovirus 12-transformed mouse cells and is restored by interferon gamma. *Proceedings of the National Academy of Sciences of the USA*, 82, 5525–5529.
Feltner, D.E., Cooper, M., Weber, J., Israel, M.A. & Thiele, C.J. (1989). Expression of class I histocompatibility antigens in neuroectodermal tumours is independent of the expression of a transfected neuroblastoma *myc* gene. *Journal of Immunology*, 143, 4292–4299.
Garrido, M.L., Perez, M., Delgado, C., Rojano, J., Algarra, I., Garrido, A. & Garrido, F. (1986). Immunogenicity of H-2 positive and H-2 negative clones of a mouse tumour, GR9. *Journal of Immunogenetics*, 13, 159–167.
Gopas, J., Rager-Zisman, B., Har-Vardi, I., Hämmerling, G.J., Bar-Eli, M. & Segal, S. (1989). NK sensitivity, H-2 expression and metastatic potential: analysis of H-2D$^k$ gene transfected fibrosarcoma cells. *Journal of Immunogenetics*, 16, 305–313.
Grand, R.J., Rowe, M., Byrd, P.J. & Gallimore, P.H. (1987). The level of

expression of class I MHC antigens in adenovirus-transformed human cell lines. *International Journal of Cancer*, 40, 213–219.
Gross, N., Beck, D. & Favre, S. (1990). *In vitro* modulation and relationship between N-*myc* and HLA class I RNA steady-state levels in human neuroblastoma cells. *Cancer Research*, 50, 7532–7536.
Harel-Bellan, A., Quillet, A., Marchiol, C., DeMars, R., Tursz, T. & Fradelizi, D. (1991). Natural killer susceptibility of human cells may be regulated by genes in the HLA region of chromosome 6. *Proceedings of the National Academy of Sciences of the USA*, 83, 5688–5692.
Hateboer, G., Timmers, H.T.M., Rustgi, A.K., Billaud, M., van 't Veer, L.J. & Bernards, R. (1993). TATA-binding protein and the retinoblastoma gene product bind to overlapping epitopes on c-Myc and adenovirus E1A protein. *Proceedings of the National Academy of Sciences of the USA*, 90, 8489–8493.
Henseling, U., Schmidt, W., Schöler, H.R., Gruss, P. & Hatzopoulos, A.K. (1990). A transcription factor interacting with the class I gene enhancer is inactive in tumorigenic cell lines which suppress major histocompatibility complex class I genes. *Molecular and Cellular Biology*, 10, 4100–4109.
Hui, K., Grosveld, F. & Festenstein, H. (1984). Rejection of transplantable AKR leukaemia cells following MHC DNA-mediated cell transformation. *Nature (London)*, 311, 750–752.
Inghirami, G., Grignani, F., Sternas, L., Lombardi, L., Knowles, D.M. & Dalla-Favera, R. (1990). Down-regulation of LFA-1 adhesion receptors by c-*myc* oncogene in human B-lymphoblastoid cells. *Science*, 250, 682–686.
Ingvarsson, S., Asker, C., Axelson, H., Klein, G. & Sumegi, J. (1988). Structure and expression of B-*myc*, a new member of the myc gene family. *Molecular and Cellular Biology*, 8, 3168–3174.
Isakov, N., Katzav, S., Feldman, M. & Segal, S. (1983). Loss of expression of transplantation antigens encoded by the H-2K locus on Lewis lung carcinoma cells and its relevance to the tumor's metastatic properties. *Journal of the National Cancer Institute*, 71, 139–145.
Jakobovitz, A., Schwab, M., Bishop, J. M. & Martin, G.R. (1985). Expression of N-*myc* in teratocarcinoma cells and mouse embryos. *Nature (London)*, 318, 188–191.
Kast, W.M. & Melief, C.J.M. (1991). Fine peptide specificity of cytotoxic T lymphocytes directed against adenovirus-induced tumours and peptide-MHC binding. *International Journal of Cancer*, Suppl. 6, 90–94.
Kast, W.M., Offringa, R., Peters, P.J., Voordouw, A.C., Meloen, R.H., van der Eb, A.J. & Melief, C.J.M. (1989). Eradication of adenovirus E1-induced tumours by E1A-specific cytotoxic T lymphocytes. *Cell*, 59, 603–614.
Katzav, S., de Baetselier, P., Tartakovsky, B., Feldman, M. & Segal, S. (1983). Alterations in major histocompatibility complex phenotypes of mouse cloned T10 sarcoma cells: association with shifts from nonmetastatic to metastatic cells. *Journal of the National Cancer Institute*, 71, 317–324.
Klein, G. (1983). Specific chromosomal translocations and the genesis of B cell derived tumours in mice and man. *Cell*, 32, 311–315.
Klein, G. (1991). Comparative action of *myc* and *bcl-2* in B cell malignancy. *Cancer Cells*, 3, 141–143.
Knuth, A., Wölfel, T., Klehmann, E., Boon, T. & Meyer zum Büschenfelde, K.H. (1989). Cytolytic T cell clones against an autologous human melanoma: specificity study and definition of three antigens by immunoselection. *Proceedings of the National Academy of Sciences of the USA*, 86, 2804–2808.
Land, H., Chen, A.C., Morgenstern, J.P., Parada, L.F. & Weinberg, R.A. (1986).

Behaviour of *myc* and *ras* oncogenes in transformation of rat embryo fibroblasts. *Molecular and Cellular Biology*, 6, 1917–1925.

Land, H., Parada, L.F. & Weinberg, R.A. (1983). Cellular oncogenes and multistep carcinogenesis. *Science*, 222, 771–778.

Larsson, L.G., Oberg, F., Stockbauer, P., Masucci, M.G. & Nilsson, K. (1992). Suppression of basal, PMA-alpha-induced, and IFN-alpha-induced, but not IFN-gamma-induced expression of HLA class I in v-*myc*-transformed U-937 monoblasts. *International Journal of Cancer*, 52, 759–765.

Lee, W.S., Kao, C.C., Bryant, G.O., Liu, X. & Berk, A.J. (1991). Adenovirus E1A activation domain binds the basic repeat in the TATA box transcription factor. *Cell*, 67, 365–376.

Lenardo, M., Rustgi, A.K., Schievella, A.R. & Bernards, R. (1989). Suppression of MHC class I gene expression by N-*myc* through enhancer inactivation. *EMBO Journal*, 8, 3351–3355.

Levine, A.J., Momand, J. & Finlay, C.A. (1991). The p53 tumour suppressor gene. *Nature (London)*, 351, 453–456.

Ljunggren, H.G. & Kärre, K. (1990). In search of the missing self – MHC molecules and NK cell recognition. *Immunology Today*, 11, 237–244.

Lobo, P.I. & Spencer, C.E. (1989). Use of anti-HLA antibodies to mask major histocompatibility complex gene products on tumour cells can enhance susceptibility of these cells to lysis by natural killer cells. *Journal of Clinical Investigation*, 83, 278–287.

Lüscher, B. & Eisenman, R.N. (1990). New light on Myc and Myb. Part 1. Myc. *Genes and Development*, 4, 2025–2035.

Maio, M., Altomonte, M., Tatake, R., Zeff, R.A. & Ferrone, S. (1991). Reduction in susceptibility to natural killer cell-mediated lysis of human FO-1 melanoma cells after induction of HLA class I antigen expression by transfection with $\beta_2$-m gene. *Journal of Clinical Investigation*, 88, 282–289.

Maitland, N.J., Brown, K.W., Poirier, V., Shaw, A.P. & Williams, J. (1989). Molecular and cellular biology of Wilms' tumour. *Anticancer Research*, 9, 1417–1426.

Marley, G.M., Doyle, L.A., Ordonez, J.V., Sisk, A., Hussain, A. & Yen, R.W. (1989). Potentiation of interferon induction of class I major histocompatibility complex antigen expression by human tumour necrosis factor in small cell lung cancer cell lines. *Cancer Research*, 49, 6232–6236.

Minafra, S., Morello, V., Glorioso, F., La Fiura, A.M., Tomasino, R.M., Feo, S., McIntosh, D. & Woolley, D.E. (1989). A new cell line (8701-BC) from primary ductal infiltrating carcinoma of human breast. *British Journal of Cancer*, 60, 185–192.

Momburg, F., Ziegler, A., Harpprecht, J., Möller, P., Moldenhauer, G. & Hämmerling, G.J. (1989). Selective loss of HLA-A or HLA-B antigen expression in colon carcinoma. *Journal of Immunology*, 142, 352–358.

Nanni, P., Colombo, M., Degiovanni, C., Lollini, P., Nicoletti, G., Parmiani, G. & Prodi, G. (1983). Impaired H-2 expression in BL6 melanoma variants. *Journal of Immunogenetics*, 10, 361–370.

Nau, M.M., Brooks, B.J., Battey, J., Sausville, E., Gazdar, A.F., Kirsch, I.R., McBride, O.W., Bertness, V., Hollis, G.F. & Minna, J.D. (1985). L-*myc*, a new *myc*-related gene amplified and expressed in human small cell lung cancer. *Nature (London)*, 318, 69–73.

Nau, M.M., Brooks, B.J., Carney, D.N., Gazdar, A.F., Battey, J.F., Sausville, E.A. & Minna, J.D. (1986). Human small cell lung cancers show amplification and expression of the N-myc gene. *Proceedings of the National Academy of Sciences of the USA*, 83, 1092–1096.

Nishimura, M.I., Stroynowski, I., Hood, L. & Ostrand-Rosenberg, S. (1988). H-2K$^b$ antigen expression has no effect on natural killer susceptibility and tumorigenicity of a murine hepatoma. *Journal of Immunology*, 141, 4403–4408.

Nouri, A.M.E., Bergbaum, A., Lederer, E., Crosby, D., Shamsa, A. & Oliver, R.T.D. (1991). Paired TIL and tumour cell line from bladder cancer: a new approach to study tumour immunology *in vitro*. *European Journal of Cancer*, 27, 608–612.

Nouri, A.M.E., Smith, M.E.F., Crosby, D. & Oliver, R.T.D. (1990). Selective and non-selective loss of immunoregulatory molecules (HLA-A, -B and -C antigens and LFA-3) in transitional cell carcinoma. *British Journal of Cancer*, 62, 603–606.

Ohlen, C., Bejarano, M.T., Gronberg, A., Torsteinsdottir, S., Franksson, L., Ljunggren, H.G., Klein, E., Klein, G. & Kärre, K. (1989). Studies of sublines selected for loss of HLA expression from an EBV-transformed lymphoblastoid cell line. Changes in sensitivity to cytotoxic T cells activated by allostimulation and natural killer cells activated by IFN or IL-2. *Journal of Immunology*, 142, 3336–3341.

Peltenburg, L.T.C., Dee, R. & Schrier, P.I. (1993). Down-regulation of HLA class I expression by *c-myc* in human melanoma is independent of enhancer A. *Nucleic Acids Research*, 21, 1179–1185.

Peltenburg, L.T.C., Steegenga, W.T., Krüse, K.M. & Schrier, P.I. (1992). *c-myc*-induced natural killer cell sensitivity of human melanoma cells is reversed by HLA-B27 transfection. *European Journal of Immunology*, 22, 2737–2740.

Peltenburg, L.T.C. & Schrier, P.I. (1994). Transcriptional suppression of HLA-B expression by c-*myc* is mediated through the core promoter elements. *Immunogenetics*, 40, 54–61.

Plaksin, D., Gelber, C., Feldman, M. & Eisenbach, L. (1988). Reversal of the metastatic phenotype in Lewis lung carcinoma cells after transfection with syngeneic H-2K$^b$ gene. *Proceedings of the National Academy of Sciences of the USA*, 85, 4463–4467.

Porgador, A., Feldman, M. & Eisenbach, L. (1989). H-2K$^b$ transfection of B16 melanoma cells results in reduced tumorigenicity and metastatic competence. *Journal of Immunogenetics*, 16, 291–303.

Quillet, A., Presse, F., Marchiol-Fournigault, C., Harel-Bellan, A., Benbunan, M., Ploegh, H.L. & Fradelizi, D. (1988). Increased resistance to non-MHC-restricted cytotoxicity related to HLA A, B expression. Direct demonstration using $\beta_2$-microglobulin-transfected Daudi cells. *Journal of Immunology*, 141, 17–20.

Redondo, M., Ruiz-Cabello, F., Concha, A., Cabrera, T., Perez-Ayala, M., Oliva, M.R. & Garrido, F. (1991). Altered HLA class I expression in non-small cell lung cancer is independent of c-*myc* activation. *Cancer Research*, 51, 2463–2468.

Rosolen, A., Whitesell, L., Ikegaki, N., Kennett, R.H. & Neckers, L.M. (1990). Antisense inhibition of single copy N-*myc* expression results in decreased cell growth without reduction of c-*myc* protein in a neuroepithelioma cell line. *Cancer Research*, 50, 6316–6322.

Ruley, E.H. (1983). Adenovirus early region E1A enables viral and cellular

transforming genes to transform primary cells in culture. *Nature (London)*, 304, 602–606.
Rustgi, A.K., Dyson, N. & Bernards, R. (1991). Amino-terminal domains of c-*myc* and N-*myc* proteins mediate binding to the retinoblastoma gene product. *Nature (London)*, 352, 541–544.
Schrier, P.I., Bernards, R., Vaessen, R.T., Houweling, A. & van der Eb, A.J. (1983). Expression of class I major histocompatibility antigens switched off by highly oncogenic adenovirus 12 in transformed rat cells. *Nature (London)*, 305, 771–775.
Schrier, P.I. & Peltenburg, L.T.C. (1993). Relationship between *myc* oncogene activation and MHC class I expression. *Advances in Cancer Research*, 60, 181–246.
Schrier, P.I., Versteeg, R., Peltenburg, L.T.C., Plomp, A.C., van 't Veer, L.J. & Krüse-Wolters, K.M. (1991). Sensitivity of melanoma cell lines to natural killer cells: a role for oncogene-modulated HLA expression? *Seminars in Cancer Biology*, 2, 73–83.
Schwab, M., Alitalo, K., Klempnauer, K.H., Varmus, H.E., Bishop, J.M., Gilbert, F., Brodeur, G., Goldstein, M. & Trent, J. (1983). Amplified DNA with limited homology to *myc* cellular oncogene is shared by human neuroblastoma cell lines and a neuroblastoma tumour. *Nature (London)*, 305, 245–248.
Schwab, M., Varmus, H.E. & Bishop, J.M. (1985). Human N-*myc* gene contributes to neoplastic transformation of mammalian cells in culture. *Nature (London)*, 316, 160–162.
Seeger, R.C., Brodeur, G.M., Sather, H., Dalton, A., Siegel, S.E., Wong, K.Y. & Hammond, D. (1985). Association of multiple copies of the N-*myc* oncogene with rapid progression of neuroblastomas. *New England Journal of Medicine*, 313, 1111–1116.
Shaw, A.P., Poirier, V., Tyler, S., Mott, M., Berry, J. & Maitland, N.J. (1988). Expression of the N-*myc* oncogene in Wilms' tumour and related tissues. *Oncogene*, 3, 143–149.
Shimizu, Y. & DeMars, R. (1989). Demonstration by class I gene transfer that reduced susceptibility of human cells to natural killer cell-mediated lysis is inversely correlated with HLA class I antigen expression. *European Journal of Immunology*, 19, 447–451.
Soong, T.W. & Hui, K.M. (1991). Identification of locus-specific DNA-binding factors for the regulation of HLA class I genes in human colorectal cancer. *International Journal of Cancer*, Suppl. 6, 131–137.
Soong, T.W. & Hui, K.M. (1992). Locus-specific transcriptional control of HLA genes. *Journal of Immunology*, 149, 2008–2020.
Soong, T.W., Oei, A.A. & Hui, K.M. (1991). Regulation of HLA class I expression in human colorectal carcinoma. *Seminars in Cancer Biology*, 2, 23–33.
Stanton, B.R., Perkins, A.S., Tessarollo, L., Sassoon, D.A. & Parada, L.F. (1992). Loss of N-*myc* function results in embryonic lethality and failure of the epithelial component of the embryo to develop. *Genes and Development*, 6, 2235–2247.
Stanton, L.W., Schwab, M. & Bishop, J.M. (1986). Nucleotide sequence of the human N-*myc* gene. *Proceedings of the National Academy of Sciences of the USA*, 83, 1772–1776.
Storkus, W.J., Alexander, J., Payne, J.A., Cresswell, P. & Dawson, J.R. (1989a). The alpha-1 alpha-2 domains of class I HLA molecules confer resistance to natural killing. *Journal of Immunology*, 143, 3853–3857.

Storkus, W.J., Alexander, J., Payne, J.A., Dawson, J.R. & Cresswell, P. (1989b). Reversal of natural killing susceptibility in target cells expressing transfected class I HLA genes. *Proceedings of the National Academy of Sciences of the USA*, 86, 2361–2364.

Storkus, W.J., Howell, D.N., Salter, R.D., Dawson, J.R. & Cresswell, P. (1987). NK susceptibility varies inversely with target cell class I HLA antigen expression. *Journal of Immunology*, 138, 1657–1659.

Storkus, W.J., Salter, R.D., Alexander, J., Ward, F.E., Ruiz, R.E., Cresswell, P. & Dawson, J.R. (1991). Class I-induced resistance to natural killing – identification of nonpermissive residues in HLA-A2. *Proceedings of the National Academy of Sciences of the USA*, 88, 5989–5992.

Storkus, W.J., Salter, R.D., Cresswell, P. & Dawson, J.R. (1992). Peptide-induced modulation of target cell sensitivity to natural killing. *Journal of Immunology*, 149, 1185–1190.

Suen, T.C. & Hung, M.C. (1991). c-*myc* reverses *neu*-induced transformed morphology by transcriptional repression. *Molecular and Cellular Biology*, 11, 354–362.

Sugio, K., Nakagawara, A. & Sasazuki, T. (1991). Association of expression between N-*myc* gene and major histocompatibility complex class I gene in surgically resected human neuroblastoma. *Cancer*, 67, 1384–1388.

Tanaka, K., Isselbacher, K.J., Khoury, G. & Jay, G. (1985). Reversal of oncogenesis by the expression of a major histocompatibility complex class I gene. *Science*, 228, 26–30.

Tanaka, K., Hayashi, H., Hamada, C., Khoury, G. & Jay, G. (1986). Expression of major histocompatibility complex class I antigens as a strategy for the potentiation of immune recognition of tumour cells. *Proceedings of the National Academy of Sciences of the USA*, 83, 8723–8727.

Taniguchi, K., Kärre, K. & Klein, G. (1985). Lung colonization and metastasis by disseminated B16 melanoma cells: H-2 associated control at the level of the host and the tumour cell. *International Journal of Cancer*, 36, 503–510.

Thiele, C.J., Reynolds, C.P. & Israel, M.A. (1985). Decreased expression of N-*myc* precedes retinoic acid-induced morphological differentiation of human neuroblastoma. *Nature (London)*, 313, 404–406.

Traversari, C., van der Bruggen, P., Luescher, I.F., Lurquin, C., Chomez, P., van Pel, A., de Plaen, E., Amarco-Stesek, A. & Boon, T. (1992a). A nonapeptide encoded by human gene MAGE-1 is recognized on HLA-A1 by cytolytic T lymphocytes directed against tumour antigen-MZ2-E. *Journal of Experimental Medicine*, 176, 1453–1457.

Traversari, C., van der Bruggen, P., van den Eynde, B., Hainaut, P., Lemoine, C., Ohta, N., Old, L. & Boon, T. (1992b). Transfection and expression of a gene coding for a human melanoma antigen recognized by autologous cytolytic T lymphocytes. *Immunogenetics*, 35, 145–152.

Vaessen, R.T., Houweling, A., Israel, A., Kourilsky, P. & van der Eb, A.J. (1986). Adenovirus E1A-mediated regulation of class I MHC expression. *EMBO Journal*, 5, 335–341.

van den Eynde, B., Hainaut, P., Herin, M., Knuth, A., Lemoine, C., Weynants, P., van der Bruggen, P., Fauchet, R. & Boon, T. (1989). Presence on a human melanoma of multiple antigens recognized by autologous CTL. *International Journal of Cancer*, 44, 634–640.

van 't Veer, L.J., Beijersbergen, R.L. & Bernards, R. (1993). N-*myc* suppresses major histocompatibility complex class I gene expression through down-regulation of the p50 subunit of NF-κB. *EMBO Journal*, 12, 195–200.

van der Bruggen, P., Traversari, C., Chomez, P., Lurquin, C., de Plaen, E., van den Eynde, B., Knuth, A. & Boon, T. (1991). A gene encoding an antigen recognized by cytolytic T lymphocytes on a human melanoma. *Science*, 254, 1643–1647.

Vasavada, R., Eager, K.B., Barbanti-Brodano, G., Caputo, A. & Ricciardi, R.P. (1986). Adenovirus type 12 early region 1A proteins repress class I HLA expression in transformed human cells. *Proceedings of the National Academy of Sciences of the USA*, 83, 5257–5261.

Versteeg, R., Krüse-Wolters, K.M., Plomp, A.C., van Leeuwen, A., Stam, N.J., Ploegh, H.L., Ruiter, D.J. & Schrier, P.I. (1989a). Suppression of class I human histocompatibility leukocyte antigen by c-*myc* is locus-specific. *Journal of Experimental Medicine*, 170, 621–635.

Versteeg, R., Noordermeer, I.A., Krüse-Wolters, K.M., Ruiter, D.J. & Schrier, P.I. (1988). c-*myc* down-regulates class I HLA expression in human melanomas. *EMBO Journal*, 7, 1023–1029.

Versteeg, R., Peltenburg, L.T.C., Plomp, A.C. & Schrier, P.I. (1989b). High expression of the c-*myc* oncogene renders melanoma cells prone to lysis by natural killer cells. *Journal of Immunology*, 143, 4331–4337.

Versteeg, R., van der Minne, C., Plomp, A.C., Sijts, A., van Leeuwen, A. & Schrier, P.I. (1990). N-*myc* expression switched off and class I HLA expression switched on after somatic cell fusion of neuroblastoma cells. *Molecular and Cellular Biology*, 10, 5416–5423.

Viel, A., Maestro, R., Toffoli, G., Grion, G. & Boiocchi, M. (1990). C-*myc* over-expression is a tumour-specific phenomenon in a subset of human colorectal carcinomas. *Journal of Cancer Research*, 116, 288–294.

Wallich, R., Bulbuc, N., Hämmerling, G.J., Katzav, S., Segal, S. & Feldman, M. (1985). Abrogation of metastatic properties of tumour cells by *de novo* expression of H-2K antigens following H-2 gene transfection. *Nature (London)*, 315, 301–305.

Weinberg, R.A. (1989). Oncogenes, anti-oncogenes and the molecular bases of multistep carcinogenesis. *Cancer Research*, 49, 3713–3721.

Whyte, P., Buchkovich, K.J., Horowitz, J.M., Friend, S.H., Raybuck, M., Weinberg, R.A. & Harlow, E. (1988). Association between an oncogene and an anti-oncogene: the adenovirus E1A proteins bind to the retinoblastoma gene product. *Nature (London)*, 334, 124–129.

Wölfel, T., Klehmann, E., Müller, C.A., Schütt, K.H., Meyer zum Büschenfelde, K.H. & Knuth, A. (1989). Lysis of human melanoma cells by autologous cytolytic T cell clones. Identification of human histocompatibility leukocyte antigen A2 as a restriction element for three different antigens. *Journal of Experimental Medicine*, 170, 797–810.

Yang, B.S., Geddes, T.J., Pogulis, R.J., de Crombrugghe, B. & Freytag, S.O. (1991). Transcriptional suppression of cellular gene expression by c-*myc*. *Molecular and Cellular Biology*, 11, 2291–2295.

Zimmerman, K.A., Yancopoulos, G.D., Collum, R.G., Smith, R.K., Kohl, N.E., Denis, K.A., Nau, M.M., Witte, O.N., Toran-Allerand, D., Gee, C.E., Minna, J.D. & Alt, F.W. (1986). Differential expression of *myc* family genes during murine development. *Nature (London)*, 319, 780–783.

# 14

# Mechanisms of tumour-cell killing and the role of MHC antigens in experimental model systems

ROGER F. L. JAMES
*University of Leicester*

## 14.1 Introduction

Even before 1975, when the role of the MHC antigens as a 'guidance mechanism' for the immune system was discovered (Zinkernagel & Doherty, 1975), it was known that some tumour cells expressed abnormally low levels of MHC antigens and/or $\beta_2$-m (Nilsson, Evsin & Welsh, 1974). In addition, there was ample experimental evidence that tumour cells could be 'recognized' and eliminated by the immune system, although early work had failed to take into account allogeneic recognition of transplantation antigens (Foley, 1953). By the 1950s it had been clearly shown that chemically induced, radiation-induced and spontaneously occurring tumours could express tumour antigens that could initiate and lead to 'tumour rejection', namely the tumour-associated transplantation antigens, TATAs (Gross, 1943). It was only in the late 1970s or early 1980s that these two strands of research could be put together and they have subsequently led to important advances in the clinical treatment of cancer.

These advances are related to the great insight of Paul Ehrlich, who postulated that the immune system acted as a surveillance system to detect changes within the body caused either by normal pathological events or by invading organisms (Ehrlich, 1909). It is now clear that the primary function of CTLs is to monitor cell surfaces for abnormal peptides presented by MHC class I antigens. T cells reactive to normal self peptides will have been made tolerant either by clonal deletion or clonal anergy. CTLs are extremely sensitive to changes within the environment of potential target cells and can detect as few as 200 foreign peptides on a single target cell (Christinck *et al.*, 1991). These may include not only expression of foreign or abnormal proteins, derived from viruses or by mutation, but also those produced by altered processing to generate different self peptides. In fact, it is now possible to conceive of at least three separate ways in which tumour cells, following transformation, could give rise to a functional CTL target. Firstly, a mutation converts a non-permissive normal cellular peptide into a peptide permissive for MHC binding. Secondly, a mutation converts a permissive MHC-binding peptide that is not normally recognized by the immune system into one that is recognized. Thirdly, a normally silent gene is activated to produce a permissive MHC-binding peptide that is recognized by CTLs (Boon, 1993).

Given that tumour-specific CTLs have been shown to exist, MHC class I presentation then becomes an essential recognition element. It, therefore, becomes a selective advantage for a tumour cell to lose expression of class I antigens and, as would be predicted, this can often be a selective loss (Festenstein & Schmidt, 1981; Hui, Grosveld & Festenstein, 1984; Bernards, 1987). That is, only the class I antigens involved in TATA presentation are lost. As will be described later, many viruses have evolved in such a way as to down-regulate MHC class I expression during and subsequent to the infection and the transforming process as this clearly gives them a selective advantage (Lu *et al.*, 1991; see Chapter 6 for details).

In addition, what role does MHC class II play in this process? In general, although there are exceptions (Erb *et al.*, 1990), class II does not act as a restriction element for CTLs. However, it has been noted that MHC class II-positive tumours often, although not always, have a better prognosis (Natali *et al.*, 1983; Esteban *et al.*, 1990) and this may be related to the normal functional role of class II-restricted $T_H$ cells in potentiating immune responses (Grey & Chesnut, 1985).

CTLs may not be the only effector cells involved in the destruction of tumours, as NK and LAK cells have been shown to be important *in vivo* and *in vitro* in tumour cell lysis (Trinchieri, 1989). The role of MHC antigens here is still being elucidated, but it seems clear that NK cells have a role in eliminating cells that show loss or reduced expression of MHC antigens (Karre *et al.*, 1986). Therefore, two balancing options for the elimination of tumour cells exist, one dependent on expression and the other dependent on the lack of expression of MHC class I antigens.

In this chapter, the possible therapeutic effects of 'transfecting' i.e. 'transferring' cloned DNA encoding MHC class I and II genes into tumour cells will be described. To a large extent this is still an experimental procedure and results obtained in experimental animals will predominate, but current preliminary clinical trials of transfection of MHC genes into human tumours may well confirm the promising results obtained so far with animals.

## 14.2 Mechanisms of tumour-cell killing

This section will deal with mechanisms whereby tumour cells can be killed *in vitro* and *in vivo*. As the topic of this book is the MHC, cytolytic mechanisms related to the MHC will receive most attention, namely those involving NK cells (which includes LAK and other cytokine-induced killer cells) and CTLs. Antibody-mediated mechanisms (e.g. ADCC) can also play a role but will not be considered here.

### *14.2.1 Natural killer cells*

One of the most fundamental aspects of the immune system is its ability to recognize friend from foe. Such discrimination can, in principle, depend upon either the recognition of the presence of non-self molecules or, alternatively, the recognition of an absence of an array of self molecules. The immune system has evolved so that B cells and T cells can recognize foreign antigens by their Ig and TCR molecules. Antibodies can recognize antigens in their soluble non-denatured form and T cells are targeted to foreign antigen presented on normal cells by MHC/peptide binding, as described in Chapter 1. This mechanism, although highly efficient, leaves a pathogenic hole which has to be filled. Firstly, the recruitment of antigen-specific cells from a small, limited number of precursor cells inevitably takes a period of time (certainly days) during which a virulent organism could kill its host. Secondly, the nature of pathogenic evolution would lead to the development of organisms that could avoid MHC binding and, thereby, recognition. Although this was, to some extent, overcome by the development of MHC polymorphism, it could not deal with organisms capable of down-regulating MHC expression (e.g. the adenoviruses, see Chapter 8). Therefore, the rather enigmatic NK cells, described first in 1975 (Kiessling, Klein & Wigzell, 1975), seem to fulfil the second role, as described above, of detecting the absence of certain MHC antigens (Karre, 1985).

NK cells are defined as cells that can spontaneously kill certain susceptible target cells in a manner that is not restricted by classical MHC products. NK cell activity is associated with the large granular lymphocyte (LGL) subpopulation of lymphocytes (about 15% of total lymphocytes in peripheral blood) and the cells are phenotypically distinct from T or B lymphocytes in that they are negative for CD3/TCR and membrane immunoglobulin (mIg). They can be positively identified by expression of CD16 (Fc$\gamma$RIIa) and CD56 (NK1.1). They can recognize and reject tumour cells, virus-infected cells and haematopoetic cells transplanted from other individuals (Karre, 1985) but how they do this is not entirely clear.

The nature of the receptor involved in target recognition by NK cells and its natural ligand have taken a long time to be characterized. However, the use of blocking mAbs and new advances in molecular cloning seem to have resolved some of these problems. Ly-49 was first identified as an antigen present on the cell surface of EL-4 cells (Nagasawa *et al.*, 1987) and was subsequently shown to be involved with MHC class I alloantigen recognition by IL-2-activated NK cells (Karlhofer, Ribaudo & Yokoyama, 1992). The gene was subsequently cloned and shown to be a member of a multigene family on chromosome 6 known as the 'mouse NK complex' (Yokoyama *et al.*, 1991). The Ly-49 gene encodes a protein consisting of two homodimer subunits (45–50 kDa) and is a type II glycoprotein with an extracellular lectin-like carboxy-terminal domain (Chan & Takei, 1989).

Karlhofer and colleagues have performed a number of elegant experiments using Ly-$49^+$ and Ly-$49^-$ NK cells from the H-$2^b$ B6 strain of mouse to elucidate the natural target ligand of the NK receptor (Karlhofer *et al.*, 1992). They tested Ly-$49^+$ (present on about 20% splenic NK cells) and Ly-$49^-$ NK cells on a large panel of tumour cell targets. They found that H-$2^b$ cells were sensitive while H-$2^d$ and H-$2^k$ tumour cells were resistant to Ly-$49^+$ NK cells. Transfection of $D^d$ (but not $K^d$ or $L^d$) made sensitive cells resistant. In contrast, Ly-$49^-$ NK cells lysed all targets. The most attractive hypothesis is that a family of Ly-49-like NK receptor isoforms exists, each of which is expressed on a subset of NK cells.

There are two basic models of how the absence of class I molecules can be perceived by NK cells. Firstly, the receptor–ligand interaction (i.e. Ly-49–MHC) inhibits NK cell activation, and, secondly, MHC expression masks recognition of the natural ligand. The experiments by Karlhofer *et al.* (1992) helped to distinguish between these two possibilities because they showed that lysis of $D^d$-transfected target cells by Ly-$49^+$ NK cells occurred if antibodies against either Ly-49 or $D^d$ were added. This is most easily explained by the receptor inhibition model, although other explanations for these results have been suggested (Karre, 1992). Interestingly, for an antibody against H-$2D^d$ to reverse resistance to NK lysis, it has to be specific for the $\alpha_1/\alpha_2$ domain, which is the peptide-binding part of the H-2 heavy chain (see Chapter 1). It has also been shown that pre-incubation of target cells with exogenous peptides that bind to surface MHC class I antigens enhanced their recognition by NK cells (Chadwick & Miller, 1992).

Although the results outlined above seem to fit most closely with the 'missing self' hypothesis for NK recognition, one paradox does remain. The Ly-$49^+$ NK cells were obtained from H-$2^b$ mice, yet they lyse H-$2^b$ tumour cells and lysis is inhibited by introduction of the foreign H-$2D^d$ gene, whereas mice of the H-$2^d$ haplotype do not express the Ly-49 molecule. The results literally favour a 'missing foreign' rather than a 'missing self' interpretation, which would make very little physiological sense. Perhaps the most logical explanation is that Ly-$49^+$ NK cells react not only with $D^d$ but also with H-$2^b$

class I antigens to which a specific self peptide(s) is bound. The peptide may be lost from tumour cells, explaining the sensitivity of $H\text{-}2^b$ tumour cells to $Ly\text{-}49^+$ NK cells. Indeed Karlhofer *et al.* (1992) showed that normal $H\text{-}2^b$ cells were resistant to $Ly\text{-}49^+$ NK lysis. Presumably the $D^d$ molecule is being recognized in the sense of $H\text{-}2^b$ plus self peptide and delivering a negative signal.

In summary, NK cells bear a range of receptors capable of recognizing autologously encoded ligands bound to the class I peptide-binding groove and this interaction delivers a negative signal. Therefore, either loss of the class I antigen from the cell surface or loss of the class I bound ligand would render cells NK sensitive. This might occur because a virus has switched off biosynthesis of particular proteins or because a viral peptide displaces the NK ligand from the class I groove.

### *14.2.2 MHC restricted cytotoxicity*

CTLs were first identified as causing lysis of allogeneic target cells *in vitro* in mixed lymphocyte reactions (Cerottini & Brunner, 1974). It was shown that they contributed in major part to the defence against viral infections (Gardner, Bowern & Blanden, 1974) and, in a classical series of experiments, CTLs were shown to kill virus-infected autologous target cells that shared MHC class I haplotypes (Zinkernagel & Doherty, 1979). Since then, many studies have demonstrated virus-specific CTLs *in vitro* and *in vivo* (McMichael *et al.*, 1983) including oncogenic virus-specific CTL (Plata *et al.*, 1987; Jiang & Flyer, 1992). Tumour-specific CTLs have also been obtained from mice bearing tumours that were not caused by viruses (Boon, 1983) and the targets of these cells are the TATA antigens mentioned earlier. As has become clear more recently, these antigens need not be tumour specific and they have become more frequently called the tumour-rejection antigens (TRA).

The multitude of evidence (reviewed by McMichael, 1992) implicating the presence of CTLs in tumour rejection suggests that they must have an important role to play in identifying and eliminating abnormal cells. These CTLs are generally found to be of the $CD4^-CD8^+$ phenotype although $CD4^+$ CTLs have been demonstrated in *in vitro* systems (Erb *et al.*, 1990) and may be induced in viral infections such as measles because of exogenous rather than endogenous delivery to target cells.

The mechanism by which CTLs recognize MHC–peptide antigens was described in Chapter 1, but the generation of effective anti-tumour CTLs is dependent on a number of factors. These include efficient generation of $T_H$

cells either by direct or indirect antigen presentation (Townsend & Allison, 1993), effective peptide transport for MHC loading (Townsend *et al.*, 1988), optimal accessory molecule (Gregory *et al.*, 1988; Colombo *et al.*, 1992) and cytokine production and, perhaps most obviously, expression of MHC class I at the cell surface (Browning & Bodmer, 1992). This also explains the results obtained with the mouse Hepa-1 tumour, which is an H-2K$^b$-loss mutant of the BW7756 tumour (Nishimura & Ostrand-Rosenberg, 1991). Although it is readily rejected in syngeneic mice, the parental tumour is highly malignant and, superficially, this would appear to be a classic example of the role of NK killing of tumour cells. However, Nishimura & Ostrand-Rosenberg (1991) clearly showed that NK cells were not involved and that killing was mediated by the classical CD4$^-$CD8$^+$ CTLs, which, apparently, cannot be generated against the parental (BW7756) tumour although they are effective against BW7756 once generated. From the model outlined above, it can be speculatively proposed that the NK repertoire in the host strain (C57L/J, H-2$^b$) does not depend upon an inactivation signal delivered by K$^b$ plus peptide. How tumour-specific CTLs can be generated against a loss mutant and not against the parental line is not clear but is likely to result from the presentation of a newly expressed TRA by H-2D$^b$.

### 14.2.3 Other mechanisms of tumour cell killing

NK- and CTL-mediated lysis of tumour cells *in vitro* and *in vivo* is related to and associated with MHC expression on target cells. However, there is no doubt that tumour cell targets can be killed by mechanisms that do not involve MHC expression/restriction (Elliott *et al.*, 1989). For instance, when PBLs or splenocytes are incubated *in vitro* with IL-2, cells develop that are capable of killing almost all target cells, including those resistant to NK cell activity. This LAK cell activity is not restricted by MHC and cannot be blocked by anti-MHC antibodies. However, these cells, like NK cells, will only kill tumour cells and not normal cell targets. LAK activity is primarily associated with LGL cells but, as these experiments are usually performed on uncloned populations, CD3$^+$ cells often acquire LAK activity after prolonged exposure to IL-2. Other lymphokines such as TNF, IL-1 and IL-4 (often in conjunction with low-dose IL-2) can also be used to generate LAK activity, perhaps by up-regulating the IL-2R $\alpha$ chain (Clark *et al.*, 1988). While this loss of MHC restriction may be an *in vitro* artifact, such cells have also been shown to be effective at tumour cell killing *in vivo*.

T cells bearing $\gamma/\delta$ TCRs have been shown to be cytolytic in a non-MHC-restricted way, and although this can be associated with the development of

NK-like activity (i.e. the presence of CD56), other γ/δ T cells have been shown to be restricted by the non-polymorphic MHC-related molecules such as Qa-1 (Vidovic *et al.*, 1989).

### *14.2.4 Summary*

It is clear that MHC expression and the level of tumour immunogenicity are inextricably linked. Both NK cell and T cell effectors can eliminate tumour cells. Enhanced H-2 expression can by-pass NK effectors while reduced levels can ameliorate CTL effector function. As the two systems work in concert, they should, in theory, be very effective and generally are, but, nevertheless, tumour cells can still evade recognition. The study by Perez *et al.* (1990) is a good example. They studied the metastatic capacity of clones derived from a chemically induced fibrosarcoma GR9 of BALB/c mice. In essence, subcutaneous injection with $H\text{-}2^{low}$ phenotypes gave few metastases while those with an $H\text{-}2^{high}$ phenotype gave a much greater number, consistent with a role for NK cells in tumour cell killing (which was confirmed *in vitro*). In contrast, intravenous injection of tumour cells gave rise to the opposite results, with the $H\text{-}2^{high}$ clones giving rise to fewer metastases. This was confirmed by analysing the phenotype of the cells found in the metastatic nodules, which had a marked loss of H-2K and H-2D expression. Interestingly, H-2K was generally down-regulated and H-2D up-regulated. Up-regulation of H-2K with IFN-γ reduced metastatic capacity in the intravenous system. This can be most easily explained by the balancing role of the NK and CTL effector mechanisms, both of which have a finite repertoire.

In the B16 melanoma system (Taniguchi, Karre & Klein, 1985), intravenous inoculation of H-2-negative cells gave no metastases unless either NK cells were depleted, H-2 expression was up-regulated by IFN-γ or the cell number of the inoculate was increased. This would be entirely consistent with the GR9 system and the suggestion would be that early tumour escape (low cell numbers) would depend on up-regulation of an NK protective H-2 antigen (K, D or L depending on strain and, therefore, repertoire of NK effectors) and a later down-regulation of H-2 loci depending, again, on the restriction elements required for CTL killing.

It would, of course, be satisfying if the balance between H-2 expression, NK/CTL killing and tumorigenesis was a consistent and universal phenomenon. However, in the $H\text{-}2^b$ mouse BW7756 tumour system (Nishimura & Ostrand-Rosenberg, 1991), it was shown that the reduced immunogenicity of a naturally occurring $H\text{-}2K^b$ loss mutant (Hepa-1) was not, as would be expected, the result of increased susceptibility to NK lysis but of an increased susceptibility to H-2D$^b$-restricted CD4$^-$CD8$^+$ CTLs. There may, therefore, be no universal rules, and resistance to tumour growth can be a function of antigen expression, NK repertoire, TCR repertoire, MHC haplotype and any number of other factors that influence tumour spread, e.g. adhesion molecule expression (Gabius *et al.*, 1987; Johnson *et al.*, 1989), antigen processing (Zou *et al.*, 1992) or the induction of T suppressor ($T_S$) rather than $T_H$ cells (Leshem & Kedar, 1990). Even in the GR9 system, Algarra *et al.* (1991) showed that metastasis derived from cloned H-2-positive (GR9.G2) or H-2-negative (GR9.B9) cells could be of either phenotype. This emphasizes the complex nature of the interaction between the immune system and tumour development *in vivo*.

## 14.3 Tumour immunogenicity

Perhaps one of the most exciting aspects of tumour immunology to be developed over the past few years has been the identification of tumour antigens recognized by CTLs in the P815 mouse mastocytoma tumour system (Boon, 1983). Boon and his colleagues, in this pioneering work, have suggested three separate mechanisms that could give rise to tumour antigens generating an anti-tumour T cell response. Firstly, mutations in a normal gene product can give rise to peptides that are capable of binding to MHC while their normal counterparts are not. Secondly, a mutation in a normally expressed gene may allow T cell recognition and activation. Thirdly, a normally silent gene can be activated in a subset of tumour cells allowing MHC binding and recognition by (non-tolerized) CTLs. The results obtained by Boon's group are very definitive because of the molecular approach taken but are built on a large data base of animal experiments which show that many normal and viral proteins can act as TRAs. For instance, Kast & Melief (1991) showed the induction of CTLs to the E1A oncoprotein of adenovirus type 5 in mice immunized with syngeneic cells transformed with the adenovirus early region 1 (E1). They also showed that an H-2$^b$-binding peptide was involved in immune rejection of E1-induced (syngeneic) tumours. Viral oncoproteins have a well-defined role as TRAs, as well as their cellular counterparts (Peace *et al.*, 1991), and cellular oncogenes may also act by influencing the expression of other cellular proteins that can themselves act as tumour antigens (Torigoe *et al.*, 1991). Therefore, it has been conclusively shown that tumour cells can be recognized by the immune system and, by definition, that MHC antigens have a pivotal role.

## 14.4 Restoration of MHC expression and tumorigenicity

Most of the work published on MHC loss in tumour systems has concentrated on class I antigens, since most tumours arise from cells that do not naturally express class II. In addition, class I antigens are intimately involved with cell-mediated anti-tumour responses (both NK and CTL mediated). However, current methods of gene transfer (transfection) are very efficient and allow cells to be 'induced' to express either a naturally lost (syngeneic) class I antigen or, for that matter, any syngeneic or allogeneic class I or II antigen. This is a powerful system for studying the effect of a particular MHC antigen in any given tumour system and it may also open the way for inducing anti-tumour immune responsiveness, i.e. a tumour vaccine (Anonymous, 1989).

### *14.4.1 MHC class I restoration*

There is now overwhelming evidence that many tumour cell types manifest a loss of MHC class I antigen expression (Festenstein & Schmidt, 1981; Bernards, 1987; Hammerling *et al.*, 1987; Tanaka *et al.*, 1988; Elliott *et al.*, 1989; Algarra *et al.*, 1991). This generally confers a growth advantage *in vivo* because of the essential role of CTLs in recognition and killing. Most animal tumours studied are established cell lines derived from tissues of diverse origin, where cell transformation has been either virally or non-virally induced. MHC class I genes are developmentally regulated (Ozato, Wan & Orrison, 1985) and their expression can be up-regulated by IFNs ($\alpha$, $\beta$ or $\gamma$), often in combination with TNF-$\alpha$ (Rosa, Cochet & Fellous, 1986), and suppressed by certain viruses and oncogenes (Gogusev *et al.*, 1988). These topics are extensively described in Chapters 3, 4 and 6. This down-regulation and, indeed, complete loss of MHC class I gene expression is often restricted to only certain loci, e.g. H-2K is often lost in leukaemic cell lines derived from the AKR mouse, which is susceptible to the Gross leukaemia virus (Hui *et al.*, 1984). These locus-specific loss 'mutants' can be made susceptible to CTL killing *in vitro* and *in vivo* by the up-regulation or the reintroduction of genes coding for the syngeneic MHC class I antigens (Hui, 1989).

The fundamental causes underlying these MHC class I loss mutations can be the complete deletion of either the class I heavy chain gene or, more often, the $\beta_2$-m gene (Kaklamanis & Hill, 1992). Defects in binding of transcription factors are common, such as KBF1 and KBF2, which bind to the enhancer A regulatory element of class I genes (Singer & Maguire, 1990; see Chapters 4 and 13). In the AKR mouse thymoma system, low levels of H-2K mRNA are present and this can be attributed to low levels of KBF1 binding to the 5′-flanking enhancer region. As transfected H-2K$^k$ genes are expressed (see below), the defect lies not in the production of KBF1 but in a *cis*-acting mechanism preventing transcription of the endogenous MHC gene.

This is not the only mechanism described for dysfunction of MHC class I expression. For instance, in the non-class I expressing fibrosarcoma and lung carcinoma lines described by Klar & Hammerling (1989), expression can be induced by IFN-γ and these tumours would appear to have defects in genes necessary for class I assembly. These are probably similar to the defects associated with peptide loading via the proteasome (peptide transporter genes) as described by Townsend *et al.* (1989) for the RMA-S cell line (see Chapter 1).

A number of methods have been used to increase tumour immunogenicity by enhancement of the expression of MHC class I antigens. These include treating tumour cells with various lymphokines such as IFNs and TNF (Tanaka *et al.*, 1988). In addition, Gorelik *et al.* (1985) produced immunogenic variants of BL6 melanoma cells by treatment with the mutagen *N*-methyl-*N*′-nitro-*N*-nitrosoguanidine and the DNA hypomethylating agent 5-azacytidine. All the immunogenic variants showed enhanced expression of MHC class I products. Treatment of an H-2-deficient murine lung carcinoma cell line with dimethyl sulphoxide also led to enhanced MHC class I expression and led to the cells becoming susceptible to T cell-mediated cytotoxicity (Bahler & Lord, 1985; Blieden *et al.*, 1991). While this approach has been informative, it has the drawback that other antigens could also be up-regulated and it is difficult to assign the positive effects to MHC expression alone. DNA-mediated gene transfer or transfection (Malissen, 1986) is the most direct way in which these effects can be separated. It involves the introduction of cloned MHC genes into eukaryotic tumour cells and can be mediated by a number of methods including calcium phosphate precipitation, lipofection, electroporation or the use of retroviral vectors (James & Grosveld, 1986). However, a note of caution should be added. To obtain demonstrable exogenous gene expression, strong promoters (such as the SV40 early promoter) are often necessary and, as plasmid integration is an essentially random process, the transfection process itself can modify the expression of endogenous MHC and non-MHC genes. For example, it was noted that 10–15% of clones transfected with the pSV2*neo* vector alone (bearing the gene for neomycin resistance) showed altered *in vivo* growth, metastasis and MHC expression compared with controls (Kerbel *et al.*, 1987). As electroporation and retroviral vectors do not produce these non-specific mutational events to the same degree, these properties may be ascribed to the trauma associated with the calcium phosphate precipitation method for transfection (Weiss & Ziegler, 1989), which is now rarely used for such studies.

One of the earliest and, perhaps, best studied systems has been the H-2K-deficient AKR mouse (H-$2^k$) leukaemia cell lines originating from Festenstein's laboratory (Festenstein & Schmidt, 1981; see also Chapter 7). The most tumorigenic clones derived were those with the lowest H-2K expression (e.g. K36.16) while, conversely, high H-2K expression (e.g. 369) was associated with low tumorigenicity (Hui *et al.*, 1991). Since H-2K acts as the restriction element for CTL activity on AKR tumour cells, the effect of *de novo* expression of H-$2K^k$ on K36.16 cells was studied by the direct introduction of the syngeneic gene (Hui *et al.*, 1984). In contrast to the parental cells, the transformed tumour cells did not grow in syngeneic mice when injected at a relatively high dose ($10^5$/mouse). Blocking the newly expressed H-2K antigens on the transformed cells with an H-$2K^k$-specific mAb resulted in tumour growth, directly demonstrating the biological relevence of H-2K antigen expression in this model. These experiments were complemented by showing that the high-H-2K-expressing, low-tumorigenic variant (369) was transformed into a highly tumorigenic line after transfection with H-$2K^k$ anti-sense DNA (Hui *et al.*, 1991). Perhaps one of the most interesting and important aspects of this model is that mice pre-immunized with H-$2K^k$-expressing variants become resistant to a further challenge with the original (non-H-$2K^k$-expressing) K36.16 tumour. This implies that the need for H-2K expression is more important for the induction of immunity (i.e. generation of $CD4^+$ $T_H$ cells) than for the generation of tumour-specific H-2K-restricted $CD8^+$ CTLs (see below). Similar results have been found in other tumour systems (Lathe *et al.*, 1987; Feldman & Eisenbach, 1988; Tanaka *et al.*, 1988), although protection against the original (MHC class I loss) tumour has not always been found (Hammerling *et al.*, 1987). The role of NK killing in the AKR system was excluded by studying the effect of *in vitro* killing of YAC-1 cells using lymph node cells from immunized (protected) versus non-immunized mice.

### *14.4.2 Allogeneic MHC class I gene transfection*

While the use of syngeneic class I gene transfection into specific 'loss' mutant cell lines is an unequivocal and elegant system to study the role of subclasses of MHC class I on tumorigenicity *in vivo*, this approach would never be a feasible way to develop a 'tumour vaccine' (Anonymous, 1989). A tumour would have to be removed, any subclass MHC class I loss defined and replaced (by transfection) with an identical haplotype. This would be a difficult (but not impossible) task given the high degree of polymorphism exhibited by the HLA-A, HLA-B and HLA-C loci. A number of groups have, therefore, studied the fate of allogeneic MHC class I gene transfected tumours in animal models (Bahler *et al.*, 1987; Cole *et al.*, 1987; Itaya *et al.*, 1987; Hui *et al.*, 1989; Cole & Ostrand-Rosenberg, 1991; Ostrand-Rosenberg *et al.*, 1991a). Clearly this situation becomes much more complex at a molecular level, as restoring a 'hole in the repertoire' (either at the presentation or responder level) would not be the only way in which allogeneic class I antigens could work. They could simply act as strongly immunogenic determinants to initiate a response to the tumour, since up to 1% of all host T cells can be recruited to a given allogeneic MHC class I antigen (Marrack & Kappler, 1988).

In the AKR, K36.16 system, Hui *et al.* (1989) were able to show that H-2$K^b$-transfected tumour cells were non-tumorigeneic in AKR mice. Furthermore, an initial challenge with the class I allogeneic-transfected tumour cells gave protective immunity against a subsequent challenge with the wild-type (non-transfected) K36.16 tumour. Mice injected with K36.16 were even protected when given the H-2$K^b$-transfected cells two days after the initial challenge with K36.16.

Given these extremely exciting and important results, a note of caution needs to be added. Not all H-2$K^b$-expressing clones were protective and this was not wholly related to the level of H-2$K^b$ expression. Even in the syngeneic (H-2$K^k$-transfected K36.16) system, there was no direct correlation between class I expression and immunogenicity (Hui *et al.*, 1984). This could be associated with down-regulation of class I expression of some clones *in vivo*, which was not tested but may be caused by unknown factors associated with the transfected tumour. In addition, the number of cells used to challenge mice is often overlooked. For instance, 100% of AKR mice injected with $10^5$ H-2$K^k$-transfected K36.16 cells (K36.$K^k$) may survive but all mice injected with $5 \times 10^5$ K36.$K^k$ cells may die (Hui *et al.*, 1984; James *et al.*, 1991). The dynamics are also important. In the allogeneic class I system, protection can be provided to an existing non-transfected K36.16 tumour-bearing mouse if challenged with H-2$K^b$-expressing transfectants within two days, but by four days after tumour inoculation, no protection can be provided. The strain combinations used may also be important. In the allogeneic class I-transfected Sa1 tumour system (Cole & Ostrand-Rosenberg, 1991), some strains of mice were immune to tumour challenge (9/12) while others (e.g. BALB/c, DBA/2, A.TL) were not and this did not seem to be related entirely to MHC haplotype. In later experiments, it was shown that while

H-2K$^b$ transfection alone was not an effective immunoprotectant in those strains, a combination of H-2K$^b$ and H-2D$^b$ was effective (Ostrand-Rosenberg, Roby & Clements, 1991b). These are similar to the results of Mandelbeim, Feldman & Eisenbach (1992), who found that double class I transfectants were effective in immunizing in the 3LL lung carcinoma system while single transfectants were not. This probably relates to the immunological repertoire found in different strains of mice and has important implications.

Finally, when relating these animal experiments to future uses of transfected tumour cells for human 'vaccine' development (Anonymous, 1989), it is important to know whether transfected cells can be killed and still provide protection. This is not clearly recorded in the literature, as most studies have used viable tumour cells, but Hui *et al.* (1984) found that heat-killed K36.K$^k$ cells did not provide protection against a subsequent challenge with K36.16. If this were to be a general case, it may provide clinicians with an ethical dilemma at some time in the future.

### *14.4.3 MHC class II gene transfection*

The role of MHC class I in tumour cell recognition and lysis has been clearly stated, making class I gene transfection an obvious choice for enhancing the immunogenicity of tumour cells, but what of class II? As there are so many examples where class I transfectants provide protection to subsequent challenge with the untransfected parental tumour (Hui *et al.*, 1984; 1989; Lathe *et al.*, 1987; Feldman & Eisenbach, 1988; Tanaka *et al.*, 1988; Cole & Ostrand-Rosenberg, 1991; Ostrand-Rosenberg *et al.*, 1991b; Mandelbeim *et al.*, 1992) then class I must surely be acting in some way as an 'adjuvant' (perhaps by filling a hole in the repertoire) and not as the restricting element for T cell recognition.

Although requiring the co-transfection of two genes (the $\alpha$ and $\beta$ chain genes), class II transfection and expression was shown to be feasible in 1983 (Malissen *et al.*, 1983). Why not then use class II, which in theory could be more effective, as TRA could be presented directly to T cells? One possible problem could be the need to have endogenous expression of invariant chain to obtain functional assembly (Peterson & Miller, 1990). However, it has subsequently been shown that absence of invariant chain may allow endogenously (viral?) produced antigens to be associated with class II and thus enhance the effectiveness of class II-transfected tumours to present TRAs (Roche & Cresswell, 1990). In the K36.16 AKR tumour system, James *et al.* (1991) were able to show that H-2E$^k$ gene transfection (K36.E$^k$ cells) was actually more effective in conferring immunity to K36.16 than was K36.$K^k$. That is, doses of > $10^5$ K36.$K^k$ cells gave significant mortality while AKR mice could survive a challenge of $4 \times 10^6$ K36.$E^k$ cells. Also, K36.$E^k$ cells could protect mice from a simultaneous challenge with K36.16 cells, whereas K36.$K^k$ could not.

Essentially similar results were obtained by Ostrand-Rosenberg, Thakur & Clements (1990) in the Sa1 tumour system, where H-2$A^k$ provided effective immunity against the parental Sa1 tumour. They were then able to show that removal of the cytoplasmic domain of the H-2$A^k$ heavy chain (important in intracellular signalling) negated the effect, suggesting a role for direct presentation by the transfected class II antigen (Ostrand-Rosenberg *et al.*, 1991b). This is supported by their work on B7 transfection, which suggests that direct presentation of TRAs by tumour cells may be an important way of triggering anti-tumour immunity (Baskar *et al.*, 1993). The transfection approach to trigger anti-tumour immunity is open to many other cell surface and secreted molecules (e.g. IL-2) not covered in this chapter (reviewed by Pardoll, 1993).

## 4.5 Conclusions

The prospect of developing systems for treating patients with modified tumour cells (following initial surgery or therapy) as a way of stimulating the immune system to rid the body of residual tumour cells is very exciting. The ground work done in animal systems has now led to clinical treatment programmes that have shown positive results (Boon, 1992). If the success of animal experiments can be translated into clinical medicine, then tumour immunotherapy may one day be commonly used to prevent tumour recurrence.

## References

Algarra, I., Gaforio, J.J., Garrido, A., Mialdea, M.J., Perezo, M. & Garrido, F. (1991). Heterogeneity of MHC class I antigens in clones of methylcholanthrene-induced tumors. Implications for local growth and metastases. *International Journal of Cancer*, Suppl. 6, 73–81.

Anonymous (1989). Tumour cell vaccines: has their time arrived? [Editorial] *Lancet*, ii, 955–957.

Bahler, D., Frelinger, J., Harwell, L. & Lord, E.M. (1987). Reduced tumorigenicity of a spontaneous mouse lung carcinoma following H-2 gene transfection. *Proceedings of the National Academy of Sciences of the USA*, 84, 4562–4565.

Bahler, D. & Lord, E.M. (1985). Enhanced immune recognition of H-2 antigen-deficient murine lung carcinoma cells following treatment with dimethyl sulfoxide. *Cancer Research*, 45, 6362–6365.

Baskar, S., Ostrand-Rosenberg, S., Nabavi, N., Nadler, L.M., Freeman, G.J. & Glimcher, L.H. (1993). Constitutive expression of B7 restores immunogenicity of tumor cells expressing truncated major histocompatibility complex class II molecules. *Proceedings of the National Academy of Sciences of the USA*, 90, 5687–5690.

Bernards, R. (1987). Suppression of MHC gene expression in cancer cells. *Trends in Genetics*, 3, 298–301.

Blieden, T.M., McAdam, A.J., Foresman, M.D., Cerosaletti, K.M., Frelinger, J.G. & Lord, E.M. (1991). Class I MHC expression in the mouse lung carcinoma, Line 1: a model for class I inducible tumors. *International Journal of Cancer*, Suppl. 6, 82–89.

Boon, T. (1983). Antigenic tumor cell variants obtained with mutagens. *Advances in Cancer Research*, 39, 121–151.

Boon, T. (1992). Toward a genetic analysis of tumor-rejection antigens. *Advances in Cancer Research*, 58, 179–220.

Boon, T. (1993). Teaching the immune system to fight cancer. *Scientific American*, 268, 32–39.

Browning, M.J. & Bodmer, W.F. (1992). MHC antigens and cancer: implications for T cell surveillance. *Current Opinion in Immunology*, 4, 613–618.

Cerottini, J.-C. & Brunner, K.T. (1974). Cell mediated cytotoxicity, allograft rejection and tumour immunity. *Advances in Immunology*, 18, 67–132.

Chadwick, B.S. & Miller, R.G. (1992). Hybrid resistance *in vitro*. Possible role of both MHC class I and self peptides in determining the level of target cell sensitivity. *Journal of Immunology*, 148, 2307–2313.

Chan, P.Y. & Takei, F. (1989). Molecular cloning and characterization of a novel murine T cell surface antigen, YE1/48. *Journal of Immunology*, 142, 1727–1736.

Christinck, E.R., Luscher, M.A., Barber, B.H. & Williams, D.B. (1991). Peptide binding to class I MHC on living cells and quantitation of complexes required for CTL lysis. *Nature (London)*, 352, 67–70.

Clark, W.R., Ostergaard, H., Gormann, K. & Torbett, B. (1988). Molecular mechanisms of CTL-mediated lysis: a cellular perspective. *Immunological Reviews*, 103, 37–52.

Cole, G.A., Clements, V.K., Garcia, E.P. & Ostrand-Rosenberg, S. (1987). Allogeneic H-2 antigen expression is insufficient for tumor rejection. *Proceedings of the National Academy of Sciences of the USA*, 84, 8613–8617.

Cole, G.A. & Ostrand-Rosenberg, S. (1991). Rejection of allogeneic tumour is not determined by host responses to MHC class I molecules and is mediated by $CD4^{-}$ $CD8^{+}$ T lymphocytes that are not lytic for the tumor. *Cellular Immunology*, 134, 480–491.

Colombo, M., Modesti, A., Parmiani, G. & Forni, G. (1992). Local cytokine availability elicits tumor rejection and systemic immunity through granulocyte–T-lymphocyte cross talk. *Cancer Research*, 52, 4853–4857.

Ehrlich, P. (1909). Ueber denjetzigen Stand der Karzinomforschung. *Nederlandsch tijdschrift voor geneeskunde*, i, 273–290.

Elliott, B.E., Carlow, D.A., Rodrick, A.-M. & Wade, A. (1989). Perspectives on the role of MHC antigens in normal and malignant cell development. *Advances in Cancer Research*, 53, 181–245.

Erb, P., Grogg, D., Troxler, M., Kennedy, M. & Flufi, M. (1990). $CD4^{+}$ T cell-mediated killing of MHC class II-positive antigen presenting cells. I.

Characterization of target cell recognition *in vivo* of *in vitro* activated $CD4^+$ killer T cells. *Journal of Immunology*, 144, 790–796.

Esteban, F., Ruiz-Cabello, F., Concha, A., Perez-Ayala, M., Sanchez-Rozas, J. & Garrido, F. (1990). HLA-DR expression is associated with excellent prognosis in squamous cell carcinoma of the larynx. *Clinical and Experimental Metastases*, 8, 319–322.

Feldman, M. & Eisenbach, L. (1988). Genes controlling the metastatic phenotype. *Cancer Surveys*, 7, 555–572.

Festenstein, H. & Schmidt, W. (1981). Variation in MHC antigenic profiles of tumour cells and its biological effects. *Immunological Reviews*, 60, 85–127.

Foley, E.J. (1953). Antigenic properties of methylcholanthrene-induced tumours in mice of the strain of origin. *Cancer Research*, 13, 835–837.

Gabius, H.-J., Bandlow, G., Schirrmacher, V., Nagel, G.A. & Vehmeyer, K. (1987). Differential expression of endogenous sugar-binding proteins (lectins) in murine tumour-model systems with metastatic capacity. *International Journal of Cancer*, 39, 643–648.

Gardner, I., Bowern, N.A. & Blanden, R.V. (1974). Cell-mediated cytotoxicity against ectromelia virus infected target cells. II. Identification of effector cells and analysis of mechanisms. *European Journal of Immunology*, 4, 68–72.

Gogusev, J., Teutsch, B., Morin, M.T., Mongiat, F., Haguenau, F., Suskind, G. & Rabotti, G.F. (1988). Inhibition of HLA class I antigen and mRNA expression induced by Rous sarcoma virus in transformed human fibroblasts. *Proceedings of the National Academy of Sciences of the USA*, 85, 203–210.

Gorelik, E., Peppoloni, S., Overton, R. & Herberman, R.B. (1985). Increase in H-2 antigen expression and immunogenicity of BL6 melanoma cells treated with N-methyl-N′-nitro-N-nitrosoguanidine. *Cancer Research*, 45, 5341–5347.

Gregory, C.D., Murray, R.J., Edwards, C.F. & Rickinson, A.B. (1988). Down regulation of cell adhesion molecules LFA-3 and ICAM-1 in Epstein–Barr virus-positive Burkitt's lymphoma underlies tumour cell escape from virus-specific T cell surveillance. *Journal of Experimental Medicine*, 167, 1811–1824.

Grey, H.M. & Chesnut, R. (1985). Antigen processing and presentation to T cells. *Immunology Today*, 6, 101–106.

Gross, L. (1943). Intradermal immunization of C3H mice against a sarcoma that originated in an animal of the same line. *Cancer Research*, 43, 125–132.

Hammerling, G.J., Klar, D., Pulm, W., Momburg, F. & Moldenhauer, G. (1987). The influence of major histocompatibility complex class I antigens on tumor growth and metastases. *Biochimica et Biophysica Acta*, 907, 245–259.

Hui, K.M. (1989). Re-expression of major histocompatibility complex (MHC) class I molecules on malignant tumour cells and its effect on host tumour interaction. *BioEssays*, 11, 22–26.

Hui, K., Grosveld, F. & Festenstein, H. (1984). Rejection of transplantable AKR leukaemia cells following MHC DNA-mediated cell transformation. *Nature (London)*, 311, 750–752.

Hui, K.M., Sim, T., Foo, T.-T. & Oei, A.-A. (1989). Tumour rejection mediated by transfection with allogeneic class I histocompatibility gene. *Journal of Immunology*, 143, 3835–3843.

Hui, K.M., Sim, B.C., Foot, T. & Oei, A.-A. (1991). Promotion of tumour growth by transfecting antisense DNA to suppress endogenous $H\text{-}2^k$ MHC gene expression in AKR mouse thymoma. *Cellular Immunology*, 136, 80–94.

Itaya, T., Yamagiwa, S., Okada, F., Oikawa, T., Kuzumaki, N., Takeichi, N., Hosokawa, G. & Jay, G. (1987). Xenogenization of a mouse lung carcinoma

(3LL) by transfection with an allogeneic class I major histocompatibility complex gene (H-2L$^d$). *Cancer Research*, 47, 3136–3140.

James, R.F.L., Edwards, S., Hui, K.M., Bassett, P.D. & Grosveld, F. (1991). The effect of class II gene transfection on the tumorigenicity of the H-2K negative mouse leukaemia cell line K36.16. *Immunology*, 72, 213–218.

James, R.F.L. & Grosveld, F. (1986). DNA mediated gene transfer into mammalian cells. In *Techniques in Molecular Biology II*, ed. J. Walker & W. Gaastra, pp. 187–202. Kent: Croom Helm.

Jiang, D. & Flyer, D.C. (1992). Immune response to Moloney murine leukaemia virus: non-viral, tumor-associated antigens fail to provide *in vivo* tumor protection. *Journal of Immunology*, 148, 974–980.

Johnson, J.P., Stade, B.G., Holzmann, B., Schwable, W. & Riethmuller, G. (1989). *De novo* expression of inter-cellular adhesion molecule 1 in melanoma correlates with increased risk of metastasis. *Proceedings of the National Academy of Sciences of the USA*, 86, 641–644.

Kaklamanis, L. & Hill, A. (1992). MHC loss in colorectal tumours: evidence for immunoselection. *Cancer Surveys*, 13, 155–171.

Karlhofer, F.M., Ribaudo, R.K. & Yokoyama, W.Y. (1992). MHC class I alloantigen specificity of Ly49$^+$ IL-2 activated natural killer cells. *Nature (London)*, 358, 66–70.

Karre, K. (1985). Role of target histocompatibility antigens in regulation of natural killer cell activity: a re-evaluation and a hypothesis. In *Mechanisms of NK Mediated Cytotoxicity*. ed. D. Callewert & R.B. Herberman, pp. 81–91. Orlando: Academic Press.

Karre, K. (1992). Natural killing: an unexpected petition for pardon. *Current Biology*, 2, 613–615.

Karre, K., Ljunggren, H.G., Piontek, K.G. & Kiessling, R. (1986). Selective rejection of H-2 deficient lymphoma variants suggests alternative immune defence strategy. *Nature (London)*, 319, 675–678.

Kast, W.M. & Melief, C.J. (1991). Fine peptide specificity of cytotoxic T lymphocytes directed against adenovirus-induced tumours and peptide-MHC binding. *International Journal of Cancer*, Suppl. 6, 90–94.

Kerbel, R.S., Waghorne, C., Man, M.S., Elliott, B.E. & Breitman, M.L. (1987). Alteration of the tumorigenic and metastatic properties of neoplastic cells is associated with the process of calcium phosphate-mediated DNA transfection. *Proceedings of the National Academy of Sciences of the USA*, 84, 1263–1267.

Kiessling, R., Klein, E. & Wigzell, H. (1975). Natural killer cells in the mouse. *Journal of Immunology*, 5, 112–116.

Klar, D. & Hammerling, C.L. (1989). Induction of assembly of MHC class I heavy chains with $\beta_2$-microglobulin by interferon-gamma. *EMBO Journal*, 8, 475–481.

Lathe, R., Kieny, M.P., Gerlinger, P., Clertant, P., Guizani, I., Cuzin, F. & Chambon, P. (1987). Tumour prevention and rejection with recombinant vaccine. *Nature (London)*, 326, 878–880.

Leshem, B. & Kedar, E. (1990). Cytotoxic T lymphocytes reactive against a syngeneic murine tumour and their specific suppressor T cells are both elicited by *in vitro* allosensitization. *Journal of Experimental Medicine*, 171, 1057–1071.

Lu, Y., Blair, D.C., Segal, S., Shih, T.Y. & Clanton, D.J. (1991). Tumorigenicity, metastases and suppression of MHC class I expression in murine fibroblasts transformed by mutant v-*ras* deficient in GTP binding. *International Journal of Cancer*, 6, 45–53.

Malissen, B. (1986). Transfer and expression of MHC genes. *Immunology Today*, 7, 106–112.

Malissen, B., Steinmetz, M., McMillan, M., Pierres, M. & Hood, L. (1983). Expression of I-$A^k$ class II genes in mouse L cells after DNA-mediated gene transfer. *Nature (London)*, 305, 440–443.

Mandelbeim, O., Feldman, M. & Eisenbach, L. (1992). H-2K double transfectants of tumor cells as anti-metastatic cellular vaccines in heterozygous recipients. Implications for the T cell repertoire. *Journal of Immunology*, 148, 3666–3673.

Marrack, P. & Kappler, J. (1988). The T-cell repertoire for antigen and MHC. *Immunology Today*, 9, 308–310.

McMichael, A. (1992). Cytotoxic T lymphocytes and immune surveillance. *Cancer Surveys*, 13, 5–21.

McMichael, A.J., Gotch, F.M., Noble, G.R. & Beare, P.A. (1983). Cytotoxic T cell immunity to influenza. *New England Journal of Medicine*, 309, 13–17.

Nagasawa, R., Gros, J., Kanagawa, O., Townsend, K., Lanier, L.L., Chiller J. & Allison, J.P. (1987). Identification of a novel T cell surface disulfide-bonded dimer distinct from the alpha/beta antigen receptor. *Journal of Immunology*, 138, 815–824.

Natali, P.G., Giacomini, P., Bigotti, A., Imai, K., Nicotra, M.R., Ng, A.K. & Ferrone, S. (1983). Heterogeneity in the expression of HLA and tumour-associated antigens by surgically removed and cultured breast carcinoma cells. *Cancer Research*, 43, 660–668.

Nilsson, K., Evsin, P.E. & Welsh, K.I. (1974). Production of $\beta_2$-microglobulin in normal and malignant cell lines and peripheral lymphocytes. *Transplantation Reviews*, 21, 53–84.

Nishimura, M.I. & Ostrand-Rosenberg, S. (1991). Mouse Hepa-1 tumor is rejected by H-$2D^b$ restricted CTL despite decreased MHC class I antigen expression. *Cellular Immunology*, 136, 414–424.

Ostrand-Rosenberg, S., Roby, C.A. & Clements, V.K. (1991b). Abrogation of tumorigenicity by MHC class II antigen expression requires the cytoplasmic domain of the class II molecule. *Journal of Immunology*, 147, 2419–2422.

Ostrand-Rosenberg, S., Roby C., Clements, V.K. & Cole, G.A. (1991a). Tumor-specific immunity can be enhanced by transfection of tumor cells with syngeneic MHC-class II genes or allogeneic MHC-class I genes. *International Journal of Cancer*, Suppl. 6, 61–68.

Ostrand-Rosenberg, S., Thakur, A. & Clements, V. (1990). Rejection of mouse sarcoma cells following transfection of MHC class II genes. *Journal of Immunology*, 114, 4068–4071.

Ozato, K., Wan, J. & Orrison, B. (1985). Mouse major histocompatibility class I gene expression begins at mid-somite stage and is inducible in earlier stage embryos by interferon. *Proceedings of the National Academy of Sciences of the USA*, 82, 2427–2432.

Pardoll, D.M. (1993). New strategies for enhancing the immunogenicity of tumors. *Current Opinions in Immunology*, 5, 719–725.

Peace, D.J., Chen, W., Nelson, H. & Cheever, M.A. (1991). T cell recognition of transforming proteins encoded by mutated *ras* proto-oncogenes. *Journal of Immunology*, 146, 2059–2065.

Perez, M., Algarra, I., Ljunggren, H., Caballero, A., Mialdea, M.J., Gaforio, J.J., Klein, G., Karre, K. & Garrido, F. (1990). A weakly tumorigenic phenotype with high MHC class-I expression is associated with high metastatic potential after surgical removal of the primary murine fibrosarcoma. *International Journal of Cancer*, 46, 258–261.

Peterson, M. & Miller, J. (1990). Invariant chain influences the immunological recognition of MHC II molecules. *Nature (London)*, 345, 172–174.

Plata, F., Langlade-Demoyen, P., Abastado, J.P., Berbar, T. & Kourilsky, P. (1987). Retrovirus antigens recognised by cytolytic T lymphocytes activate tumor rejection *in vivo*. *Cell*, 48, 231–237.

Roche, P.A. & Cresswell, P. (1990). Invariant chain association with HLA-DR molecules inhibits immunogenic peptide binding. *Nature (London)*, 345, 615–618.

Rosa, F.M., Cochet, M.M. & Fellous, M. (1986). Interferon and major histocompatibility complex genes: a model to analyse eukaryotic gene regulation. *Interferon*, 7, 47–52.

Singer, D.S. & Maguire, J.E. (1990). Regulation of the expression of class I MHC genes. *Critical Reviews in Immunology*, 10, 235–257.

Tanaka, K., Yoshioka, T., Bieberich, C. & Jay, G. (1988). Role of major histocompatibility complex class I antigens in tumor growth and metastases. *Annual Review of Immunology*, 6, 359–380.

Taniguchi, K., Karre, K. & Klein, G. (1985). Lung colonisation and metastases by disseminated B16 melanoma cells: H-2 associated control at the level of the host and the tumor cell. *International Journal of Cancer*, 36, 503–508.

Torigoe, T., Sato, N., Takashima, T., Cho, J.-M., Tsuboi, N., Qi., W., Hara, I., Wada, Y., Takahashi, N. & Kikuchi, K. (1991). Tumor rejection antigens on BALB3T3 cells transformed by activated oncogenes. *Journal of Immunology*, 147, 3251–3258.

Townsend, A., Bastin, J., Gould, K., Brownlee, G., Andrew, M., Coupar, B., Boyle, D., Chan, S. & Smith, G. (1988). Defective presentation to class I restricted cytotoxic T lymphocytes in vaccinia infected cells is overcome by enhanced degradation of antigen. *Journal of Experimental Medicine*, 168, 1211–1224.

Townsend, A., Ohlin, C., Bastin, J.L., Ljunggren, H.G., Foster, L. & Karre, K. (1989). Association of class I major histocompatibility heavy and light chains induced by viral peptides. *Nature (London)*, 340, 443–448.

Townsend, S.E. & Allison, J.P. (1993). Tumor rejection after direct co-stimulation of $CD8^+$ T cells by B7-transfected melanoma cells. *Science*, 259, 368–370.

Trinchieri, G. (1989). Biology of natural killer cells. *Advances in Immunology*, 47, 187–376.

Vidovic, D., Roglic, M., McKune, K., Guerder, S., MacKay, C. & Dembic, Z. (1989). Qa-1 restricted recognition of foreign antigen by γδ T-cell hybridoma. *Nature (London)*, 340, 646–648.

Weiss, E.H. & Ziegler, A. (1989). HLA genes and function. *Immunology Today*, 10, S16–S18.

Yokoyama, W.M., Ryan, J.C., Hunter, J.J., Smith, H.R.C., Stark, M. & Seaman, W.E. (1991). cDNA cloning of mouse NKR-PM and genetic linkage with Ly-49. Identification of a natural killer cell gene complex on mouse chromosome 6. *Journal of Immunology*, 147, 3229–3236.

Zinkernagel, R.M. & Doherty, P.C. (1975). H-2 compatibility requirement for T cell mediated lysis of target cells infected with lymphocytic choriomeningitis virus. *Journal of Experimental Medicine*, 141, 1427–1436.

Zinkernagel, R.M. & Doherty, P.C. (1979). MHC-restricted cytotoxic T cells: studies on the biological role of polymorphic major transplantation antigens determining T cell restriction – specificity, function and responsiveness. *Advances in Immunology*, 27, 51–177.

Zou, J.-P., Shimizu, J., Ikegame, K., Yamamoto, N., Ono, S., Fujiwara, H. & Hamaoka, T. (1992). Tumor-bearing mice exhibit a progressive increase in tumor antigen-presenting cell function and a reciprocal decrease in tumor antigen-responsive $CD4^+$ T cell activity. *Journal of Immunology*, 148, 648–655.

# 15

# Manipulation of MHC antigens by gene transfection and cytokine stimulation: a possible approach for pre-selection of suitable patients for cytokine therapy

AHMAD M. E. NOURI
*The Royal London Hospital*

## 15.1 Introduction

### *15.1.1 MHC antigens and the immune system*

Since the report by Mitchison in the early 1950s demonstrating that cell-mediated, rather than humoral, immunity played a greater role in tumour rejection (Mitchison, 1953), its primacy in tumour rejection has become an increasingly accepted mechanism by most immunologists. This has mainly been attributed to specific immunity involving MHC antigens as restriction element and to a lesser extent the non-specific immunity involving LAK/NK activity (Jabrane-Ferrat *et al.*, 1990; Mule *et al.*, 1984).

It has long been established that the MHC antigens are an individual's fingerprint and they exist in two forms, the class I and class II antigens. Zinkernagel & Doherty (1979) showed that they act as associative molecules for presentation of non-self antigens to CTLs and $T_H$ cells, respectively. The critical role of CTLs for regulating resistance to viral infection (McMichael *et al.*, 1977) as well as in graft and tumour rejection in experimental models (Hui, Grosveld & Festenstein, 1984; Wallich *et al.*, 1985) has previously been reported. $T_H$ cells have been shown to act mainly as an immune amplifier, since their stimulation results in the production of a series of immunoreactive cytokines, such as IL-2, which in turn are critical for initiating immune responses, including activation of CTLs (Greenberg *et al.*, 1988). There is increasing realization of the critical role of the so-called immune surveillance system not only for dealing with processes resulting in generation of non-self proteins derived from viruses or by mutation but also by altered processing, leading to generation of different self antigens. Support for such an immune surveillance theory has come from both clinical and laboratory observations.

The most significant clinical observations include the spontaneous tumour regression that has been seen occasionally in different tumours and is thought to be immunologically mediated (Oliver, Nethersell & Bottomley, 1989); the effectiveness of IL-2 in a small minority of melanoma patients, resulting in complete and durable remission (Rosenberg *et al.*, 1987); the therapeutic efficacy of BCG for treating superficial bladder cancer patients and its efficacy, correlating with the degree of class II antigen expression on urothelium and the release of cytokines such as IFN-γ in the patient's urine (Prescot *et al.*, 1990); the effectiveness of cytokines such as IFN-α, which is known to be an efficient MHC antigen inducer (Nouri *et al.*, 1992), in treating some tumours (Oliver, 1991); and the decreased frequency of loss of MHC antigens in immunosuppressed individuals compared with an unsuppressed control group (List *et al.*, 1991).

In the laboratory context, it has been shown that tumour-infiltrating lymphocytes (TILs) isolated from melanoma tumours show MHC-restricted, CTL-mediated killing activity against autologous tumour cells (Itoh, Platsoucas & Blach, 1988), a phenomenon that is seen more frequently in patients responding to cytokine therapy than in a non-responder group (Aebersold *et al.*, 1991). Melanoma-derived TILs show restricted T cell receptor rearrangement, demonstrating their restricted oligoclonality (Morita *et al.*, 1992). Reintroduction of a non-expressed class I gene into class I-negative tumours resulted in the change of their phenotype from tumorigenic to rejectable tumours (Hui *et al.*, 1984; Wallich *et al.*, 1985). The final piece of evidence in support of the immune surveillance theory has been the report that the introduction of immunoreactive cytokine gene, like those giving rise to IL-2, IL-4 and IFN-γ, into tumours by gene transfection is effective in controlling tumour growth (Fearon *et al.*, 1990; Gansbacher *et al.*, 1990a,b).

### *15.1.2 MHC class I antigens and tumours*

An increasing number of reports have demonstrated abnormal expression of MHC antigens in a variety of human tumours. Class I antigen abnormalities have been shown in a substantial proportion of colon adenocarcinomas (Rees *et al.*, 1988; Smith, Bodmer & Bodmer, 1988; Momburg *et al.*, 1989), breast carcinomas (Fleming *et al.*, 1981), Burkitt's lymphomas (Masucci *et al.*, 1987) and carcinomas of the larynx (Esteban *et al.*, 1990), cervix (Connor & Stern, 1990) and ovary (Ferguson, Moore & Fox, 1985). Using the peroxidase anti-peroxidase (PAP) staining technique, the pattern of MHC antigen expression on human tumour tissue sections has been analysed in this laboratory. This technique was chosen not only because of its sensitivity but also because each section provides test (tumour) and control (tumour stroma) tissue on the same sample. Using the mAb W6/32 (which detects a monomorphic determinant on all $\beta_2$-m-associated class I antigens) as the first antibody, 23 of 52 (45%) transitional carcinomas (TCC) of the bladder, 2 of 2 (100%) Wilms' and 6 of 13 (46%) renal carcinomas showed some degree of abnormality, ranging from low expression on a small percentage of tumour cells to complete absence of all class I antigens from all tumour cells (Nouri *et al.*, 1990).

Since the presentation of non-self antigens to T cells occurs via individual polymorphic MHC antigens (Hammerling *et al.*, 1987; Crowley *et al.*, 1991; Pandolfi *et al.*, 1991), tissue sections from a subgroup of bladder cancer patients were stained with a panel of mAbs directed against polymorphic class determinants on HLA-A2, A3, Bw4 and Bw6 class I antigens and the results were compared with those obtained with mAb W6/32 (Fig 15.1). The percentage of abnormal cases increased from 45% (using W6/32) to 84% (using polymorphic mAbs), indicating that analysis of MHC antigens with the W6/32 mAb provides an underestimate of the frequency of MHC abnormality in these tissues. This can clearly be seen from Table 15.1, which summarizes the staining patterns in 18 bladder tumours.

This is an important observation and may explain the failure, in several studies, to correlate the extent of MHC abnormality as observed using monomorphic mAbs and the clinical behaviour of the tumour. Nonetheless, results obtained by Ottesen *et al.* (1987) and Tomita *et al.* (1990a), as well as those obtained in this laboratory (Nouri *et al.*, 1990) in bladder tumours and those of van Duinen *et al.* (1988) in melanoma, have indicated that there might be some degree of correlation between these two parameters. A similar conclusion has also been reached by other groups studying breast (Concha *et al.*, 1991), larynx (Esteban *et al.*, 1989) and aneuploid ovarian cancers (Moore, Fowler & Olafsson, 1990). In addition, an analysis of melanoma tumours demonstrated that visceral metastases often express lower levels of class I antigens than primary tumours (Brocker *et al.*, 1985), suggesting that reduction in class I expression might be associated with increased malignancy. However, these views are not universally accepted and reports from Wintzer *et al.* (1990), analysing breast cancer, and Durrant *et al.* (1987), studying colorectal carcinoma, have shown no such correlation. Further clarification of these conflicting reports will be resolved when a more comprehensive collection of mAbs against polymorphic class I antigens becomes available.

### *15.1.3 MHC class II antigens and tumours*

In general, the expression of class II antigens on cells other than B cells and antigen-presenting cells like macrophages is thought to be a sign of on-going

(a)

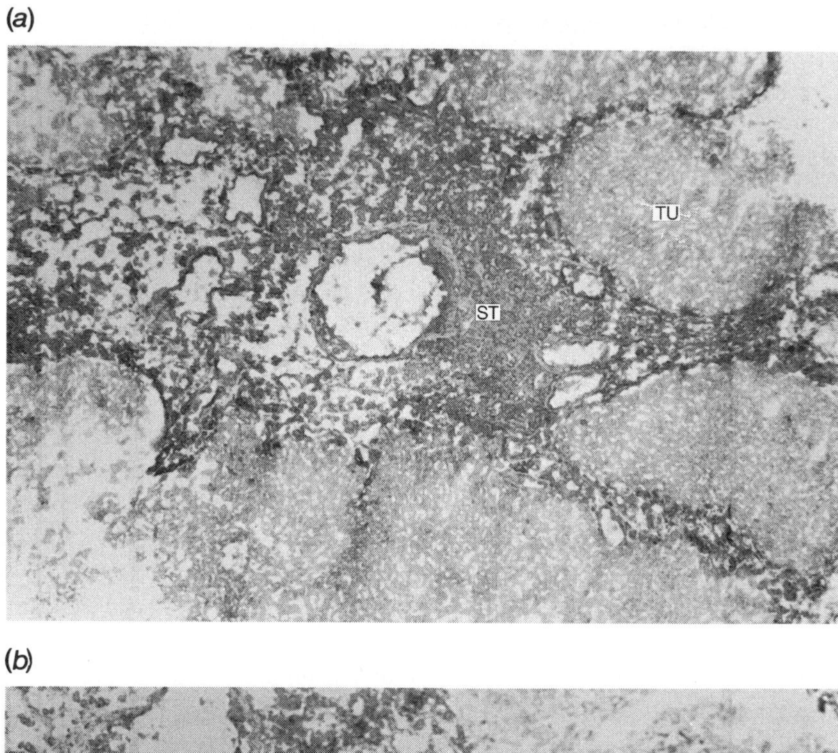


(b)

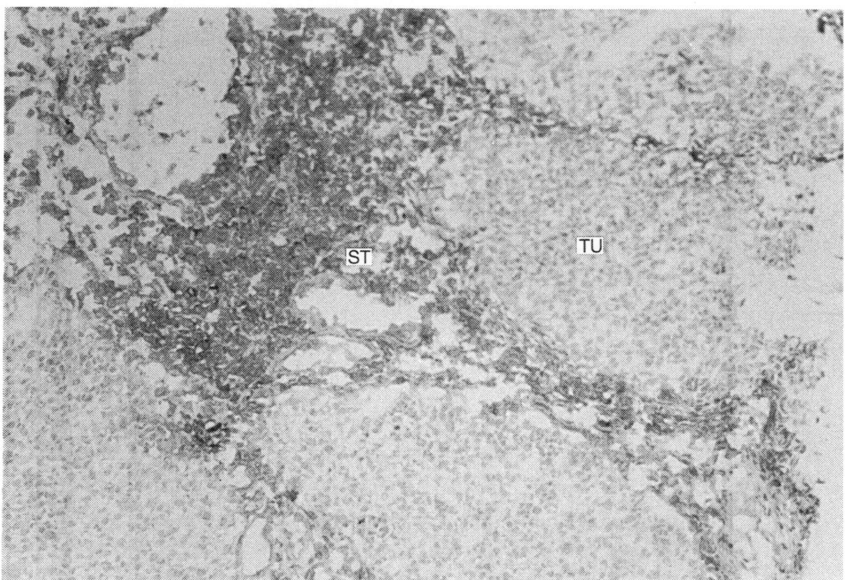


Fig. 15.1. Staining pattern on a tissue section of an individual with transitional carcinoma of the bladder using (*a*) mAbs against all β2-m-associated HLA-A, HLA-B and HLA-C (W6/32) and (*b*) Bw6 126.39 antigens. ST and TU denote tumour stroma and tumour, respectively.

Table 15.1. *Expression of HLA-A, HLA-B and HLC-C on bladder tumours*

| Patient | Tumour type | Mature A, B, C | A, B, C heavy chain | Light chain ($\beta_2$-m) (BBM.1) | A2 + B17 (MA2.1) | A2 + Aw69 (BB7.2) | A3 (GAPA3) | Bw4 (116.5.28) | Bw6 (126.39) |
|---|---|---|---|---|---|---|---|---|---|
| 1 | N | 4s | 4s | 4s | – | – | – | – | 4s |
| 2 | N | 4s | 4s | 4s | – | – | – | – | 4s |
| 3 | N | 4s | 4s | 4s | 4s | 4s | 4m | – | 4s |
| 4 | I | 4s | 4m | 4s | – | – | – | – | 4m |
| 5 | I | 4s | 4m | 4s | 4m | 4w | – | – | 3m |
| 6 | N | 4s | 4m | 4s | 4s | 4w | 4m | 4w | 4m |
| 7 | I | 4s | 0 | 4s | 4m | 3w | – | – | 4w |
| 8 | I | 4s | 4w | 4s | 4s | 4s | 4w | n.d. | 4s |
| 9 | I | 4s | 4w | 4s | – | – | 4m | – | 4s |
| 10 | N | 4m | 0 | 4s | 4m | 4w | – | – | 4w |
| 11 | N | 4w | 0 | 4w | 0 | 0 | – | 0 | 44w |
| 12 | N | 4w | 4w | 4w | – | – | 4w | 0 | – |
| 13 | I | 3m | 0 | 4m | – | – | – | 4w | 0 |
| 14 | N | 2s | 1s | 2s | – | – | – | 2w | 1s |
| 15 | I | 2s | 1m | 3m | 3s | n.d. | – | – | – |
| 16 | N | 1w | 0 | 1w | – | – | – | 0 | 0 |
| 17 | I | 1w | 0 | 1w | – | n.d. | – | 0 | – |
| 18 | I | 0 | 0 | 0 | 0 | 0 | 0 | 0 | 0 |

Assessment of stromal or tumour cell staining. –, No staining; 0, stroma positive, but tumour negative; 1–4, stroma and tumour positive with increasing amounts of tumour cells staining positively: 1 ($< 10\%$), 2 ($\geqslant 10\%$ to $< 50\%$), 3 ($\geqslant 50\%$ to $< 95\%$), 4 ($\geqslant 95\%$).

Strength of antigen expression: s, strong; m, moderate; w, weak.

I, invasive tumour; N, non-invasive tumour; n.d., not done.

immunological reactions. Cohen *et al.* (1987) reported that the presence of these surface antigens in melanoma patients is a good prognostic factor. Similarly, Prescot *et al.* (1989) argued that the expression of such antigens on bladder urothelium following BCG therapy correlated with a clinical response. In contrast, Norazmi *et al.* (1989), analysing colon carcinoma, argued that the presence of these antigens correlated with poor prognosis. In our studies using the PAP staining technique, 19 of 25 bladder tumours investigated were positive for class II antigens using the HB55 mAb and this ranged from positive staining on a small percentage of tumour cells to strong expression on all the tumour cells (3 of 25 cases).

The precise explanation of why cells that are normally negative for class II antigens become positive is unexplained. One possible hypothesis is that the detection of putative tumour-specific antigen(s) by infiltrating T cells results in T cell activation, releasing cytokines that in turn result in the induction of these antigens. An additional level of complexity regarding the expression of class II antigens has been reported by Guerry *et al.* (1984). They demonstrated that melanoma tumour cell lines overexpress non-presenting class II antigens and this defect could be corrected by transfection of the normal class II gene (Alexander, Lee & Guerry, 1991).

Similar to the findings of Smith *et al.* (1988) in colorectal carcinoma, one out of nine bladder tumour biopsies in our studies showed a strong presence of HLA-DR antigens on normal urothelium (Fig. 15.2) adjacent to HLA-DR-negative tumour cells, indicating the potential of some tumours for defective class II antigen expression. This defect might represent an additional step for tumour cells to escape immunological detection, particularly considering the increasing awareness of cytolytic activity of $T_H$ cells against tumour target cells (Hou *et al.*, 1992).

## 15.2 MHC antigen expression in tumours and infiltration of T cells into tumours

A report from Itoh *et al.* (1988) indicated that TILs isolated from melanoma patients could be propagated *in vitro* in the presence of IL-2 without losing specific MHC-restricted cytotoxicity against autologous tumour cells. This was followed by a study by Rosenberg *et al.* (1990) showing that melanoma-derived TILs, which had been transfected with a neomycin-resistance gene, could be detected in the patient's blood circulation 200 days after injection and at the site where the tumour underwent rejection for up to 70 days. Taking these finding together with the reports of Dayan & Marshall (1964), Pomerance (1972) and Tsujihashi *et al.* (1989), which demonstrated that the

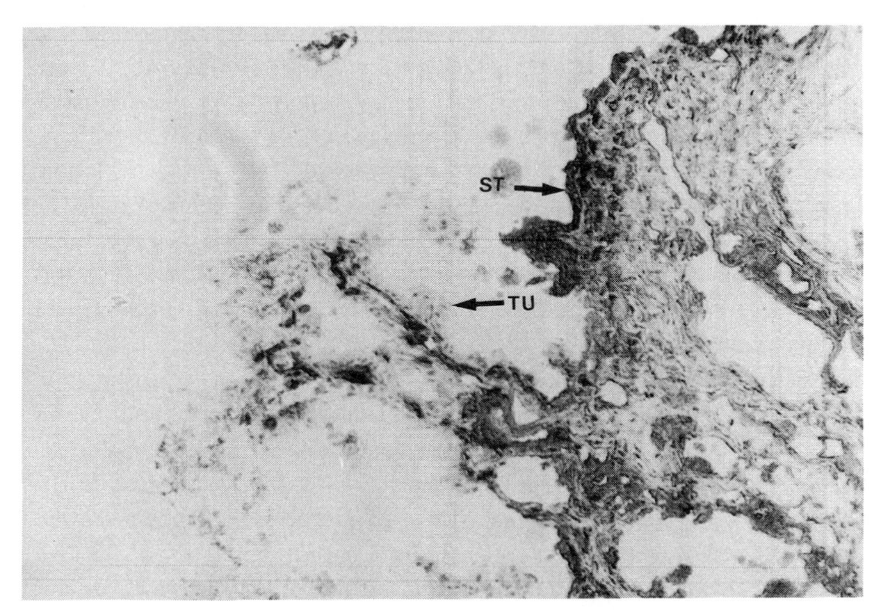


Fig. 15.2. Class II antigen expression (detected by HB55 mAb) on a tissue section of an individual with transitional cell carcinoma. TU and ST denote tumour and normal bladder urothelium, respectively.

degree of T cell infiltration is a good prognostic factor, we focussed our attention on whether, firstly, TIL could be isolated and expanded from different tumours and, secondly, if there was any correlation between the expression of MHC antigens and the success of TIL development. TILs have so far been successfully established from 7 of 18 (39%) bladder, 10 of 15 (66%) kidney and 10 of 13 (77%) testis tumours (Nouri *et al.*, 1993a), indicating that TILs could be isolated and expanded from a significant proportion of different tumours. The low rate of success in bladder compared with testis and kidney might be because of the diathermic damage to bladder tumours that occurs during tissue collection.

Tomita *et al.* (1990b) reported that, in renal carcinoma, TILs infiltrated more frequently into tumours expressing normal levels of MHC class I antigens. In our studies (Nouri *et al.*, 1991a) using W6/32 and BBM.1 (detects $\beta_2$-m) mAbs on a subgroup of bladder tumours, positive staining was seen in 6 of 6 (100%) tumours that developed TILs but only in 6 of 13 (46%) patients whose tumours failed to generate TILs ($p < 0.02$, Fisher's exact test). These findings indicate that T cells infiltrate and possibly proliferate at tumour sites where there is normal expression of class I antigens, which in turn might act as accessory molecules for presentation of putative tumour specific-antigen(s). No TILs were developed from 54% of tumours that had apparently normal class I expression detected using the W6/32 mAb. A more comprehensive investigation using mAbs against polymorphic determinants on class I antigens might explain this discrepancy. This is consistent with the relatively rare occasion where isolated TILs showed MHC-restricted killing against autologous tumour cells, taking into account the frequency of MHC abnormality in tumours like bladder carcinoma.

A conventional [$^3$H]thymidine-incorporation assay (Nouri *et al.*, 1991a) was used to investigate the proliferative response of TILs to IL-2 alone and IL-2 plus conditioned medium. Conditioned medium was prepared by activating fresh mononuclear cells for 2 hours with the mitogen phytohaemagglutin (PHA) and, following three washes, the cells were incubated for a further 48 hours. The cell-free supernatant was then removed and used as conditioned medium (CM). TILs from two individuals (FB and FS), expanded in the presence of IL-2 and CM (5%, v/v) over a 10 day period in culture, showed a 64-fold and 24-fold increase in cell numbers, respectively. This is indicative of the potential of IL-2 to stimulate TIL expansion to sufficient numbers for both *in vitro* specific cytotoxicity studies (Nouri *et al.*, 1991b) and possible clinical trials (Topalian *et al.*, 1987).

## 15.3 Cytolytic activity and phenotypic profile of TILs

The cytolytic activity of TILs against various tumour targets was investigated using a conventional $^{51}$Cr release assay or a colorimetric technique (Nouri *et al.*, 1991b; Hussain, Nouri & Oliver, 1993). The killing activity of TILs was at its maximum during the first three weeks of culture in all cases tested. At an effector/target (E/T) ratio of 25/1 the percentage of killing by TILs from one individual (Wil) expanded in IL-2 for more than two weeks and tested against allogeneic tumour cell lines Daudi, Molt 4, U937 and K562 was 60%, 39%, 16% and 14%, respectively (Fig. 15.3). The killing potential of these TILs was also tested against autologous (Wil) and allogeneic (SKV14, epithelial tumour line) tumour targets and resulted in 15% and 7% killing, respectively, demonstrating the lack of effective MHC-restricted cytotoxicity against autologous tumour cell targets.

It is now well-established that one consequence of T cell activation is the expression of class II antigens on their cell surface (Nouri *et al.*, 1993a). TILs isolated from six bladder tumour biopsies were cultured for more than 30 days with IL-2 and analysed for cell-surface antigen expression by flow cytometry (Table 15.2). In all cases, more than 69% and 70% of the cells were positive for CD3 and class II antigens (detected with HB55 mAb),

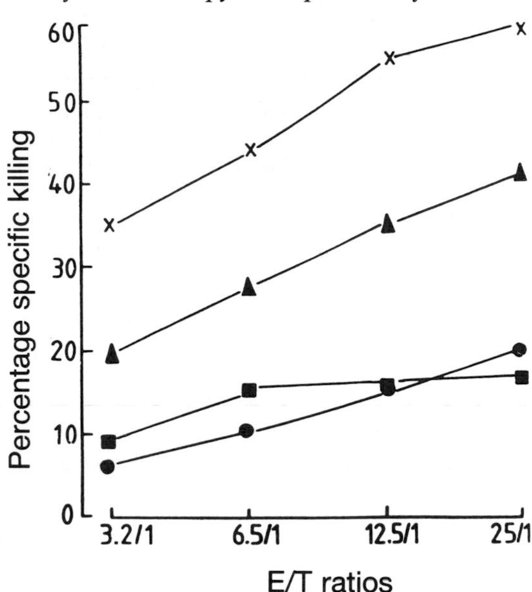


Fig. 15.3. Cytolytic activity of TILs from one individual against different allogeneic tumour cell lines: Daudi (X); Molt 4 (▼); U937 (■); K562 (●). E/T ratio: effector/target ratio.

respectively, indicating the activated nature of the T cells. While the percentage of CD8-positive cells remained above 29%, in two of six patients, CD4-positive cells constituted a small minority of the total cell population. This indicates that the long-term expansion of T cells with IL-2, in the absence of stimulating antigen, results in the preferential expansion of the CTL population, a phenomenon that has previously been reported by Topalian *et al.* (1987).

## 15.4 Expression of MHC antigens in tumour cell lines

The pattern of MHC antigen expression on cell lines established from different tumours was investigated using peroxidase staining and this was compared with results obtained with tumour biopsies. In addition, other techniques including radioactive surface binding (RD) (Nouri *et al.*, 1992), isoelectric focussing (IEF) (Nouri *et al.*, 1992), dot blot (DB) (Gillott *et al.*, 1993) and immunoprecipitation (IP) were employed to locate and quantify these antigens on tumour cells.

There have been many reports describing the pattern of MHC class I antigens on human tumour cell lines and their behaviour following cytokine

Table 15.2. *Cell surface phenotype of TILs from different individuals*

| | Cells positive for (%)[a] | | | | |
|---|---|---|---|---|---|
| Patient | All class I[b] | All class II[c] | CD4 | CD8 | CD3 |
| FS | 98 | 95 | 45 | 50 | 87 |
| Wil | 98 | 98 | 2 | 33 | 80 |
| JF | 95 | 83 | 23 | 40 | 89 |
| FB | 100 | 70 | 4 | 30 | 87 |
| AW | 89 | 86 | 27 | 36 | 83 |
| LR | 100 | 93 | 24 | 29 | 69 |
| Mean | 96.6 | 87.5 | 20.8 | 36.3 | 82.5 |

[a] Percentage of TILs positive for various phenotypes in tumour biopsies from six individuals using mAbs on cells after 30 days in culture.
[b] All class I detected with W6/32 mAb.
[c] All class II detected with HB55 mAb.

stimulation (Basham *et al.*, 1985; Burrone *et al.*, 1985). In our study, we used four carcinoma cell lines established in our laboratory, two bladder carcinoma lines, Wil and Fen, and two testicular carcinoma lines, Ha and Lan (Nouri *et al.*, 1992), plus 17 cell lines collected from external sources (mainly from ATCC).

### *15.4.1 MHC class I antigens*

The degree of cell surface MHC class I antigen expression on different lines was established using the RD technique and the results are presented in Table 15.3. A value below 150 c.p.m. corresponded to background and indicated a lack of class I expression. Cell lines Fen, Ha and Tera I showed no binding and the level of binding for the remaining lines ranged from $240 \pm 49$ to $2064 \pm 407$ c.p.m., demonstrating, firstly, that there was a significant minority of cell lines which were completely negative for these antigens and, secondly, that there was a wide variation of antigen expression among the lines.

Following stimulation for 48 hours with the optimum concentration of IFN-γ (100 U/ml, using conditions established in earlier experiments) the pattern of class I expression fell into three categories: group one remained negative (Ha, Fen and Tera I); group two expressed high levels of antigens spontaneously and showed no further up-regulation with IFN-γ (for example J82, TccSup and T24); and group three showed significant up-regulation with IFN-γ (Tera II, Ep2102 and T47D). Similar results were observed when the

Table 15.3. *Class I and II antigen surface expression on cells from various cell lines before and after IFN-γ treatment*

| | Class I | | Class II | |
|---|---|---|---|---|
| | Untreated | IFN-γ | Untreated | IFN-γ |
| RT4 | 566 ± 161 | 772 ± 123*** | 72 ± 22 | 212 ± 92** |
| Scaber | 920 ± 109 | 999 ± 58*** | 92 ± 12 | 619 ± 24* |
| RT112 | 1020 ± 57 | 902 ± 64*** | 73 ± 14 | 261 ± 28** |
| T24 | 1786 ± 91 | 2162 ± 234*** | 60 ± 12 | 57 ± 18*** |
| J82 | 2064 ± 407 | 2284 ± 236*** | 53 ± 19 | 1021 ± 122* |
| TccSup | 1680 ± 67 | 1619 ± 198*** | 39 ± 7 | 110 ± 53* |
| Wil | 1208 ± 67 | 1194 ± 113*** | 44 ± 6 | 1227 ± 118* |
| Fen | 64 ± 5 | 76 ± 16*** | 58 ± 23 | 1337 ± 34* |
| SKV14 | 1627 ± 288 | 2146 ± 118*** | 43 ± 18 | 1484 ± 183* |
| 5637 | 1138 ± 110 | 1150 ± 204*** | 65 ± 3 | 1292 ± 143* |
| MCF7 | 1186 ± 77 | 1080 ± 59*** | 62 ± 14 | 960 ± 114* |
| T47D | 1105 ± 77 | 1438 ± 182** | 93 ± 17 | 1575 ± 174* |
| Tera I | 63 ± 12 | 65 ± 30*** | 40 ± 9 | 677 ± 90* |
| Tera II | 240 ± 49 | 1560 ± 139* | 59 ± 34 | 93 ± 26*** |
| Ep2102 | 660 ± 108 | 1222 ± 149* | 92 ± 72 | 130 ± 32*** |
| Lan | 799 ± 76 | 680 ± 32*** | 48 ± 8 | 89 ± 22*** |
| Ha | 84 ± 14 | 60 ± 37*** | 54 ± 12 | 1005 ± 20* |

Expression was determined by the RD technique and results are expressed as mean ± SD (c.p.m.) of three replicates.
Probabilities: *, 0.001; **, 0.01; ***, non-significant.

cells were stimulated with IFN-α (Table 15.4). These results indicate that if the therapeutic efficacy of cytokines like IFN-α for treating cancer patients is mediated via up-regulation of MHC antigens, tumours falling into the group three category ought to benefit most from such therapy, as the up-regulation of these antigens makes the tumours more 'visible' to CTLs.

In order to assess the nature of the MHC defect, three cell lines were subjected to more detailed analyses. IEF was carried out on paired TIL and tumour cell lines of one individual (Wil). Following pulse-labelling of the two cell types with [$^{35}$S]methionine, they were lysed and, using W6/32 mAb as precipitating antibody, all class I antigens were isolated and separated by IEF (Fig. 15.4). The pattern of protein species in TILs clearly showed that the individual was HLA-A2, HLA-A3, HLA-B44 and HLA-B7 (HLA-B7 typing was confirmed using standard serological techniques because of the indistinguishable nature of HLA-A3 and HLA-37 bands with the IEF technique). No HLA-B44 band could be seen despite the presence of both HLA-A2 and HLA-A3 bands. Using the peroxidase technique with a mAb

Table 15.4. *Effects of combination of IFNs on induction of class I and II antigens on tumour cell lines*

| | Untreated | IFN-γ | IFN-α | IFN-γ + IFN-α |
|---|---|---|---|---|
| *Class I* | | | | |
| Tera II | 177 ± 6 | 1731 ± 27 | 1326 ± 54 | 1572 ± 144 |
| Ha | 74 ± 12 | 74 ± 10 | 76 ± 10 | 76 ± 15 |
| T24 | 2092 ± 299 | 2635 ± 357 | 2173 ± 132 | 2500 ± 228 |
| TccSup | 1590 ± 105 | 1689 ± 271 | 1641 ± 321 | 1770 ± 216 |
| *Class II* | | | | |
| Tera II | 60 ± 10 | 95 ± 29 | 56 ± 8 | 71 ± 11 |
| Ha | 76 ± 15 | 1166 ± 185 | 83 ± 15 | 949 ± 67 |
| T24 | 60 ± 12 | 71 ± 11 | 51 ± 9 | 59 ± 17 |
| TccSup | 39 ± 7 | 60 ± 7 | 41 ± 13 | 71 ± 8 |

Results are expressed as mean ± SD in c.p.m. Interferons were added at the beginning of the 48 h culture: IFN-γ, 100 U/ml; IFN-α, 1000 U/ml.

against HLA-B7, no staining was observed on Wil tumour cells, confirming the complete absence of B locus antigens in this tumour cell line. If the putative tumour-specific antigen(s) was presented to CTLs in the context of HLA-B locus antigens, the absence of these antigens would prevent effective tumour cell recognition and killing. To this end attempts are now in progress to correct the gene expression defect by gene transfection and to establish how such correction influences the specific killing by autologous TILs.

Exposure of this cell line to IFN-γ did not result in a significant increase in the total cell surface expression of class I antigens assayed by the RD technique using W6/32 mAb, although both HLA-A2 and HLA-A3 antigens were significantly up-regulated (Table 15.5). Under the conditions in which IFN-γ induced class II antigens, no induction of the absent B7 antigen occurred. This might be a novel observation, indicating that whilst IFN-γ had no significant effect on class I antigens using antibody against monomorphic class I antigens (detected with W6/32) it resulted in a significant up-regulation of polymorphic HLA-A2 and HLA-A3 antigens, as described also by Hakem *et al.* (1989).

A second cell line studied in detail using the IP technique was the Fen cell line (derived from a bladder tumour). The cells were cultured in the presence or absence of IFN-γ for 48 hours, and, following pulse-labelling, cell lysates were precipitated with mAbs HC10 (which detects free class I heavy chains) and BBM. 1 (which detects $\beta_2$-m). The resulting immuno-precipitates were separated by gel electrophoresis (Fig. 15.5). A complete

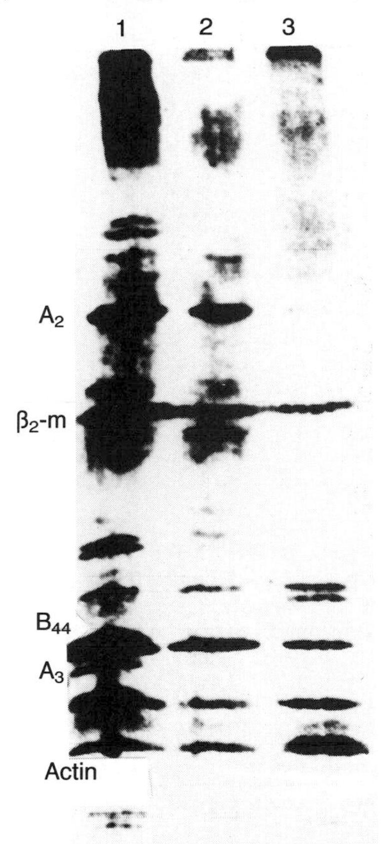


Fig. 15.4. Isoelectric focussing of tumour cells and TILs from one individual, Wil. Lane 1, Wil tumour; lane 2, Wil TILs; lane 3, HLA-A3-positive control (Hom-2 cell line).

Table 15.5. *Expression of class II and monomorphic and polymorphic class I antigens by Wil line with and without IFN-γ stimulation*

| | Untreated | Treated | *p* values |
|---|---|---|---|
| Class I | 1208 ± 67 | 1194 ± 113 | NS |
| Class II | 59 ± 3 | 1227 ± 118 | 0.0001 |
| HLA-A2 | 421 ± 87 | 773 ± 49 | 0.001 |
| HLA-A3 | 851 ± 59 | 1138 ± 128 | 0.01 |
| HLA-B7 | 61 ± 4 | 73 ± 7 | NS |

Results are expressed as mean ± SD of three replicates (c.p.m.). NS denotes not significant.

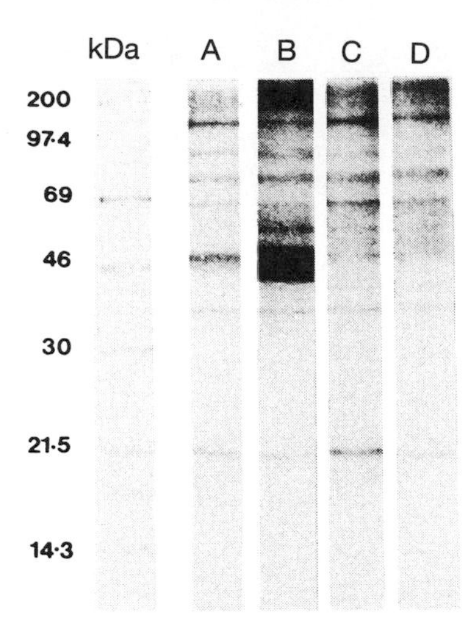


Fig. 15.5. Immunoprecipitation profile of Fen cells cultured in the presence and absence of IFN-$\gamma$. Fen cells were lysed before (lanes A and C) and after (lanes B and D) IFN-$\gamma$ stimulation for 48 hours. Cell lysates were precipitated with mAb BBM.1 (lanes A and B), detecting $\beta_2$-m, and mAb HClO (lanes C and D), detecting class I free heavy chain.

absence of the $\beta_2$-m species was observed whether the cells were stimulated or not. A similar observation has been reported by Momburg & Koch (1988). The low levels of free heavy chain present in unstimulated cells were greatly up-regulated by IFN-$\gamma$ stimulation, indicating that the absence of class I antigens in the Fen cell line was caused by the absence of $\beta_2$-m. This was further investigated by transfection of the cells with the $\beta_2$-m gene (a generous gift from Dr E. J. Baas). The levels of surface binding with W6/32 mAb using the RB technique before and after transfection were $132 \pm 20$ and $2\,000 \pm 48$ c.p.m., respectively, and these values remained unchanged following IFN stimulation (Table 15.6).

In a subgroup of cell lines, attempts were also made to investigate whether a combination of cytokines could provide the necessary signal for correction of missing MHC antigens. As can be seen from Table 15.4 in none of the three positive lines (Tera II, T24 and TccSup) did addition of IFN-$\alpha$ to IFN-$\gamma$-stimulated cells result in a significant increase in the levels of class I antigens; even class I-negative cell lines such as Ha remained negative. Consistent with the findings of Avila-Carino *et al.* (1988), these results indicate that for each cell line there is a finite maximum of cell surface class I antigens. A combination of IFNs could not result in additive class I expression

Table 15.6. *Inducibility of MHC antigens by interferons on Fen cells before and after transfection of the $\beta_2$-m gene*

| | Class I | Class II |
|---|---|---|
| *Before transfection* | | |
| Untreated | 132 ± 20 | 186 ± 11 |
| IFN-γ | 144 ± 14 | 2391 ± 134 |
| IFN-α | 136 ± 15 | 240 ± 26 |
| *After transfection* | | |
| Untreated | 2000 ± 48 | 108 ± 11 |
| IFN-γ | 2161 ± 156 | 2184 ± 113 |
| IFN-α | 2037 ± 136 | 106 ± 15 |

Results of cells treated with IFN-γ (100 U/ml) and IFN-α (1000 U/ml) and cultured for 48 hours are expressed in c.p.m. as mean ± SD of three replicates.

and such combinations could not induce the expression of absent class I antigens on class I-negative cell lines.

Further confirmation of these results was established using the DB technique. This technique has been modified in our laboratory to provide an accurate and rapid assessment of inducible antigens without the use of radioisotopes (Gillott *et al.*, 1993). Lysates from Fen and Fen transfected with the $\beta_2$-m gene were cultured in the presence and absence of IFN-γ. As can be seen from Fig. 15.6, the Fen plus $\beta_2$-m cell line showed positive results with

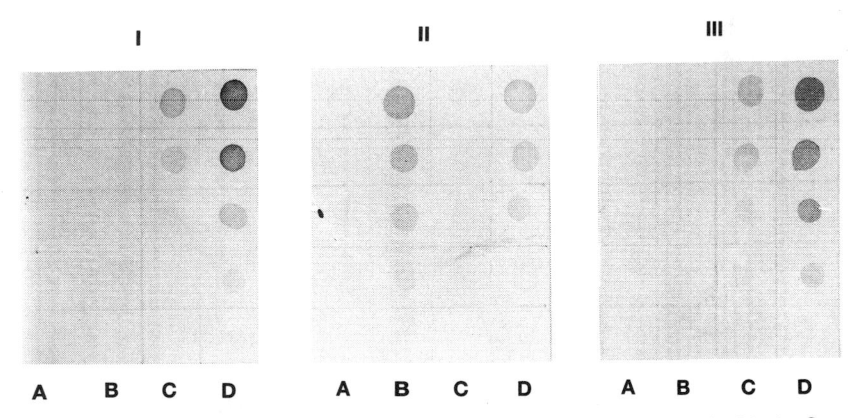


Fig. 15.6. Dot blot analysis of Fen (lanes A and B) and Fen transfected with the $\beta_2$-m gene (lanes C and D). Blots I were stained with mAb W6/32 (class I antigens), II with mAb HC10 (free class I heavy chains) and III with mAb BBM.1 (free $\beta_2$-M). Lanes A and C are unstimulated cells, lanes B and D cells stimulated with IFN-γ.

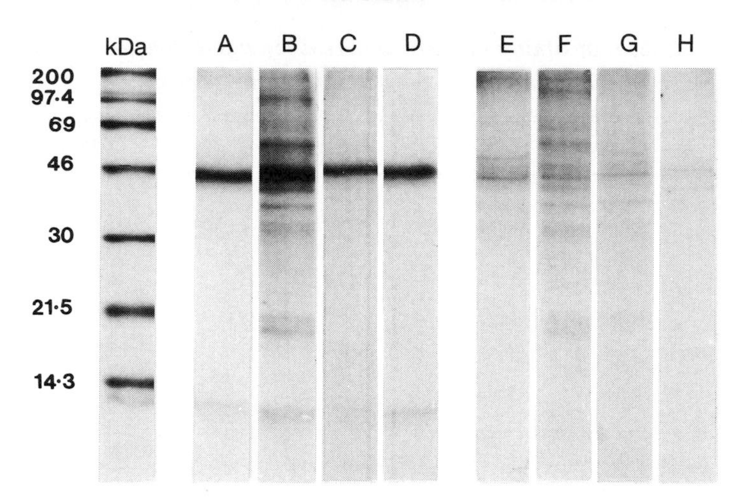


Fig. 15.7. Pattern of expression of class I and class II antigens in four tumour cell lines. IP profile of class I antigens, precipitated with mAb W6/32, (lanes A–D), and class II antigens, precipitated with mAb HB55 (lanes E–H). Lanes A and E, EP2102; lanes B and F, T24; lanes C and G, Tera II; and lanes D and H, Tcc Sup.

W6/32 (I), HC10 (II) and BBM.1 (III). Untransfected Fen cells, whilst completely negative for $\beta_2$-m and class I antigens, were positive for free class I heavy chains (detected by HC10), particularly after IFN-$\gamma$ stimulation. The absence of $\beta_2$-m on transfected cells indicated that all the $\beta_2$-m polypeptides became engaged in forming class I antigens since the BBM.1 mAb can only recognize free $\beta_2$-m molecules.

### *15.4.2 MHC class II antigens*

Using the RD technique, none of the lines investigated expressed class II antigens spontaneously (Tables 15.3 and 15.4) while IFN-$\gamma$, but not IFN-$\alpha$, stimulation resulted in their induction. The cells fell into two groups: group one, in which there was no induction (T24, TccSup. Tera II, Ep2102 and Lan) and group two, where the cells were induced with varying intensities of antigen expression ranging from $212 \pm 92$ to $1575 \pm 174$ c.p.m. (Tables 15.3 and 15.4), results that are consistent with findings reported by Watanabe & Jacob (1991). Confirmation of these results was obtained by the IP technique. As shown in Fig. 15.7, while all four cell lines Ep2102, T24, Tera II and TccSup showed a normal pattern of expression of class I antigens (free heavy chain and $\beta_2$-m polypeptides) when precipitated with W6/32, none of the lines showed class II antigen polypeptides when precipitated with HB55 mAb (which detects HLA-DR antigens).

Table 15.7. *Effects of combination of IFNs on induction of class I and II antigens by tumour cell lines*

| | Ep2102 | Fen | RT112 |
|---|---|---|---|
| *Class I* | | | |
| Untreated | 347 ± 52 | 69 ± 26 | 1056 ± 28 |
| IFN-γ | 566 ± 33 | 62 ± 16 | 1421 ± 138 |
| IFN-γ + IFN-β (1000 U/ml) | 713 ± 112 | 60 ± 14 | 1413 ± 14 |
| IFN-γ + IFN-β (2000 U/ml) | 772 ± 74 | 56 ± 9 | 1294 ± 40 |
| *Class II* | | | |
| Untreated | 52 ± 8 | 62 ± 10 | 66 ± 7 |
| IFN-γ | 40 ± 17 | 1411 ± 110 | 208 ± 45 |
| IFN-γ + IFN-β (1000 U/ml) | 56 ± 37 | 670 ± 128 | 164 ± 20 |
| IFN-γ + IFN-β (2000 U/ml) | 37 ± 6 | 542 ± 60 | 115 ± 13 |

Results of cells cultured for 48 hours are expressed as means ± SD in c.p.m. IFN-γ (100 U/ml) and IFN-β (1000 or 2000 U/ml) were added at the start of the culture period.

IFN-β alone did not have any effect on either class I or class II antigens, but when added to IFN-γ-stimulated cells resulted in a significant inhibition of class II but not class I antigen expression (Table 15.7), an observation that is consistent with the findings of Ling, Warren & Yogel (1985) using murine macrophages and those of Shaw *et al.* (1985). These results raise two very important issues. Firstly, the notion that, similar to results obtained in bladder tumour sections, there is a significant minority of tumour cell lines that fails to express class II antigens in response to cytokine stimulation. Secondly, the fact that combinations of some cytokines may lead to down-regulation of MHC antigens and this might be an important issue to be considered when cancer patients are treated with combinations of cytokines.

## 15.5 Non-MHC-restricted tumour cell killing

Non-specific killing or NK activity was first described by Kiessling *et al.* (1975). Carlson & Wegman (1977) defined this as a residual killing activity in an MHC-restricted killing (CTL) system that could not be inhibited by antibodies specific for MHC antigens. The identification of IL-2 led to the definition of a second non-MHC-restricted killing phenomenon, the so-called LAK activity (Grimm *et al.*, 1982). However, Lange *et al.* (1991) argued that these activities may be the same, the difference having more to do with the activation or differentiation stage of the cells. Here, non-MHC-restricted killing activity is referred to as LAK/NK.

The idea that MHC antigens had no effect on the efficiency of LAK/NK killing has been challenged by a number of investigators and led to the suggestion that there is an inverse relationship between these two parameters, with a complete absence of MHC antigens on tumour cell targets acting as a stimulatory signal for the LAK/NK activity (Stern *et al.*, 1982; Karre *et al.*, 1986; Gorelik, Gunji & Herberman, 1988; Lobo & Spencer, 1989; Shimizu & DeMars, 1989; Maziarz *et al.*, 1990). Storkus *et al.* (1989) have pursued this and argued that this is restricted to certain class I polymorphic antigens (HLA-A3, HLA-B7 and HLA-B27). However, this is not a universally accepted conclusion and other investigators have argued that there is no correlation between these two phenomena (Leiden *et al.*, 1989; Pena *et al.*, 1989; Stam *et al.*, 1989; Aosi *et al.*, 1991).

We performed a number of experiments to resolve this controversy and our overall conclusion was that there are cell surface molecules other than MHC antigens on the tumour cell target dictating the efficiency of LAK/NK killing. This was based on three sets of results. Firstly, there was no inverse correlation between LAK/NK killing and the levels of class I antigen expression when cell lines with varying degrees of class I antigens were used as targets. In the case of T47D and RT112 lines, although both lines display equal intensities of class I antigens, their respective sensitivity to LAK/NK killing was 59 and –9%, respectively (Table 15.8). Secondly, stimulation of target cells with IFNs, which are known to increase the expression of MHC antigens (Nouri *et al.*, 1992), if anything increased rather than decreased target cell susceptibility to LAK/NK killing (Fig. 15.8). Thirdly, if MHC antigens participate in LAK/NK killing, one would expect a decrease in target susceptibility following the introduction of class I antigens by gene transfection into a class I-negative tumour target cell. As can be seen from Table 15.9, cell surface class I expression in transfected and non-transfected Fen cells was $2000 \pm 48$ and $132 \pm 20$ c.p.m., respectively. The corresponding values for LAK/NK at E/T ratios of 20/1 and 10/1 were 45%, 36% and 51%, 45%, respectively, indicating no significant change in the target susceptibility following the introduction of class I antigens. At the lowest E/T ratio of 1/1, however, the transfected cells were slightly less sensitive than the non-transfected cells and this needs to be explored further.

A working hypothesis is that the over-riding factor controlling the non-MHC-restricted LAK/NK killing is a receptor–target interaction independent of the mechanism regulating class I antigen expression (Nouri *et al.*, 1993b). Under circumstances where the receptor expression is low, certain class I antigens, as demonstrated by Storkus *et al.* (1991), would have a conformational masking effect, hence becoming the limiting factor for the killing.

Table 15.8. *LAK/NK activity of IL-2-activated mononuclear cells on cell lines expressing different intensities of class I antigens*

| | | LAK/NK killing (E/T ratios) | | |
|---|---|---|---|---|
| Cell lines | Class I | 20/1 | 10/1 | 5/1 |
| J82 | 2064 ± 407 | 2 | 1 | 1 |
| SKV14 | 1627 ± 288 | 11 | 17 | −2 |
| Wil | 1208 ± 67 | 14 | 2 | n.d. |
| MCF7 | 1186 ± 77 | 28 | 22 | 16 |
| A431 | 1121 ± 67 | 15 | −2 | 7 |
| 5637 | 1118 ± 110 | −1 | 1 | −2 |
| T47D | 1105 ± 77 | 59 | 43 | 31 |
| RT112 | 1026 ± 57 | −9 | 1 | −16 |
| Scaber | 920 ± 109 | 0 | −1 | −20 |
| Lan | 749 ± 76 | 2 | 7 | 16 |
| Ep2102 | 706 ± 54 | 36 | 35 | 33 |
| RT4 | 566 ± 61 | −20 | −17 | 10 |
| Tera II | 207 ± 26 | 34 | 28 | 18 |
| Fen | 86 ± 12 | 77 | 62 | 52 |
| Tera I | 70 ± 21 | 26 | 20 | 11 |
| $r$ | | −0.23 | −0.30 | −0.47 |
| $p$ | | 0.39 | 0.27 | 0.08 |

Results are expressed as mean ± SD (c.p.m.) of three replicates for class I antigens. Effector/target ratio (E/T ratio) denotes sensitivity to LAK/NK killing as a percentage. The Spearman correlation value ($r$) and significance level ($p$) is given between the class I antigens and the degree of killing. n.d. denotes not done.

However, when the receptors are in excess, LAK/NK lysis occurs whatever the level of class I antigens.

## 15.6 Conclusions

The understanding of MHC antigen expression is surely a critical issue if tumours are to be recognized and rejected by CTLs. Clinical trials using various cytokines have only been beneficial in a small minority of melanoma and renal carcinoma patients (Oliver & Nouri, 1991), where MHC-restricted killing of tumour cells by autologous TILs has been demonstrated. Therefore, it is imperative to identify this small minority of tumours and study them in great depth. The level of class I antigens on tumour biopsies and the behaviour of tumour cells following cytokine stimulation (once there is a sufficient number of primary tumour cells) might prove to be useful for pre-selection of this small minority.

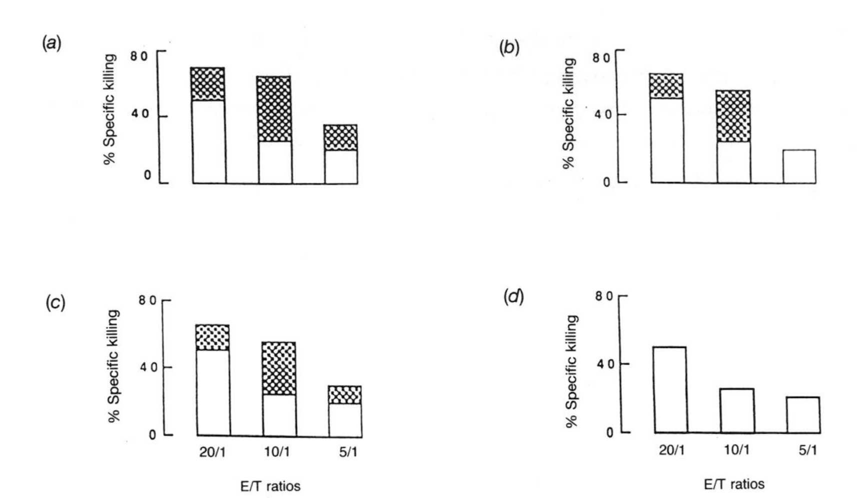

Fig. 15.8. Treatment of the Fen tumour cell line with IFN increased susceptibility to the cytotoxic activity of IL-2-activated LAK/NK cells. Target cells were stimulated with IFN 24 hours prior to the cytoxicity assay. (*a*) IFN-$\gamma$, 100 U/ml; (*b*) IFN-$\alpha$, 1000 U/ml; (*c*) IFN-$\beta$, 2000 U/ml; (*d*) unstimulated cells.

Table 15.9. *LAK/NK killing on Fen cell line before and after* $\beta_2$*-m gene transfection*

| | Non-transfected cells | Transfected cells |
|---|---|---|
| *E/T ratio (% killing)* | | |
| 20/1 | 36 | 45 |
| 10/1 | 45 | 51 |
| 5/1 | 41 | 52 |
| 2.5/1 | 26 | 48 |
| 1/1 | 31 | 10 |
| *Class I antigens* | $132 \pm 20$ | $2000 \pm 48$ |

Effector/target ratios (E/T ratio) denotes sensitivity to LAK/NK killing activity, expressed as percentage killing.
Class I antigen levels expressed as mean ± SD of three replicates (c.p.m.).

In addition, the correction of absent MHC antigens by gene transfection and expansion of the TILs *in vitro* will provide the necessary tools for analysing putative tumour-specific antigen(s). Such gene correction, possibly in combination with transfection of cytokine genes such as those for IL-2, IL-4 or IL-12, could also act as a powerful tumour vaccine and might be a fruitful approach for interfering with tumour escape mechanism(s).

## Acknowledgements

This work was supported in part by the Department of Medical Oncology, The Royal London Hospital, Imperial Cancer Research Fund, Grand Metropolitan, Mercury and Barclays Foundation. I am indebted to Professor R. T. D. Oliver for his scientific contributions and support and Dr G. E. Blair for his meticulous editing of the manuscript. I would also like to thank Mrs R. F. Hussain, Mr D. J. Gillott, Dr M. Mansouri and Mrs R. Qurachi for their contributions and clinical colleagues, particularly Mr B. Jenkins, Mr C. Fowler and Mr A. Paris, for providing clinical materials.

## References

Aebersold, P., Hyatt, C., Johnson, S., Hines, K., Korcak, L., Sanders, M., Lotze, M., Yang, J., Topalian, S. & Rosenberg, S.A. (1991). Lysis of autologous melanoma cells by tumour infiltrating lymphocytes: association with clinical response. *Journal of the National Cancer Institute*, 83, 932–937.

Alexander, M.A., Lee, W. & Guerry, D. (1991). Retroviral vector transfection of class II positive human metastatic melanoma cell line with a matched HLA-DR β1 gene restores its capacity to present antigen. *Proceedings of the American Association of Cancer Research*, 32, abstract 1413.

Aosi, F., Ohlan, C., Ljunggren, H.G., Frankkson, L., Ploegh, H., Townsend, A., Karre, K. & Stauss, H.J. (1991). Different types of allospecific CTL clones identified by their ability to recognize peptide loading defective target cells. *European Journal of Immunology*, 21, 2767–2771.

Avila-Carino, J., Torsteinsdottir, S., Bejarano, M.T., Klein, G., Klein, E. & Masucci, M.G. (1988). Combined treatment with interferon (IFN)-gamma and tumour necrosis factor (TNF)-alpha up-regulates the expression of HLA class I determinants in Burkitt's lymphoma lines. *Cellular Immunology*, 117, 303–311.

Basham, T.Y., Nikoloff, B.J., Merigan, T.C. & Morhenn, V.B. (1985). Recombinant gamma interferon differentially regulated class II antigen expression and biosynthesis on cultured normal human keratinocytes. *Journal of Interferon Research*, 88, 393–400.

Brocker, E.B., Sutter, L., Bruggen, J., Ruiter, D.J., Macher, E. & Sorg, C. (1985). Phenotypic dynamics of tumour progression in human malignant melanoma. *International Journal of Cancer*, 36, 29–35.

Burrone, O.R., Kefford, F.R., Gilmore, D. & Milstein, C. (1985). Stimulation of HLA-A,B,C by IFN-α. The derivation of Molt 4 variants and differential expression of HLA-A,B,C subset. *EMBO Journal*, 4, 2855–2858.

Carlson, G. & Wegman, T. (1977). Rapid *in vivo* destruction of semi-syngeneic cells by non-immunised mice as a consequence of non-identity at H2. *Journal of Immunology*, 118, 2130–2137.

Cohen, P.J., Lotz, M.T., Roberts, J.R. & Rosenberg, S.A. (1987). Immunopathology of sequential tumour biopsies in patients on IL-2. *American Journal of Pathology*, 129, 208–216.

Concha, A., Cabrera, T., Ruiz-Cabello, F. & Garrido, F. (1991). Can the HLA phenotype be used as a prognostic factor in breast carcinoma? *International Journal of Cancer*, Suppl.6, 146–169.

Connor, M.E. & Stern, P.L. (1990). Loss of MHC class I expression in cervical carcinomas. *International Journal of Cancer*, 46, 1029–1035.

Crowley, N.J., Darrow, T.L., Quinn-Allen, M.A. & Seigler, H.F. (1991). MHC-restricted recognition of autologous melanoma by tumour-specific cytotoxic T cells. Evidence for restriction by a dominant HLA-A allele. *Journal of Immunology*, 146, 1692–1699.

Dayan, A.D. & Marshall, A.M.E. (1964). Immunological reaction in man against certain tumours. *Lancet*, ii, 1102–1103.

Durrant, L.G., Ballantyne, K.C., Armitage, N.C., Robins, R.A., Marksman, R., Hardcastle, J.D. & Baldwin, R.A. (1987). Quantitation of MHC antigen expression on colorectal tumours and its association with tumour progression. *British Journal of Cancer*, 56, 425–432.

Esteban, F., Concha, A., Huelin, C., Perez-Ayala, M., Pedrinaci, S., Ruiz-Cabello, F. & Garrido, F. (1989). Histocompatibility antigens in primary and metastatic squamous cell carcinoma of the larynx. *International Journal of Cancer*, 43, 436–440.

Fearon, E.R., Pardoll, D.M., Itaya, T., Golumbek, P., Levitsky, H.I., Simons, J.W., Karasuyama, H. & Vogelstein, B. (1990). Interleukin-2 production by tumour cells bypass T helper function in generation of an anti-tumour response. *Cell*, 60, 397–403.

Ferguson, A., Moore, M. & Fox, H. (1985). Expression of MHC products and leukocyte differentiation antigens in gynaecological neoplasms: an immunohistological analysis of the tumour cells and infiltrating leukocytes. *British Journal of Cancer*, 52, 551–563.

Fleming, K.A., McMichael, A., Morton, J.A., Woods, J. & McGee, J.O'D. (1981). Distribution of HLA class I antigens in normal human tissue and in mammary cancer. *Journal of Clinical Pathology*, 34, 779–784.

Gansbacher, B., Bannerji, R., Daniels, B., Zier, K., Cronin, K. & Gilboa, E. (1990a). Retroviral vector-mediated γ-interferon gene transfer into tumour cells generates potent and long lasting anti-tumour immunity. *Cancer Research*, 50, 7820–7825.

Gansbacher, B., Zier, K., Daniel, B., Cronin, K., Bannerji, R. & Gilboa, E. (1990b). Interleukin-2 gene transfer into tumour cells abrogates tumorigenicity and induces protective immunity. *Journal of Experimental Medicine*, 172, 1217–1224.

Gillott, D.J., Nouri, A.M.E., Compton, S.J. & Oliver, R.T.D. (1993). Accurate and rapid assessment of MHC antigen up-regulation following cytokine stimulation as a possible tool for pre-selecting cancer patients for cytokine therapy. *Journal of Immunological Methods*, 165, 231–239.

Gorelik, E., Gunji, Y. & Herberman, R.B. (1988). H-2 antigen expression and sensitivity of BL16 melanoma cells to natural killer cell cytotoxicity. *Journal of Immunology*, 140, 2096–2102.

Greenberg, P.D., Klarnet, J.P., Kern, D.E. & Cheever, M.A. (1988). Therapy of disseminated tumours by adoptive transfer of specifically immune T cells. *Progress in Experimental Tumour Research*, 32, 104–127.

Grimm, E.A., Mazumder, A., Zhang, H.Z. & Rosenberg, S.A. (1982). The lymphokine activated killer cell phenomenon: lysis of NK resistant fresh solid tumour cells by IL-2-activated autologous human peripheral blood lymphocytes. *Journal of Experimental Medicine*, 155, 1823–1841.

Guerry, D. 4th, Alexander, M.A., Herlyn, M.F., Zehngebot, L.M., Mitchell, K.F., Zmijewski, C.M. & Lusk, E.J. (1984). HLA-DR histocompatibility leukocyte antigens permit cultured human melanoma cells from early but not advanced disease to stimulate autologous lymphocytes. *Journal of Clinical Investigation*, 73, 267–271.

Hakem, R., Bouteiller, P.L., Barad, M., Trujillo, M., Mercier, P., Wietzerbin, J. & Lemonnier, F.A. (1989). IFN-γ mediated differential regulation of the expression of HLA-B7 and HLA-A3 class I genes. *Journal of Immunology*, 142, 297–305.

Hammerling, G.J., Klar, D., Plum, W., Momburg, F. & Moldenhauer, G. (1987). The influence of major histocompatibility complex class I antigens on tumour growth and metastasis. *Biochimica et Biophysica Acta*, 907, 245–249.

Hou, S., Doherty, P.C., Zijlstra, M., Jaenisch, R. & Katz, J.M. (1992). Delayed clearance of Sendai virus in mice lacking class 1 MHC-restricted $CD8^+$ T cells. *Journal of Immunology*, 149, 1319–1325.

Hui, K., Grosveld, F. & Festenstein, H. (1984). Rejection of transplantable AKR leukaemia cells following MHC DNA-mediated transformation. *Nature (London)*, 311, 750–752.

Hussain, R.F., Nouri, A.M.E. & Oliver, R.T.D. (1993). A new approach for measurement of cytotoxicity using a colorimetric assay. *Journal of Immunological Methods*, 160, 89–96.

Itoh, K., Platsoucas, C.D. & Blach, C.M. (1988). Autologous tumour specific cytotoxic T lymphocytes in the infiltrate of human metastatic melanomas: activation by interleukin-2 and autologous tumour cells and involvement of the T cell receptor. *Journal of Experimental Medicine*, 168, 1419–1441.

Jabrane-Ferrat, N., Calvo, F., Faille, A., Lagabrielle, J.F., Boisson, N., Quillet, A. & Fradelizi, D. (1990). Recombinant gamma interferon provokes resistance of human breast cancer cells to spontaneous and IL-2 activated non-MHC restricted cytotoxicity. *British Journal of Cancer*, 61, 558–562.

Karre, K., Ljunggren, H.G., Piontek, G. & Kiessling, R. (1986). Selective rejection of H-2 deficient lymphoma variants suggests alternative immune defence strategy. *Nature (London)*, 319, 675–677.

Kiessling, R., Klein, E. & Wigzell, H. (1975). Natural killer cells in the mouse. I. Cytotoxic cells with specificity for mouse Moloney leukemia cells. Specificity and distribution according to genotype. *European Journal of Immunology*, 5, 112–116.

Lange, A., Fetting, R., Jazwiec, B., Moniewska, A., Ennen, J., Ernst, M. & Flad, H.D. (1991). Augmentation of interleukin-2-activated cytotoxicity after treatment of cells with inhibitors of interferon production. *Cellular Immunology*, 133, 285–294.

Leiden, J.M., Karpinsky, B.A., Gottschalk, L. & Kornbluth, J. (1989). Susceptibility of natural killer mediated cytolysis is independent of the level of target cell class I HLA expression. *Journal of Immunology*, 142, 2140–2147.

Ling, P.D., Warren, M.K. & Vogel, S.N. (1985). Antagonistic effect of interferon-β on the interferon-γ-induced expression of Ia antigen in murine macrophages. *Journal of Immunology*, 135, 1857–1863.

List, A.F., Grogan, T.M., Spier, C.M. & Miller, T.P. (1991). Tumour infiltrating T lymphocytes (TIL) response is deficient in B cell NHL arising in immunosuppressed (IC) host. *American Society for Clinical Oncology*, abstract 940.

Lobo, P. & Spencer, C.E. (1989). Use of anti-HLA antibodies to mask major histocompatibility complex gene products on tumour cells can enhance susceptibility of these cells to lysis by natural killer cells. *Journal of Clinical Investigation*, 83, 278–287.

Masucci, M.G., Torsteindottir, S., Colombani, J., Brautbar, C., Klein, E. & Klein, G. (1987). Down-regulation of class I HLA antigens and Epstein–Barr virus encoded latent membrane protein in Burkitt's lymphoma lines. *Proceedings of the National Academy of Sciences of the USA*, 84, 4567–4571.

Maziarz, R.T., Mentzer, S.J., Burakoff, S.J. & Faller, D.V. (1990). Distinct effects of interferon-$\gamma$ and MHC class I surface antigen levels on resistance of the K562 tumour cell line to natural killer-mediated lysis. *Cellular Immunology*, 130, 329–338.

McMichael, A.J., Ting, A., Zweerink, H.J. & Askonas, B.A. (1977). HLA-restriction of cell-mediated lysis of influenza virus-infected human cells. *Nature (London)*, 270, 524–526.

Mitchison, A. (1953). Passive transfer of transplantation immunity. *Proceedings of the Royal Society, Series B*, 142, 72–75.

Momburg, F. & Koch, S. (1988). Abrogation of $\beta_2$-microglobulin but not of HLA-A,B,C heavy chain expression at the transcriptional level in human colon carcinoma cells. *Immunobiology*, 178, 128–133.

Momburg, F., Ziegler, A., Harprecht, J., Moller, P., Moldenhauer, G., Hammerling, G.J. (1989). Selective loss of HLA-A or HLA-B antigen expression in colon carcinoma. *Journal of Immunology*, 142, 352–358.

Moore, D.H., Fowler, W.C. & Olafsson, K. (1990). Class I histocompatibility antigen expression: a prognostic factor for aneuploid ovarian cancers. *Gynecological Oncology*, 38, 458–461.

Morita, T., Salmeron, M.A., Moser, R.P., Ross, M.I. & Itoh, H. (1992). Oligoclonal expansion of V$\beta$8$^+$ cells in interleukin-2-activated T cells residing in subcutaneous metastatic melanoma. *Clinical and Experimental Metastasis*, 10, 69–76.

Mule, J.J., Shu, S., Schwarz, S.L. & Rosenberg, S.A. (1984). Adoptive immunotherapy of established pulmonary metastases with LAK cells and recombinant interleukin-2. *Science*, 225, 1487–1489.

Norazmi, M., Hohmann, A.W., Skinner, J.M. & Bradley, J. (1989). Expression of MHC class I and II antigens in colonic carcinomas. *Pathology*, 21, 248–253.

Nouri, A.M.E., Bergbaum, A., Lederer, E., Crosby, D., Shamsa, A. & Oliver, R.T.D. (1991b). Paired tumour infiltrating lymphocytes (TIL) and tumour cell lines from bladder cancer: a new approach to study tumour immunology *in vitro*. *European Journal of Cancer*, 27, 608–612.

Nouri, A.M.E., Dos Santos, A.V.L., Crosby, D. & Oliver, R.T.D. (1991a). Correlation between class I antigen expression and the ability to generate tumour infiltrating lymphocytes from bladder tumour biopsies. *British Journal of Cancer*, 64, 996–1000.

Nouri, A.M.E., Hussain, R.F., Dos Santos, A.V.L., Gillott, D.J. & Oliver, R.T.D. (1992). Induction of MHC antigens by tumour cell lines in response to interferons. *European Journal of Cancer*, 28, 1110–1115.

Nouri, A.M.E., Hussain, R.F., Dos Santos, A.V.L., Mansouri, M. & Oliver, R.T.D. (1993b). Intensity of class I antigen expression on human tumour cell lines and its relevance to the efficiency of non-MHC restricted killing. *British Journal of Cancer*, 67, 1223–1228.

Nouri, A.M.E., Hussain, R.F., Oliver, R.T.D., Handy, A.M., Bartkova, I. & Bodmer, G.J. (1993a). Immunological paradox in testicular tumours: the presence of a large number of activated T cells despite the complete absence of MHC antigens. *European Journal of Cancer*, 29, 1895–1899.

Nouri, A.M.E., Smith, M.E.F., Crosby, D. & Oliver, R.T.D. (1990). Selective and non-selective loss of immunoregulatory molecules (HLA-A,B,C antigens and

LFA-3) in transitional cell carcinoma. *British Journal of Cancer*, 62, 603–606.

Oliver, R.T.D. (1991). New views on rejection mechanisms and their relevance to interleukin-2 as a treatment for renal cell cancer. *European Journal of Cancer*, 27, 1168–1172.

Oliver, R.T.D., Nethersell, A.B.W. & Bottomley, J.M. (1989). Unexplained spontaneous regression and alpha interferon as treatment for metastatic renal cell carcinoma. *British Journal of Urology*, 63, 128–131.

Oliver, R.T.D. & Nouri, A.M.E. (1991). T cell immune response to cancer in humans and its relevance for immunodiagnostics and therapy. *Cancer Surveys*, 13, 173–204.

Ottesen, S.S., Kieler, J. & Christensen, B. (1987). Changes in HLA-A,B,C expression during spontaneous transformation of human urothelial cells *in vitro*. *European Journal of Cancer and Clinical Oncology*, 23, 991–994.

Pandolfi, F., Boyle, L.A., Trentin, L., Kurnick, J.T., Isselbacher, K.J. & Gattoni-Celli, S. (1991). Expression of the HLA-A2 antigen in human melanoma cell lines and its role in T cell recognition. *Cancer Research*, 51, 3164–3170.

Pena, J., Solana, R., Alonso, M.C., Santamaria, M., Serrano, R., Ramirez, R. & Carracedo, J. (1989). MHC class I expression on tumour cells and their susceptibility to NK lysis. *Journal of Immunogenetics*, 16, 407–411.

Pomerance, A. (1972). Pathology and progression following total cystectomy for carcinoma of bladder. *British Journal of Urology*, 44, 451–456.

Prescot, S., James, K., Busuttil, J., Hargreave, T.B., Chisholm, G.D. & Smyth, J.F. (1989). HLA-DR expression by high grade superficial bladder cancer treated with BCG. *British Journal of Urology*, 63, 264–269.

Prescot, S., James, K., Hargreave, T.B., Chisholm, D.G. & Smyth, J.F. (1990). Radio-immunoassay detection of interferon gamma in urine after intravesical Evans BCG therapy. *Journal of Urology*, 144, 1248–1252.

Rees, R.C., Buckle, A.M., Gelsthorpe, K., James, V., Potter, C.W., Rogers, K. & Jacob, G. (1988). Loss of polymorphic A and B locus HLA antigens in colon carcinoma. *British Journal of Cancer*, 57, 374–377.

Rosenberg, S.A., Aebersold, P., Cornetta, K., Kasid, A., Morgan, R.A., Moen, R., Karson, E.M., Lotze, M.T., Yang, J.C. & Topalian, S.C. (1990). Gene transfer into humans – immunotherapy of patients with advanced melanoma, using tumour infiltrating lymphocytes modified by retroviral gene transduction. *New England Journal of Medicine*, 323, 570–578.

Rosenberg, S.A., Lotze, M.T., Muul, L.M., Chang, A.E., Avis, F.P., Leitman, S., Lineham, W.M., Robertson, C.N., Lee, R.E. & Rubin, J.T. (1987). A progress report on the treatment of 157 patients with advanced cancer using lymphokine-activated killer cells and interleukin-2 or high dose interleukin-2 alone. *New England Journal of Medicine*, 316, 898–903.

Shaw, A.R.E., Chan, J.K.W., Reid, S. & Seebaler, J. (1985). HLA-DR synthesis, induction and expression in HLA-DR negative carcinoma cell lines of diverse origins by interferon γ but not by interferon β. *Journal of the National Cancer Institute*, 74, 1261–1268.

Shimizu, Y. & DeMars, R. (1989). Demonstration by class I gene transfer that reduced susceptibility of human cells to natural killer cell-mediated lysis is inversely correlated with human HLA-class I antigen expression. *European Journal of Immunology*, 19, 447–451.

Smith, M.E.F., Bodmer, W.F. & Bodmer, J.G. (1988). Selective loss of HLA-A,B,C locus products in colorectal adenocarcinomas. *Lancet*, i, 823–824.

Stam, N.J., Kast, W.M., Voordouw, A.C., Pastoors, L.B., van der Hoeven, F.A., Melief, C.J. & Ploegh, H.L. (1989). Lack of correlation between levels of MHC class I antigens and susceptibility to lysis of small cell lung carcinoma (SCLC) by natural killer cells. *Journal of Immunology*, 142, 4113–4117.

Stern, P., Gidlund, M., Orn, A. & Wigzell, H. (1982). Natural killer cells mediate lysis of embryonal carcinoma cells lacking MHC. *Nature (London)*, 285, 341–342.

Storkus, W.J., Alexander, J., Payne, J.A., Dawson, J.R. & Cresswell, P. (1989). Reversal of natural killing susceptibility in target cells expressing transfected class I HLA genes. *Proceedings of the National Academy of Sciences of the USA*, 86, 2361–2364.

Storkus, W.J., Salter, R.D., Alexander, J., Ward, F.E., Ruiz, R.E., Cresswell, P. & Dawson, J.R. (1991). Class I-induced resistance to natural killing: identification of non-permissive residues in HLA-A2. *Proceedings of the National Academy of Sciences of the USA*, 88, 5989–5992.

Tomita, Y., Matsumoto, Y., Nishiyama, T. & Fujiwara, M. (1990a). Reduction of major histocompatibility complex class I antigens on invasive and high grade transitional cell carcinoma. *Journal of Pathology*, 162, 157–164.

Tomita, Y., Nishiyama, T., Fujiwara, M. & Sato, S. (1990b). Immunohistochemical detection of major histocompatibility complex antigens and quantitative analysis of tumour infiltrating mononuclear cells in renal cell cancer. *British Journal of Cancer*, 62, 354–359.

Topalian, S.L., Muul, L.M., Solomon, D. & Rosenberg, S.A. (1987). Expansion of human tumour infiltrating lymphocytes for use in immunotherapy trials. *Journal of Immunological Methods*, 102, 127–141.

Tsujihashi, H., Uejima, S., Akiyama, T. & Kurita, T. (1989). Immunohistochemical detection of tissue infiltrating lymphocytes in bladder tumours. *Urologia Internationalis*, 44, 5–9.

van Duinen, S.G., Ruitter, D.J., Broker, E.B., van der Velde, S.C., Welvaart, K. & Ferrone, S. (1988). Level of HLA antigens in loco-regional metastases and clinical course of the disease in patients with melanoma. *Cancer Research*, 48, 1019–1025.

Wallich, R., Bulbuc, N., Hammerling, G.J., Katzav, S., Sehgal, S. & Feldman, M. (1985). Abrogation of metastatic properties of tumour cells: *de novo* expression of H-2K antigens following H-2 gene transfection. *Nature (London)*, 315, 289–295.

Watanabe, Y. & Jacob, C.O. (1991). Regulation of MHC class II antigen expression. *Journal of Immunology*, 146, 899–905.

Wintzer, H.O., Benzing, M. & von Kleist, S. (1990). Lack of prognostic significance of $\beta_2$-microglobulin, MHC class I and class II antigen expression in breast carcinomas. *British Journal of Cancer*, 62, 289–295.

Zinkernagel, R.M. & Doherty, P.C. (1979). MHC cytotoxic T cells: studies on the biological role of polymorphic major transplantation antigens determining T cell restriction specificity. *Advances in Immunology*, 27, 51–77.

# 16

# Overexpression of MHC proteins in pancreatic islets: a link between cytokines, viruses, the breach of tolerance and insulin-dependent diabetes mellitus?

MARTA VIVES-PI, NURIA SOMOZA, FRANCESCA VARGAS and RICARDO PUJOL-BORRELL
*University Autonoma of Barcelona*

## 16.1 Introduction

The expression of MHC products by pancreatic islet cells has been extensively studied because these cells are the targets of the destructive autoimmune response that leads to insulin-dependent diabetes mellitus (IDDM) and also because they constitute a set of distinct cell populations ideal for the study of peripheral tolerance by gene targeting in transgenic (tg) mice. Many of the studies described in this chapter were first prompted by concepts previously proposed in the aberrant class II expression hypothesis of endocrine autoimmunity (Bottazzo *et al.*, 1983).

### *16.1.1 The aberrant class II expression hypothesis of endocrine autoimmunity*

Aberrant (actually ectopic) HLA class II expression, i.e. the expression of HLA class II by cells of lineages that are normally class II negative, was first detected in the thyroid follicular cells of glands resected from patients suffering from autoimmune thyrotoxicosis, or Graves' disease (Hanafusa *et al.*, 1983). This finding, together with the demonstration that thyroid follicular cells can be induced to express HLA class II antigens *in vitro* (Pujol-Borrell *et al.*, 1983), led to the formulation of the aberrant class II expression hypothesis. This is based on the assumption that the lack of active immune response to endocrine cells and other scarce and differentiated cells present in peripheral tissues is the result of their low expression of HLA proteins, which makes them 'invisible' to T cells (immunological silence). Essentially the hypothesis proposes that the expression of class II proteins by endocrine cells could lead to the effective presentation of their specific autoantigens to the T cells. The interaction between class II-positive endocrine cells and $CD4^+$

T lymphocytes would play a central role in triggering, driving and/or maintaining the autoimmune response. This model requires an initiating factor capable of provoking the local release of cytokines responsible for the induction of a transient expression of class II by endocrine cells and for attracting T cells. The larger the number of T lymphocytes recruited, the greater would be the probability that some autoreactive cells may recognize the class II-positive endocrine cells. The transition from a limited autoimmune response to an overt clinical disease would be slow and dependent on the functional reserve of the organ as well as the capacity of other regulatory circuits of the immune system to curb the reaction (Pujol-Borrell & Todd, 1987).

The finding of ectopic class II expression in thyroid autoimmune glands led to the search for other examples of class II expression in autoimmune diseases. Many were found, but the functional repercussions of ectopic class II expression may well be very different depending on the organ and the allele (Pujol-Borrell & Todd, 1987). Considerable work was dedicated to investigating whether this hypothesis was applicable to IDDM, which forms a paradigm of endocrine autoimmune diseases since in IDDM the destructive process is confined to a single cell type. Two independent approaches were taken, firstly, to examine HLA expression in samples of pancreatic tissue from diabetic patients and, secondly, to study the regulation of MHC gene expression by pancreatic islet cells cultured *in vitro*. These studies were carried out in humans (in normal and pathological samples), as well as in normal rodents and in two strains of rodents that develop a spontaneous autoimmune diabetes.

### *16.1.2 Islet cell MHC expression in IDDM*

Normal human endocrine islet cells constitutively express low levels of class I but no class II antigens (Pujol-Borrell *et al.*, 1986). Marked changes in MHC expression were detected in IDDM. Before describing these changes, some key features of IDDM will be reviewed. IDDM is a relatively common disease that results from the selective destruction of the insulin-producing cells of the islets of Langerhans. In most cases, the disease becomes clinically manifest before the age of 40 with peaks at 6 and 12 years. Around 20% of patients have a family history of IDDM and there is a clinical association with other 'organ-specific autoimmune diseases', e.g. Hashimoto's thyroiditis, primary myxedema, Graves' disease, autoimmune adrenalitis and pernicious anaemia. IDDM patients often have serum autoantibodies to several islet cell antigens as well as to thyroid and gastric parietal cell antigens.

There is a strong association of IDDM with certain HLA class II alleles, particularly DR3, DQw8 and DR4 (Nepom, 1990).

## 16.2 Pathology of the islets in IDDM

The two most characteristic features of the islets of Langerhans in type 1 diabetes are β cell depletion and infiltration by mononuclear cells, i.e. insulitis (Meyenburg, 1940). Although important, β cell depletion is not complete at clinical onset: around 15% of β cells remain and some persist for up to six years (Foulis & Stewart, 1984). Approximately one third of the islets are affected by 'insulitis'. This inflammatory process is mainly seen in pancreatic specimens taken at autopsy from newly diagnosed IDDM patients (Gepts, 1965; Junker *et al.*, 1977; Gepts & De May, 1978). In the pancreas of such patients, there are three types of islet: (a) islets that contain no residual β cells but have normal or excess numbers of the other endocrine cell types, i.e. glucagon-secreting α cells, somatostatin-secreting δ cells and pancreatic polypeptide-secreting PP cells; (b) islets with insulitis; this affects insulin-containing islets and is often associated with a reduction in the number of β cells; (c) islets that are histologically normal and contain insulin (Foulis, Liddle & Farquharson, 1986). By far the most common islets are those described in (a) above. To find the pancreatic lobes where the insulitis process is active can be difficult and this helps to understand the relatively recent recognition of insulitis as the hallmark of IDDM. Bottazzo *et al.* (1985) carried out the first detailed immunopathological study of the insulitis process and found that the majority of mononuclear cells were $CD8^+$ T lymphocytes. NK and $CD4^+$ lymphocytes were also present, but in smaller numbers. A proportion of T lymphocytes expressed HLA-DR antigens, indicating that they were activated. Macrophages were not seen.

### *16.2.1 HLA expression in the diabetic pancreas*

In the study by Bottazzo *et al.* (1985), HLA class I expression was strikingly increased in many islets. Foulis & Farquharson (1986), using fixed tissue and antibodies that recognized formalin-resistant determinants on HLA antigens (Epenetos, Bobrow & Adams, 1985), extended this observation to another 14 similar pancreases. Class I antigens were markedly increased in the islets with insulitis and also in many of the islets that contained a normal proportion of β cells, but not in most islets devoid of insulin-producing cells. This suggested that class I overexpression preceded insulitis. In both studies described above, ectopic expression of class II was also detected and found to be restricted to the β cells. Hanafusa *et al.* (1990) confirmed these findings in small pancreatic biopsies performed under laparoscopy from seven newly diagnosed diabetic patients: class I overexpression was found in four of the seven patients and ectopic class II expression in the β cells in one patient. In a study from Finland, a remarkable overexpression of class I was seen in many islets, but ectopic class II expression in β cells was not detected (Hanninen *et al.*, 1992). In one pancreas having similar characteristics recently examined in great detail in our laboratory, we found that 58% of the islets overexpressed class I (Fig. 16.1) and about 20% of insulin-positive islets showed ectopic class II expression in the endocrine cells, although class II expression was not confined to the β cells (Somoza *et al.*, 1992). One aspect of the results from all these studies is particularly remarkable: islets that had lost their β cells did not show ectopic class II expression or lymphocytic infiltration, indicating that these three features are probably closely associated.

### *16.2.2 Expression of adhesion molecules in the islets*

The migration of lymphocytes follows two patterns: (a) recirculating T cells bind to 'homing receptors' (carbohydrate molecules expressed in particular endothelial cells, i.e. high endothelial venules, HEV, on the lymph nodes) through specialized surface receptors such as LAM-1 (leukocyte adhesion molecule 1, a member of the L-selectin family); (b) lymphocytes move into inflamed tissues because locally produced cytokines, particularly IFN-γ, TNF-α and IL-1, activate the capillary endothelial cells, which then express increased amounts of adhesion molecules, such as ICAM-1, and become very adhesive (Dustin *et al.*, 1986; Pober *et al.*, 1987). This second type of mechanism may be operating in the diabetic pancreas, since the endothelial cells express very high levels of ICAM-1 both inside and around the islets (Hanafusa *et al.*, 1990; Hanninen *et al.*, 1992; N. Somoza and R. Pujol-Borrell, unpublished results). Lampeter *et al.* (1992) have reported high levels of circulating ICAM-1 and L-selectin in recent-onset diabetes and first-degree relatives of IDDM patients. This is remarkable, given the modest dimension of the inflammatory process in the islets. Elevated levels of circulating adhesion molecules had been previously described during on-going inflammation or extensive tissue damage (Rothelin *et al.*, 1991). Some adhesion molecules, such as ICAM-1 and LFA-3, which are ligands of LFA-1 and CD2, respectively, are important in the immune response not only because they help the lymphocytes to migrate to the site of inflammation but also

(*a*)

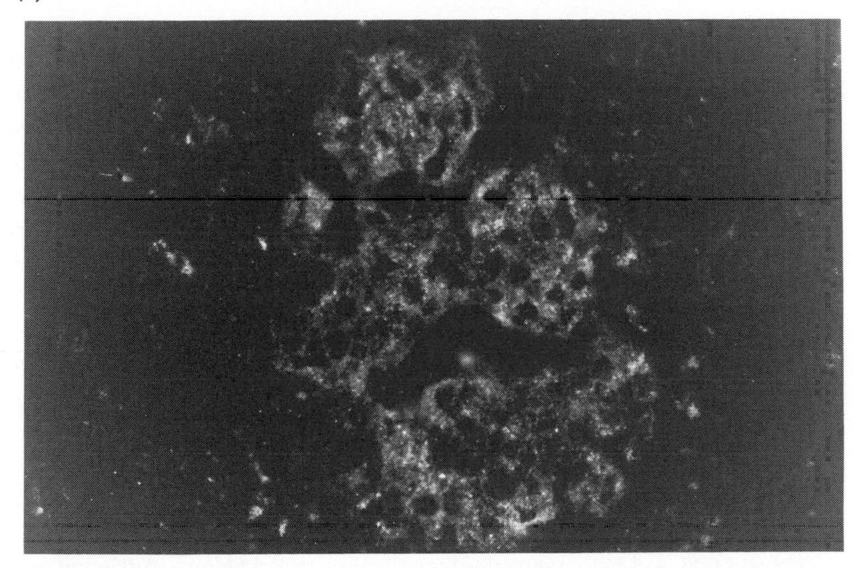

(*b*)

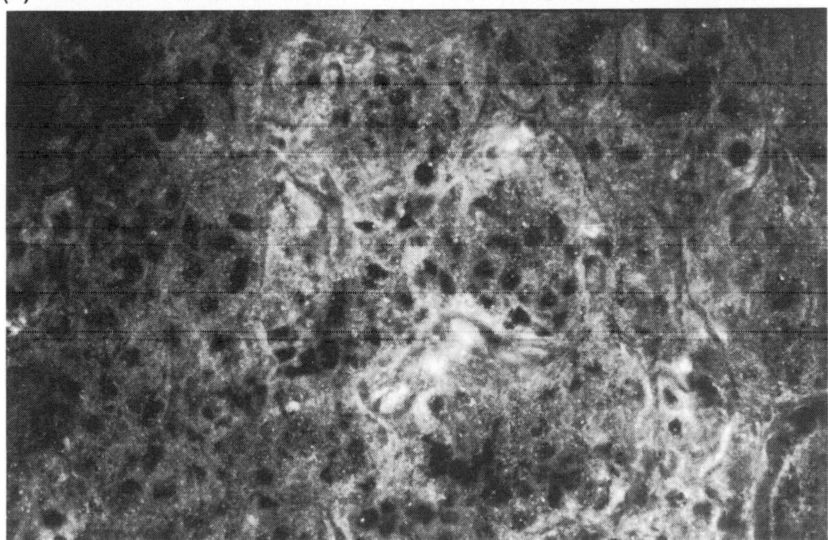

Fig. 16.1. Double immunofluorescent antibody staining of a cryostat section of a newly diagnosed diabetic for (*a*) glutamic acid decarboxylase (GAD) and (*b*) for HLA class I. Note the increase in class I expression throughout the whole islet. (30×.)

because they mediate adhesion between lymphocytes and target cells and may provide activation signals for the effector cells (see below and Chapter 1).

As expected, β cells do not normally express ICAM-1, but IFN-γ and TNF-α both strongly induce ICAM-1 (Campbell *et al.*, 1988a; Vives *et al.*, 1991). Islet cells spontaneously express ICAM-1 several hours after plating in culture, a phenomenon also observed in thyrocytes (Tolosa *et al.*, 1992). This should be taken into account when pre-culturing islets for transplantation experiments. In basal conditions or after treatment with cytokines, β cells do not express LFA-3, or at least not in amounts detectable by flow cytometry.

### *16.2.3 Second signals*

T cells require other signals for activation in addition to the main signal generated by the interaction of the TCR with the MHC–peptide complex (Schwartz, 1992; see also Chapter 1). These other signals, originally termed co-stimulation, are normally provided by the APC. One of the main molecules responsible for co-stimulation has been identified: this is B7, which binds to CD28 and CTLA-4 in lymphocytes (see Section 16.8.4 below). Adhesion molecules, in particular LFA-1 and CD2, may also play a role in T cell activation. It has been shown that the binding of ICAM-1 to LFA-1 is followed by a $Ca^{2+}$ influx (Carrera *et al.*, 1988), while CD2 can mediate antigen-independent activation (Bockenstedt *et al.*, 1988; Denning *et al.*, 1988). Some cytokines could also play the role of a second signal. Stimulation through the TCR in the absence of co-stimulatory signals results in T cell paralysis (Lamb *et al.*, 1983). This mechanism provides a very appealing approach to the induction of tolerance, which is being actively investigated (Lenschow *et al.*, 1992).

### *16.2.4 Cytokines*

Attempts have been made to identify the cytokines present in the diabetic pancreas. So far, the most conclusive results are those of Foulis, Farquharson & Meager (1987), who detected IFN-α in the β cells of diabetic pancreases (even in those that were morphologically normal) and IFN-γ in 45% of the lymphocytes infiltrating the islets (Foulis, McGill & Farquharson, 1991). Using PCR and cytokine-specific primers, it has been possible to detect IFN-α, IFN-β and IL-6 in the pancreas of a newly diagnosed diabetic patient (Somoza *et al.*, 1994).

## 16.3 Induction of MHC expression in pancreatic islets cultured *in vitro*

MHC class I products are constitutively expressed at low levels in human pancreatic islet cells (Baekkeskov *et al.*, 1981) Stimulation *in vitro* for 24–

48 hours with IFN-α, β or γ induces a rapid and marked increase in their expression (Pujol-Borrell *et al.*, 1986). The induction of HLA class II expression in human islet cells requires the combination of IFN-γ plus TNF-α, which contrasts with most types of cell where IFN-γ is a sufficient stimulus (Pujol-Borrell *et al.*, 1987). In this sense, the regulation of MHC expression in β cells is similar to that of neurons, oligodendrocytes (Mauerhoff *et al.*, 1988) and Sertoli cells (Tokuda, Kasahara & Levy, 1990). In mice, the situation is more complex, since while some strains of mice also require IFN-γ plus TNF-α for the induction of class II in the islet cells (Campbell *et al.*, 1988b), in others IFN-γ is active by itself (Wright *et al.*, 1988). Results from *in vitro* studies showed that the islets from diabetes-prone BB rats were readily induced to express class II antigens *in vitro* by stimulation with rat IFN-γ, alone, while islets from the non-diabetic-prone animals of the same strain were resistant to IFN-γ (Walker *et al.*, 1986). It is plausible that in the diabetes-prone animals, in which insulitis is very prominent and starts relatively early in life, the β cells had already been exposed to cytokines *in vivo* (including IFN-γ and TNF-α) before being stimulated *in vitro* with IFN-γ.

Cells of the RINm5F rat insulinoma cell line have also been induced to express class II antigens by incubation with recombinant rat IFN-γ (Varey *et al.*, 1988). The threshold of class II induction seems to be lower in this cell line than in normal rat islet cells. This also applies to the human islet cell line HP62, which was generated by transfection of human islet cells with a plasmid vector containing the SV40 early region (Soldevila *et al.*, 1991). These differences are not unexpected since transformed cell lines often behave quite differently from the parental lineage in terms of HLA expression.

In trying to simulate the viral infection postulated to trigger IDDM, human islet cells in culture have been directly infected with reovirus (Campbell *et al.*, 1988c) or attenuated mumps virus (R. Pujol-Borrell & I. Todd, unpublished observations) in culture. Both types of infection induced an increase in class I but not class II expression. By contrast, in a human insulinoma cell line, infection by measles or mumps viruses induced both class I and class II expression (Cavallo *et al.*, 1992). This increase in class I or II expression is probably the consequence of cytokine production by the infected islet cells.

The fact that the combination of IFN-γ and TNF-α given *in vitro* induces the simultaneous and parallel overexpression of ICAM-1 and HLA class I and II antigens in β cells is somehow in contradiction with the finding that in the islets of diabetic patients only HLA class I and sometimes class II are overexpressed while ICAM-1 is only moderately expressed (Hanninen *et al.*, 1992; Somoza *et al.*, 1994). This implies that factors other than IFN-γ and

TNF-α are probably involved in the modulation of HLA expression in the islet cells of the diabetic pancreas.

### *16.3.1 Are there tissue-specific, strain-dependent differences in class II inducibility?*

In view of the discrepancies in the literature regarding the induction of class II in the islet cells of mice, Leiter *et al.* (1989) carried out a systematic study of the kinetics of MHC class I and class II induction *in vitro*, in macrophages and islet cells from different strains of mice. Their results pointed to marked interstrain differences: in CBA mice neither macrophages nor islet cells expressed class II constitutively and class II was only inducible in macrophages; in non-obese non-diabetic mice (NON is a strain related to NOD (non-obese diabetic) mice, but not prone to diabetes) both islet cells and macrophages constitutively expressed class II antigens; finally, in NOD mice, macrophages expressed class II antigens constitutively while islets were class II antigens negative but responded readily to induction by IFN-γ. The analysis of NOD × CBA $F_1$ mice suggested the existence of genetic elements that act *in trans* to inhibit the induction of class II in the islets. That constitutive expression of class II in islet cells protects NON from diabetes would be in agreement with the results obtained from the pIns-I-E transgenic mice in which transgene-dependent constitutive expression of class II products in the islet cells has been shown to be tolerogenic (see Section 16.8.1 below). Similar findings, i.e. tissue-specific, strain-dependent differences of class II inducibility, have been reported in experimental encephalomyelitis (Massa, ter Meulen & Fontana, 1987). G. Obiols, L. Santamaria, M. Sospedra, E. Tolosa, C. Roura & R. Pujol-Borrell (unpublished results), working with primary cultures of human thyrocytes stimulated with IFN-γ, have also found interindividual, tissue-specific differences of inducibility that seem intrinsic. All these findings, together with the description of allelic variants of functional relevance in the transcriptional regulatory sequences of the DQ genes (Andersen *et al.*, 1991), suggest that one component of disease predisposition conferred by HLA haplotypes may well be located in some of these regulatory sequences.

### *16.3.2 Is there defective class I expression in diabetes?*

Low class I expression has been detected on splenocytes of the NOD mouse and in the lymphocytes of diabetic patients (Faustman *et al.*, 1992). The reduced expression of MHC class I seems to be secondary to a defect in the peptide transporters (TAPs, see Chapters 1 and 2) that transfer cytoplasmic peptides to the endoplasmic reticulum where they bind and stabilize class I heterodimer (van Kaer *et al.*, 1992). In NOD mice, the reduction of class I expression results in low responses in autologous mixed lymphocyte reactions (AMLR). Interestingly, the $\beta_2$-m 'knock out' mice, i.e. mice whose endogenous $\beta_2$-m gene has been inactivated and which, as a consequence, do not express HLA class I, also develop a diabetic syndrome with insulitis, albeit late in life (Faustman *et al.*, 1992). It has been proposed that defective class I expression would interfere with the process of selection, allowing many autoreactive T cells to escape deletion and thus favouring autoimmunity. It should be noted that the results that indicated defective class I expression in NOD mice were soon challenged by several groups (Gaskins, Monaco & Leiter, 1992).

## 16.4 Viruses in human type 1 diabetes

The possible involvement of viruses in causing or triggering IDDM has been postulated for many years. At present, the evidence is rather indirect and can be grouped as epidemiological and serological on the one hand and experimental on the other, i.e. animal models and *in vitro* studies. This evidence will be briefly reviewed because it may help to understand the immunopathology of the diabetic pancreas.

### *16.4.1 Epidemiological and serological evidence*

Clinical and epidemiological observations, many of them dating back to the 1960s, pointed to the possible aetiological role of common viruses in IDDM. Seasonal variations in the incidence of new cases of IDDM (with peaks in the autumn and winter months) and the existence of peaks at the age when children join or move school suggested that common viruses, e.g. rhinovirus or Coxsackie, could be involved in the pathogenesis of IDDM (Gamble *et al.*, 1969; Durruty, Ruiz & Garcia de los Rios, 1979; Gamble, 1980). Elevated titres of IgM antibody to Coxsackie virus in newly diagnosed IDDM patients supported this concept (King *et al.*, 1983). However, follow-up studies of first-degree relatives of diabetic children who have themselves subsequently become diabetic demonstrated that clinical IDDM is preceded by a period of several years during which autoantibodies are present, while the $\beta$ cell mass declines slowly. It is, therefore, highly unlikely that viruses can cause IDDM by acute damage of the islet cells, perhaps only in very exceptional cases (Yoon *et al.*, 1979). Acute viral infection can, of course, trigger clinical symptoms in an individual with a low insulin reserve. It is still quite plausible, however, that a chronic or latent infection of the islets will be found to be the primary cause, no matter how indirectly, of the autoimmune process that leads to IDDM.

### *16.4.2 Difficulties in detecting viruses in the human islets*

The deep anatomical location of the pancreas and the risk of inducing a chemical pancreatitis by the leakage of pancreatic enzymes to the peritoneum has deterred most investigators from performing pancreatic biopsies. Therefore, the available data that suggest a viral infection in the islets of diabetic patients is indirect. The most convincing data include:

1. The presence of IFN-α (a product of virally infected cells) in human islet β cells in diabetics (Foulis *et al.*, 1987).
2. An increased frequency of IDDM in patients with congenital rubella syndrome (Forrest, Menses & Burgess, 1971).
3. The detection of CMV sequences in DNA extracted from lymphocytes of IDDM patients (Pak *et al.*, 1988) and the anti-islet reactivity of a mAb generated by immunization with CMV (Pak *et al.*, 1990).
4. The detection of retroviruses in the islets of NOD mice (see below).
5. The involvement of retroviruses in other human autoimmune diseases, although this is still controversial, e.g. the presence of *gag* sequences in the DNA of lymphocytes and thyrocytes of patients with Graves' disease (Ciampolillo *et al.*, 1989; Lagaye *et al.*, 1992); the reports of retroviruses in AIDS-associated systemic lupus erythematosus (De Clerck *et al.*, 1988), multiple sclerosis (Reddy *et al.*, 1989) and Sjögren's syndrome (Talal *et al.*, 1990). It has been difficult, however, to establish a definite link between viruses and autoimmune disease mainly because of the failure to detect viral gene products in the target organ, although they have been demonstrated in animal models (see Section 16.5).

Several mechanisms have been proposed by which viruses could induce autoimmunity: (a) molecular mimicry/crossreactivity, (b) modification of the T cell repertoire, and (c) induction of class II expression in the target cells. In view of the experiments with tg mice (see below), the possibility that a virus infecting the β cells induces second signals capable of stimulating both anti-viral and autoimmune responses should also be considered.

There have been many failed attempts to demonstrate viral gene products in the islets of the diabetic pancreas. Among the published studies, it is worth mentioning that Foulis *et al.* (1990) could not detect enteroviral capsid protein VP1 in pancreases from diabetics, while VP1 was detected in the myocytes of 12 out 20 patients who died of acute Coxsackie virus myocarditis. We have tested for the presence of adenoviruses and retroviruses in fresh specimens of two diabetic pancreases by nested PCR using degenerate primers, with negative results. The inoculation of homogenized tissue from two diabetic pancreases, into human embryonic fibroblasts, Hep-2 and Vero cell cultures did not induce cytopathic changes. Transmission electron microscopy did not reveal any virus particles or cytopathic changes. Further investigations are still being performed on this material (N. Somoza, & R. Pujol-Borrell, unpublished results).

## 16.5 The NOD mouse

The NOD strain of mouse presents a spontaneous diabetes that closely resembles human type 1 diabetes (Tochino, 1986). These mice develop diabetes at an early age as a consequence of autoimmune and selective β cell destruction. The symptoms are, as in human diabetes, weight loss, ketoacidosis, polyuria and polydipsia. The incidence of the disease depends on the colony, but it is higher in females (72–100%) than in males (39–50%). NOD mice have a genetic pre-disposition to the disease conferred by at least six genes (Todd *et al.*, 1991), one of which is located in the MHC region and has high homology to human DQw8. T lymphocytes, macrophages and NK cells all seem to be important for the development of insulitis. Treatment with silica (a toxic agent for macrophages) prevents the disease (Charlton, Balcelj & Mandel, 1988); however, diabetes can be transferred to healthy animals by lymphocytes (Bendelac *et al.*, 1987; Miller *et al.*, 1988; Haskins *et al.*, 1989).

It is controversial whether there is ectopic class II expression in the β cells of NOD mice (Motojima *et al.*, 1986; Tarui, Tochino & Nonaka, 1986; Hanafusa *et al.*, 1987; Signore *et al.*, 1987). These mice have structural alterations in the Ia genes (Acha-Orbea & McDevitt, 1988), i.e. I-E products are not expressed because of a deletion in the promoter of the α chain (Hattori, Buse & Jackson, 1986), and it is not, therefore, surprising that the detection of class II antigens has been problematic. Formby (1989) has shown by flow cytometry and double immunofluorescence that there is ectopic expression of class II (I-$A^k$) in the β cells of NOD mice, which preceded insulitis. Class II positive β cells were also capable of stimulating the proliferation of autologous $CD4^+$ lymphocytes. The results of Janeway's group in their work with autoreactive T cell clones also suggested that β cells can directly stimulate T cells (Reich *et al.*, 1989). The tg NOD mice described by Lund *et al.* (1990) incorporated transgenes encoding either a modified I-A β chain (Pro-56 gives rise to a different conformation) or a normal I-E α chain. Both modifications protected the mice from IDDM, while islets presented minimal infiltration, but this was not associated with a deletion of any Vβ family of T cells. The mechanism of protection remains unclear, but it may be a change in the T cell repertoire.

Electron microscopy studies on NOD mice have demonstrated retrovirus-like particles in the β cells but not in the α, δ and PP cells (Fujita *et al.*, 1984; Leiter, 1985). Nakagawa *et al.* (1992) detected *gag* protein in the islets of NOD mice. Retroviruses are, however, common in mice and these observations must be evaluated carefully.

## 16.6 The BB rat

BB rats develop a diabetic syndrome that resembles IDDM in humans. Diabetes is linked to the MHC and also to a genetic trait that determines a marked lymphopenia (Mordes & Rossini, 1987). Dean *et al.* (1986) carried out a systematic immunohistopathological study on cryostat sections of pancreases from a cohort of 96 of these animals at different stages preceding the onset of the disease and found that, although class II expression was increased in the capillary of the islets very early, the β cells did not express class II antigens until the insulitis process was well advanced.

## 16.7 Virus-induced diabetes in rodents

One variant of encephalomyocarditis virus (EMCV) selectively infects the islet β cells of specific strains of rodents. In DBA/2 mice, diabetes appears to result directly from viral injury to the β cells, whereas in BALB/c mice, both viral and immunopathological (but not autoimmune) mechanisms seem to play a role. In contrast to the spontaneous models of diabetes, the MHC genes do not determine susceptibility (Craighead, Huber & Sriram, 1990). Diabetogenic strains of Coxsackie B4 virus also produce a diabetic syndrome in susceptible mice, which resembles human diabetes (Webb *et al.*, 1976). Guberlski *et al.* (1991) observed an unexpectedly high incidence of diabetes in a colony of diabetes resistant (DR) BB/Wor rats and found that it was related to infection by a parvovirus known as Kilham's rat virus. Viral antigen was not detected in pancreatic islet cells and β cell cytolysis was not seen until after insulitis, which suggests that the viral infection had triggered an autoimmune response to the islets.

The different mechanisms by which the infection of β cells by a virus may result in loss of tolerance to islet cell antigens have been recently explored using tg mice (see Section 16.8.3 below).

## 16.8 Mice expressing transgenes in β cells

The identification of tissue-specific regulatory sequences in the 5′ flanking region of the insulin gene has made it possible to generate DNA constructs in which the rat insulin promoter directs the transcription of a chosen gene. This regulatory element has been designated pInspro (insulin promoter) or RIP (rat insulin promoter). When introduced into the mouse genome, generating a tg animal, these constructs induce a strong and selective expression of the transgene in the β cells. Several groups have used this approach to test the effect of *de novo* expression of MHC products (Pujol-Borrell & Bottazzo, 1988; Harrison *et al.*, 1989), viral proteins and cytokines in β cells (see Table 16.1). These experiments have changed some of the prevailing ideas on peripheral tolerance.

### *16.8.1 Expression of isogeneic or allogeneic MHC products*

Sarvetnick *et al.* (1988) introduced the pInsproI–$A_{\alpha}^{d}$ and pInsproI–$A_{\beta}^{d}$ genes into a mouse strain that was syngeneic for these genes. They found that the tg mice (actually the $F_1$ animals carrying the transgenes for both the I-A α and β chains) became diabetic, requiring insulin at 8 weeks of age. The islets showed clear class II expression and insulin depletion but no insulitis. There was no evidence of autoimmunity to the islet cells. Lo *et al.* (1988) used similar constructs (InsI–$E_{\alpha}^{d}$ and InsI–$E_{\beta}^{b}$) but introduced them into mice that lacked endogenous I-E expression and which, therefore, experience the transgene products as allogeneic. Here again the tg mice developed IDDM but there was no insulitis or any other evidence of immunological response to the islets. The β cells were strongly stained for class II but only faintly for insulin. Mixed lymphocyte culture experiments demonstrated that the tg mice were tolerant to the MHC antigens coded by the transgene.

It is of particular interest that the mechanism underlying tolerance to I-$E^{b}$ products in this line of tg mice is different from that found in mice naturally expressing I-$E^{b}$. In the normal I-$E^{b}$ positive mice, T cells carrying the TCR encoded by the Vβ17 genes cannot be found, even if mice of this strain have the Vβ17 genes in their genome. It has been shown that TCRs encoded by Vβ17 recognize mainly I-E and this has been considered to be a demonstration of clonal deletion. By contrast, InsI-$E^{b}$-positive tg mice possess in

Table 16.1. *Mice expressing transgenes in the β cells*

| Constructs | Insulitis | Autoimmunity | IDDM | References |
|---|---|---|---|---|
| *MHC products* | | | | |
| pInspro I-A$^d$ | No/minimal | No | Yes | Sarvetnick *et al.* (1988) |
| pInspro I-E$^b$ | No | No | Yes | Lo *et al.* (1988) |
| Inspro H2-K$^b$ | No | No | Yes | Allison *et al.* (1988) |
| *Cytokines* | | | | |
| pInspro IFN-γ | Yes | Yes | Yes | Sarvetnick *et al.* (1990) |
| Inspro IFN-α | ND | ? | Yes | Stewart *et al.* (1993) |
| Inspro TNF-β | Yes | No | No | Picarella *et al.* (1992) |
| Inspro TNF-α | Yes | No | No | Higuchi *et al.* (1992) |
| *Viral antigens* | | | | |
| Inspro TAg (SV40) | Yes | Yes | No | Adams *et al.* (1987) |
| Inspro HA | Yes | Yes | Yes | Roman *et al.* (1990) |
| Inspro LCMV | No | No | No | Oldstone *et al.* (1991) |
| Inspro LCMV + viral infection | Yes | Yes | Yes | Oldstone *et al.* (1991) |
| *Multiple* | | | | |
| InsproH-2K$^b$ + TCR anti-K$^b$ | No | No | No | Schonrich *et al.* (1991) |
| Inspro LCMV GP + TCR anti-LCMV | No | No | No | Ohashi *et al.* (1991) |
| Inspro LCMV GP + TCR anti-LCMV + viral infection | Yes | Yes | Yes | Ohashi *et al.* (1991) |
| InsproH-2K$^b$ + TCR anti-K$^b$ + Inspro-IL-2 | Yes | Yes | Yes | Heath *et al.* (1992) |
| pInsI-E + pInsB7 | Yes | Yes | Yes | D. Guerder and R. A. Flavell, personal communication |
| pInsTNF + pInsB7 | Yes | Yes | Yes | D. Guerder and R. A. Flavell, personal communication |
| pInsB7 | No | No | No | Harlan *et al.* (1993) |
| pInsB7 + STZ | Yes | Yes | Yes | Harlan *et al.* (1993) |

their repertoire lymphocytes that express receptors encoded by the Vβ17 genes but still they are tolerant to I-$E^b$ (Parham, 1988; Burkly *et al.*, 1989). Subsequent experiments have demonstrated that class II I-$E^b$-positive islets from these tgs were not rejected when transplanted into I-E-negative animals. However, when the islet recipients were primed with I-$E^b$-positive spleen cells prior to transplantation, they were able to respond and destroy the I-E-positive tg islet cells. This suggests that class II antigens were recognized by the effector cells but class II-positive β cells were themselves apparently incapable of initiating an immune response by naive, non-tolerant T lymphocytes *in vivo*. To assess further the rules that determine the outcome of presentation by class II-positive islet cells, *in vitro* experiments were carried out using I-E-restricted T cell clones specific for a herpes virus peptide. The pre-culture of I-$E^b$-positive islets from tg mice with these T cell clones rendered them unresponsive to the virus peptide even when the antigen was presented by conventional APC (Markmann *et al.*, 1988). This led to the conclusion that antigen presentation by class II-bearing β cells results in subsequent T cell paralysis. Experiments carried out by Miller's group originated from a slightly different rationale but led to similar conclusions (Allison *et al.*, 1988). In view of the striking increase of class I expression in the islets of newly diagnosed diabetic patients and the similar changes induced *in vitro* by infecting islet cells with reoviruses, these authors designed a tg mouse to test the effect of early intense class I expression in islet cells. The transgene was a construct of the rat insulin II gene promoter with H-$2K^b$ and was introduced into different strains of mice, some of which were syngeneic and some allogeneic. A high proportion of the tg animals developed IDDM but there were no signs of immune reaction to the strongly class I-positive islet cells. Thymectomy did not influence the development of IDDM and their peripheral T cells were tolerant to the allogeneic class I products expressed by the β cells. Thymic T cells were, however, not tolerant and peripheral tolerance ceased when the β cells expressing the allogeneic class I disappeared late in the life of these animals (Morahan, Allison & Miller, 1989). These results suggest again that β cells present antigens in a 'tolerogeneic' manner.

One question that was not addressed in the first MHC tgs, but which is now often raised in view of the disparity of results, is to what extent the peptides occupying the presenting cavity of MHC antigens are also important in allogeneic situations. Obviously the array of peptides available for presentation is very different in the thymus and in the islets. This is particularly relevant in tg mice carrying allogeneic MHC products in their β cells. A pitfall of these experiments is that some of these tg mice may be 'leaky', and transcripts of the transgenes have been detected in the thymus.

Another question raised by these experiments with tg mice is the cause of the insulin secretory failure in these animals. The most likely explanation is that massive production of MHC molecules is deleterious *per se* to the β cells *in vivo*. Parham (1988) proposed that insulin secretion is impeded because the hormone forms complexes with the MHC antigens in the pre-secretory cytoplasmic compartments of the β cells. This argument was based on the high avidity of MHC molecules for peptides. Fragments derived from insulin could easily be trapped in the peptide-binding groove of MHC antigens (see Chapter 1). As a consequence, the secretion of the hormone would be greatly disrupted. In this context, it is relevant to remember that tg animals incorporate many tandem copies of the transgene, randomly integrated in the genome. It is possible that a massive expression of MHC antigens will be by itself incompatible with the normal production of insulin.

### *16.8.2 Expression of cytokines*

Sarvetnick *et al.* (1990) also developed tg mice, with a construct where the IFN-γ gene was under the control of the mouse insulin promoter. The pIns–IFN-γ tg mice developed IDDM and pancreatic inflammation with ectopic class II expression in the islets. Subsequent transplantation experiments demonstrated that these animals had developed an autoimmune response to the islets. A tg mouse expressing IFN-α in the β-cells has also been generated; such mice develop a syndrome closely resembling IDDM (Stewart *et al.*, 1993).

Two lines of tg mice expressing TNF-β and TNF-α in the β cells have been described (Higuchi *et al.*, 1992; Picarella *et al.*, 1992). A considerable leukocytic inflammatory infiltrate consisting of B lymphocytes and of $CD4^+$ and $CD8^+$ T cells was seen in animals from both lines, but neither of them progressed to diabetes. There was also a reduction of the total endocrine cell mass, islet fibrosis and growth of intraislet ductules (indicating, perhaps, islet regeneration). These two tgs demonstrate that islet inflammation caused by TNF expression is not diabetogenic *per se*. The difference between the IFN-γ and TNF tg mice may indicate that IFN-γ is more efficient in generating (either directly or indirectly) second signals for the activation of T cells.

It is relevant to mention here that, in contrast to the results obtained with tg mice, the capacity of class II-positive endocrine cells to present exogenous and endogenous antigens to T cells *in vitro* has been demonstrated using murine (Charriere & Michel-Bechet, 1982; Salamero & Charriere, 1985) and human thyrocytes (Londei *et al.*, 1984; Londei, Bottazzo & Feldmann, 1985; Grubeck-Loebenstein *et al.*, 1988; Wataru *et al.*, 1988). In all these experiments, class II expression was induced on the endocrine cells by cytokines, as in the pIns–IFN-γ tg mice.

### *16.8.3 Mice expressing viral transgene products in the islets*

The expression of viral neo-antigens on the surface of β cells could be the first event in the process that leads to diabetes. The effect of selective expression of viral antigens in the β cells has been tested by several groups using the tg approach. Adams, Alpert & Hanahan (1987), in a key experiment, introduced as transgene a plasmid construct of the rat insulin promoter linked to SV40 T antigen (a viral oncogene). The tg mice developed hyperplasia of the islets, insulinoma-like tumours and, in some cases, pancreatitis with 'autoantibodies' to the SV40 T antigens, but never IDDM. The 'autoimmune' response occurred only when the transgene began to be expressed in the islets of adult animals, while neonatal expression resulted in tolerance. In another model, the expression of influenza virus haemagglutinin in the pancreatic β cells caused lymphocyte infiltration, destruction of insulin-producing cells, autoantibodies in sera and, finally, diabetes (Roman *et al.*, 1990). Oldstone *et al.* (1991) created lines of tg mice expressing lymphocytic choriomeningitis virus (LCMV) nucleoprotein (NP) or glycoprotein (GP) in the β cells. In contrast to the previous experiment, they did not develop diabetes, but when the animals were challenged by a true viral infection there was progressive insulitis with selective β cell destruction. This indicates that a viral gene introduced in the germline and expressed in β cells had not reinduced real tolerance. Another group generated a variant of this model in which a line of tg mice bearing the LCMV-GP in the β cells was bred with a tg mouse carrying a TCR specific for LCMV-GP (Ohashi *et al.*, 1991). Again, these mice did not develop diabetes unless they were infected with LCMV, indicating that self-reactive T cells may remain unresponsive until appropriately activated.

### *16.8.4 Double and treble pIns transgenic mice incorporating second signals*

Schonrich *et al.* (1991) generated a double tg line of mice that incorporated the pIns–H-2$K^b$ construct and the rearranged α and β genes of a TCR that recognized H-2$K^b$. T cells expressing high levels of the tg TCR were deleted intrathymically, but some $K^b$-specific T cells persisted in the periphery and were apparently indifferent to the expression of $K^b$ by the islet cells. When this tg strain was crossed with tg mice expressing IL-2 in the pancreatic β cells, the triple tg mice developed overt diabetes with insulitis (Heath *et al.*, 1992). These results indicate that autoreactive T cells that ignore self antigens may cause autoimmune diabetes when provided with help in the form of IL-2.

D. Guerder & R. A. Flavell (personal communication) generated four different tg mice expressing the human B7 genes in β cells. These mice did not develop diabetes, but when crossed with the tg mice that express I-E in the islets, the hybrid line developed insulitis and diabetes by 12 weeks of age. In addition, the mice resulting from crossing the B7 tg and the TNF tg mice also developed diabetes. A second line of tg mice that expresses B7 on pancreatic β cells has been described (Harlan *et al.*, 1993). These mice did not develop diabetes, but when they were injected with a single subdiabetogenic dose of streptozotocin, 50% of the transgenics developed hyperglycaemia while none of the non-transgenic litter mates was affected. It seems, therefore, that endocrine cells expressing class II can, under certain circumstances, be ignored by T cells or elicit an active immune response. The current explanation for this discordant outcome is based on the second-signal hypothesis (Janeway, 1989). Experiments carried out many years ago with T cell clones specific for the haemagglutinin of influenza virus had shown the possibility of inducing anergy by incomplete presentation (Lamb *et al.*, 1983). Other experiments injecting MHC class II-positive L cell transfectants (Madsen *et al.*, 1988) or formalin-fixed APCs (Jenkins & Schwartz, 1987) led to similar conclusions. The second signal required for the activation of T cells could be IL-2, as this lymphokine overcame T cell paralysis *in vitro* in the experiments by Markmann *et al.* (1988), but other cytokines normally provided by conventional APCs may also be involved (Jenkins, Ashwell & Schwartz, 1988). However, at present, the best-established pathway for the second signal uses CD28 and CTLA4, whose ligand is B7, a transmembrane protein expressed mainly in B cells. It is quite plausible that different pathways intervene in presentation by different types of APC. Another set of molecules able to provide or enhance signals for the activation of T cells are adhesion molecules, such as ICAM-1 and LFA-3, which are present in lymphokine-treated islet cells but not in untreated islet cells of these tg mice (see above).

These considerations have implications for the ectopic class II expression hypothesis of autoimmunity. If the ectopic class II expression observed in the pathological tissue results from the action of cytokines secreted by T lymphocytes recognizing a virus, i.e. IFNs and TNF, then β cells would probably express the full array of membrane glycoproteins that participate in antigen presentation. In addition, it can reasonably be assumed that the micro-environment would contain other cytokines capable of providing a second signal for the activation of autoreactive T cells. Therefore, it is likely that in this scenario, presentation by the class II-positive β cells would activate, rather that tolerize, class II-restricted T cells. If, on the contrary, ectopic class II expression results from the direct effect of viral infection through an intracellular pathway, it remains uncertain whether a second signal of activation for the T cells would be available. Finally, if ectopic class II expression is only a late event, presentation by class II-positive β cells may have little impact on the autoimmune process, apart from perhaps maintaining it.

## 16.9 Conclusions

The study of MHC expression in β cells has led to far more complex issues than were first envisaged when the aberrant class II expression hypothesis of autoimmunity was first formulated at a time when the complexity of antigen presentation and T cell activation were not yet known. The phenomenon of overexpression of MHC antigens on target cells in autoimmune diseases has been well documented, although some uncertainty remains in animal models of diabetes as well as the β cell specificity of MHC expression in humans. Whether MHC expression is secondary to the action of cytokines released by T cells responding to a viral infection or already involved in the autoimmune response is not yet known (Fig. 16.2). Another possibility is that MHC expression is directly induced by a non-cytopathic virus infecting the β cells (Fig. 16.3). The role of class II expression in the pathogenesis of IDDM is still under investigation. Partly as a consequence of the hypothesis mentioned above, β cells have become the paradigm peripheral cell for the study of tolerance. The numerous lines of tg mice that have been described, expressing a variety of products in the islets, have not yet established what is the relative role of immunological silence (or ignorance), anergy and suppression in the maintenance of peripheral tolerance. However, these transgenic animal models have begun to throw light on these important problems, which lie at the frontier of modern immunology.

## Acknowledgements

Work in the author's laboratory is supported by grants from the 'Fondo de Investigaciones Sanitarias' (Project 90E1262–6-E), from Fundación Ramon

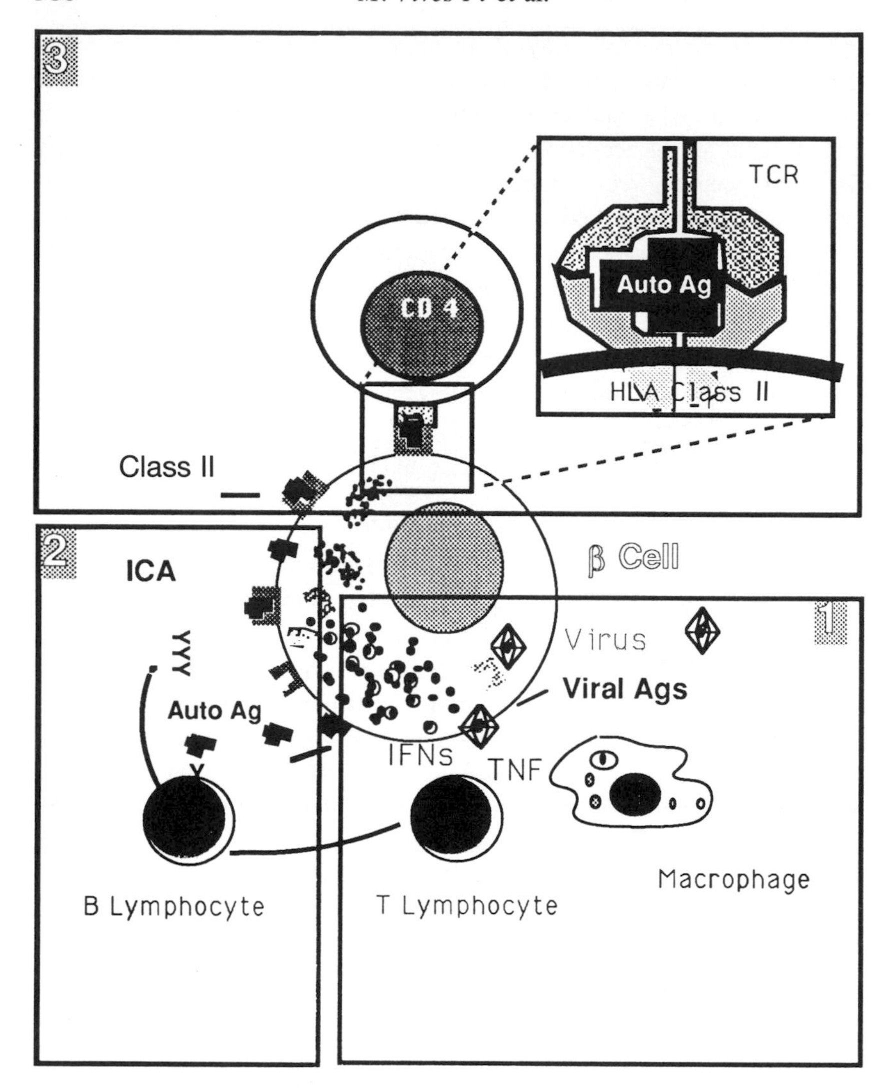


Fig. 16.2. A mechanism proposed to explain the induction of class II expression in β cells and the initiation of an autoimmune response to β cells.

1. A virus, which may reach the pancreas through the ductal system, selectively infects the β cells. T lymphocytes responding to the viral antigens provide help for autoreactive B lymphocytes and secrete cytokines.
2. The B lymphocytes produce islet cell antibodies (ICA). IFN-γ plus TNF-α/β produced by T lymphocytes induce HLA class II expression in the β cells.
3. Class II-positive β cells are recognized by autoreactive $CD4^+$ T cells and through this initial interaction the autoimmune response is triggered.

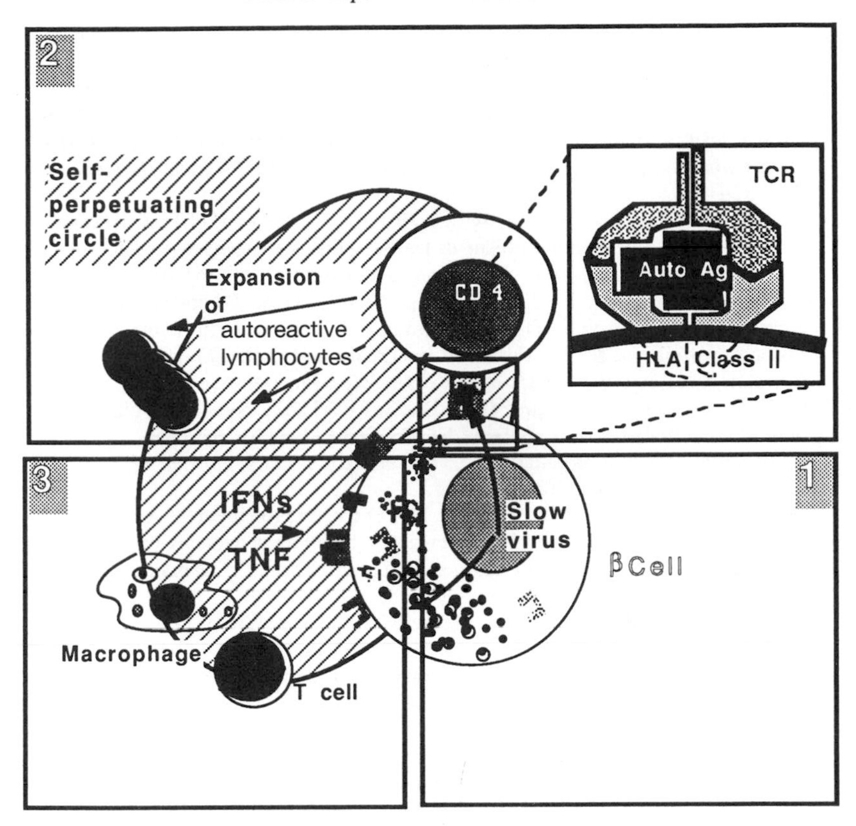


Fig. 16.3. A proposed mechanism of class II induction in β cells.

1. The selective infection of Æ cells by a slow virus results in the direct induction of class II expression.
2. Autoreactive lymphocytes would recognize autoantigen(s) in the β cells and secrete lymphokines resulting in further induction of class II.
3. The autoreactive T cells responding to the β cell autoantigen(s) secrete cytokines' which will spread further the induction of class II expression in other β cells. This will result in the amplification of the response, which will enter a self-perpetuating circle.

Areces, CAYCIT (Project PB860/86) and Eureka (Project EU 286). We express our gratitude to Professor Màrius Foz for his continuous support and encouragement and to all the members of the Immunology Unit for their help.

## *References*

Acha-Orbea, H. & McDevitt, H.O. (1988). The first external domain of the non obese diabetic mouse class II I-A beta chain is unique. *Proceedings of the National Academy of Sciences of the USA*, 84, 2435–2439.

Adams, T.E., Alpert, S. & Hanahan, D. (1987). Non-tolerance and autoantibodies to a transgenic self antigen expressed in pancreatic beta cells. *Nature (London)*, 325, 223–228.

Allison, J.A., Campbell, I.L., Morahan, G., Mandel, T.E., Harrison, L.C. & Miller, J.F.A.P. (1988). Diabetes in transgenic mice resulting from over-expression of class I histocompatibility molecules in pancreatic β cells. *Nature (London)*, 333, 529–533.

Andersen, C.L., Beaty, J.S., Nettles, W., Seyfried, C.E., Nepom, G.T. & Nepom, B.S. (1991). Allelic polymorphism in transcriptional regulatory regions of HLA-DQB genes. *Journal of Experimental Medicine*, 173, 181–192.

Baekkeskov, S., Kanatsuna, T., Klareskog, L., Nielsen, D.A., Peterson, P.A., Rubenstein, A.H., Steiner, D.F. & Lernmark, A. (1981). Expression of major histocompatibility class I antigens on pancreatic islet cells. *Proceedings of the National Academy of Sciences of the USA*, 78, 6456–6460.

Bendelac, A., Carnaud, C., Boitard, C. & Bach, J.F. (1987). Syngeneic transfer of autoimmune diabetes from diabetic NOD mice to healthy neonates. Requirement for both populations $L3T4^+$ and $Lyt2^+$ cells. *Journal of Experimental Medicine*, 166, 823–832.

Bockenstedt, L., Goldsmith, M.A., Dustin, M., Olive, D., Springer, T.A. & Weiss, A. (1988). The CD2 ligand LFA-3 activates T cells but depends on the expression and function of the antigen receptor. *Journal of Immunology*, 141, 1904–1911.

Bottazzo, G.F., Dean, B.M., McNally, G.M., McKay, E.H., Swift, P.G.F. & Gamble, D.R. (1985). *In situ* characterization of autoimmune phenomena and expression of HLA molecules in the pancreas in diabetic insulitis. *New England Journal of Medicine*, 313, 353–360.

Bottazzo, G.F., Pujol-Borrell, R., Hanafusa, T. & Feldmann, M. (1983). Hypothesis: role of aberrant HLA-DR expression and antigen presentation in the induction of endocrine autoimmunity. *Lancet*, ii, 1115–1119.

Bottazzo, G.F., Todd, I., Mirakian, R., Belfiore, A. & Pujol-Borrell, R. (1986). Organ-specific autoimmunity: a 1986 overview. *Immunological Reviews*, 94, 137–169.

Burkly, L., Lo, D., Kanagawa, O., Brinster, R.L. & Flavell, R.A. (1989). T cell tolerance by clonal anergy in transgenic mice with non lymphoid expression of MHC class II I-E. *Nature (London)*, 342, 564–566.

Campbell, I.L., Cutri, A., Wilkinson, D., Boyd, A.W. & Harrison, L.C. (1988a). ICAM-1 is induced on isolated endocrine islets by cytokines but not by reovirus infections. *Proceedings of the National Academy of Sciences of the USA*, 86, 4282–4286.

Campbell, I.L., Harrison, L.C., Ashcroft, R. & Jack, I. (1988b). Reovirus infection enhances expression of class I on human pancreatic beta and rat insulinoma cells. *Diabetes*, 37, 362–365.

Campbell, I.L., Oxbrow, L., Kay, T. & Harrison, L.C. (1988c). Regulation of MHC protein expression in pancreatic beta cells by IFN-gamma and TNF-alpha. *Molecular Endocrinology*, 2, 101–107.

Carrera, A.C., Rincón, M., Sánchez-Madrid, F., Lopez-Botet, M. & de Landazuri, M.O. (1988). Triggering of co-mitogenic signals in T cell proliferation by anti LFA-1 (CD18, CD11a), LFA-3 and CD7 monoclonal antibodies. *Journal of Immunology*, 141, 1919–1924.

Cavallo, M.G., Baroni, M.G., Toto, A., Gearing, A.J.H., Forsey, T. & Andreani, D. (1992). Viral infection induces cytokine release by beta islet cells. *Immunology*, 75, 664–668.

Charlton, B., Balcelj, A. & Mandel, T.E. (1988). Administration of silica particles or anti-Lyt2 antibody prevents β-cell destruction in NOD mice given cyclophosphamide. *Diabetes*, 37, 930–935.

Charriere, J. & Michel-Bechet, M. (1982). Syngeneic sensitization of mouse lymphocytes on monolayers of thyroid epithelial cells. III. Induction of thyroiditis by thyroid sensitized T lymphoblasts. *European Journal of Immunology*, 12, 421–425.

Ciampolillo, A., Marini, V., Mirakian, R., Buscema, M., Schultz, T., Pujol-Borrell, R. & Bottazzo, G.F. (1989). Retroviral-like sequences in Graves' disease: implications for human autoimmunity. *Lancet*, i, 1096–1100.

Craighead, J.E., Huber, S.A. & Sriram, S. (1990). Animal models of picornavirus-induced autoimmune disease: their possible relevance to human disease. *Laboratory Investigation*, 63, 432–446.

De Clerck, L., Couttenye, M., De Broe, M. & Stevens, W. (1988). Acquired immunodeficiency syndrome mimicking Sjögren's syndrome and systemic lupus erythematosus. *Arthritis and Rheumatism*, 31, 272–275.

Dean, B.M., Walker, R., Bone, A.J., Baird, J.D. & Cooke, A. (1986). Prediabetes in the spontaneously diabetic BB/E rat: lymphocytic subpopulations in the pancreatic infiltrate and expression of rat MHC class II in the endocrine cells. *Diabetologia*, 28, 464–466.

Denning, S.M., Dustin, M.L., Springer, T.A., Singer, K.H. & Hynes, D.F. (1988). Purified lymphocyte function-associated antigen 3 (LFA-3) activates human thymocytes via the CD2 pathway. *Journal of Immunology*, 141, 2980–2985.

Durruty, P., Ruiz, F. & Garcia de los Rios, M. (1979). Age at diagnosis and seasonal variation in the onset of insulin-dependent diabetes in Chile (Southern hemisphere). *Diabetologia*, 17, 357–360.

Dustin, M.L., Rothlein, R., Bhan, A.K., Dinarello, C.A. & Springer, T.A. (1986). Induction by IL-1 and interferon-gamma: tissue distribution, biochemistry and function of a natural adherence molecule (ICAM-1). *Journal of Immunology*, 137, 245–254.

Epenetos, A.A., Bobrow, L.G. & Adams, T.E. (1985). A monoclonal antibody that detects HLA-D region antigen in routinely fixed, wax-embedded sections of normal and neoplastic lymphoid tissues. *Journal of Clinical Pathology*, 38, 12–17.

Faustman, D., Li, X., Lin, H.Y., Fu, Y., Eisenbarth, G., Avruch, J. & Guo, J. (1992). Linkage of faulty major histocompatibility complex class I to autoimmune diabetes. *Science*, 254, 1756–1761.

Formby, B. (1989). Mechanisms of autoimmunity in the NOD diabetic mouse. *Diabetologia*, 32, 488A–489A.

Forrest, J.M., Menses, M.A. & Burgess, J.A. (1971). High frequency of diabetes mellitus in young adults with congenital rubella. *Lancet*, ii, 332–334.

Foulis, A.K. & Farquharson, M.A. (1986). Aberrant expression of HLA DR antigens in insulin containing beta cells in recent onset type 1 (insulin-dependent) diabetes mellitus. *Diabetes*, 35, 1215–1226.

Foulis, A., Farquharson, M., Cameron, S., McGill, M., Schonke, H. & Kandolf, R. (1990). A search for the presence of the enteroviral capsid protein VP1 in pancreases of patients with type 1 (insulin-dependent) diabetes and pancreases and hearts of infants who died of coxsackieviral myocarditis. *Diabetologia*, 33, 290–298.

Foulis, A.K., Farquharson, M.A. & Meager, A. (1987). Immunoreactive alpha-interferon in insulin-secreting beta cells in type 1 diabetes mellitus. *Lancet*, ii, 1423–1427.

Foulis, A.K., Liddle, C.N. & Farquharson, M.A. (1986). The histopathology of the pancreas in type 1 (insulin-dependent) diabetes: a 25 year review of deaths in patients under 20 years of age in the United Kingdom. *Diabetologia*, 29, 267–279.

Foulis, A.K., McGill, M. & Farquharson, M.A. (1991). Insulitis in type 1 (insulin dependent) diabetes mellitus in man – macrophages, lymphocytes and interferon-γ containing cells. *Journal of Pathology*, 165, 97–103.

Foulis, A.K. & Stewart, J.A. (1984). The pancreas in recent-onset type I (insulin-dependent) diabetes mellitus: insulin content of islets, insulitis and associated changes in the exocrine acinar tissue. *Diabetologia*, 26, 456–461.

Fujita, H., Fujino, H., Nonaka, K., Tarui, S. & Tochino, Y. (1984). Retrovirus-like particles in pancreatic β cells of NOD (non-obese diabetic) mice. *Biomedical Research*, 5, 67–70.

Gamble, D.R. (1980). The epidemiology of insulin-dependent diabetes with particular reference to virus infection to its aetiology. *Epidemiological Reviews*, 2, 49–70.

Gamble, D.R., Kinsley, M.L., Fitzgerald, M.G., Bolton, R. & Taylor. K.W. (1969). Viral antibodies in diabetes mellitus. *British Medical Journal,* 3, 627–630.

Gaskins, H.R., Monaco, J.J. & Leiter, E.H. (1992). Expression of intra-MHC transporter (Ham) genes and class I antigens in diabetes-susceptible NOD mice. *Science,* 256, 1826–1828.

Gepts, W. (1965). Pathologic anatomy of the pancreas in juvenile diabetes mellitus. *Diabetes,* 14, 619–633.

Gepts, W. & De May, J. (1978). Islet cell survival determined by morphology and immunocytochemical study of the islets of Langerhans in juvenile diabetes mellitus. *Diabetes,* 27 (suppl.), 251–261.

Grubeck-Loebenstein, B., Londei, M., Greenall, C., Pirich, K., Kassal, H., Waldhause, W. & Feldmann, M. (1988). Pathogenic relevance of HLA class II expressing thyroid follicular cells in nontoxic goiter and in Graves' disease. *Journal of Clinical Investigation,* 81, 1608–1614.

Guberlski, D.L., Thomas, V.A., Shek, W.R., Like, A.A., Handler, E.S., Rossini, A.A., Wallace, J.E. & Welsh, R.M. (1991). Induction of type I diabetes by Kilham's rat virus in diabetes resistant BB/Wor rats. *Science,* 254, 1010–1013.

Hanafusa, T., Fujino-Kurihara, H., Miyazaki, A., Yamada, K., Nakajima, H., Miyagawa, J., Kono, N. & Tarui, S. (1987). Expression of class II MHC antigens on pancreatic beta cells in NOD mouse islets. *Diabetologia,* 30, 104–108.

Hanafusa, T., Miyazaki, A., Miyagawa, J., Tamura, S., Inada, M., Yamada, K., Shinji, Y., Katsura, H., Yamagata, K., Itoh, N., Asakawa, H., Nakagawa, C., Otsuka, A., Kawata, T., Kono, S. & Tarui, S. (1990). Examination of islets in the pancreas biopsy specimen from newly diagnosed type 1 (insulin-dependent) diabetic patients. *Diabetologia,* 33, 105–111.

Hanafusa, T., Pujol-Borrell, R., Chiovato, L., Russell, R.C.G., Doniach, D. & Bottazzo, G.F. (1983). Aberrant expression of HLA-DR antigen on thyrocytes in Graves' disease: relevance for autoimmunity. *Lancet,* ii, 1111–1115.

Hanninen, A., Jalkanen, S., Salmi, M., Toikkanen, S., Nikolakaros, G. & Simell, O. (1992). Macrophages, T cell receptor usage and endothelial cell activation in the pancreas at the onset of insulin-dependent diabetes mellitus. *Journal of Clinical Investigation,* 90, 1901–1910.

Harlan, D.M., Hengartner, H., Huang, M.L., Kang, Y.H., Abe, R., Moreadith, R.W., Pircher, H., Gray, G.S., Ohashi, P.S. & Freeman, G.J. (1993). Mice expressing both B7-1 and viral glycoprotein on pancreatic beta cells along with glycoprotein-specific transgenic T cells develop diabetes due to a breakdown of T-lymphocyte unresponsiveness. *Proceedings of the National Academy of Sciences of the USA*, 91, 3137–3141.

Harrison, L.C., Campbell, I.L., Allison, J. & Miller, J.F.A.P. (1989). MHC molecules and β cell destruction. Immune and non-immune mechanisms. *Diabetes,* 36, 815–818.

Haskins, K., Portas, M., Bergman, B., Lafferty, K. & Bradley, B. (1989). Pancreatic islet-specific T cell clones from non obese diabetic mice. *Proceedings of the National Academy of Sciences of the USA*, 86, 8000–8004.

Hattori, M., Buse, J.B. & Jackson, R.A. (1986). The NOD mouse: recessive diabetogenic gene in the major histocompatibility complex. *Science,* 231, 733–735.

Heath, W.R., Allison, J., Hoffmann, M.W., Schonrich, G., Hammerling, G., Arnold, B. & Miller, J.F.A.P. (1992). Autoimmune diabetes as a consequence of locally produced interleukin-2. *Nature (London)*, 359, 547–549.

Higuchi, Y., Herrera, P., Muniesa, P., Huarte, J., Belin, D., Ohashi, P., Aichele, P., Orci, L., Vassalli, J.D. & Vassalli, P. (1992). Expression of a tumor necrosis factor α transgene in murine pancreatic β cells results in severe and permanent insulitis without evolution towards diabetes. *Journal of Experimental Medicine*, 176, 1719–1731.

Janeway, C. (1989). Immunogenicity signals 1, 2, 3 . . . and O. *Immunology Today*, 110, 283–285.

Jenkins, M.K., Ashwell, J. & Schwartz, R.H. (1988). Allogeneic non spleen cells restore the responsiveness of normal T cell clones stimulated with antigens and chemically modified antigens presenting cells. *Journal of Immunology*, 140, 3324–3330.

Jenkins, M.K. & Schwartz, R.H. (1987). Antigen presentation by chemically modified splenocytes induces antigen specific T cell unresponsiveness *in vitro* and in *vivo*. *Journal of Experimental Medicine*, 165, 302–319.

Junker, K., Egeberg, J., Kromann, H. & Nerup, J. (1977). An autopsy study of the islet of Langerhans in acute onset juvenile diabetes mellitus. *Acta Pathologica et Microbiologica Scandinavica*, 85, 699–706.

King, M.L., Shaikh, A., Bidwell, D., Voller, A. & Banatvala, J.E. (1983). Coxsackie B-virus-specific IgM responses in children with insulin-dependent diabetes mellitus, *Lancet*, i, 1397–1399.

Lagaye, S., Vexiau, P., Morozov, V., Guenebant-Clandet, V., Tobalz-Tapiero, J., Canivet, M., Cathelinean, G., Peries, J. & Emanoil-Ravier, R. (1992). Human spumaretrovirus-related sequences in the DNA of leukocytes from patients with Graves' disease. *Proceedings of the National Academy of Sciences of the USA*, 89, 10070-100074.

Lamb, J.R., Skidmore, B.J., Green, N., Chillez, J.M. & Feldmann, M. (1983). Induction of tolerance in influenza virus immune T lymphocytes clones with synthetic peptides of influenza haemagglutinin. *Journal of Experimental Medicine*, 157, 1434–1447.

Lampeter, E.R., Kishimoto, T.K., Rothlein, R., Mainolfi, E.A., Bertrams, J., Kolb, H. & Martin, S. (1992). Elevated levels of circulating adhesion molecules in subjects at risk for IDDM. *Diabetes*, 44, 1668–1671.

Leiter, E.H. (1985). Type C retroviral production by pancreatic beta cells. Association with accelerated pathogenesis in C3H-db (diabetic) mice. *American Journal of Pathology*, 119, 22–32.

Leiter, E.H., Christianson, G.J., Serreze, D.V., Ting, A.T. & Worthen, S.M. (1989). MHC antigen induction by IFN-γ on cultured mouse pancreatic β cells and macrophages. Genetic analysis of strain differences and discovery of an 'occult' class I-like antigen in NOD/It mice *Journal of Experimental Medicine*, 170, 1243–1262.

Lenschow, D.J., Zeng, Y., Thistlethwaite, J.R., Montag, A., Brady, W., Gibson, M.G., Linsley, P.S. & Bluestone, J.A. (1992). Long-term survival of xenogeneic pancreatic islet grafts induced by CTLA41g. *Science*, 257, 789–792.

Lo, D., Burkly, L.C., Widera, G., Cowing C., Flavell, R.A., Palmiter, R.D. & Brinster, R.L. (1988). Diabetes and tolerance in transgenic mice expressing class II molecules in pancreatic beta cells. *Cell*, 53, 159–168.

Londei, M., Bottazzo, G.F. & Feldmann, M. (1985). Human T cell clones from thyroid autoimmune thyroid glands: specific recognition of autologous thyroid cells. *Science*, 228, 85–89.

Londei, M., Lamb, J.R., Bottazzo, G.F. & Feldmann, M. (1984). Epithelial cells expressing aberrant MHC class II determinants can present antigen to cloned human T cells. *Nature (London)*, 312, 639–641.

Lund, T., O'Reilly, L., Hutchings, P., Kanagawa, O., Simpson, E., Gravely, R., Chandler, P., Dyson, J., Picard, J.K., Edwards, A., Kioussis, D. & Cooke, A. (1990). Prevention of insulin-dependent diabetes mellitus in non-obese diabetic mice by transgenes encoding modified I-A β chain or normal I-E α chain. *Nature (London)*, 345, 727–729.

Madsen, J.C., Superina, R.A., Wood, K.J. & Morris, P.J. (1988). Immunological unresponsiveness induced by recipient cells transfected with donor MHC genes. *Nature (London)*, 332, 161–164.

Markmann, J., Lo, D., Naji, A., Palmiter, R.D., Brinster, R.L. & Hner-Katz, E. (1988). Antigen presenting function of class II MHC expressing beta cells. *Nature (London)*, 336, 476–479.

Massa, P.T., ter Meulen, V. & Fontana, A. (1987). Hyperinducibility of Ia antigen on astrocytes correlates with strain-specific susceptibility to experimental autoimmune encephalomyelitis. *Proceedings of the National Academy of Sciences of the USA*, 84, 4219–4223.

Mauerhoff, T., Pujol-Borrell, R., Mirakian, R. & Bottazzo, G.F. (1988). Differential expression and regulation of major histocompatibility complex (MHC) products in neural and glial cells of the human fetal brain. *Journal of Neuroimmunology*, 18, 271–289.

Meyenburg, H.V. (1940). Veber 'insulitis' bei Diabetes. *Schweizerische Medizinische Wochenschrift*, 21, 554.

Miller, B.J., Appel, M.C., O'Neil, J.J. & Wicker, L.S. (1988). Both Lyt2$^+$ and L3T4$^+$ T cell subsets are required for the transfer of diabetes in nonobese diabetic mice. *Journal of Immunology*, 140, 52–58.

Morahan, G., Allison, J. & Miller, J.F.A.P. (1989). Tolerance of class I histocompatibility antigens expressed extrathymically. *Nature (London)*, 339, 622–624.

Mordes, J.P. & Rossini, A.A. (1987). Keys to understanding autoimmune diabetes mellitus: the animal models of insulin-dependent diabetes mellitus. In *Baillière's Clinical Immunology and Allergy*, Vol. I, ed. D. Doniach & G.F. Bottazzo, pp.29–52, London: Ballière Tindall.

Motojima, K., Mullen, Y., Azama, A. & Wicker, L. (1986). Ia$^+$ lymphocytes and Ia$^+$ beta cells in NOD mouse islets. *Diabetes*, 35 (Suppl.1), 69A.

Nakagawa, C., Hanafusa, T., Miyagawa, J., Yutsudo, M., Nakajima, H., Yamamoto, K., Tomita, K., Kono, N., Hakura, A. & Tarui, S. (1992).

Retrovirus *gag* protein p30 in the islets of non-obese diabetic mice: relevance for pathogenesis of diabetes mellitus. *Diabetologia*, 35, 614–618.

Nepom, G.T. (1990). A unified hypothesis for the complex genetics of HLA associations with IDDM. *Diabetes*, 39, 1153–1157.

Ohashi, P.S., Oehen, S., Buerki, K., Pircher, H., Ohashi, C.T., Odermatt, B., Malissen, B., Zinkernagel, R.M. & Hengartner, H. (1991). Ablation of tolerance and induction of diabetes by virus infection in viral antigen transgenic mice. *Cell*, 65, 305–317.

Oldstone, M.B.A., Nerenberg, M., Southern, P., Price, J. & Lewicki, H. (1991). Virus infection triggers insulin-dependent diabetes mellitus in a transgenic model: role of anti-self (virus) immune response. *Cell*, 65, 319–331.

Pak, C.Y., Cha, C.Y., Rajotte, R.V., McArthur, R.G. & Yoon, J.W. (1990). Human pancreatic islet cell specific 38 kilodalton autoantigen identified by cytomegalovirus-induced monoclonal islet cell antibody. *Diabetologia*, 33, 569–572.

Pak, C.Y., Eun, H.M., McArthur, R.G. & Yoon, J.W. (1988). Association of cytomegalovirus infection with autoimmune type 1 diabetes. *Lancet*, ii, 1–4.

Parham, P. (1988). Immunology: intolerable secretion in tolerant transgenic mice. *Nature (London)*, 333, 500–503.

Picarella, D.E., Krattz, A., Li, C.B., Ruddle, N.H. & Flavell, R.A. (1992). Insulitis in transgenic mice expressing tumor necrosis factor β (lymphotoxin) in the pancreas. *Proceedings of the National Academy of Sciences of the USA*, 89, 10036–10040.

Pober, J.S., Lapierre, L.A., Stolpen, A.H., Brock, T.A., Springer, T.A., Fiers, W., Bevilacqua, M.P., Mendrick, D.L. & Gimbrone. M.A. (1987). Activation of cultured human endothelial cells by recombinant lymphotoxin: comparison with tumor necrosis factor and interleukin-1 species. *Journal of Immunology*, 138, 3319–3324.

Pujol-Borrell, R. & Bottazzo, G.F. (1988). Puzzling diabetic transgenic mice: a lesson for human type 1 diabetes? *Immunology Today*, 9, 303–306.

Pujol-Borrell, R., Hanafusa, T., Chiovato, L. & Bottazzo, G.F. (1983). Lectin-induced expression of DR antigen on human cultured follicular thyroid cells. *Nature (London)*, 303, 71–73.

Pujol-Borrell, R. & Todd, I. (1987). Inappropriate HLA class II in autoimmunity: is it a primary phenomenon? In *Ballières's Clinical Immunology and Allergy*, Vol. 1, ed. D. Doniach & G.F. Bottazzo, pp. 1–27. London: Ballière Tindall.

Pujol-Borrell, R., Todd, I., Doshi, M., Bottazzo, G.F., Sutton, R., Gray, D., Adolf, G.R. & Feldmann, M. (1987). HLA class II induction in human islet cells by interferon-gamma plus tumor necrosis factor or lymphotoxin. *Nature (London)*, 326, 304–306.

Pujol-Borrell, R., Todd, I., Doshi, M., Gray, D., Feldmann, M. & Bottazzo, G.F. (1986). Differential expression and regulation of MHC products in the endocrine and exocrine cells of the human pancreas. *Clinical Experimental Immunology*, 65, 128–139.

Reddy, E.P., Sandberg-Wollheim, M., Mettus, R.V., Ray, P.E., de Freitas, E. & Koprowski, H. (1989). Amplification and molecular cloning of HTLV-I sequences from DNA of multiple sclerosis patients. *Science*, 243, 529–533.

Reich, E.P., Sherwin, N.S., Kanagawa, O. & Janeway, C.A. (1989). An explanation for the protective effect of MHC class II I-E molecule in murine diabetes. *Nature (London)*, 341, 326–328.

Roman, L.M., Simons, L.F., Hammer, R.E., Sambrook, J.F. & Gething, M.J.

(1990). The expression of influenza virus hemagglutinin in the pancreatic β cells of transgenic mice results in autoimmune diabetes. *Cell*, 61, 383–396.

Rothelin, R., Mainolfi, E.A., Czaijkowski, M. & Martin, S.D. (1991). A form of circulating ICAM-1 in human serum. *Journal of Immunology*, 147, 3788–3793.

Salamero, J. & Charriere, J. (1985). Syngeneic sensitization of mouse lymphocytes on monolayers of thyroid epithelial cells. VII. Generation of thyroid specific cytotoxic effector cells. *Cellular Immunology*, 91, 111–118.

Sarvetnick, N., Liggitt, D., Pitts, S.L., Hansen, S.E. & Stewart, T.E. (1988). Insulin-dependent diabetes mellitus induced in transgenic mice by ectopic expression of class II MHC and interferon gamma. *Cell*, 52, 773–782.

Sarvetnick, N., Shizuru, J., Liggit, D., Martin, L., McIntire, B., Gregory, A., Parslow, T. & Stewart, T. (1990). Loss of pancreatic islet tolerance induced by beta cell expression of interferon-gamma. *Nature (London)*, 346, 844–847.

Schonrich, G., Kalinke, U., Momburg, F., Malissen, M., Schmitt-Verhulst, A.M., Malissen, B., Hamerling, G.J. & Arnold, B. (1991). Down-regulation of T cell receptors on self-reactive T cells as a novel mechanism for extrathymic tolerance induction. *Cell*, 65, 293–304.

Schwartz, R. (1992). Costimulation of T lymphocytes: the role of CD28, CTLA-4 and B & BB1 in interleukin-2 production and immunotherapy. *Cell*, 71, 1065–1068.

Signore, A., Cooke, A., Pozzilli, P., Butcher, E., Simpson, E., Beverley, P.C.L. (1987). Class II, IL-2 receptor positive cells in the pancreas of NOD mice. *Diabetologia*, 30, 902–905.

Soldevila, G., Buscema, M., Marini, V., Sutton, R., James, R., Bloom, S.R., Robertson, R.P., Mirakian, R., Pujol-Borrell, R. & Bottazzo, G.F. (1991). Transfection with SV40 gene of human pancreatic endocrine cells. *Journal of Autoimmunity*, 4, 381–396.

Somoza, N., Vargas, F., Martí, M., Roura, C., Vives, M., Usac, E.F., Soldevila, G., Ariza, A., Bragado, R., Jaraquemada, D. & Pujol-Borrell, R. (1992). Immunohistopathologic and molecular studies on the pancreas of a newly diagnosed type-1 diabetic patient. *Diabetologia*, 35 (Suppl. 1), A41.

Somoza, N., Vargas, F., Roura-Mir, C., Vives-Pi, M., Fernández-Figueras, M.T., Ariza, A., Gomis, R., Bragado, R., Martí, M., Jaraquemada, D. & Pujol-Borrell, R. (1994). Pancreas in recent onset IDDM: changes in HLA, adhension molecules and autoantigens, restricted T cell receptor Vb usuage and cytokine profile. *Journal of Immunology*, 153, 1360–1377.

Stewart, T.A., Hultgren, B., Huang, X., Pitts-Meek, S., Hully, J. & MacLachlan, N.J. (1993). Induction of type I diabetes by interferon α in transgenic mice. *Science* 260, 1942–1946.

Talal, N., Dauphinee, M., Dang, H., Alexander, S., Hart, D. & Garry, R. (1990). Detection of serum antibodies to retroviral proteins in patients with primary Sjögren's syndrome (autoimmune exocrinopathy). *Arthritis and Rheumatism*, 33, 774–781.

Tarui, S., Tochino, Y. & Nonaka, K. (ed.) (1986). Insulitis and type I diabetes. In *Lessons from the NOD Mouse*. Tokyo: Academic Press.

Tochino, Y. (1986). Discovery and breeding of the NOD mouse. Insulitis and type I diabetes. In *Lessons from the NOD Mouse*, ed. S. Tarui, Y. Tochino, K. Nonaka', pp. 3–10. Academic Press: Tokyo.

Todd, J.A., Aitman, T.J., Cornall, R.J., Ghosh, S., Hall, J.R.S., Hearne, C.M., Knight, A.M., Love, J.M., McAleer, M.A., Prins, J.B., Rodrigues, N., Lathrop, M., Pressey, A., DeLarato, N.H., Peterson, L.B. & Wicker, L.S. (1991).

Genetic analysis of autoimmune type one diabetes mellitus in mice. *Nature (London)*, 351, 542–547.

Tokuda, N., Kasahara, M. & Levy, R.B. (1990). Differential regulation and expression of MHC and Hysix gene products on mouse testicular Leydig and Sertoli cell line. *Journal of Autoimmunity*, three, 457–472.

Tolosa, E., Roura, C., Martí, M., Belfiore, A. & Pujol-Borrell, R. (1992). Induction of inter-cellular adhesion molecule-one but not of lymphocyte function-associated antigen-three in thyroid follicular cells. *Journal of Autoimmunity*, five, 119–135.

van Kaer, L., Ashton-Rickardt, P.G., Ploegh, H.L. & Tonegawa, S. (1992). *TAPone* mutant mice are deficient in antigen presentation, surface class I molecules and CD4–8+ T cells. *Cell*, 71, 1205–1214.

Varey, A.-M., Lydyard, P., Dean, B., van der Meide, P.H., Baird, J.D. & Cooke, A. (1988). Interferon-gamma induces class II MHC antigens on RINmfiveF cells. *Diabetes*, 37, 209–212.

Vives, M., Soldevila, G., Alcalde, L., Lorenzo, C., Somoza, N. & Pujol-Borrell, R. (1991). Adhesion molecules in human islet cells: *de novo* induction of ICAM-1 but not of LFA-three. *Diabetes*, 40, 1382–1390.

Walker, R., Cooke, A., Bone, A.J., Dean, B.M., van der Meide, P. & Baird, J.D. (1986). Induction of MHC class II antigen on pancreatic beta cells isolated from BB/E rats. *Diabetologia*, 29, 749–751.

Wataru, H., Lahat, N., Platzer, M., Schmidt, S. & Davies, T.F. (1988). Activation of MHC restricted rat T cells by cloned syngeneic thyrocytes. *Journal of Immunology*, 141, 1098–1102.

Webb, S.R., Loria, R.M., Madge, & Kibrick, S. (1976). Susceptibility of mice to group B coxsackie virus is influenced by the diabetic gene. *Journal of Experimental Medicine*, 143, 1239–1248.

Wright, J.R., Epstein, H.R., Hauptfeldt, V. & Lacy, P.E. (1988). Tumour necrosis factor enhances interferon induced la antigen expression on murine islet parenchymal cells. *American Journal of Pathology*, 130, 427–430.

Yoon, J.W., Austin, M., Onodera, T. & Notkins, A.B. (1979). Virus-induced diabetes mellitus: isolation of a virus from the pancreas of a child with diabetic ketoacidosis. *New England Journal of Medicine*, 300, 1173–1179.

# 17

# The role of cytokines in contributing to MHC antigen expression in rheumatoid arthritis

FIONULA M. BRENNAN
*The Kennedy Institute of Rheumatology, London*

## 17.1 Introduction

Rheumatoid arthritis (RA) is a chronic inflammatory disease with autoimmune features chiefly affecting the synovial joints. In common with many other autoimmune diseases, there is emerging evidence to suggest that susceptibility to RA is determined by several genes (Wordsworth & Bell, 1991). However, despite the evidence of a genetic predisposition (in particular HLA-DR4 or HLA-DR1), it is clear that the genetic component is probably relatively small since concordance rates in monozygotic twins are no more than 30%, suggesting that other factors are involved (Wordsworth & Bell, 1991; Rigby *et al.*, 1991; Silman, 1991).

The target tissue for the autoimmune and inflammatory response in RA is the synovial joint. In healthy individuals, the synovial membrane is virtually acellular, but in RA it is infiltrated extensively by large numbers of cells from the blood, including activated T cells, macrophages and plasma cells (Janossy *et al.*, 1981; Fφrre, Dobloug & Natvig, 1982; Burmester *et al.*, 1982). In addition, there is proliferation of fibroblasts in the lining layer of the synovial membrane. Many of these infiltrating and resident cells including endothelial cells, fibroblasts and T cells normally express little, if any, HLA class II antigens but are activated in the RA synovium and express abundant HLA class II antigens (Førre *et al.*, 1982; Klareskog *et al.*, 1982; Burmester *et al.*, 1987). These cells may act as antigen-presenting cells and be involved in the stimulation of autoreactive T cells. Thus augmented expression of HLA class II has functional relevance as it is an essential requirement for antigen presentation. However, increased expression of adhesion molecules such as ICAM-1 may also be of importance as it enhances cell–cell interactions.

## 17.2 HLA class II expression in RA synovium

Since HLA class II expression is regulated by cytokines, the analysis of what cytokines are present in the RA synovial joint and how they are regulated should yield information regarding the pathogenesis of this disease. This chapter seeks to review the expression of HLA class II antigens in RA and the regulation of such molecules by cytokines found within the synovial joint.

The expression of HLA class II antigens (mRNA and protein) in RA tissue has been investigated intensively by a number of workers, using immunohistological techniques on sections of tissue and by investigating the expression of HLA class II antigens on cultured synovium using both molecular and biochemical techniques. Most immunohistological studies have focussed on HLA-DR expression, which is found to be increased on macrophages (Janossy *et al.*, 1981; Poulter *et al.*, 1981; Burmester *et al.*, 1982), vascular endothelium (Klareskog *et al.*, 1982; Klareskog, Johnell & Hulth, 1984; Palmer *et al.*, 1985), 'dendritic-like' cells (Winchester & Burmester, 1981) and T cells (Duke *et al.*, 1982). However, such findings were also observed in other forms of synovitis, including osteoarthritic synovitis (Lindblad *et al.*, 1983; Lindblad & Hedfors 1985). The expression of other HLA class II antigens such as HLA-DP and HLA-DQ, which can also function as 'presentation' molecules, has been studied less intensively. Barkley *et al.* (1989) showed that HLA-DR and HLA-DQ expression was comparable between RA and reactive arthritis, but that HLA-DQ expression on endothelium and interstitial cells of RA compared with reactive arthritis was increased. These studies indicate that the particular HLA class II antigens expressed and their chronicity may both be important features that contribute to the disease process in RA. Indeed, data also from our laboratory (Kissonerghis, Maini & Feldmann, 1988) investigating HLA class II expression at the mRNA level showed that active RA synovial tissue expressed high mRNA levels of all HLA class II $\alpha$ chain transcripts (HLA-DR, HLA-DP and HLA-DQ). Furthermore, these levels were maintained if these RA synovial cells were cultured *in vitro* in the absence of any exogenous stimulus for up to six days. The persistence of HLA class II expression was not an extended half-life of the $\alpha$ chain mRNA, since blocking transcription with actinomycin D in such RA synovial cells resulted in a mRNA half-life of 30 minutes, which was similar to that seen in monocytic cells stimulated with IFN-$\gamma$.

The mechanism involved in chronic expression of HLA class II in the RA cultures and also presumably within the synovium is not clear, but it is likely to involve both soluble mediators and cell–cell interactions. Indeed, HLA class II expression rapidly declines on synovial joint macrophages if the

CD3$^+$-positive lymphocytes are removed from the synovial joint cell cultures by panning or by complement-mediated lysis (F. Brennan, unpublished observation). Furthermore, if the synovial fibroblasts are 'grown-out' from the RA synovial cultures, they too lose HLA class II expression as the other cells in the culture dish die. In summary, these studies suggest that there is continual HLA class II synthesis in active RA joints, which then produce molecules capable of maintaining the class II expression. The possible identity of these molecules is covered in more detail in the next section. However, in RA, unlike in Graves' disease (discussed in Chapter 3), the exact nature of the most important antigen-presenting cell is not known since both infiltrating cells and tissue resident cells in the joint are HLA class II positive and could potentially present autoantigen.

## 17.3 Cytokine regulation of HLA class II expression

### *17.3.1 IFN-γ*

IFN-γ is considered to be the most important stimulus for induction of HLA class II expression (Korman *et al.*, 1985) and, as such, is a likely candidate for induction of class II in autoimmune target tissue. This was confirmed for thyrocytes (discussed in Chapter 3) but not for other tissues, such as pancreatic β cell islet in which a synergy between TNF-α and IFN-γ was required (Pujol-Borrell *et al.*, 1987; see Chapter 16). It was pertinent, therefore, to assess the evidence for endogenous IFN-γ in RA tissue regulating HLA class II expression and to determine the effect of exogenous IFN-γ on RA synovial tissue HLA class II antigen expression *in vitro*.

Using a combination of molecular and biochemical approaches, we have investigated cytokine expression in RA synovial tissue (Table 17.1). In freshly isolated RA synovial tissue, a low level of mRNA for IFN-γ was observed; however, this increased as the cells were cultured *in vitro* (Buchan *et al.*, 1988a). In contrast, a negligible amount of immunoreactive protein was detected using sensitive immunoassays (Buchan *et al.*, 1988a; Brennan *et al.*, 1989a). The absence of other T cell cytokines in abundance, including lymphotoxin (LT) and IL-2 has also been described by others (Firestein *et al.*, 1988), which has led Firestein & Zvaifler (1990) to suggest that the pathogenic changes in RA are caused by the activity of macrophages and fibroblasts within the joint, the role of T cells being minimal. It is conceivable, however, that small amounts of IFN-γ (produced by activated T cells) are present but undetectable in the assays used. To investigate this further,

Table 17.1. *Summary of cytokines produced spontaneously by RA synovial cells*

| Cytokine | mRNA | Protein |
|---|---|---|
| IL-1α | Yes | Yes |
| IL-1β | Yes | Yes |
| TNF-α | Yes | Yes |
| LT | Yes | (+/–)[a] |
| IL-2 | Yes | (+/–)[a] |
| IL-3 | No | No |
| IL-4 | ? | No |
| IFN-γ | Yes | (+/–)[a] |
| IL-6 | Yes | Yes |
| GM-CSF | Yes | Yes |
| IL-8/NAP-1 | Yes | Yes |
| RANTES | Yes | ? |
| G-CSF | Yes | ? |
| M-CSF | No | Yes |
| TGF-β | Yes | Yes |
| EGF | Yes | Yes |
| TGF-α | No | No |
| PDGF-A | Yes | Yes |
| PDGF-B | Yes | Yes |
| IL-10 | Yes | Yes |

NAP-1, neutrophil-activating peptide; RANTES, regulated on activation, normal T expressed and secreted.
[a] Demonstration of protein has proved difficult.

we compared the HLA-DR expression in synovial mononuclear cells cultured for two days in the presence of a neutralizing antibody to the interferon-γ receptor with cells incubated in the presence of an isotype control antibody (Fig. 17.1). The results showed that blocking the IFN-γ receptor in one of the four RA synovial cultures resulted in a significant decrease in HLA-DR expression, as determined by flow cytometry, despite the fact that immunoreactive IFN-γ was not detectable in any of the culture supernatants. Interestingly, the proportion of T lymphocytes in this RA synovial mononuclear cell culture was higher than the others and, although secreted IFN-γ protein was not found, positive staining was found in T cells from cytospin secretions of this culture (F. Brennan, unpublished observations).

Addition of recombinant IFN-γ to such cultures also yielded variable results. Where the expression of HLA-DR was initially low (Fig. 17.2*a*), this could be further enhanced by addition of exogenous IFN-γ to the cultures. In the majority of synovial cultures (illustrated in Fig 17.2*b*), however, addition

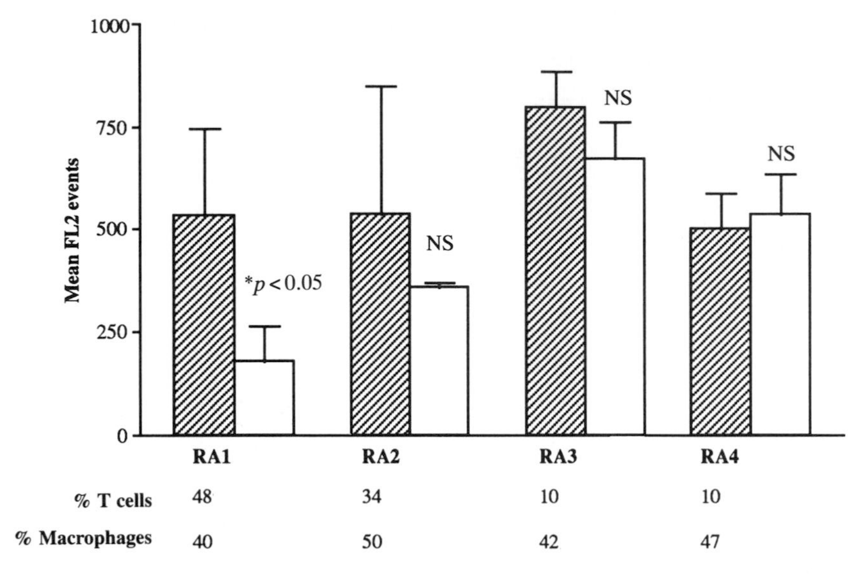


Fig. 17.1. Role of IFN-γ in maintaining HLA class II expression in RA mononuclear cells. RA synovial joint mononuclear cells were cultured for two days in RPMI 1640 and 10% foetal calf serum in the presence (clear bar) or absence (hatched bar) of a mAb against the IFN-γ receptor. HLA-DR expression was determined with a monoclonal anti-DR-PE (phycoerythrin) conjugate (Becton Dickinson) and analysed by flow cytometry. Results are expressed as mean FL2 events (triplicate cultures) ± SD. NS, not significant.

of IFN-γ had no effect, suggesting that class II antigen expression was already maximal. In conclusion, from these studies it is unlikely that IFN-γ plays a major role in the induction and/or maintenance of HLA class II antigens in RA synovium; however, it may be involved in synergy with other molecules (discussed in the next section).

The lack of evidence for IFN-γ as a prime candidate in inducing HLA class II expression in RA synovial joints is also borne out by immunohistological analysis in which the presence of IFN-γ has been difficult to demonstrate. Furthermore, the observation that a HLA class II-inducing factor found in RA joint samples is not neutralized by antibody to IFN-γ suggests that other cytokines are involved in this mechanism.

### *17.3.2 GM-CSF*

In contrast to the paucity of T cell-derived cytokines, there is an abundance of other cytokines, derived principally from activated macrophages in the joint (Table 17.1). These include proinflammatory cytokines, such as IL-1

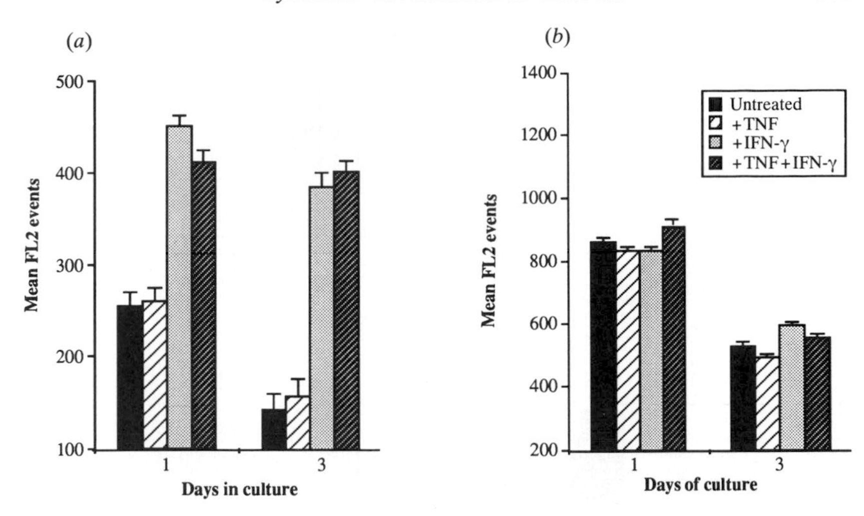


Fig. 17.2. Modulatory effect of IFN-γ and TNF-α on HLA class II antigen expression on RA synovial joint mononuclear cells.

RA mononuclear cells, RA5α (*a*) and RA6α (*b*), were cultured *in vitro* for two days in the presence of IFN-α (500 U/ml), TNF-α (500 U/ml) or a combination of both cytokines. HLA class II expression was determined after three and five days in culture using a monoclonal anti-DR-PE conjugate and analysed by flow cytometry. Results are expressed as mean FL2 events (triplicate cultures) ± SD.

and TNF-α (Buchan *et al.*, 1988b; Brennan *et al.*, 1989b), GM-CSF (Haworth *et al.*, 1991), IL-6 (Hirano *et al.*, 1988), IL-8 (Brennan *et al.*, 1990a) and TGF-β (Brennan *et al.*, 1990b). Of these cytokines, the haemopoietic factor GM-CSF was of particular interest since it induced HLA-DR expression on peripheral blood monocytes in the absence of IFN-γ (Fig. 17.3) and also synergized with IFN-γ to increase HLA class II expression further (Chantry *et al.*, 1990). The presence of GM-CSF in RA synovial cultures has been investigated principally by Firestein and colleagues. It was initially described in RA synovial tissue as an HLA class II-inducing factor that was blocked by antibodies to GM-CSF. Immunoreactive and biologically active GM-CSF was then demonstrated in both RA synovial cultures (Alvaro-Gracia, Zvaifler & Firestein, 1989) and RA synovial fluid (Xu *et al.*, 1989), and synovial macrophages were identified by *in situ* hybridization as the source of GM-CSF (Alvaro-Gracia *et al.*, 1991). These studies, therefore, indicate that GM-CSF is a likely candidate in induction of HLA class II expression in RA synovial tissue. However, GM-CSF does not induce HLA-DQ on human monocytes (Gerrard, Dyer & Mostowski, 1990) or cultured human umbilical vein endothelial cells (G. Howells, unpublished observations). Furthermore, although GM-CSF may be involved in HLA-DR expression on macrophage-

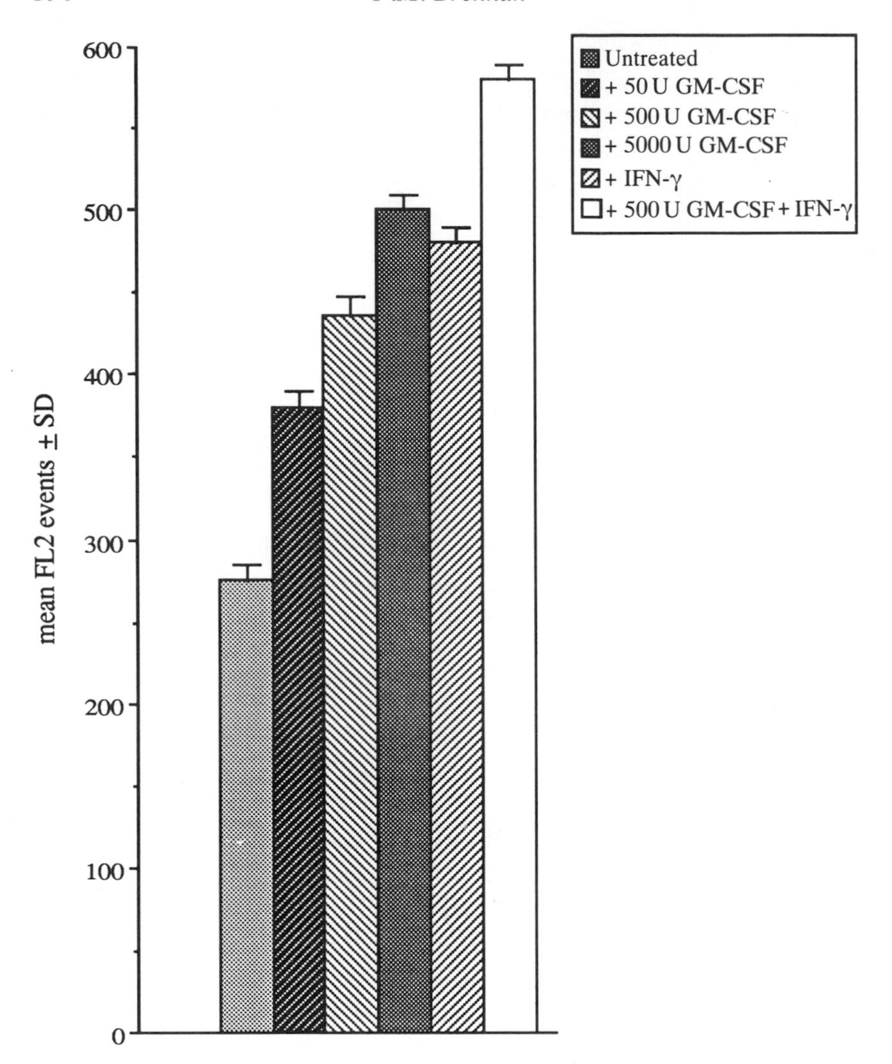


Fig. 17.3. Induction of surface HLA class II expression on human monocytes. Peripheral blood-enriched monocytes were cultured in the presence of increasing doses of GM-CSF (50–50 000 U/ml), IFN-γ (500 U/ml) or a mixture of both for three days. The cells were then harvested and surface HLA-DR expression determined by immunofluorescence and flow cytometry. Data are expressed as mean FL2 events ± SD after subtraction of background binding.

like 'type A' synoviocytes (Chantry *et al*., 1990), it is unlikely to be involved in HLA-DR expression on the 'type B' synoviocytes, or fibroblasts, since these cells do not respond to GM-CSF and presumably lack the α chain and/or the common β chain of the GM-CSF receptor. In contrast, IFN-γ has been shown to induce HLA-DR on many different cell types including fibroblasts (Rosa & Fellous 1984). Therefore, it remains to be determined which (novel) molecule(s) and/or mechanisms are involved in HLA class II expression on RA synovial fibroblasts and to what extent IFN-γ is involved in this process.

### *17.3.4 TNF-α*

In contrast to T cell-derived cytokines, TNF-α is another proinflammatory cytokine produced in large amounts by RA synovial cultures. TNF-α was identified as an important pivotal molecule in the pathogenesis of RA, based on both *in vitro* (Brennan *et al.*, 1989b; Brennan, Maini & Feldmann, 1992) studies and *in vivo* studies in animal models (Williams, Feldmann & Maini, 1992). These showed that removal of TNF-α in RA synovial cultures inhibits the production of the equally proinflammatory cytokine IL-1 (Brennan *et al.*, 1989b) and GM-CSF (Haworth *et al.*, 1991) and its removal in DBA/1 mice with collagen-induced arthritis reduced the severity of the disease. Although the proinflammatory effects of TNF-α are well documented, its role in induction of HLA class II antigens is less clear. TNF-α by itself does not induce HLA class II expression, but it does synergize with IFN-γ in this function (Portillo *et al.*, 1989). It was, therefore, of interest to establish the effect of TNF-α depletion on HLA class II expression in the RA cultures (Fig. 17.4). In this experiment, synovial membrane mononuclear cells were cultured in the presence of neutralizing antibodies to TNF-α, GM-CSF, IFN-γ (anti-IFN-γ receptor) or a combination of all three antibodies. The results showed that, as previously described, neutralization of IFN-γ had no effect, whereas neutralization of GM-CSF had some inhibitory effect on HLA class II expression and removal of TNF-α had an even greater effect. When TNF-α, GM-CSF and IFN-γ were all inhibited, a further reduction in HLA class II expression was observed. This modulation of HLA class II antigens by TNF-α in these RA synovial membrane cultures is unlikely to be direct and is either mediated by induction of GM-CSF or, possibly, by up-regulation of adhesion molecules such as ICAM-1 and/or vascular cell adhesion molecule (V-CAM) adhesion molecules, which are both expressed at increased levels on activated macrophages within the synovial membrane (Koch *et al.*, 1991).

### *17.3.5 IL-4*

Interleukin-4 is produced by activated T cells and, like other T cell-derived cytokines in the RA synovial joint, is not found at significant levels in RA

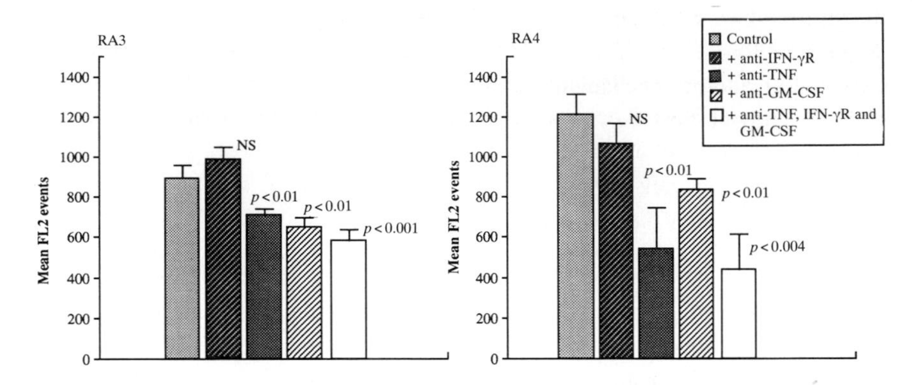


Fig. 17.4. Complexity of cytokine involvement in maintenance of class II expression on RA synovial joint mononuclear cells. Cells isolated from RA synovial joints (RA3, RA4) were cultured for five days in the presence of neutralizing antibodies against IFN-γ receptor, TNF-α, GM-CSF or with a combination of all three antibodies. Cells were harvested and HLA class II expression determined by immunofluorescence and flow cytometry. Data are expressed as mean FL2 events (triplicate cultures) ± SD after subtraction of background binding. NS, not significant.

synovium (Table 17.1). IL-4 displays many different biological properties, which can have both stimulatory and inhibitory effects on immune cells. Thus, IL-4 acts as a growth factor for T and B cells (Defrance *et al.*, 1987; Spits *et al.*, 1987), induces HLA class II expression on B cells and monocytes (Noeele *et al.*, 1984; Gerrard *et al.*, 1990) and induces IL-6 production by endothelial cells (Howells *et al.*, 1991). These effects could exacerbate inflammation and are, therefore, proinflammatory. Conversely, IL-4 also has anti-inflammatory potential, such as inhibition of LPS-induced IL-1, TNF-α, IL-6 and PGE2 production in human monocytes (Essner *et al.*, 1989; Hart *et al.*, 1989; te Velde *et al.*, 1989). Therefore, although IL-4 has the potential to induce HLA class II expression, it is very much dependent on the target cell and its activation status. Furthermore, many of the effects of IL-4, such as induction of class II antigens on B cells, are inhibited in the presence of IFN-γ (Mond *et al.*, 1985). Clearly the overall effect on HLA class II expression of a multifunctional cytokine such as IL-4 will depend on the target cell, stage of differentiation and activation status, as well as the influence of other cytokines in the environment. Interestingly, based on its anti-inflammatory effects on cytokine production by RA synovial cells, the use of IL-4 as a therapeutic agent in RA has been suggested. Clearly this would have to be viewed in the light of its T and B cell activation properties, in addition to the effect on HLA class II expression.

### *17.3.6 IL-13*

IL-13, or P600 as it was formerly known, is a cytokine that has been recently cloned from a human T cell library and exhibits a number of properties that overlap with those of IL-4 (McKenzie *et al.*, 1993; Minty *et al.*, 1993). On monocytes and B cells it will up-regulate HLA class II and CD23 antigen expression, whereas on T cells (unlike IL-4) it has no biological action. On monocytes, it inhibits the production of proinflammatory cytokines, such as TNF and IL-1, induced by lipopolysaccharides (LPS) in a manner analogous to IL-4. The presence of IL-13 in RA synovial tissue and the possible modulatory effects on HLA class II expression remain to be determined.

### *17.3.7 IL-10*

IL-10, or CSIF, cytokine synthesis-inhibitory factor, as it was formerly described in the mouse (Fiorentino, Bond & Mosmann, 1989), was cloned from a cDNA library constructed from human $CD4^+$ T cells (Vieira *et al.*, 1991). It exists as a native, biologically active homodimer of molecular mass 39 kDa. The properties and effects of IL-10 on human cells have been recently reviewed (de Waal Malefyt *et al.*, 1992). IL-10 is produced by a variety of cells, including T cells, B cells and monocytes. On peripheral blood mononuclear cells, IL-10 inhibits IFN-γ, GM-CSF, LT, TNF-α and IL-3 production and antigen-driven T cell proliferation. IL-10 also strongly down-regulates HLA-DR, HLA-DP and HLA-DQ expression and abolishes the up-regulation of HLA class II expression on monocytes induced by IFN-γ and IL-4. The down-regulatory effects of IL-10 on HLA class II expression may be responsible for the inhibition of antigen presentation by human monocytes. Clearly the immunoregulatory properties of IL-10 are very potent. The presence and immunomodulatory effect of IL-10 in RA synovial tissue has recently been investigated. IL-10 has been detected in both RA and osteoarthritis (OA) synovium (F. Brennan, unpublished data) and the further addition of exogenous IL-10 to RA synovial cell cultures inhibited TNF-α production and HLA class II expression. Clearly the immunomodulatory properties of IL-10 in chronic inflammatory disease will receive much attention in the near future.

### *17.3.8 TGF-β*

The TGF-βs were originally described as cytokines that were capable of conferring a transformed phenotype on a number of non-transformed fibroblast lines but had diverse effects on growth and development (Massague, 1987). Three isoforms of TGF-β have been described in mammalian cells, with TGF-$\beta_1$ being the most well studied. It is produced by cells of the immune system, including macrophages (Assoian *et al.*, 1987), B cells (Kehrl *et al.*, 1986) and T cells (Alvarez-Mon *et al.*, 1986) and inhibits a number of immune functions including T and B cell proliferation (Shalaby & Ammann, 1988), immunoglobulin production, NK cell generation (Rook *et al.*, 1986) and cytokine production by peripheral blood mononuclear cells (Espevik *et al.*, 1987; Chantry *et al.*, 1989).

The regulatory effects of TGF-β on cytokine production have been studied and these show that TGF-β, like IL-4 and IL-10, inhibits TNF-α and IL-1 production in LPS-stimulated monocytes (Chantry *et al.*, 1989). Furthermore, TGF-β can inhibit both constitutive and IFN-γ-induced HLA class II expression on human monocytes (Czarniecki *et al.*, 1988). Since increased production of proinflammatory cytokines such as IL-1 and TGF-α are characteristic of RA synovial tissue and HLA class II expression is over-expressed on a range of cell types in the synovium, we investigated both whether TGF-β was present in RA synovium and what immunomodulatory effects were observed after the addition of exogenous TGF-β to RA synovial cultures.

TGF-β is initially released as a latent, inactive molecule that is subsequently activated by proteolytic cleavage. However, abundant active TGF-β was found in synovial fluid from patients with both RA and non-RA inflammatory diseases (Brennan *et al.*, 1990b). Furthermore, synovial mononuclear cells isolated from the RA joint produced both TGF-$\beta_1$ and TGF-$\beta_2$ mRNA and protein in culture. Therefore, despite the abundance of this immunomodulatory cytokine, excessive cytokine production and HLA class II expression were still observed. The production of TNF-α and the expression of HLA class II antigens was then examined in RA mononuclear cells that were cultured for two days with ng/ml TGF-β1 or TGF-β2. Unlike in peripheral blood monocyte cultures, neither cytokine production (Fig. 17.5) nor HLA class II expression (Fig. 17.6) was inhibited upon addition of TGF-β. A possible explanation may lie in the observation that activated monocytes are refractory to the inhibitory effects of TGF-β (Chantry *et al.*, 1989) and that in order for inhibition of either cytokine production or HLA class II expression to occur the inhibitory cytokine needs to be added to the cells before they are activated. Clearly cells isolated from RA synovium are already in an activated state and are refractory to the inhibitory effects of TGF-β. Therefore, although HLA class II expression on RA synovium is not down-regulated by

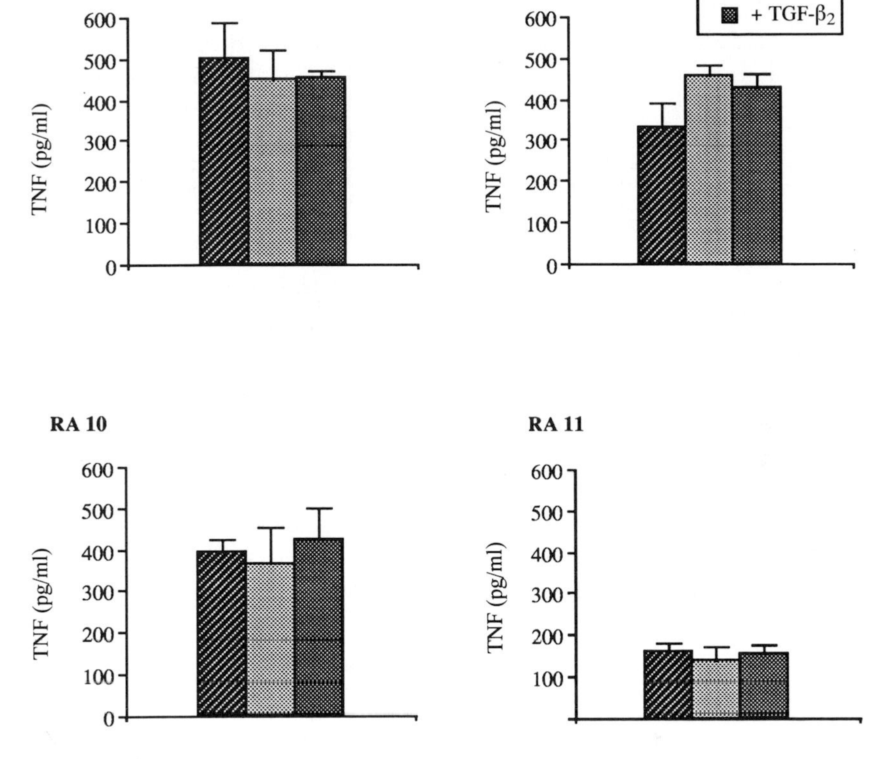


Fig. 17.5. TNF-α production by RA synovial joint mononuclear cells in culture: effect of TGF-β. Cells in four cultures (RA8–RA11) were grown *in vitro* for two days in RPMI 1640/10% FCS with and without 10 ng/ml TGF-$\beta_1$/$\beta_2$. Supernatants were harvested and assayed for TNF-α production by ELISA.

TGF-β, it does not exclude the possibility that TGF-β has some immunoregulatory role *in vivo*.

## 17.4 Conclusions

On the basis of immunohistological techniques and from *in vitro* studies on RA synovial membrane mononuclear cells, it is clear that augmented HLA class II expression on 'resident' and 'infiltrating' cells in the synovial joint is a feature of RA. As HLA class II expression is a striking feature of virtually all autoimmune diseases in humans and HLA class II antigen expression is

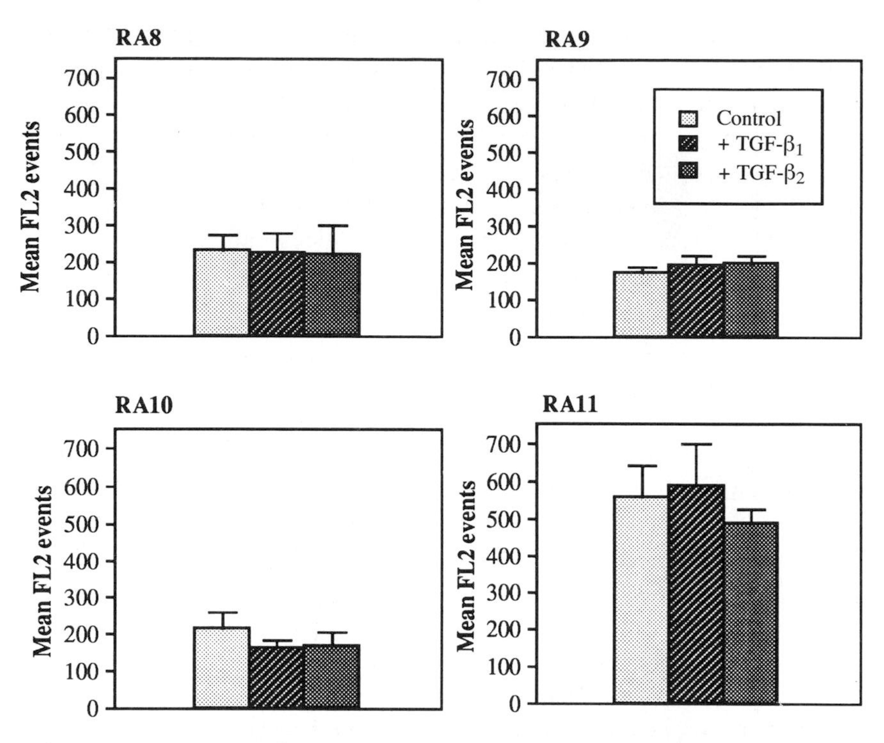


Fig. 17.6. Effect of TGF-β on HLA class II expression on RA synovial joint mononuclear cells in culture (RA8–RA11, as in Fig. 17.5). Cells were cultured *in vitro* for two days in RPMI 1640/10% FCS with and without 10 ng/ml TGF-$\beta_1/\beta_2$. Cells were harvested and HLA class II expression determined by immunofluorescence and flow cytometry. Results are expressed as mean FL2 events (triplicate cultures) ± SD after subtraction of background binding.

essential for antigen presentation to T cells, it has been postulated that these two features (abnormal class II expression and antigen presentation to autoreactive T cells) have a major role in the pathogenesis of human autoimmune diseases such as RA. This is supported by the observation that T cells recognizing an autoantigen, collagen type II, were cloned from a synovial membrane of a patient with severe RA (Londei *et al.*, 1989). However, although collagen type II may be one of the autoantigens involved, it is not known how many RA patients have collagen type II-specific T cells. Furthermore, as many different cells in the joint express abundant HLA class II antigens, it is not known which antigen-presenting cell is the most important *in vivo*.

Since the HLA class II antigens in the RA synovial joint are expressed chronically, the identification of the factors that initiate and/or maintain expression has been investigated. These studies have indicated that cytokines such as GM-CSF, TNF-α and to some extent IFN-γ may be involved. However, it is also clear that these cytokines, either alone or in combination, do not account for the chronic expression of all HLA-DR, HLA-DP and HLP-DQ antigens on the different cell types within the joint. Furthermore, other cytokines such as TGF-β and IL-10, which are present in the RA synovial joint, are known to inhibit HLA class II expression. Therefore, the mechanism and molecules involved that maintain chronic class II expression in autoimmune disease remain to be established. One possibility is that the mediators involved are cell membrane molecules that have cytokine-like effects, such as the ligands of the NGF/TNF receptor superfamily. These ligands include CD40-L, CD30-L, CD27-L and have now been cloned and expressed. They are membrane-associated molecules that exert cytokine-like effects upon engagement of their respective receptors, such as growth stimulation and activation. What is not clear at this stage is whether such ligands modulate the expression of cell-associated antigens such as HLA class II antigens. If these, or other cell membrane-associated ligands, do modulate HLA class II expression, this would enable modulation of class II to occur directly when cells interact in the absence of secreted cytokines. Therefore, in sites of inflammation such as RA synovium where cells express elevated levels of adhesion molecules and adhere strongly to each other and/or the extracellular matrix, the cell–cell interactions themselves and the concomitant engagement of membrane ligands with their receptors may be sufficient to maintain the chronic features of disease, including HLA class II expression and cytokine production. The challenge for the future will be to establish which mechanisms are involved in local class II expression and antigen presentation in the synovial joint. Once the mechanism and cell types involved in class II expression in RA are identified, this would allow the emergence of new therapies directed at modulating class II expression more selectively and, more particularly, enable the down-regulation of HLA class II expression on cells within the synovial joint.

## References

Alvarez-Mon, M., Derynck, R., Sporn, M.B. & Fauci, A.S. (1986). Production of transforming growth factor β by human T lymphocytes and its potential role in the regulation of T cell growth. *Journal of Experimental Medicine*, 163, 1037–1050.

Alvaro-Gracia, J.M., Zvaifler, N.J., Brown, C.B., Kaushansky, L. & Firestein, G.S. (1991). Cytokines in chronic inflammatory arthritis. VI. Analysis of the synovial cells involved in granulocyte-macrophage colony stimulating factor production and gene expression in rheumatoid arthritis and its regulation by IL-1 and TNF-α. *Journal of Immunology*, 146, 3365–3371.

Alvaro-Gracia, J.M., Zvaifler, N.J. & Firestein, G.S. (1989). Cytokines in chronic inflammatory arthritis IV. Granulocyte/macrophage colony stimulating factor-mediated induction of class II MHC antigen on human monocytes: a possible role in rheumatoid arthritis. *Journal of Experimental Medicine*, 170, 865–875.

Assoian, R.K., Leurdelys, B.E., Stevenson, H.C., Miller, P.J., Madtes, D.K., Raines, E.W., Ross, R. & Sporn, M.J. (1987). Expression and secretion of type β transforming growth factor by activated human macrophages. *Proceedings of the National Academy of Sciences of the USA*, 84, 6020–6024.

Barkley, D., Allard, S., Feldmann, M. & Maini, R.N. (1989). Increased expression of HLA-DQ antigens by interstitial cells and endothelium in the synovial membrane of rheumatoid arthritis patients compared with reactive arthritis patients. *Arthritis and Rheumatism*, 32, 955–963.

Brennan, F.M., Chantry, D., Jackson, A.M., Maini, R.N. & Feldmann, M. (1989a). Cytokine production in culture by cells isolated from the synovial membrane. *Journal of Autoimmunity*, 2 (Suppl.), 177–186.

Brennan, F.M., Chantry, D., Jackson, A., Maini, R.N. & Feldmann, M. (1989b). Inhibitory effect of TNF-α antibodies on synovial cell interleukin-1 production in rheumatoid arthritis. *Lancet*, ii, 244–247.

Brennan, F.M., Chantry, D., Turner, M., Foxwell, B., Maini, R. & Feldmann, M. (1990b). Detection of transforming growth factor-beta in rheumatoid arthritis synovial tissue: lack of effect on spontaneous cytokine production in joint cell cultures. *Clinical and Experimental Immunology*, 81, 278–285.

Brennan, F.M., Maini, R.N. & Feldmann, M. (1992). TNF-α – a pivotal role in rheumatoid arthritis. *British Journal of Rheumatology: Reviews in Clinical Science*, 31, 293–298.

Brennan, F.M., Zachariae, C.O.C., Chantry, D., Larsen, C.G., Turner, M., Maini, R.N., Matsushima, K. & Feldmann, M. (1990a). Detection of interleukin-8 biological activity in synovial fluids from patients with rheumatoid arthritis and production of IL-8 mRNA by isolated synovial cells. *European Journal of Immunology*, 20, 2141–2144.

Buchan, G.S., Barrett, K., Fujita, T., Taniguchi, T., Maini, R.N. & Feldmann, M. (1988a). Detection of activated T cell products in the rheumatoid joint using cDNA probes to interleukin 2, IL-2 receptor and interferon γ. *Clinical and Experimental Immunology*, 71, 295–301.

Buchan, G., Barrett, K., Turner, M., Chantry, D., Maini, R.N. & Feldmann, M. (1988b). Interleukin-1 and tumour necrosis factor mRNA expression in rheumatoid arthritis: prolonged production of IL-1α. *Clinical and Experimental Immunology*, 73, 449–455.

Burmester, G.R., Dimitriu-Bona, A., Waters, S.J. & Winchester, R.J. (1982). Identification of three major synovial lining cell populations by monoclonal antibodies directed to Ia antigens and antigens associated with monocytes/macrophages and fibroblasts. *Scandinavian Journal of Immunology*, 17, 69–70.

Burmester, G.R., Jahn, B., Rohwerr, P., Zacher, J., Winchester, R.J. & Kalden, J.R. (1987). Differential expression of Ia antigens by rheumatoid synovial lining cells. *Journal of Clinical Investigation*, 80, 595–604.

Chantry, D., Turner, M., Abney, E.R. & Feldmann, M. (1989). Modulation of cytokine production by transforming growth factor β. *Journal of Immunology*, 142, 4295–4300.

Chantry, D., Turner, M., Brennan, F.M., Kingsbury, A. & Feldmann, M. (1990). Granulocyte-macrophage colony stimulating factor induces both HLA-DR expression and cytokine production by human monocytes. *Cytokine*, 2, 60–67.

Czarniecki, C.W., Chiu, H.H., Wong, G.W., McCabe, S.M. & Palladino, M.A. (1988). Transforming growth factor-β1 modulates the expression of class II histocompatibility antigens on human cells. *Journal of Immunology*, 140, 4217–4223.

de Waal Malefyt, R., Yssel, H., Roncarolo, M.-G., Spits, H., & de Vries, J.E. (1992). Interleukin-10. *Current Opinion in Immunology*, 4, 314–320.

Defrance, T., Vanbervliet, B., Aubry, J.P., Takebe, Y., Arai, N., Miyajima, A., Yokota, T., Lee, F., Arai, K.I., de Vries, J.E. & Banchereau, J. (1987). B cell growth-promoting activity of recombinant human IL-4. *Journal of Immunology*, 139, 1135–1141.

Duke, O., Panayi, G.S., Janossy, G. & Poulter, L.W. (1982). An immunohistological analysis of lymphocyte subpopulations and their microenvironment in the synovial membranes of patients with rheumatoid arthritis using monoclonal antibodies. *Clinical and Experimental Immunology*, 49, 22–30.

Espevik, T., Figari, I.S. Shalaby, M.R., Lackdoes, G.A., Lewis, G.D., Shepard, H.M. & Palladino, M.A. Jr (1987). Inhibition of cytokine production by cyclosporin A and transforming growth factor beta. *Journal of Experimental Medicine*, 166, 571–576.

Essner, R., Rhoades, K., McBridge, W.H., Morton, D.L. & Economou, J.S. (1989). IL-4 downregulates IL-1 and TNF gene expression in human monocytes. *Journal of Immunology*, 142, 3857–3861.

Fiorentino, D.F., Bond, M.W. & Mosmann, T.F. (1989). Two types of mouse helper T cells IV. Th2 clones secrete a factor that inhibits cytokine production by Th1 clones. *Journal of Experimental Medicine*, 170, 2081–2095.

Firestein, G.S., Xu, W.D., Townsend, K., Broide, D., Alvaro-Gracia, J., Glasebrook, A. & Zvaivler, N.J. (1988). Cytokines in chronic inflammatory arthritis. I. Failure to detect T cell lymphokines (IL-2 and IL-3) and presence of CSF-1 and a novel mast cell growth factor in rheumatoid arthritis. *Journal of Experimental Medicine*, 168, 1578–1586.

Firestein, G.S. & Zvaifler, N.J. (1990). How important are T cells in chronic rheumatoid synovitis? *Arthritis and Rheumatism*, 33, 768–773.

Førre, Ø., Dobloug, J.H. & Natvig, J.B. (1982). Augmented numbers of HLA-DR positive T lymphocytes in the synovial tissue of patients with rheumatoid arthritis. *In vivo*-activated T lymphocytes are potent stimulators in the mixed lymphocyte reaction. *Scandinavian Journal of Immunology*, 15, 227–233.

Gerrard, T.L., Dyer, D.R. & Mostowski, H.S. (1990). IL-4 and granulocyte macrophage colony stimulating factor selectively increase HLA-DR and HLA-DP antigens but not HLA-DQ antigens on human monocytes. *Journal of Immunology*, 144, 4670–4674.

Hart, P.H., Vitti, G.F., Burgess, D.R., Whitty, G.A., Piccoli, D.S. & Hamilton, J.H. (1989). Potential anti-inflammatory effects of interleukin-4: suppression of human monocyte tumour necrosis factor α, interleukin-1 and prostaglandin $E_2$. *Proceedings of the National Academy of Sciences of the USA*, 86, 3803–3807.

Haworth, C., Brennan, F.M., Chantry, D., Turner, M., Maini, R.N. & Feldmann, M. (1991). Expression of granulocyte-macrophage colony stimulating factor (GM-CSF) in rheumatoid arthritis: regulation by tumour necrosis factor α. *European Journal of Immunology*, 21, 2575–2579.

Hirano, T., Matsuda, T., Turner, M., Miyasaka, N., Buchan, G., Tang, B., Sato, K., Shimizu, M., Maini, R., Feldmann, M. & Kishimoto, T. (1988). Excessive production of interleukin 6/B cell stimulatory factor-2 in rheumatoid arthritis. *European Journal of Immunology*, 18, 1797–1801.

Howells, G., Pham, P., Taylor, D., Foxwell, B. & Feldmann, M. (1991).

Interleukin-4 induces IL-6 production by human endothelial cells: synergy with interferon-γ. *European Journal of Immunology*, 21, 97–101.

Janossy, G., Panayai, G., Duke, O., Bofill, M., Poulter, L.W. & Goldstein, G. (1981). Rheumatoid arthritis: a disease of T-lymphocyte/macrophage immunoregulation. *Lancet*, ii, 839–841.

Kehrl, J.H., Roberts, A.B., Wakefield, S.J., Jakowlew, S., Sporn, M.B. & Fauci, A.S. (1986). Transforming growth factor β is an important immunomodulatory protein for human B lymphocytes. *Journal of Immunology*, 137, 3855–3860.

Kissonerghis, M., Maini, R.N. & Feldmann, M. (1988). High rate of HLA class II mRNA synthesis in rheumatoid arthritis joints and its persistence in culture: down regulation by recombinant IL-2. *Scandinavian Journal of Immunology*, 29, 73–82.

Klareskog, L., Forsum, U., Scheynius, A., Kabelitz, D. & Wigzell, H. (1982). Evidence in support of a self perpetuating HLA-DR dependent delayed type cell reaction in rheumatoid arthritis. *Proceedings of the National Academy of Sciences of the USA*, 79, 3632–3636.

Klareskog, L., Johnell, O. & Hulth, A. (1984). Expression of HLA-DR and HLA-DQ antigens on cells within the cartilage–pannus junction in rheumatoid arthritis. *Rheumatology International*, 4, 11–15.

Koch, A.E., Burrows, J.C., Haines, G.K., Carlos, T.M., Harlan, J.M. & Leibovich, S.J. (1991). Immunolocalisation of endothelial and leukocyte adhesion molecules in human rheumatoid and osteoarthritic synovial tissues. *Laboratory Investigation*, 64, 313–320.

Korman, A.J., Boss, J.M., Spies, T., Sorrentino, R., Okado, K. & Strominger, J.L. (1985). Genetic complexity and expression of human class II histocompatibility antigens. *Immunological Reviews*, 85, 45–86.

Lindblad, S. & Hedfors, E. (1985). Intra-articular variation in synovitis: local macroscopic signs of inflammatory activity are significantly correlated. *Arthritis and Rheumatism*, 28, 977–986.

Lindblad, S., Klareskog, L., Hedfors, E., Forsum, U. & Sundstrom, C. (1983) Phenotypic characterisation of synovial tissue cells *in situ* in different types of synovitis. *Arthritis and Rheumatism*, 26, 1321–1332.

Londei, M., Savill, C., Verhoef, A., Brennan, F., Leech, Z.A., Duance, V., Maini, R.N. & Feldmann, M. (1989). Persistence of collagen type II specific T cell clones in the synovial membrane of a patient with RA. *Proceedings of the National Academy of Sciences of the USA*, 86, 636–640.

Massague, J. (1987). The TGF-β1 family of growth and differentiation factors. *Cell*, 49, 437–438.

McKenzie, A.N.J., Culpepper, J.A., de Waal Malefyt, R., Briere, F., Punnonen, J., Aversa, G., Sato, A., Dang, W., Cocks, B.G., Menon, S., de Vries, J.E., Banchereau, J. & Zurawski, G. (1993). Interleukin 13, a T cell derived cytokine that regulates human monocyte and B-cell function. *Proceedings of the National Academy of Sciences of the USA*, 90, 3735–3739.

Minty, A., Chalon, P., Derocq, J.M., Dumont, X., Guillemot, J.C., Kaghad, M., Labit, C., Leplatoid, P., Kiauzun, P., Miloux, B., Minty, C., Casellas, P., Loison, G., Lupker, J., Shire, D., Ferrara, P. & Caput, D. (1993). Interleukin-13 is a new human lymphokine regulating inflammatory and immune responses. *Nature (London)*, 362, 248–250.

Mond, J.J., Finkelmann, F.D., Sharma, C., Ohara, J. & Serrate, S. (1985). Recombinant interferon gamma inhibits the B cell proliferative response stimulated by soluble but not a sepharose bound anti-immunoglobulin antibody. *Journal of Immunology*, 135, 2513–2517.

Noeele, R., Krammer, P.H., Ohara, J., Uhr, J.W. & Vitetta, E.S. (1984). Increased expression of Ia antigens on resting B cells: an additional role for B cell growth factor. *Proceedings of the National Academy of Sciences of the USA*, 81, 6149–6153.

Palmer, D.G., Selvendran, Y., Allen, C., Revell, P.A. & Hogg, N. (1985). Features of synovial membrane identified with monoclonal antibodies. *Clinical and Experimental Immunology*, 59, 529–538.

Portillo, G., Turner, M., Chantry, D. & Feldmann, M. (1989). Effect of cytokines on HLA-DR and IL-1 production by a monocytic tumour, THP-1. *Immunology*, 66, 170–175.

Poulter, L.W., Duke, O., Hobbs, S., Janossy, G. & Panayi, G. (1981). Histochemical discrimination of HLA-DR positive cell populations in the normal and arthritic synovial lining. *Clinical and Experimental Immunology*, 48, 381–388.

Pujol-Borrell, R., Todd, I., Doshi, M., Bottazzo, G.F., Sutton R., Gray, D., Adolf, G.R. & Feldmann, M. (1987). HLA class II induction in human islet cells by interferon-γ plus tumour necrosis factor or lymphotoxin. *Nature (London)*, 326, 304–306.

Rigby, A.S., Silman, A.J., Voelm, L., Gregory, J.C., Ollier, W.E.R., Kahan, M.A., Nepom, G.T. & Thomson, G. (1991). Investigating the HLA component in rheumatoid arthritis. An additive (dominant) mode of inheritance is rejected, a recessive model is preferred. *Genetic Epidemiology*, 8, 153–175.

Rook, A.H., Kehrl, J.H., Wakefield, L.M., Roberts, A.B., Sporn, M.B., Burlington, D.B., Lane, H.C. & Fauci, A.S. (1986). Effects of TGF-β on the functions of NK cells: depressed cytolytic activity and blunting of IFN-γ responsiveness. *Journal of Immunology*, 136, 3916–3920.

Rosa, F. & Fellous, M. (1984). The effect of gamma-interferon on MHC antigens. *Immunology Today*, 5, 261–262.

Shalaby, M.R. & Ammann, A.J. (1988). Suppression of immune cell function *in vitro* by recombinant human TGF-β. *Cellular Immunology*, 112, 343–350.

Silman, A.J. (1991). Is rheumatoid arthritis an infectious disease? *British Medical Journal*, 303, 200–201.

Spits, H., Yssel, H., Takebe, Y., Arai, N., Yokota, T., Lee, F., Arai, K.I., Blanchereau, J. & de Vries, J.E. (1987). Recombinant interleukin-4 promotes growth of human T cells. *Journal of Immunology*, 139, 1142–1147.

te Velde, A.A., Huijbens, R.J.F., Heije, K., de Vries, J.E. & Figdor, C.G. (1989). Interleukin-4 (IL-4) inhibits secretion of IL-1β, tumour necrosis factor-α and IL-6 by human monocytes. *Blood*, 76, 1392–1397.

Vieira, P., de Wall Malefyt, R., Dang, M.N., Johnson, K.E., Kastelein, R., Fiorentino, D.F., de Vries, J.E., Roncarolo, M.G., Mossmann, T.R. & Moore, K.W. (1991). Isolation and expression of human cytokine synthesis inhibitory factor (CSIF/IL-10) cDNA clones: homology to Epstein–Barr virus open reading frame BCRFI. *Proceedings of the National Academy of Sciences of the USA*, 88, 1172–1176.

Williams, R.O., Feldmann, M. & Maini, R.N. (1992). Anti-TNF ameliorates joint disease in murine collagen-induced arthritis. *Proceedings of the National Academy of Sciences of the USA*, 89, 9784–9788.

Winchester, R.J. & Burmester, G.R. (1981). Demonstration of Ia antigens on certain dendritic cells and on a novel elongate cell found in human synovial tissue. *Scandinavian Journal of Immunology*, 14, 439–444.

Wordsworth, B.P. & Bell, J.L. (1991). Polygenic susceptibility in rheumatoid arthritis. *Annals of Rheumatic Disease*, 50, 343–346.

Xu, W.D., Firestein, G.S., Taetle, R., Kaushansky, K. & Zvaifler, N.J. (1989). Cytokines in chronic inflammatory arthritis. II. Granulocyte-macrophage colony stimulating factor in rheumatoid synovial effusions. *Journal of Clinical Investigation*, 83, 876–882.

# 18

# Expression of an MHC antigen in the central nervous system: an animal model for demyelinating diseases

LIONEL FEIGENBAUM, TADURU SREENATH
and GILBERT JAY
*Jerome H. Holland Laboratory, Rockville*

## 18.1 Introduction

Demyelinating disorders of the CNS, while bound together by a common process involving damage to or improper laying-down of the myelin sheath, do not share a common aetiology. This is demonstrated by recent advances, which have enabled the classification of these conditions into three groups based on disease aetiology: genetic, viral and autoimmune.

Pelizaeus–Merzbacher disease exemplifies the genetic component of certain demyelination conditions. This X-linked recessive disorder affecting CNS myelination in children is attributed to deficient biosynthesis of the proteolipid protein (PLP) (Koeppen *et al.*, 1987). Lack of PLP, one of the most abundant components of the myelin sheath, contributes to physical symptoms such as spastic extremities and lack of purposeful movement and head control.

The viral aetiology of demyelinating diseases is seen in progressive multifocal leukoencephalopathy (PML). Originally described as a complication of leukaemia and Hodgkin's disease (Aström, Mancall & Richardson, 1958), PML develops mostly in immunocompromised patients with chronic diseases and, more recently, in individuals with AIDS (Levy, Bredsen & Rosenblum, 1985; Niedt & Schinelle, 1985). Symptoms in PML reflect the widespread destruction of the CNS white matter. Neurological deficits include dementia, confusion, aphasia, hemiparesis and ataxia. The viral aetiology of PML has been demonstrated by isolation of the papovavirus JC from the brains of diseased patients (Padgett *et al.*, 1971) and by the demonstration that transgenic mice with part of the JC viral genome develop a neurological disorder that involves dysmyelination of the CNS (Small *et al.*, 1986; Feigenbaum, Hinrichs & Jay, 1992).

The autoimmune aetiology of post-vaccinal encephalomyelitis is shown by a perivenular inflammatory reaction and widespread demyelination in the white matter of compromised individuals. This condition results from rabies vaccine injection following culture in CNS tissues.

By far the most common of demyelinating disorders is multiple sclerosis (MS). Although its clinical course is markedly variable, its most common form is recurrent attacks with frequent pathological involvement of the brain, spinal cord and optic nerve (Antel & Arnason, 1983). Macroscopically, MS lesions are represented by white matter plaques. Microscopically, these plaques show areas of demyelination containing intact axons, unless the lesions are severe. The number of oligodendrocytes is normal in early stages of disease but decreases as involvement progresses.

Although MS shares a common process with other demyelinating disorders, it is a much more complicated condition. It has no specific aetiology. Instead, it is suggested that its cause lies in an interplay between genetic, viral and autoimmune aetiologies. Any effort to develop an animal model for MS must take into account all three factors.

## 18.2 Genetic susceptibility to MS

Epidemiological evidence for genetic susceptibility to MS is partly derived from family studies. The risk of acquiring MS is 5 to 25 times higher among first degree relatives; even second- and third-degree relatives are at a higher risk compared with the general population (Sadovnick & McLeod, 1981).

Twin studies constitute a classical way to determine a genetic link to disease. The strongest evidence to date emanates from studies looking at MS concordance in monozygotic (MZ) and dizygotic (DZ) twins (Currier & Eldridge, 1982; Ebers *et al.*, 1986). A concordance of 26% for MZ twins and 2.3% for DZ twins was observed. A significantly higher concordance in MZ twins strongly suggests a genetic component to the disease. However, since only one fourth of the MZ twins are concordant, it is more likely that the genetic component is polygenic or that environmental factors may also play a role in disease susceptibility.

The most studied genetic loci in human MS are genes in the MHC. Early studies have demonstrated a strong association of MHC in Northern American and Northern European MS populations with class I HLA-A3 and HLA-B7 and class II HLA-DR2 and HLA-Dw2 (Bertram & Kuwert, 1982). However, in other ethnic groups, no association with these specific MHC class I and class II antigens has been found. If an association is present, it is with different HLA alleles; in Japan the most common HLA is DR6 (Naito *et al.*, 1978), and in Italy and Jordan it is DR4 (Kurdi *et al.*, 1977; Marrosu *et al.*, 1988).

Further evidence for involvement of MHC class I and class II in susceptibility to MS is seen in the association of specific HLA alleles with differing clinical course of the disease; HLA-A1, HLA-B8 and HLA-DR3 are typically associated with progressive MS, whereas HLA-A3, HLA-B7 and HLA-DR2 are more common in the relapsing form of the disease (Hammond *et al.*, 1988). The polygenic nature of susceptibility to MS can partly be explained by the involvement of more than one locus in the MHC. Alternatively, TCR recognition of peptides presented by antigen-presenting cells could be the basis for a second level of genetic susceptibility. The TCR is composed of an α and a β chain encoded by two gene clusters. Studies determining β-chain haplotype in sibling pairs concordant for relapsing remitting MS demonstrated a higher proportion of shared haplotypes in involved individuals (Seboun *et al.*, 1989). Identical studies on TCR α-chain alleles have also documented increased risk groups with certain haplotypes (Oskenberg *et al.*, 1989).

## 18.3 Viral involvement in MS

Evidence that environmental factors are contributing to the development of MS is found in epidemiological and biological studies. Careful examination of the distribution pattern of populations with MS suggests an increase in the prevalence of the disorder with an increase in the distance of the geographic location from the equator (Gorelick, 1989). Although these data have implications for genetic susceptibility, they also indicate environmental factors, such as viral infections, in the development of disease. Individuals migrating before puberty from a high-risk to a low-risk zone, or conversely, acquire the risk of developing disease of the host country (Alter, Leibowitz & Speer, 1962; Kurtzke, Dean & Botha, 1970). For example, the prevalence of MS in first generation immigrants to the UK from the Indian subcontinent and the West Indies is similar to UK natives (Elian & Dean, 1990). A second line of evidence is found in twin studies, which suggest a substantial environmental effect is required for the early onset of MS (Ebers *et al.*, 1986). The early onset of disease among MZ twins is associated with more birth anoxia, unusual infantile and childhood infections and major operations compared with unaffected twins. That the environment plays a role in the onset of MS is further supported by observations in the Faeroe Islands, where no case of MS had been reported before the occupation by British troops during World War II (Kurtzke & Hyllested, 1979; Fischman, 1981). Later, about 24 patients with MS were reported between 1943 and 1960. The incidence of MS dropped back to zero after 1975. Analysis of this study suggests that the disease epidemic was related to the widespread infections transmitted to native inhabitants between the age of 13 and 26 years.

The contribution of viral infections to MS pathology comes from studies of established MS populations with infectious diseases such as scarlet fever, rheumatic chorea, pneumonia and encephalitis, which all may damage the nervous system (Currier & Eldridge, 1982; Sibley, Bamford & Clark, 1983). In recent years, several viruses, including retrovirus, coronavirus, orthomyxovirus, paramyxovirus and poxvirus, have been implicated in MS (Burks *et al.*, 1980; Johnson *et al.*, 1984; Koprowski *et al.*, 1985; Hemachudha *et al.*, 1987). The viral aetiology of MS is more specifically suggested in a study of a cluster-focus of the disease in Henribourg, Saskatchewan, where 8 of 283 children having contracted measles over a 15-year period developed MS later in life (Hader, Irvine & Schiefer, 1990).

It is not clear from any of these studies how injury to the CNS triggers MS. Two mechanisms have been postulated. Direct viral infection of oligodendrocytes may cause damage to the cell and the resulting myelin breakdown may induce a secondary immune response. Alternatively, it may induce a primary autoimmune reaction against certain components of the oligodendrocyte or the myelin, leading to secondary demyelination (Waksman, 1983).

## 18.4 Autoimmune aspects of MS

MS has been associated with various immune mechanisms. It is not clear whether abnormalities in the cerebrospinal fluid, such as increased IgG, IgA, IgM, myelin basic protein (MBP), as well as proteolytic fragments of MBP (McFarlin & McFarland, 1982), are either causative or indicative of the inflammatory response in the CNS. The occurrence of inflammation in the CNS with relatively selective attacks on the myelin sheath suggests that demyelination induced by specific myelin epitopes may also cause sensitization to other myelin-associated antigens. It has been postulated that aberrant regulation of autoantigen-specific T cells results in demyelination in MS patients. However, the autoencephalitogen has not been isolated and our knowledge of the autoimmune component of MS is mostly derived from animal models, such as experimental allergic encephalomyelitis (EAE). EAE is induced in several animal species by autoimmunization with constituents of myelin and is characterized by either a monophasic illness with perivascular inflammation and minimal demyelination (Raine, 1983) or, in certain laboratory animals such as the SJL strain of mice or strain 13 guinea pigs, by chronic relapsing diseases with increasing demyelination. MBP is the major encephalitogenic agent in EAE, and the amino acid sequences responsible for disease induction, in various mouse strains, have been determined (Zamvill *et al.*, 1985; Zamvill & Steinman, 1990). PLP is a second major encephalitogen that results in inflammatory demyelination in selected mouse strains (Sobel *et al.*, 1990).

The fact that EAE can be adoptively transferred by immune cells in rats (Paterson, 1960) and guinea pigs (Stone, 1961) but not by serum suggests a cell-mediated mechanism of aetiology. The demyelinating lesions of MS patients consist of inflammatory infiltrates that include macrophages, lymphocytes and plasma cells (Prineas & Wright, 1978). Subsequent experiments involving the isolation of encephalitogenic T cell lines have demonstrated that T lymphocytes play a central role in mediating events of EAE (Ortiz-Ortiz & Wergle, 1976; Ben-Nun, Wekerle & Cohen, 1981). The $CD4^+$ $T_H$ cell appears to play an important role in disease progression through MHC-restricted antigen-presenting cells (Zamvill & Steinman, 1990). Although much attention has been focussed on the role of T cells in disease progression, primary demyelination could also result from autoantibodies. Evidence for this mechanism comes from EAE studies demonstrating that demyelination can occur with intravenous injection of myelin-specific antibodies in animals with a breached blood–brain barrier (Schluesener *et al.*, 1987).

## 18.5 An animal model for MS

The blood–brain barrier acts to exclude macromolecules and cells of immune origin, thus rendering the CNS an immune-privileged organ system. Perhaps as a result of the lack of immune cells from the systemic circulation, neuronal, glial and endothelial cells in the CNS do not express significant levels of MHC class I and class II antigens (Head & Griffin, 1985). The appearance of lesions in the CNS white matter of MS patients is concordant with a loss of function of the blood barrier (Harris *et al.*, 1991) and a gain of expression of MHC class I and class II on a variety of cell types, including astrocytes, vascular endothelial cells, infiltrating B cells and macrophages (Sobel *et al.*, 1984; Traugott, Raine & McFarlin, 1985; Traugott, Scheinberg & Raine, 1985).

A role for MHC class I and class II in the development of MS is suggested in studies designed to inhibit EAE; in mice, the use of either peptides that bind and block the MHC antigen-binding site (Lamont *et al.*, 1990) or mAbs to the MHC antigens resulted in an altered presentation of the encephalitogenic epitopes to $CD4^+$ cells (Steinman *et al.*, 1981). IFN-$\gamma$ is an immunomodulating cytokine capable of up-regulating the expression of MHC class I and

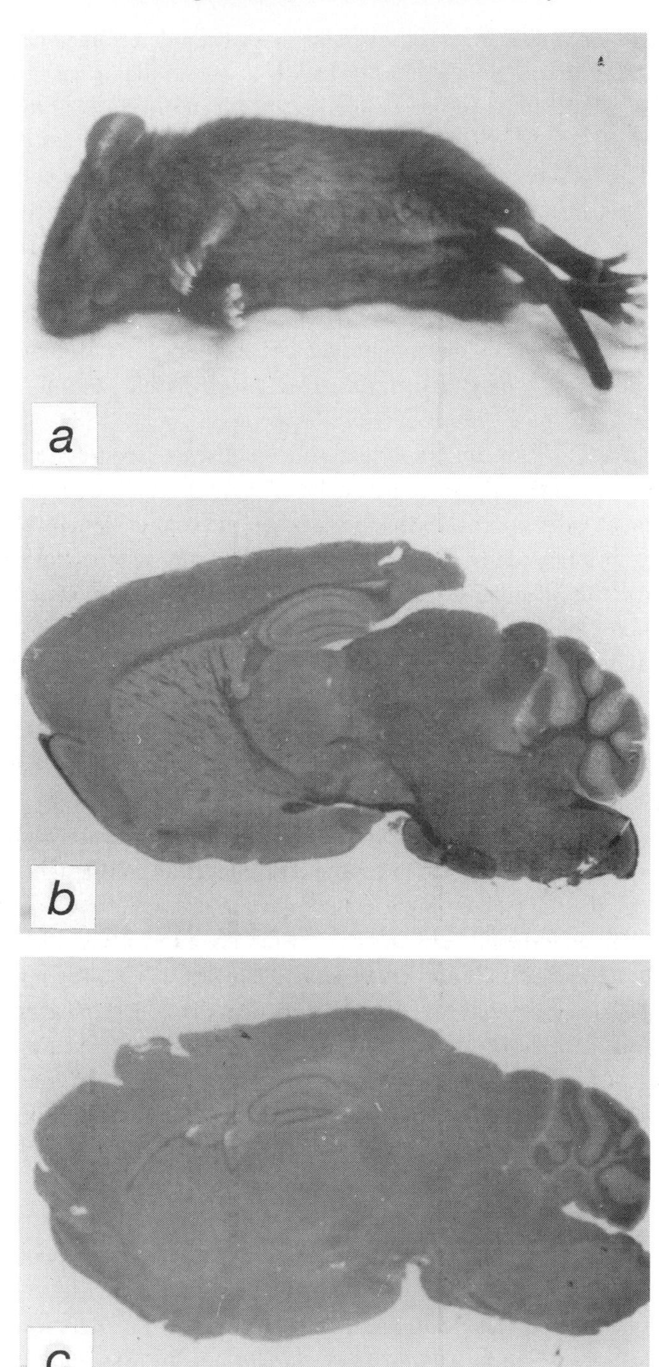
a
b
c

class II, which are elevated in lymphocytes of patients with MS (Hirsch, Panitch & Johnson, 1985). In addition, clinical trials using IFN-$\gamma$ as a therapeutic agent against MS have demonstrated that 7 of 18 patients treated with this agent displayed exacerbations of symptoms within a month of treatment (Panitch *et al.*, 1987; Traugott & Lebon, 1988).

These findings have led to the hypothesis that an increase in expression of MHC antigens in the CNS may play a pivotal role in the development of MS. To test this theory, we have up-regulated the expression of an MHC class I gene in the CNS using a transgenic mouse system. The use of this *in vivo* system is necessary and beneficial since it will mimic the cell-to-cell interactions necessary for development of disease, which cannot be reproduced in an *in vitro* setting. It also enables us to demonstrate the role of a single gene in the disease process. Demyelination in MS may result from up-regulation of the MHC antigen on oligodendrocytes. Targeting of expression of the class I $K^b$ gene to oligodendrocytes was achieved by using the MBP regulatory region, which shows absolute specificity for these cells (Katsuki *et al.*, 1988). A gene construct containing the MBP regulatory sequences directing the expression of the class I $K^b$ gene (MBP–$K^b$) was introduced into the germline of mice by pronuclear microinjection (Yoshioka, Feigenbaum & Jay, 1991). To facilitate the generation of transgenic animals, a partial H-$2^b$ haplotype was provided by C57BL6 males breeding with CD-1 superovulated females. A backcross of transgenic animals to C57BL6 mice ensured a syngeneic background for the $K^b$ gene in the resulting mice.

Two out of five transgenic mice generated did not transmit the transgene as a result of extreme mosaicism. An additional mouse died within 28 days following birth, after displaying severe tremors and occasional shaking. The remaining two founder mice, C2 and C7, developed to a reproductive age without any obvious pathology and successfully passed the transgene to their offspring. Both of these mice were genetically mosaic. Two to three weeks following birth, transgenic mice from the C7 line displayed a characteristic shaking phenotype, accompanied by extensor seizures with rigid limbs and extended toes (Fig. 18.1*a*). The seizures, while frequent, never lasted more than a minute. By four to five weeks after birth, approximately half of the transgenic mice died while the rest had sparser seizures and progressively

Fig. 18.1. Gross examination and microscopical analysis of control and transgenic mice. (*a*) A 21-day-old C7 transgenic mouse during an episode of extensor seizure. (*b*) LFB-stained mid-sagittal section of the brain of a four-week-old control mouse. Areas of staining, especially the hippocampus and the cerebellum represent myelination of these structures. (*c*) LFB-stained mid-sagittal section from a C7 transgenic mouse with no staining in the hippocampus and cerebellum.

lost their tremor phenotype, so that by eight weeks of age they appeared normal. Mice from the C2 line also developed tremors with extensor seizures, although at a lower frequency; 8 of 22 mice developed this phenotype. They showed weaker shaking and fewer seizures. All the mice from line C2 recovered from their phenotype in a time frame similar to the surviving mice from the C7 line. Based upon what appeared to be neurological deficits in mice from the C2 and C7 lines, a systematic analysis of the CNS and the peripheral nervous system (PNS) was performed.

Luxol fast blue (LFB) stains myelin-associated lipids and was used to determine the extent of myelination in the transgenic mice. Staining of a mid-sagittal section of the brain from a normal four-week-old mouse revealed large amounts of myelin in the cerebellum and hippocampal area (Fig. 18.1*b*). In comparison, an age-matched transgenic mouse from the C7 line displayed virtually no staining, suggesting extensive hypomyelination (Fig. 18.1*c*). A higher magnification of the hippocampal area again showed strong staining of white matter tracks in the control mouse (Fig. 18.2*a*); an age-matched four-week-old transgenic mouse from the C7 line showed very little staining in the same area (Fig. 18.2*b*). The white matter in the cerebellum, surrounded by the granular layer of the grey matter, showed very strong staining in the non-transgenic mouse (Fig. 18.2*c*). However, virtually no staining was seen in a similar location in a four-week-old C7 transgenic mouse that displayed strong tremors and frequent extensor seizures. A large decrease in LFB staining was observed not only in the brain but also in the white matter of the spinal cord. The myelinated white matter located on the outer boundary of the spinal cord (Fig. 18.2*e*) showed no signs of myelination in the sample from a transgenic mouse of the C7 line (Fig. 18.2*f*). A similar analysis performed on affected mice from the C2 line showed decreased staining in all the areas described above; however, the extent of hypomyelination was not as extensive as in the C7 line, reflecting the weaker phenotype observed in these mice.

To characterize further the hypomyelination of the CNS observed in these MBP-$K^b$ mice, electron microscopy of the optic nerve from control and transgenic animals with a neurological phenotype was performed. An electron-dense structure surrounding individual axons was observed in a three-week-old control mouse. This structure, representing the myelin sheath, was thick and rarely interrupted (Fig. 18.3*a*). In an age-matched transgenic mouse with severe shaking and tonic seizures, an extensive area with hypomyelinated axons was seen. On closer examination, remnants of axons with intracellular organelles scattered in the tissue milieu were detected (Fig. 18.3*b*, arrowheads). This suggested active demyelination and occasional axonal cell

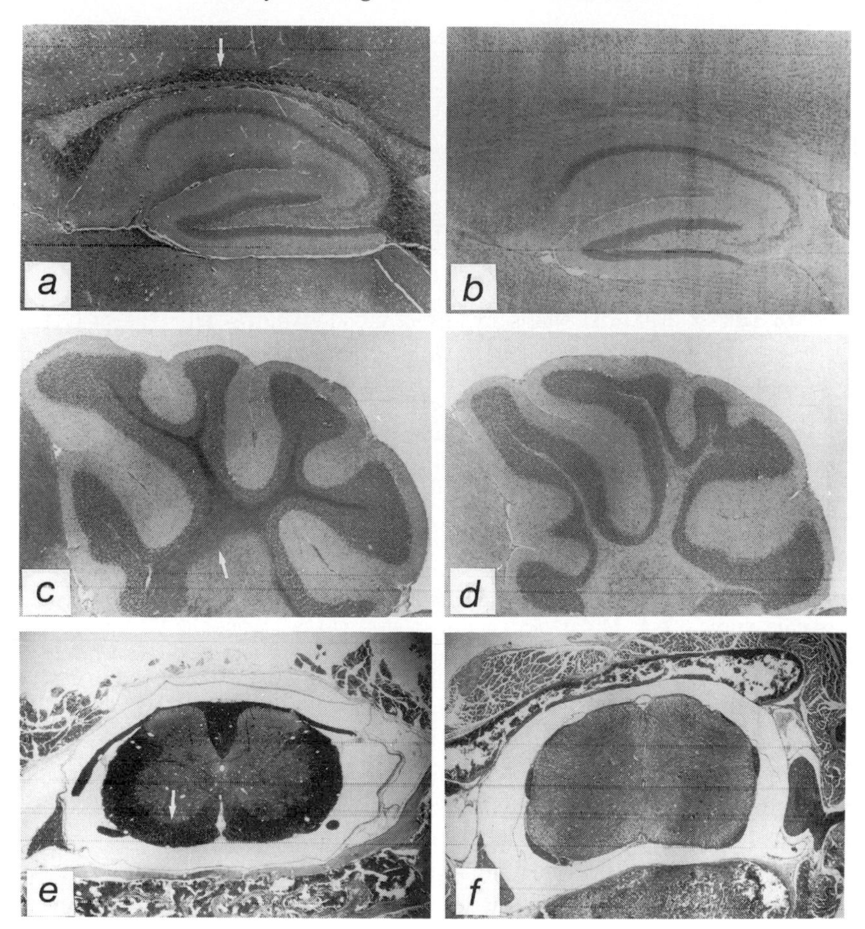


Fig. 18.2. Microscopical analysis of control and transgenic mice at four weeks of age. (*a*) LFB-stained section of the hippocampus of a control mouse. Areas of intense staining are indicated (arrow). (*b*) LFB-stained section of the hippocampus of a transgenic mouse showing no staining where staining would normally be observed. (*c*) LFB-stained section of the cerebellum of a control mouse, with intense staining of the white matter (arrow). (*d*) LFB-stained section of the cerebellum of a C7 transgenic mouse with no staining of the white matter. (*e*) LFB-stained section of the spinal cord of a control mouse, with intense staining of the white matter (arrow). (*f*) LFB-stained section of the spinal cord of a C7 transgenic mouse with no staining of the white matter.

death in the CNS of these mice. Although axonal cell death is not usually seen in MS patients, extreme cases do display this feature. This transgenic system will most likely mimic extreme cases of MS since the signal for pathogenesis, in this case the expression of the $K^b$ gene, is not localized but

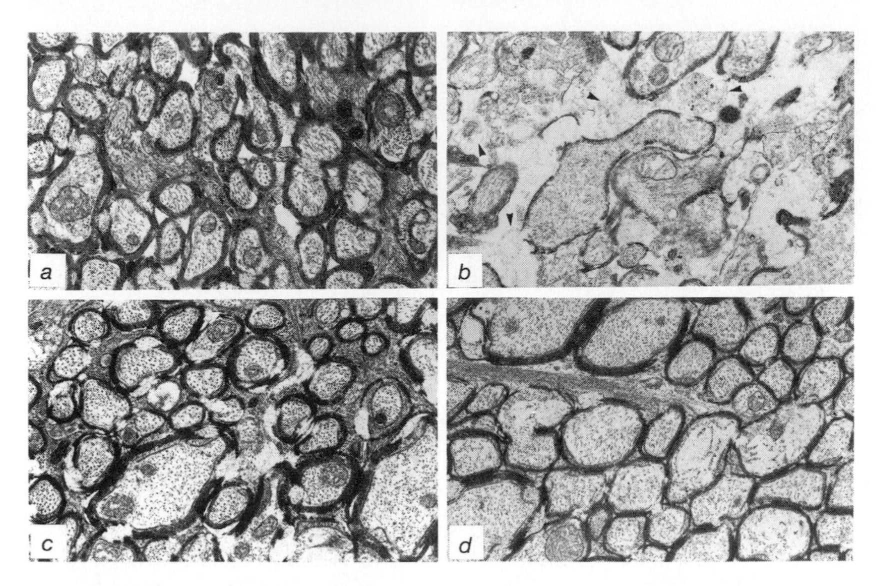


Fig. 18.3. Electron micrographs of sections of the optic nerve from control and transgenic mice. (*a*) A 24-day-old control mouse; (*b*) a 24-day-old C2 transgenic littermate. Areas showing remnants of myelinated axons are indicated (arrowheads). (*c*) A 16-week-old control mouse; (*d*) a 16-week-old C2 transgenic litter mate. Magnification, × 2332.

instead present in a generalized fashion affecting most oligodendrocytes. Axons that showed myelination in the transgenic mouse had a myelin sheath less than half the thickness of normal axons (Fig. 18.3*b*).

The recovery from tremors and tonic seizures was explained by remyelination in a four-month-old transgenic mouse that no longer showed any neurological symptoms. Compared with an age-matched litter mate (Fig. 18.3*c*), the transgenic mouse also had a myelin sheath around every axon (Fig. 18.3*d*), although not quite as thick as in the control. Electron microscopic analysis of the sciatic nerve did not reveal any abnormalities in either of the transgenic lines. The onset of pathology and its remission in the MBP-$K^b$ mice correlated perfectly with the developmental profile of MBP expression. Significant MBP expression started at day 5 post-partum and peaked at day 18, the time at which the tremor and tonic seizures were most severe in the MBP-$K^b$ mice. The level of expression then fell to approximately one fourth its peak level during adult life (Zeller *et al.*, 1984), corresponding to the disappearance of the phenotype. This suggests that a threshold level of expression of the $K^b$ gene is necessary for demyelination to occur. The variability in the extent of the observed neurological phenotype between and within transgenic lines most likely reflects the level of expression of the transgene.

These experiments demonstrate that expression of the MHC class I $K^b$ gene on oligodendrocytes is sufficient to cause demyelination. The pathology observed is not a result of loss of function of the oligodendrocytes, since normal levels of MBP mRNA are observed in the transgenic mice (Yoshioka *et al.*, 1991).

Although demyelination is observed in the MBP-$K^b$ mice, immune infiltration, a hallmark of MS, was not observed, suggesting that the presence of autoreactive T cells and B cells is not necessary for active demyelination. Instead, the expression of an MHC class I antigen on the cell surface of oligodendrocytes may suffice to interfere with the proper organization of the myelin sheath leading to autodegradation and demyelination. The assumption that MHC class I is expressed on the cell surface is derived from previous transgenic and transfection studies that have suggested that $\beta_2$-m, a protein necessary for transport of the class I antigens to the cell surface, is present and not limiting in all cells tested (Tanaka *et al.*, 1985; Yoshioka *et al.*, 1987).

The absence of a secondary immune response in the transgenic animals may be explained by the lack of an appropriate genetic background in the C57BL6 mice. Studies of genetic susceptibility to EAE have demonstrated that C57BL6 mice are resistant. However, SJL mice are highly susceptible to EAE. Expression of the MBP–$K^b$ gene in a SJL mouse background by backcrossing of our transgenic lines would most likely provide the genetic factors needed for an autoimmune response. This in turn may contribute to a secondary immune response and worsen the disease that has been observed.

That demyelination can result from expression of MHC antigens on oligodendrocytes offers an attractive explanation for the apparent lack of specificity of the viral aetiology of MS. A common mechanism for the upregulation of expression of MHC antigens in the CNS, in response to a multitude of viral agents (Burks *et al.*, 1980; Johnson *et al.*, 1984; Koprowski *et al.*, 1985; Hemachudha *et al.*, 1987), may well be the activation of IFN-$\gamma$ secretion by infiltrating T cells. This hypothesis can easily be tested by targeting the expression of this lymphokine to the CNS. An alternative interpretation of these results is a direct activation of MHC antigens in response to viral infection of oligodendrocytes. Evidence for this theory is found in *in vitro* experiments demonstrating the ability of human T-lymphotropic virus type I to activate MHC class I genes (Sawada *et al.*, 1990).

Although genetic susceptibility to MS has implicated MHC class I genes and this animal model demonstrates their involvement in disease, it is important to demonstrate that MHC class II genes are also capable of inducing a similar pathology, in keeping with results from epidemiological studies of MS patients and biological investigations of EAE.

## References

Alter, M., Leibowitz, V. & Speer, J. (1962). Risk of multiple sclerosis related to age at immigration to Israel. *Archives of Neurology*, 15, 234–237.

Antel, J.P. & Arnason, B.G.W. (1983). Multiple sclerosis and other demyelinating diseases. In *Principles of Internal Medicine*, ed. R.G. Petersdorf, R.D. Adams, E. Braunwald, K.J. Isselbacher, J.B. Martin & J.D. Wilson, pp. 2098–2104. New York: McGraw-Hill.

Aström, K.E., Mancall, E.L. & Richardson, E.P. Jr (1958). Progressive multifocal leuko-encephalopathy. *Brain*, 81, 93–111.

Ben-Nun, A., Wekerle, H. & Cohen, I.R. (1981). The rapid isolation of clonable antigen specific T lymphocyte lines capable of mediating autoimmune encephalomyelitis. *European Journal of Immunology*, 11, 195–199.

Bertram, J. & Kuwert, E. (1982). HLA antigen frequencies in multiple sclerosis. *European Journal of Neurology*, 7, 74–79.

Burks, J.S., Devald, L.D., Jankovski, L.D. & Gerges, J.C. (1980). Two coronaviruses isolated from the central nervous system tissue of two multiple sclerosis patients. *Science*, 209, 933–934.

Currier, R.D. & Eldridge, R. (1982). Possible risk factors in multiple sclerosis as found in a national twin study. *Archives of Neurology*, 39, 140–144.

Ebers, G.C., Bulman, D.E., Sadovnick, A.D., Paty, D.W., Warren, S., Hader, W., Murray, T.J., Seland, T.P., Duquette, P., Grey, T., Nelson, R., Nicolle, M. & Brunet, D. (1986). A population-based study of multiple sclerosis in twins. *New England Journal of Medicine*, 315, 1638–1642.

Elian, M.N. & Dean, G. (1990). Multiple sclerosis among United Kingdom born children of immigrants from the Indian subcontinent, Africa and the West Indies. *Journal of Neurology, Neurosurgery and Psychiatry*, 53, 906–911.

Feigenbaum, L., Hinrichs, S.H. & Jay, G. (1992). JCV and SV40 enhancers and transforming proteins: role in determining tissue specificity and pathogenicity in transgenic mice. *Journal of Virology*, 66, 1176–1182.

Fischman, H.R. (1981). Multiple sclerosis: a two-stage process? *American Journal of Epidemiology*, 114, 244–252.

Gorelick, P.B. (1989). Clues to the mystery of multiple sclerosis. *Multiple Sclerosis*, 85, 125–134.

Hader, W.J., Irvine, D.G. & Schiefer, H.B. (1990). A cluster-focus of multiple sclerosis at Henribourg, Saskatchewan. *Canadian Journal of Neurological Science*, 17, 391–394.

Hammond, S.R., English, D., Dewtit, C., Maxwell, I.C., Milligen, J.S., Stewart-Wynne, E.G., McLeod, J.G. & McCall, J. (1988). The clinical profile of MS in Australia: a comparison between medium-frequency and high-frequency prevalence zones. *Neurology*, 38, 980–986.

Harris, J.O., Frank, J.O., Patronas, N., McFarlin, D.E. & McFarland, H.F. (1991). Serial gadolinium-enhanced magnetic resonance imaging scans in patients with early, relapsing-remitting multiple sclerosis: implications for clinical trial and natural history. *Annals of Neurology*, 29, 548–555.

Head, J.R. & Griffin, W.S.T. (1985). Functional capacity of solid tissue transplant

in brain: evidence for immunological privilege. *Proceedings of the Royal Society of London, Series B*, 224, 375–387.
Hemachudha, T., Griffin, D.E., Giffels, J.J., Johnson, R.T., Moser, A.B. & Phanuphak, P. (1987). Myelin basic protein as an encephalitogen in encephalomyelitis and polyneuritis following rabies vaccination. *New England Journal of Medicine*, 316, 269–374.
Hirsch, R.L., Panitch, H.S. & Johnson, K.P. (1985). Lymphocytes from multiple sclerosis patients produce elevated levels of gamma interferon *in vitro*. *Journal of Clinical Immunology*, 5, 386–389.
Johnson, R.T., Griffin, D.E., Hirsh, R.L., Wolinsky, J.S., Roestenbeck, S., DeSoriano, I.L. & Vaisberg, A. (1984). Measles encephalomyelitis – clinical and immunologic studies. *New England Journal of Medicine*, 310, 137–141.
Katsuki, M., Kimura, M., Yokohama, M., Kobayashik, K. & Nomura, T. (1988). Conversion of normal behavior to shiverer by myelin basic protein antisense cDNA in transgenic mice. *Science*, 241, 593–595.
Koeppen, A.H., Ronca, N.A., Greenfield, E.A. & Hans, M.B. (1987). Defective biosynthesis of proteolipid protein in Pelizaeus–Merzbacher disease. *Annals of Neurology*, 21, 159–170.
Koprowski, H., Defreitas, E.C., Harper, M.E., Sandberg-Wolham, M., Sheremata, W.A., Robert-Guroff, M., Saxinger, C.W., Feinberg, M.B., Wong-Staal, F. & Gallo, R.C. (1985). Multiple sclerosis and human T cell lymphotropic retroviruses. *Nature (London)*, 318, 154–160.
Kurdi, A., Ayesh, I., Addallah, A., Maayota, U., McDonald, W.I., Compston, D.A.S. & Batchelor, J.R. (1977). Different B-lymphocytes alloantigens associated with multiple sclerosis in Arabs and Northern Europeans. *Lancet*, i, 1123–1125.
Kurtzke, J.F., Dean, G. & Botha, D.P. (1970). A method for estimating the age at immigration of white immigrants to South Africa, with an example of its importance. *South African Medical Journal*, 44, 663–669.
Kurtzke, J.F. & Hyllested, K. (1979). Multiple sclerosis in Faeroe Islands. I. Clinical and epidemiological features. *Annals of Neurology*, 5, 6–21.
Lamont, A.G., Sette, A., Fujinami, R., Colon, S.M., Miles, C. & Greg, H.M. (1990). Inhibition of experimental autoimmune encephalomyelitis induction in SJL/J mice by using a peptide with high affinity for $IA^s$ molecules. *Journal of Immunology*, 145, 1687–1693.
Levy, A.M., Bredsen, D.E. & Rosenblum, M.L. (1985). Neurological manifestations of the acquired immunodeficiency syndrome (AIDS): experience at UCSF and review of the literature. *Journal of Neurosurgery*, 62, 475–495.
Marrosu, H.G., Muntoni, F., Munu, M.R., Spinicci, G., Pichelda, M.P., Goddi, F., Cossu, P. & Pirastu, M. (1988). Sardinian multiple sclerosis is associated with HLA-DR4: a serological and molecular analysis. *Neurology*, 38, 749–753.
McFarlin, D.E. & McFarland, H.F. (1982). Multiple sclerosis. *New England Journal of Medicine*, 307, 1183–1188.
Naito, S., Kuroiwa, Y., Itoyama, T., Tsubaki, T., Horikawa A., Susazuki, T., Noguchi, S., Ohtsuki, S., Tokumi, H., Myatake, T., Takahata, N., Kawanami, S. & McMichael, A.J. (1978). HLA and Japanese MS. *Tissue Antigens*, 12, 19–24.
Niedt, G.W. & Schinelle, R.A. (1985). Acquired immunodeficiency syndrome: clinico-pathology study of 56 autopsies. *Archives of Pathology and Laboratory Medicine*, 109, 727–734.
Ortiz-Ortiz, L. & Wergle, W.O. (1976). Cellular events in the induction of

experimental allergic encephalomyelitis in rats. *Journal of Experimental Medicine*, 144, 604–616.

Oskenberg, J.R., Sherritt, M., Begovich, A.B., Erlich, H.A., Bernard, C.C., Cavelli-Sforza, L.L. & Steinman, L. (1989). T cell receptor V alpha and C alpha alleles associated with multiple sclerosis and myasthenia gravis. *Proceedings of the National Academy of Sciences of the USA*, 86, 988–992.

Padgett, B.L., Walker, D.L., ZuRhein, G.M., Eckroade, R.J. & Dessel, B.H. (1971). Cultivation of papova-like virus from human brain with progressive multifocal leukoencephalopathy. *Lancet*, i, 1257–1260.

Panitch, H.S., Hirsch, R.L., Haley, A.S. & Johnson, K.P. (1987). Exacerbation of multiple sclerosis in patients treated with gamma-interferon. *Lancet*, i, 893–895.

Paterson, P.Y. (1960). Transfer of allergic encephalomyelitis by means of lymph node cells. *Journal of Experimental Medicine*, 111, 119–133.

Prineas, J.W. & Wright, R.G. (1978). Macrophages, lymphocytes and plasma cells in the perivascular compartment in chronic multiple sclerosis. *Laboratory Investigation*, 38, 408–421.

Raine, C.S. (1983). Multiple sclerosis and chronic relapsing EAE: comparative ultrastructural neuropathology. In *Multiple Sclerosis*, ed. J.F. Hallpike, C.W.M. Adams & W.W. Tourtellotte, pp. 413–478. Baltimore: Williams & Wilkins.

Sadovnick, A.D. & McLeod, P.M.J. (1981). The familial nature of multiple sclerosis: empiric recurrence risk for first-, second-, and third-degree relatives of patients. *Neurology*, 31, 1039–1041.

Sawada, M., Suzumura, A., Yoshida, M. & Marumachi, T. (1990). Human T-leukemia virus type I transactivation induces class I major histocompatibility complex antigen expression in glial cells. *Journal of Virology*, 64, 4002–4006.

Schluesener, H.J., Sobel, R.A., Linington, C. & Weiner, H.L. (1987). A monoclonal antibody against a myelin oligodendrocyte glycoprotein induces relapses and demyelination in central nervous system autoimmune disease. *Journal of Immunology*, 139, 4016–4021.

Seboun, E., Robinson, M.A., Doolittle, T.H., Ciulla, T.A., Kindt, J.J. & Hauser, S.L. (1989). A susceptibility locus for multiple sclerosis is linked to the T cell receptor β chain complex. *Cell*, 57, 1095–1100.

Sibley, W.A., Bamford, C.R. & Clark, K. (1983). Triggering factors in multiple sclerosis. In *Diagnosis of Multiple Sclerosis*, ed. C.E. Poster & D. Paty, pp. 14–24. New York: Thieme-Stratton.

Small, J.A., Scangos, G.A., Cork, L., Jay, G. & Khoury, G. (1986). The early region of human papovavirus JC induces dysmyelination in transgenic mice. *Cell*, 46, 13–18.

Sobel, R.A., Blanchette, B.W., Bhan, A.K. & Colvin, R.B. (1984). The immunopathology of experimental allergic encephalomyelitis II. Endothelial cell Ia increases prior to inflammatory cell infiltration. *Journal of Immunology*, 32, 2402–2407.

Sobel, R.A., Tushy, V.K., Lu, Z.J., Laursen, R.A. & Lees, M.B. (1990). Acute experimental allergic encephalomyelitis in SJL/J mice induced by a synthetic peptide of myelin proteolipid protein. *Journal of Neuropathology and Experimental Neurology*, 49, 468–479.

Steinman, L., Rosenbaum, J.T., Sriram, S. & McDewitt, H.O. (1981). *In vivo* effects of antibodies to immune response gene products: prevention of experimental allergic encephalomyelitis. *Proceedings of the National Academy of Sciences of the USA*, 78, 7111–7114.

Stone, S.H. (1961). Transfer of allergic encephalomyelitis by lymph node cells in inbred guinea pigs. *Science*, 134, 619–621.

Tanaka, K., Isselbacher, K.J., Khoury, G. & Jay, G. (1985). Reversal of oncogenesis by the expression of a major histocompatiblity complex class I gene. *Science*, 228, 26–30.

Traugott, U. & Lebon, P. (1988). Multiple sclerosis: involvement of interferons in lesion pathogenesis. *Annals of Neurology*, 24, 243–251.

Traugott, U., Raine, C.S. & McFarlin, D.E. (1985). Acute experimental allergic encephalomyelitis in the mouse: immunopathology of the developing lesion. *Cellular Immunology*, 91, 240–244.

Traugott, U., Scheinberg, L.C. & Raine, C.S. (1985). On the presence of Ia-positive endothelial cells and astrocytes in multiple sclerosis and its relevance to antigen presentation. *Journal of Neuroimmunology*, 28, 1–14.

Waksman, B.H. (1983). Viruses and immune events in pathogenesis of multiple sclerosis. In *Viruses and Demyelinating Diseases*, ed. C.A. Mims, M.L. Cuzner & R.E. Kelly, pp. 155–165. New York: Academic Press.

Yoshioka, T., Bieberich, C., Scangos, G. & Jay, G. (1987). A transgenic MHC class I antigen is recognized as self and acts as a restriction element. *Journal of Immunology*, 139, 3861–3867.

Yoshioka, T., Feigenbaum, L. & Jay, G. (1991). Transgenic mouse model for central nervous system demyelination. *Molecular and Cellular Biology*, 11, 5479–5486.

Zamvill, S., Nelson, P., Trotter, J., Mitchell, D., Knobler, R., Fritz, R. & Steinman, L. (1985). T-cell clones specific for myelin basic protein induce chronic relapsing paralysis and demyelination. *Nature (London)*, 317, 355–358.

Zamvill, S. & Steinman, L. (1990). The T lymphocyte in experimental allergic encephalomyelitis. *Annual Review of Immunology*, 8, 579–621.

Zeller, N.K., Hunkeler, M.J., Campagnoni, A.T. & Lazzarini, R.A. (1984). Characterization of mouse myelin basic protein messenger RNAs with a myelin basic protein cDNA clone. *Proceedings of the National Academy of Sciences of the USA*, 81, 18–22.

# Index

References in italic denotes illustrations.


abbreviations xv–xviii
Abelson leukaemia virus 175
*abl* oncogene, and MHC regulation 175–6
activation 111
  domain
  pathways 169–70
  protein stimulation, and lymphoid neoplasia 170
activator protein, and transcription 111
active transport, class I peptide 3–4
adenovirus
  class I murine repression 83, 203–10
    post-translational regulation 211
    transcriptional regulation 210–11
  conserved regions
    CR1–3 *196*, 197
    CR3 and NRE 208
    CRE2 206–7, 208
  degradation protection 199
  expression of genome 195–200
  genome 195, *196*
  in IDDM 371
  limited 201–2
  MHC, effect on 136
  molecular oncogenicity 202–3
  molecular pathology 193–203
  negative regulatory element (NRE) 208
  oncogenicity 192, 201–2
  in rodent cells 200–2
  serotype 193
  structure 193, 195
  transformation 192, 200–3
    comparison Ad5 and Ad12 201
adenovirus early genes
  E1A 192, 195, *196*, 197, *198*
  in Ad5 206
  in Ad12 206, *210*
    binding proteins 197
    conserved regions 197
    down-regulation of class I 83, 205–10, 296
    functions and domain structure *198*
    immortalizing function 201
    ISG transcription inhibition 218
    *myc* homology 299
    oncogenic determinant region 203
    promoters 197
    rodent 201
    serotype, and oncogenicity 202
  E1B 199, 201
  E2A 200
  E2B 200
  E2F induction 197
  E3
    14.7 kDa protein 219–20
    19.0 kDa protein 211–13
    allelic differences 215–16
    binding site 215
    class I binding 213, 214
    CTL lysis protection 216–17
    down-regulation of class I 211
    gene 192
    heavy chain blocking 215
    polymorphism 214
  E6/7 recognition, by CTL 245
adenovirus late genes 195, 200
adenovirus subgenera
  Ad2
    E3 19 kDa glycoprotein 211–13
    transport block 141–2
  Ad5
    class I expression 296–7
    expression of E1A/E1B 201
      *see also* adenovirus early genes
    IFN inhibition 218

ISG transcription inhibition 218
non-oncogenic 202
Ad12
class I expression 203–5, 296–7
immortalization 201
oncogenic region 203
oncogenicity 202, 204
tumorigenicity 204–5
Ad40/41 213–14
adhesion, antigen-independent 17–18
adhesion molecules 17–19, *65*
in BL immunosurveillance escape 266–7
and CMV infection 279, 287
in insulitis 364, 366
and leucocyte adhesion 288
ligands 17
up-regulation, BL 265
viral interference 142
*see also* CD antigens; ICAM-1; LFA-1; LFA-3
adrenal hyperplasia, congenital 36
AIDS, class II 180
AKR mouse leukaemia cell line
class I 326–7
class II 327–8
H-2K deficient 325
alanine-rich sequence, in adenoviruses 202–3
allele specificity, class I peptide binding 15–16
allele-specific peptide motifs, class I 10, 11
allogeneic class I transfection 325–7
AMP *see* cAMP
anti-Ia antibodies, and immune disease remission 108
anti-inflammatory function of IL-10 52–3
antibody-dependent cytotoxic cellular response (ADCC) in HIV 180
antigen
class I/II *7*
degradation 143
expression
down-regulation 134, 135
excessive 134
flowchart 138
normal mechanism 133–4
site of viral interference 138, 139, 140–3
virus subverted 134
*see also* MHC
inhibition, peptide competition 141, 142
internalized 13–14
lymphocyte-specific 153–4
modulation
non-viral 144–5
retroviral 150
novel, on AKR/Gross cell lines 167
presentation 3
'tolerogenic' 375
class I/II *7*
processing, contradictory cytokine effects 64
recognition, T cell 234
recycling, virally modified 142–3
response to viruses
AKR/Gross leukaemia 164–8
assembly inhibition 140
defective endogenous viral sequences 168–9
MCF virus 158
Moloney murine leukaemia 151–60
RadLV 160–4
retroviral modulation 150
tolerization 3
transport inhibition 141–2
*see also* class I; class II
antigen-presenting cells (APC)
adhesion molecules *65*
co-stimulators
B7 in insulitis 366
second signal response 378
professional/non-professional 64–5
thyrocyte, differential ability 66
ATP, in peptide transport 3, 9
autocrine regulation of class I expression 172–3
autoimmune disease 44
aberrant class II 108, 361–2
multiple sclerosis 412–13
autoimmune response in islets, virally induced 372–3
autoimmunity
aberrant class II 108, 361–2
ectopic class II expression hypothesis 378–9
initiation *380*
virally caused 49, 50, 370–1

B locus antigen, in Wil tumour cell line 346
B lymphocyte
class II induction 105
DRA promoter activity, and X2 116
function, and X box 114
infection by EBV 261
latent 262–3
maturation, and class II expression 105
phenotypic change and LMP1 264–5
proliferation, TGF-β inhibited 400
stimulatory factor-1 *see* interleukin IL-4
B7 ligand
co-stimulator in insulitis 366
in second signal response 378
bacterial antigen modulation 145
BALB/c
fibroblasts, with leukaemia virus 151–2
spleen cells, H-2 antigen expression 168

bare lymphocyte syndrome (BLS)
  cell line analysis 109–10
  RF-X omission 115
BAT2/3 *see* G genes
BB rat, IDDM 372
BCG therapy 340
*bcl–2* gene 298
  LMP1 265
BCRF1 *see* vIL-10
β cells
  class II, induction 46–7, 366–7
  expression of TNF-α, transgenic mice 376
  HLA expression in insulitis 363–4
  immune response in mouse transgenes 375
  MHC subregion expression 59
  and pre-pathological IDDM 369–70
  TNF-α/IFN-γ interaction 50–1
  'tolerogenic' antigen presentation 375
  transgenes in mice 373–9
  viral transgene product expression 377
$\beta_2$-m *see* $\beta_2$-microglobulin
biopsy, cervical cancer 244
bladder tumour 337, 339
blood–brain barrier 413
BLS *see* bare lymphocyte syndrome
bone marrow, haemopoietic cell proliferation 53
Burkitt's lymphoma (BL) 261, 263
  BL30 cells, transcriptional deficiency 272–3
  cell surface markers 263
  class I
    analysis *270*, 271
    down-regulation 267, *268*, 269
    expression 267–73
  c-*myc* 298
  CTL resistant 266–7
  distinguished from LCL 263
  EBV-negative 264
  latency I 264
  phenotypic drift 263, 264
  viral protein expression 267
bystander cells, class I expression in CMV infection 281–3

c-*fos* 116
C glycoproteins
  C2 36
    transgenic mouse studies (EAE) 414–16
  C4 36
  C7 transgenic mouse studies (EAE) 414–16
  function 36
cachectin *see* tumour necrosis factor
calnexin 6
cAMP
  and prostaglandins 56–7
  TSH stimulation 56
cancer
  and oncogenes 295
  *see also* cervical cancer; colon carcinoma; lung carcinoma; malignancy; sarcoma, tumour
capillary endothelial cells, adhesion molecule production 364, 366
cathepsins 14
CBP/CP1 *see* NF-Y
CD antigens
  CD2 17
    in CMV infection 288
  $CD4^+$ 17, 18–19
    in autoimmune response 361–2
    cell generation, and H2-K 325
    and class II 18
    CTL 2
    helper T cells 43
    tropism, in SIV 182
  $CD4^+$–$CD8^+$ CTLs, in tumour rejection 319
  $CD8^+$ 17, 18–19
    antigen recognition 104
    and class I 18
    CTL 2
    T cell antigen recognition 103–4
    T cell and insulitis 363
  CD11a/CD18 17
  CD28, in second signal pathway 378
  CD54 17
  CD58 17
  *see also* adhesion molecules; LFA
cell
  differential in class II induction, IFN-γ 46–7, 47–8
  maturation, and contradictory effect of TNF-α 51
  type
    class I expression control by oncogenes 173
    cytokine function modulation 45–6
cervical cancer 233
  biopsy 235
  pre-malignant disease 245
  *see also* human papillomavirus
α-chain transcripts *see* HLA-DP; HLD-DQ; HLA-DR
β chain, class II 4, 12
  in HIV 180
chloramphenicol acetyl transferase, chimaeric gene 156–7, 207–8, *209*
chromatin configuration, and class I factor binding 88
chromosome 6 *see* MHC
chromosome 17 *see* MHC
*cim* (class I modifier) 32
CIN lesions 244
*cis*-acting elements

class I 82–3, *85*
*myc* transcription-level control 302
cis-Golgi apparatus *see* Golgi apparatus
class I
absent in non-lymphoreticular tumours 170
accessory molecules 6
assembly 6
autocrine control 172–3
binding in HIV infection 181
and CD8+ antigen recognition 103–4
α chain 4–5
$\alpha_1$, $\alpha_2$ domains and complex formation 215
*cis*-regulatory elements *82*, *85*
cytokine mediation 59
*see also* cytokines
down-regulation 136–7
by adenovirus early genes 205–10
adenoviruses 210–11
c-*myc* 300–4
HPV 235
tumour cell implication 305
enhancer elements 81, *82*
expression
CTL-mediated lysis of tumour cells 171
developmental control 80
differential in tumours 48, 323, 344–50
and Gross MuLV 166
modulation by viruses 136–7
and NK activity 318, 319
tissue-dependent 80
trophoblast 47–8
and tumours 336–7
function 4–6
heavy chain 4–6, 29
blocking by E3 19 kDa 215
in HIV infection 179, 181
human/murine comparison 81, *82*
IFN regulation absence in MD 61
IFN-α/β stimulation 48
*in vivo* factor binding 86–8
induction, in leukaemia virus 151–3
instability, in CMV-infected cells 285
maturation, HPV infection 239–40
MHC sequences 27–8
modifier *see cim*
murine 28, 30
negative elements 83
novel antigens 167
organization 27–30
overexpression, in insulitis 363, 364
peptide-bonding site 5
and allele-specific motifs 10–12
peptide
generation 6–8
loading 10
motifs 10, 11
translation 53
translocation 8–10
pockets 5–6, 10–12
polymorphism 5–6
pseudogenes 27, 28
regulatory elements 81–4
*src* oncogene interaction 174–5
structure 3–6
surface expression,and CTL lysis tolerance 216–17
and TIL infiltration level 341–2
tissue-specific expression and footprinting 87
transcription regulation 89, 94
retinoic acid 94
TNF-α 35
transfection 325–7
transplantation antigens 1
transplantation genes 27, 27–8
*see also* HLA-A; HLA-B; HLA-C
transport block 141–2
tumours
NK sensitivity 306
recognition 295–6, 297
restoration 218, 323–5
up-regulation
HTLV-1 177–8
in human leukaemia 177–8
class II 30–3, 44, *104*
aberrant expression 108, 361–2
in AIDS 180
antigen-response involvement 103–4
assembly 12–13
bacterial modulation 145
and CD4+ antigen recognition 104
α chain 4, 12
conserved regions *113*
contradictory effects of IFN-α/β 49, 50
differential tumour cell expression 344–50
down-regulation
ectopic expression in insulitis 363–4
by EGF 54
neurotransmitter-mediated 57
enhancement, by EGF 55
expression 44, 104–7
aberrant 107–9, 361–2
cAMP-mediated TSH effect 56
chronic, in rheumatoid arthritis 403
control 55
and IL-4 51
modulation by viruses 136–7
murine, IL-10 mediated 53
prostaglandin 57
5-HT 57
tumours 337, 340, *341*
function 1, 4, 12
genes
molecular regulation 112–19

class II (*cont.*)
organization 30–4
for peptide translocation 8–9
glucocorticoid
enhancement 56
inhibition 55
IFN expression 48–9
induction 105
cell-differential, by IFN-$\gamma$ 46–7, 47–8
CSF 53
discrepancies, in IDDM 368
HTLV-1 178–9
macrophage, plasma protein 57–8
mechanism of *380*, *381*
pancreatic $\beta$ cells 46–7
interleukin effect 51–3
molecular level regulation 106–7
promoters 112
regulatory elements 112
*trans*-factors 109
transcriptional control 105
motifs 15
murine 31, 33–4
peptide-binding allele specificity 15–16
delivery 14–15
generation 13–14
polymorphism 15–16
prothymosin-enhanced 57
recycling 16–17
in rheumatoid arthritis 390, 391–2
second messenger activation 106
selective expression 104
structure 4, *5*, 12
T cell variable response 378
transfection 327–8
transport 12–13
up-regulation 105–6
class III 1–2, 34–7
murine 37
CMV *see* cytomegalovirus
CNS, MHC expression, and multiple sclerosis 415
*see also* multiple sclerosis
co-stimulation 64, 66
in insulitis 366
coeliac disease 193, 194
COL11A2 gene 33
colon carcinoma, c-*myc* 298
colony-stimulating factor 53
complement component
mapping 34
polymorphism 36
complementary genes, class II 109
conjunctivitis 193, 194
conserved regions
class II *113*
*see also* enhancer A
*see also* adenovirus conserved regions
coronavirus, effect on MHC 136
COUP-TF in Ad-transformed cells 207
Coxsackie virus, and IDDM 369
CR *see* conserved regions
CSF *see* colony-stimulating factor
CSIF *see* interleukin-10
CTL *see* cytotoxic T lymphocyte
CTLA4, in second signal pathway 378
cyclosporin A 60
cytokine
and adenovirus infection 217–20
class I 59, 80–1
bystander cells in CMV infection 283
co-stimulation 64, 366
contradictory effects in antigen processing 64
differential cell-type interaction 45–6
failure of tumour cell response 351
inhibitory effect 61
involvement in inflammatory process 60–1
MHC expression regulation 44–5, 58–9, 61, 351
*in vivo* evidence 59–61
NK-$\kappa$B activation 92
in IDDM 366
and immune surveillance 336
in insulitis 364
modulation of synthesis
IL-10 52–3
TGF-$\beta$ 400
in mouse $\beta$ cells 374
non-MHC function 45
rheumatoid arthritis
class II 392–401, 403
synovium 391–2
synergy 60
$T_H$ cells 2
*see also* GM-CSF; interferons; interleukins; tumour necrosis factor
cytokine synthesis inhibitory factor *see* IL-10
cytolysis
HPV-transfected cells 243–4
by TILs 342–3
TNF-$\alpha$, in Ad infection 219
cytomegalovirus
adhesion molecule expression 287
cell recognition 288–9
non-MHC restricted cells 289–90
class I
down-regulation 279–81
expression in bystander cells 281–3
homologue 286
instability 285
intracellular expression 283–4
synthesis 284–5
transport failure 285
gene expression 239
in the immunocompromised 278–9

infection
latent 278
late-stage, and CTL recognition 289
NK lysis 289–90
cytoplasmic partitioning, in NF-κB 92
cytotoxic T lymphocytes
action 2
function 2, 315
recognition of infected cells
adenovirus 216–17
Burkitt's lymphoma 266–7
cytomegalovirus 279, 288–9
Epstein–Barr virus 262
general mechanism 319, 335
hepatitis B 252
human papillomavirus 245
retroviruses 151, 166
role of class I H-2K 166
tumour rejection 315–16
antigens 319–20
lysis 171
mechanisms 320
recognition 246, 295, 315–16

$D^K$ protein 166
D-region, murine class I 30
degradation protection in adenovirus infection 199
deletions, class I 29
demyelination 108, 409–10
autoantibodies 413
in transgenic mice 416–17, *418*, 419
*see also* multiple sclerosis
dermatitis, HLA-DR3 preponderance 108
development, and class I expression 80
diabetes *see* IDDM
diarrhoea 193, 194
differentiation
regulation, *myc* oncogenes 298
repression 197
DM locus 32–3
DNA
class I sequence 27–8
class II sequence 31
degradation protection in adenovirus infection 199
methylation 47–8
proviral insertion
synthesis stimulation, by EGF 54
*trans*-regulation 163
transcription factors 93–4, 111
DOB subregion 32
DP subregion 31
and cytokines 58
in pancreatic β cells 59
DPA promoter, J element 118
DQ subregion 31
and cytokines 58
in pancreatic β cells 59
DQB promoter, J element 118
DQW8, and IDDM 363
DR subregion 31
and cytokines 58
and IDDM 363
in pancreatic β cells 59
DRA promoter, class II
expression, and Y element 112
inhibition, by YB-1 113
octamer motif 117–18
promoter activity, and X2 116
DZα gene 33

Eα murine gene, NF-X binding 115, 116
EAE *see* experimental allergic encephalomyelitis
early genes *see* adenovirus; human papillomavirus
EBNA1 (BL viral protein) 267
EBV *see* Epstein–Barr virus
EBV-negative BL lines 264
ectopic class II expression hypothesis of autoimmunity 378–9
EGF *see* epidermal growth factor
Ehrlich, P. 315
embryo, class I expression 80
encephalomyelitis, post-vaccinal 409–10
encephalomyocarditis virus 372
endo H 239, *240*, 244
in CMV-infected cells 285
endocrine autoimmunity 361–2
endocytic compartment, class II sampling 4
endoglycosidase H *see* endo H
endoplasmic reticulum
adenovirus infection 212–14
class I
binding to viral protein 214
chain synthesis 6
peptide binding 3, 10
human papillomavirus 239
lumenal domain 212, 214
peptide transport
deficiency 8–9
premature binding 12–13
and TAP 9
rough and adenovirus proteins 212
enhancer A 302–3
KBF binding 323
N-*myc* expression 302, 303
enhancer elements, class I 81, *82*
*see also* NF-κB
*env* protein 165, 169
environmental factors, and multiple sclerosis 411–12
epidermal growth factor 54–5
in thyroid disease 63

Epstein–Barr virus 5
  cellular phenotype modulation 264–5
  CTL response 261–2
  IFN-γ interference 135
  latency I infection 264
  latent infection 262–3
  target cells 262
  vIL-10 production 52
  *see also* Burkitt's lymphoma
experimental allergic encephalomyelitis
  aberrant class I 108
  inhibition studies 413, 415
  and multiple sclerosis 412–13

factor B, mapping 34
factor binding, class I 86–8
fibroblasts in rheumatoid arthritis 392, 396
α-foetoprotein 58
  class II regulation 107
footprinting, factor binding to class I 86–8
*fos* oncogene, and MHC regulation 176–7

G genes
  class III 35, 36–7
  products 36
G-protein family 29
*gag* genes 36–7
  protein product 165
gene expression
  enhancement, class I 83
  eukaryotic, transcriptional control 110–12
  in lymphoblastoid cell lines 261–2
glucocorticoids
  class II regulation 107
  effect on MHC 55–6
glutamate, class II down-regulation 57
glycosylation in adenovirus proteins 212–13
GM-CSF
  control of class II in rheumatoid arthritis 394–7
  inhibition, by IL-10 399
Golgi apparatus
  antigen presentation 7
  cis-Golgi
    class I peptide binding 10
    in HPV infection 239
  class I expression in CMV infection 283–4, 285
  class II 13
gp70 151
graft rejection 1
Graves' disease 56
growth acceleration, adenovirus-transformed cell 197

H genes, and integrated retrovirus 169
H-2
  expression
    A-RadLV 162
    AKR/Gross virus infection 165
    heavy chain 215
    restriction by retroviruses 151
    thymus 170
    *trans*-activation by LTR from AKV 165–6
    and tumorigenicity 304
    tumour-resistant mice 161
    up-regulation, and RadLV virus 160–1
    and *Pim-I* 163
H-2D loss, tumour 321
H-2K
  $CD4^+$ cell generation 325
  as CTL restriction element 166, 325
  gene deletion 175
  increased expression, and allogeneic rejection 153
  levels in AKR tumour-carrying mice 167–8, 325
  loss, in tumour 321
  transfection rejection 167
$H-2K^b$
  and c-*fos* mRNA transcript 176
  footprinting *87*
  promoter region as negative regulator 176
  tumour protection activity 326–7
H2TF1 element of region 1, class I
  down-regulation by *myc* 302
  factor binding in adenovirus infection 207
haemochromatosis 29
haemopoietic cell proliferation, bone marrow 53
HAL-DQ, in rheumatoid arthritis 391
HBV *see* hepatitis B virus
HC10 mAb 237, *238*, 239
HCGI-HCGVII 29
heavy chain class I 4–6
  deletion 323
  in infection
    adenovirus 215–16, 217
    cytomegalovirus 284
    Epstein–Barr virus 272
    human papillomavirus 238–9
helix-turn-helix motif 111
helper T cells 63
helper virus 175
hepadnavirus, effect on MHC 136
hepatitis B virus 251
  chronic 252
    IFN treatment 252–3
    non-response 253–7
  core protein 252
  immune system suppression 143
  ORFs 253–4
  proteins 253
  transfection experiments 253–5
  viral clearance 251–2

*see also* polymerase viral
hepatocyte destruction, in HBV 252
herpesvirus, effect on MHC 136
*see also* Epstein–Barr virus
HIV 179–81
class I 179, 180–1
effect on MHC 137, 138, 140
and 'empty' class I proteins 181
IFN-γ interference 135
HLA *see* MHC
HLA-A 28
bladder tumour 339
region II copies 82
HLA-A2 27–9
MAGE-1 recognition 296
tumour-specific peptide 295–6
in Wil tumour cell line 346, 347
HLA-A3 27–9
and multiple sclerosis 410, 411
in Wil tumour cell line 346, 347
HLA-A11-positive cell line
allele down-regulation in Burkitt's lymphoma 269
W6/32 mAb study 269–72
HLA-B 28
bladder tumour 339
down-regulation, c-*myc* 300, 302–3
and increased NK sensitivity 306
locus-specific 300–1
promoter region 300
region II copies 82
HLA-B7 in Wil tumour cell line 346, 347
HLA-C 28
bladder tumour 339
down-regulation by c-*myc*, allele specific 300
region II copies 82
HLA-D cytokine-mediated expression 59
HLA-DP, in rheumatoid arthritis 391
HLA-DP/Q inhibition, by IL-10 399
HLA-DR
accessory cells, for NK lysis 290
bladder tumour 340, *314*
eosinophil expression, CSF-mediated 53
and IDDM 108
and IL-3 51
infected keratinocytes 244–5
inhibition, by IL-10 399
pre/post-ovulatory 56
in rheumatoid arthritis 391
up-regulation by TNF 50
HLA-DR2, and multiple sclerosis 410, 411
HLA-DR3, disease preponderance 108
HLA-DR4, and multiple sclerosis 410
HLA-DR6, and multiple sclerosis 410
HLA-Dw2, and multiple sclerosis 410
HLA-E 28–9
HLA-F 29
HLA-G 29
HLA-H 29
Hmt region, murine, class I 30
Hodgkin's disease 261
hormonal control of MHC expression 55–8
hormone receptor superfamily 93–4
HPV *see* human papillomavirus
HSP70 genes 35–6
5-HT 57
human immunodeficiency viruses *see* HIV
human leukaemia viruses (HTLV) *see* leukaemia virus
human papillomavirus (HPV)
class I down-regulation 235
early genes
class I expression 237, 244
protein products 234, 236–7
recognition by CTL 245
transcription *236*
genome 233–4
immunization 234
infection 233
cell-mediated lysis 243–4
class I expression 237–40, 244, 245
and class I maturation 239–40
class II expression 240–2, 245
in keratinocytes 235–7
hXBP-1 gene 116
hypomyelination, in C7 transgenic mice 416, *417*

Ia expression, and TATA functionality 117
anti-Ia antibodies, and immune disease remission 108
ICAM-1 17, *65*
in cytomegalovirus infection 287, 289
in IDDM 367–8
in insulitis 364, 366
and lysis by CTLs 266–7
ICS (IFN consensus sequence) 81
binding protein (ICSBP)
in lymphoid tissue 90
repression effect 91
*see also* IRF
motif 84
reporter gene induction 84
sequence conservation 84
IDDM
class II 108, 363
inducibility discrepancies 368
defective
class I 368–9
TAPs 369
genetic predisposition, NOD mouse 371
HLA expression 363–4
serotype 108
islet cell MHC expression 362–3, 373, 375

IDDM (*cont.*)
pre-pathological changes 369
prevention by silica (NOD mouse) 371
viral induction, in rodents 372–3
viral involvement 369–71
without insulitis 373
*see also* insulitis
IFN *see* interferon
IL *see* interleukin
immune disease, and anti-Ia antibodies 108
immune response, and helper T cells 63
immune surveillance theory 335–6
immune system specificity 103
immunogenicity, tumour 322
immunoglobulin gene superfamily and leukaemia virus 155–6
immunological silence 361
immunoprecipitation study, Fen cell line 346, 348
immunoselection, during tumorigensis 166
immunostimulation, chronic, and leukaemogenesis 170
immunosuppression
and class II up-regulation 24
cytomegalovirus 278–9
by IL-10 52–3
infection 44
and class II enhancement 105
persistent, hepatitis 143–4
viral, and soluble mediator 163
inflammation
CNS in multiple sclerosis 412
liver in hepatitis 252
MHC involvement 60–1, 143
prolonged 67
T cell activation 66
TNF-caused, non-diabetogenic 376
*see also* insulitis
initiator sequence (INR) transcription start site 111
insulin
dependent diabetes mellitus *see* IDDM
production failure, and MHC overexpression 376
promoter 373, 374
insulitis 363
adhesion molecule production 364, 366
HLA expression 363–4
interferon
class I control 80–1
pancreas 366–7
combinations in tumour cell lines 350–1
effect on adenovirus infection 217–18
genes 89, 90
promoter 172
in IDDM 366–7, 367–8
inducibility 81
inhibition, by hepatitis B core protein 252
non-responder cell selection 254–5
regulated genes (ISG) 90, 91, 218
response
class I *cis*-acting elements 83
in hepatitis B infection 255, *256*, 257, *258*
stimulation and LAK/NK effectiveness 352, 353
Fen cell line *354*
treatment, Fen tumour cell line 346, 348–9
viral neutralization 138, 140
trophoblasts 47–8
up-regulation in HBV infection 251–2
IFN-α 48–9
class I enhancement in melanoma 60
class I/II expression in carcinoma cell lines 345, 346
class II induction 350–1
DM gene induction 33
rheumatoid arthritis *395*
species differences 49
IFN-α/β
class II regulation 107
contradictory effects on class II expression 49
HBV clearance 251
viral interference 138
IFN-β 48–9
adenovirus infection 207
autocrine control of class I 172–3
*cis*-regulatory elements *85*
class II induction 350–1
IFN-γ 46–8
class I restoration in adenovirus infection 218
class I/II expression in carcinoma cell lines 344, 345
class II regulation 44, 105–7, 115–16, 350–1
cytokine synergy 62–3
differential cellular expression 60
DRA regulation, and hXBP-1 116
glucocorticoid inhibition 55
inflammation 61
inhibition 399
human papillomavirus infection 240, 241
keratinocyte class II expression 47
leukaemia virus infection 166
macrophage class II expression 52
MHC subregion expression 58, 176
and physiological MHC maintenance 59–60
receptor, down-regulation by TGF-β 54
rheumatoid arthritis 392–4, *395*
thyrocytes class II 62
trophoblast MHC expression 47–8

TSH 56
tumour necrosis factor 50
viral inhibition 135, 138
*see also* ICS; IRF
interleukin
macrophage class II expression 52
IL-1
rheumatoid arthritis 393, 394–5
and TNF-α 397
IL-2
and LAK activity 351
in melanoma regression 336
TIL stimulation 342
IL-3
effect on class II 51
inhibition, by IL-10 399
IL-4 51–2
class II up-regulation 105–6
rheumatoid arthritis 393, 397–9, 398
target cell-dependent effects 398
IL-6
enhanced by IL-4 398
in IDDM 366
in rheumatoid arthritis 393, 395
IL-10 52–3
class II regulation 107
immunoregulatory properties 399–400
murine 53
in rheumatoid arthritis 399–400
vIL-10 52
IL-13 (P600) 399
interstitial pneumonitis 278
invariant chain 4, 12–13
degradation 13
proteolysis 14
IRE *see* ICS
IRF (IFN regulatory factor) 89–91
IFN induction 90
IRF-1, activation effect 90, 91
IRF-2, repression effect 90, 91
ISG *see* interferon regulated genes
islet cells (of Langerhans) 362
class I expression in trangeneic mice 375
MHC production 366–7
reovirus-infected 367
viral infection determination 370–1
*see also* IDDM
ISRE *see* ICS

J element 118
Jurkat cell transfection 159

$K^K$ protein 166–7
KB site 81
*see also* region I
keratinocytes
cell line MHC peptide profile 245–6
human papillomavirus infection 235–7
cell-mediated lysis 243–4
class I expression 237–40, 244, 245
class II expression 47, 240–2, 245

LAK (lymphokine activated killer) activity
in human papillomavirus infection *242*, 243–4
and MHC 352
in tumour cell lines 351–3
in tumour defence 320
large granular lymphocytes (LGL) 317
latency I (BL) 264
LCL *see* lymphoblastoid cell lines
*Leishmania* sp. 145
leucine zipper motif 111, 116
leucocyte adhesion, in CMV infection 288
leukaemia viruses
human 177–9
class I up-regulation 177–8
class II induction 178
neurological disorders 179
MoMuLV 151–60, 164
class I induction 152–3
lymphocyte-specific antigen 153–4
murine
in lymphocytes 153
promoter region 159–60
T cell activation 169–70
*trans*-repressor 160
radiation leukaemia virus 160–4
A-RadLV 162
DNA-binding protein 163
H2 expression 162
receptor structure 164
soluble mediator production 163
T cell infection 161–2
*see also* LTR
leukaemogenesis 169–70
LFA
LFA-1, c-*myc* down-regulation 303
*see also* CD18
LFA-3
adhesion molecule 266–7
CMV infection 287–9
in insulitis 364, 366
*see also* CD58; adhesion molecule ligands
ligand
adhesion molecules 17
NK cell recognition 317–18
in rheumatoid arthritis 403
LMP1, and phenotypic change in B lymphocytes 264–5
LMP2 32
LMP2/7 6, 8
LMP7 32
LTR (long terminal repeat)
endogenous retroviral 168–9

LTR (long terminal repeat) (*cont.*)
mapping 157–60
*trans*-activating in leukaemia virus 157, 165–6
lung carcinoma, c-*myc* 298
Ly-19 gene 318
lymphoblastoid cell lines
and Burkitt's lymphoma 263
Epstein–Barr virus 261
latent gene expression 261–2
lymphocyte
oncogene-mediated class I 173–4
MHC regulation, retroviral infection 153
migration in insulitis 364
*see also* T cell
lymphocyte function associated antigens *see* LFA
lymphocytic choriomeningitis virus (LCMV) 377
lymphoid neoplasia generation 170
lymphoid organs, ICSBP gene 90
lymphokine
purpose 104
in tumour defence 320
tumour treatment 324
*see also* interferon; interleukin, tumour necrosis factor
lymphoma
immunosurveillance escape 165
induction, MCF 1233 165
lymphotoxin *see* tumour necrosis factor
lysis *see* LAK; NK
lysosomes
antigen proteolysis 13–14
class II peptide generation 13

mAbs *see* monoclonal antibodies
macrophage
class II
expression, and cytokines 52
induction 53, 57–8, 105
IDDM in NOD mouse 371
in rheumatoid arthritis 392
susceptibility, human papillomavirus infection *242*, 243–4
MAGE-1 melanoma gene 296
malignancy
class I down-regulation 337
and Epstein–Barr virus 261
MCF *see* mink cell focus-inducing virus
MCP-1, *trans*-activation 170
mediator interaction, MHC expression 61–3
melanoma
B16 321
IFN-α class I enhancement 60
initiation by c-*myc* activation 305
*myc*-mediated class I down-regulation 299–300
prognosis, and class II expression 340
regression 336
MHC
as associative molecules 335
cytokine-modulated expression 44–6, 58–9, 351
CSF 53
EGF 54–5
IFN-α 48–9
IFN-β 48–9
IL-3 51
IL-10 52–3
interferon 46–8
*in vivo* evidence 59–61
TGF-β 53–4
TNF 49–51
ectopic *65*
expression
cytotrophoblast 48
failure of 109
hormonal regulation 55–8
insulin production failure 376
mediator interaction 61–3
promoters 117–19
thyrocytes 44–5
transcriptional regulation 110–12
virus-modulated 133–5, 136–7
gene division 1
immunopathological changes 44
and LAK/NK activity correlation 352
mapping 1, *2*
oncogenes suppression 152, 171
polymorphism 29
restriction 10, 134
tissue distribution 43–4
*see also* class I–III; DP; DQ; DR; HLA–D
mice
β cell transgenes 373–9
transgenic studies, for multiple sclerosis 414–19
*see also* NOD mice
$\beta_2$-microglobulin
*cis*-regulator elements *85*
expression
with class I heavy chain 84–5
failure 348
in tumour cells 323
footprinting 87–8, *87*
function 4–6
in HPV-transfected keratinocytes 238, 239
as marker for hepatocyte IFN response 257
PAM regulatory element 88
mRNA expression, IFN-γ enhanced 166
in *src*-transformed cells 174
structure 4–6
synthesis, in CMV-infected cells 284

transfection, and LAK/NK Fen line effectiveness 354
microtubule components, and adenoviral infection 213
mink cell focus-inducing virus 164–5
sequences 169
U3 region 158
'missing foreign' hypothesis, NK activity 318
'missing self' hypothesis 305
NK activity 318
MoMuLV *see* leukaemia virus
monoclonal antibodies
class I expression in keratinocytes 237–8
T cell function 17, 169–70
monocyte, IL-10 production 52
mononucleosis 261
*mos* oncogene, MHC regulation 171–4
motif
DRA octamer 117–18
helix-turn-helix 111
ICS, Tla gene 84
leucine zipper 111, 116
peptide
class I 10, 11
class II 15
recognition 113–14
X-Y 118
X3 consensus 115
mRNA *see* RNA
multiple sclerosis (MS) 410
animal model 413–20
autoimmune aspects 412–13
and class II expression 57
environmental factors 411–12
genetic susceptibility 410–11
HAL-DR2/4 preponderance 108
mimicry in transgenic mice 416–19
progressive, HLA alleles 411
relapsing 411
viral involvement 411–12
MuLV (murine leukaemia virus) *see* leukaemia viruses
Mutu-BL line, class I heavy chain expression 272
myasthenia gravis, HLA-DR3 preponderance 108
*myc*
activation, consequences 304–7
adenovirus homology 299
class I down-regulation 299–300
transcription-level control 302
c-*myc* 298
and HLA-B down-regulation 302–3, *304*
locus-specific class I down-modulation 300–4
c-*myc*/N-*myc* compared 301–2
N-*myc* 298, 299
class I down-regulation 302, *304*
HLA down-regulation 301–2
oncogenes 298–304
cross-regulation 298
structure 298
tumours, neuroendocrine origin 301
myelopathy 179

natural killer cells
activity
and class I expression 318, 319
cytomegalovirus infection 289–90
receptor inhibition model 318
tumour cell lines 351–3
adenovirus sensitivity 205
identification 317
and 'missing self' hypothesis 305
sensitivity
and HLA-B down-regulation, tumour cells 306
HPV-infected cells *242*, 243–4
and tumour class I regulation 306
tumour defence 316, 317–19
interaction with class I 306–7
negative regulatory element 208
neuroblastoma 299
*myc*-mediated class I down-regulation 300
neuroendocrine tumours 301
neurological disorders, and HTLV-1 179
neutrophil, class I polypeptide translation, CSF mediated 53
NK-κB 91–2
adenovirus infection 207
factor binding, in region I 95
physiological role 95
structure and function 92
NF-Y 112, 114
motif recognition 114
uniqueness 118–19
NGF/TNF cytokines in rheumatoid arthritis 403
NK *see* natural killer cells
NOD mouse, IDDM studies 368, 371–2
noradrenaline, class II down-regulation 57
NRE 208
nuclear hormone receptor superfamily
RAR 93, 409
RXR 93–4
heterodimer stability 94
physiological role 95
THR 93
VDR 93
*see also* retinoic acid
nuclear proteins, and class I chain unfolding 88

oligodendrocyte MHC expression, and demyelination 419

oncogene
  control of class I, and cell type 173
  mutation, and tumorigenic cell properties 295
  and non-lymphoreticular tumours 170–1
  as tumour antigens 322
  *see also* specific oncogene
oncogenicity
  adenovirus infection 297
    by class I repression 204
    hamster 204–5
    limited 201–2
    mouse 205
oncoproteins, as TRAs 322
orthomyxovirus, effect on MHC 136
OSG gene 36
ovulation, and HLA-DR expression 56

P cell stimulating factor *see* IL-3
p24 181
p53 199, 295
  suppression 234
p105-*RB* 197
p600 *see* IL-13
PAM (class I) 87–8
pancreas *see* β cells; IDDM; islet cells
PAP technique, MHC antigen expression 336–7, *338*
papillomavirus *see* human papillomavirus
papovavirus
  effect on MHC 136
  JC, and PML 409
paramyxovirus, effect on MHC 136
paraparesis, tropical spastic 179
partitioning, cytoplasmic, in NF-κB 92
Pelizaeus–Mersbacher disease 409
peptide
  active transport 3–4
  class I
    binding groove 5–6, 10–12
    bonding site 5
    loading 10
    translocation 8–10
  class II
    binding groove 12, 16
    delivery 14–15
    exchange 16–17
    generation 13
    specificity of binding 15–16
  formation, antigenic 6–8
  processing deficiency 8–9
  transport, and TAP 9
  *see also* motif
peptide-transporter genes *see* TAP
phenotypic drift 263
  Burkitt's lymphoma 264
*Pim-I* 163
pInspro *see* insulin promoter
plasma cell, class II expression 105
plasma protein, class II induction in macrophage 57–8
PLP 409
PML *see* progressive multifocal leukoencephalopathy
polymerase viral (POL) 253–4
  IFN response 255, *256*
    failure 257, *258*
  liver biopsy 255, 257
polymorphism
  class I 5–6
    human 29
  class II 15–16
  complement components 36
  in HIV 181
  and integrated retrovirus 169
  by oncogenes 173
  post-transcriptional control of class I
  transporter gene 32
PPAR *see* nuclear hormone receptor superfamily RXR
pre-initiation complex 111
pregnancy, MHC trophoblast expression 47
progressive multifocal leukoencephalopathy 409
promoter
  activation, N-Tera2 EC cells 95
  adenovirus 195, 197
  class I, adenovirus action site *210*
  eukaryotic gene transcription 110–11
  generic *110*
  HLA-B, and c-*myc* down-regulation 300
  IFN, serum-factor responsive region 172
  leukaemia virus responsive 159–60
  thymidine kinase, down-regulation 208
prostaglandin, and class II expression 57, 107
protease
  in antigenic peptide formation 6, 8
  cathepsins 14
  inhibitors, viral 141
  for invariant chain degradation 13
protein
  degradation to peptides 6–8
  nuclear and class I 88
  protein interactions 111
  transcriptional regulation 111
  viral antigens 8
protein kinase
  activation, in class II regulation 106
  inhibition, viral 218
protein X (HBV) 253
proteolipid protein, *see* PLP
proteolysis, of antigen 13–14
proteosome components, IFN-γ sensitive 141
prothymosin 57
proto-oncogene *see* oncogene

proviral insertion, preferential 163
proviral sequence, endogenous 168
pseudogenes
class I 27, 28, 29
class II 31

Q7/10 genes 84
Qa region 30, 83–4

RA *see* rheumatoid arthritis
radiation leukaemia virus (RadLV) *see* leukaemia viruses
*ras* oncogene
lymphocyte
class I 173
class II 173–4
MHC regulation 171–4
protein product 295
rat BB, IDDM development 372
*Rb* suppression 234
receptor inhibition model of NK cell activity 318
region I
binding 86
conservation 82
and NF-κB factor binding 95
region I/II (enhancer), class I 81, *82*
region II, multiple copies 82
regulatory element
class I 81–4
transcription factor accessibility 88
rejection, and H-2K gene 153
*rel* gene family 92
remyelination, in transgenic mice 418
reovirus 136
infection of islet cells 367
reporter gene
class I, RXR activation 94
enhancement, N-Tera2 EC cells 95
induction 84
*see also* chloramphenicol acetyl transferase
respiratory disease, adenovirus caused 193, 194
restriction (MHC) 134
restriction element, H2-K 325
retinitis 278
retinoic acid
$\beta_2$-m gene expression 84
class I
gene transcription induction 8 1
and region I 86
and region II 81
nuclear hormone receptor (VDR) 93
in regulation of N-Tera2 EC cells 95
retrovirus
effect on MHC 137
and IDDM 370, 371
integrated, and H gene 169
Moloney murine leukaemia 151–60
site of antigen inhibition 140
*see also* HIV; leukaemia viruses; sarcoma viruses
RF-X 115
rhabdovirus 136
rheumatoid arthritis 390
anomalous TGF-β effect 400–1, *402*
class II
cytokine control 391–401
expression 390, 391–2, 401–2
induction by GM-CSF 395
genetic component 390
HLA/DR2/4 preponderance 108
pathogenic change origin 392
RING subregion human class II 32–3
RNA
mRNA
from adenovirus early gene 195, 197, 199
antigen-encoding in leukaemia virus infection 153, 155
IFN enhanced 166
tRNA synthetase sequences 35
RNA polymerase
promoter sites, U3 159
protein–protein interaction 111
rRF-X (recombinant) 115
RXR *see* nuclear hormone receptor superfamily

sarcoma virus
murine 170–7
oncogene containing, and MHC suppression 152
second signal (messenger)
class II activation 106
insulitis 366
*see also* co-stimulation
pathway 378
transgenic mice 377–9
self-tolerance, and non-responsive T cells 66
serotonin *see* 5-HT
serum factor, in class I control in tumour cells 172–3
sialic acid 168
signalling events 17–19
silica, and IDDM prevention 371
simian immunodeficiency virus 181–2
site α (class I) 82
SIV *see* simian immunodeficiency virus
SLE *see* systemic lupus erythematosus
spleen, H-2 antigen expression 168
*src* oncogene
class I antigen interaction 174–5
MHC regulation 174–5
*Staphylococcus aureus* 145

steroid, hormones
  receptors 93
  sex, effect on MHC 56
  *see also* glucocorticoids
stress protein *see* HSP70
suppressor proteins, tumour/oncogene 295
  *see also* p53; *ras*
synovial cells, cytokine production in rheumatoid arthritis 393
synoviocytes 396
synovitis, class II expression 391
systemic lupus erythematosus 108

T cell
  activation
    and inflammation 66
    and MuLV protein up-regulation 169–70
    process 17–18
    professional/non-professional 64–6
  adhesion molecules *65*
  co-stimulation 64
  and EAE aetiology 413
  education, thymic 59–60
  leukaemia, and HTLV-1 virus 177
  non-responsive, and peripheral self-tolerance 66
  paralysis, and mouse β cell antigen presentation 375
  proliferation, TGF-β inhibited 400
  RadLV infection 161–2
  receptor
    CB3 complex 2
    domain structure 3
    receptor for RadLV 164
  role 43
  surface protein, up-regulation by MoMuLV 153–5
  variable response to class II expression 378
  *see also* cytotoxic T lymphocyte
T helper ($T_H$) cells 2
T lymphoblastoid cells, AKR/Gross virus infection 165
TAP
  polymorphism 9–10
  TAP1 3–4, 9
  TAP1/2
    deficiency, BL30 cells 272–3
    products 9
    transport function 32
  TAP2 3–4, 9
TATA antigens in tumour defence 319
  *see also* tumour-rejection antigens
TATA sequence 110–11
  class II 117
  transgenic mice 117
*tax* gene 178–9
  TCR *see* T cell, receptor
tenascin 36
tetanus toxin (TT) degradation 14
TGF-β 53–4
  anomalous effect in rheumatoid arthritis 400–1, *402*
  effect on cytokines 400
THR *see* nuclear hormone receptor superfamily
thromboxane/prostaglandin antagonism 57
thymic education of cells 59–60
thymidine kinase promoter down-regulation 208
thymus
  infected, and H-2 levels 170
  and non-oncogenetic adenovirus rejection 202
  T cell elimination 3
thyrocyte
  class II differential antigen-presenting ability 66
  class II induction
    IFN-γ 62
    TSH 62–3
  IFN-α effect 49
  MHC expression 44–5
thyroid cell line, and oncogene-controlled class I 173
thyroid disease, autoimmune
  mediator interaction 62
  MHC subregion expression 58
thyroid stimulating hormone
  cAMP-mediated effect on class II 56
  effect on thyrocytes 56, 62–3
  EGF inhibition
  synergy with IFN-γ 56, 62–3
thyroiditis, and EGF activity 63
TIL 340–2
  infiltration level
    and class I expression 341–2
    and prognosis 341
  killing activity 336
  phenotypic profile 342–3, 344
  stimulation, by IL-2 342
tissue distribution, MHC antigen 43–4
TL murine region, endogenous proviral sequences 168
Tla 83–4
  murine 28, 30
TNF, HLA-DR up-regulation 50
tumour necrosis factor (TNF) 34–5
  TNF-α 34–5
    class I transcription enhancement 218
    class II 107
    class III 32
    cytolysis in adenovirus infection 219
    human papillomavirus infection 240–1
    inhibition, by IL-10 399
    pancreas class II 367

production, transgenic mice 376
region I binding activity induction 86
removal, and IL-1 inhibition 397
in rheumatoid arthritis 395, 397
TNF-β (lymphotoxin) 50
mapping 32
TNF/IFN-γ synergy 62
togavirus, effect on MHC 136
toxic shock syndrome 142
TRA *see* tumour-rejection antigens
*trans*-activation
adenovirus early genes 197
H-2 gene 165–6
by MoMuLV 156
region mapping *156*, 157–60
MuLV, response element 160
viral, consensus sequence 159
*trans*-factor
binding, class I 85–6
class II regulation 109
*trans*-induction of class II in islet cells 368
*trans*-repressor
class I in viral infection 152–3
leukaemia virus 160
transcription
and activator protein 111
class I
developmental regulation 80
down-regulation by adenovirus 192
and *myc* 302
class II 105
IFN-γ induction 106
eukaryotic genes 110–12
factor
accessibility 88
cloning (class I) 89–95
complex components 94–5
E2F 197
H2TF1 207, 302
IRF 89–91
KBF1/2 323–4
NF-κB 91–2
NF-Y 112, 114, 118–19
RXR subgroup 93–4
p53 *see* p53
initiation class I regulation by adenovirus 206
leukaemia virus 152–3
start site 111, 112
X element 115
viral inhibition 140
transfection
allogenic class I 325–7
class II 327–8
rejection, and H-2K 167
transforming growth factor-β *see* TGF-β
in tumour treatment 324
transgenic studies, factor binding to class I 86–8
transplantation antigens 1
transplantation genes 1, 27–8
*see also* HLA-A–C
transport
failure
in BL30 cells 272–3
class I antigens in CMV infection 285
inhibition, viral 141–2, 215
viral mRNA 199
transporter gene polymorphism 32
transporters *see* TAP1/2
tRNA *see* RNA
trophoblasts 47–8
tryptophan repeat, in IRF 89
TSH *see* thyroid-stimulating hormone
tumorigenesis, immunoselection 166
tumorigenicity
adenovirus and class I repression 204–5
H-2 modulation 304
tumour
class I expression 297, 336–7
differential 344–50
down-regulation implication 305, 337
expression loss, advantages 316
under-expression as intrinsic property 305
up-regulation treatment 324
class I/II defect study 345–50
class II expression 337, 340, *341*
class II positive 316
CTL defence 315–16
cytolysis
CTL-mediated 171
*see also* TILs
defence 316–17
composite system 322
LAK activity 320
MHC-restricted cytotoxicity 319–20
NK cells 317–19
T cell receptors 320–1
Fen line study 346, 348–9
growth, and E6/7 234
IFN-γ effect on TNF-α 50
immunity, H-2E$^k$ 327–8
immunogenicity 322
initiation, by c-*myc* activation 305
malignancy 305, 337
MHC expression
class I 344–50
class II 350–1
differential 48, 343–4
natural killer sensitivity 306
neuroendocrine origin, and *myc* 301
non-lymphoreticular, and class I levels 170
protection activity, H-2k$^b$ 326–7

tumour (*cont.*)
  recognition 295–6
    by class I CTL 295
    gp70 151
  regression, spontaneous 335–6
  rejection 315
    *see also* TATA antigens
  treatment, by transfection 324
  Wil line study 345–7
  *see also* bladder tumour; Burkitt's lymphoma; cervical cancer; cancer
tumour-causing virus, and promoter region regulation 159–60
tumour-infiltrating lymphocytes *see* TIL
tumour-rejection antigens 319
  viral oncoproteins as 322
tumour vaccine 325–6, 327, 354

U3 region of retrovirus genome
  MCF virus 158
  RNA polymerase III promoter sites 159
ubiquitin 6
UL-18 gene (class I homologue) 286
URR (upstream regulatory region), HPV 234

vaccine
  transfected cell lines as 167
  tumour 325–6, 327, 354
VDR *see* nuclear hormone receptor superfamily; retinoic acid
vIL-10 52
virus
  antigen, in mouse β cells 374
  early gene *see* adenovirus early gene
  helper 175
  infection
    and autoimmune response 49, 50
    and IDDM 369–71
    induced diabetes in rodents 372–3
  interference location 138, 139, 140–3
  involvement in multiple sclerosis 411–12
  neo-antigen β cell expression 377
  oncogenes, murine 170–7
  pathogenicity, and MHC expression modulation 143–4
  peptide/antigen association prevention 141, 142
  protein, as antigenic peptide 8
  resistance, and CTLs 335
  *trans*-activation, consensus sequence 159
  *see also* individual viruses

W/Z/S elements 106–7, 112, 117
W6/32 mAb 237, *238*
  cytomegalovirus infection study 284–5
  limitations of conventional tests 337
  PAP technique 336–7, *338*
  study of HLA-A11 down-regulation 269–72

X box, function 114–15
X element 106–7, 112, *113*, 114–16
  murine, partial systemic function 117
  transgenic mice 116–17
X-Y motif 118
X1 protein-specificity 115
X2 element 116
X2-binding protein 116
X3 consensus motif 115

Y element 106–7, 112, *113*, 114
  partial systemic function 117
  transgenic mice 116–17
Y5 33
YB-1 114
YEBP *see* NF-Y

zinc finger
  CR3 197
  RXR subgroup 93
zipper motif 111
zipper protein 116